# International Physics Conference at the Anatolian Peak 2016 (IPCAP2016)



Journal of Physics: Conference Series Volume 707

Erzurum, Turkey
25 – 27 February 2016

**Editors:**

**Ridvan Durak**
**Sebahattin Tuzemen**

ISBN: 978-1-5108-2418-8
ISSN: 1742-6588

**Printed from e-media with permission by:**

Curran Associates, Inc.
57 Morehouse Lane
Red Hook, NY 12571

**Some format issues inherent in the e-media version may also appear in this print version.**



Printed by Curran Associates, Inc. (2016)



Institute of Physics
Dirac House, Temple Back
Bristol BS1 6BE UK

Phone: 44 1 17 929 7481
Fax: 44 1 17 920 0979

techtracking@iop.org

**Additional copies of this publication are available from:**

Curran Associates, Inc.
57 Morehouse Lane
Red Hook, NY 12571 USA
Phone: 845-758-0400
Fax: 845-758-2633
Email: curran@proceedings.com
Web: www.proceedings.com

# Table of contents

## Volume 707

**International Physics Conference at the Anatolian Peak (IPCAP2016)**
**25–27 February 2016, Erzurum, Turkey**

**Accepted papers received: 30 March 2016**
**Published online: 4 May 2016**

### Preface



### Papers

#### Atomic, Molecular and Plasma Physics

**Solid State and Nano Physics**

**Nuclear Physics and Nuclear Energy**



**Particle Physics**



**Optic and Lasers**

## Mathematical Physics



## Applied Physics



## Physics Education

**Astronomy and Astrophysics**

**Other topics**

# Preface

The present volume of *Journal of Physics: Conference Series* represents contributions from participants of the International Physics Conference at the Anatolian Peak (IPCAP2016) Erzurum, Turkey, 25-27 February 2016, organized by Atatürk University Physics Department that were peer-reviewed by 40 expert referees to the best professional and scientific standards.

The conference aims; to share the experiences and research results of academicians, researchers, and students (scientists) studying physics and related fields, to exchange information about the latest developments in physical science, to discuss the collaboration facilities and finally to provide scientific communication between scientists from different countries.

The Conference venue was the Atatürk University Nenehatun Cultural Center in Erzurum. IPCAP 2016 conference covers a wide spectrum of important topics on physics with 12 different issues;

1. Atomic, Molecular and Plasma Physics;
2. Solid State and Nanophysics;
3. Nuclear Physics and Nuclear Energy;
4. Particle Physics;
5. Optic and Lasers;
6. Statistic Physics;
7. Mathematical Physics,
8. Applied Physics;
9. Biophysics and Medical Physics;
10. Physics Education;
11. Astronomy and Astrophysics and
12. Other topics.

Even though we have organized the first IPCAP 2016 conference this year, about 300 papers will be presented in 4 separated halls and in 8 oral and 2 poster sessions during 3 days.

I would like to thank the Invited Speakers for their significant contributions to the conference. I would also like to thank the members of the International Advisory and Scientific Committee, the members of the Organizing, conference secretary and sponsors.

I hope the reader will enjoy this special issue and will find ideas for new bright achievements.

***Rıdvan Durak, Atatürk University, Erzurum-Turkey, March 23, 2016.***

# Board and committees

### Conference Chairs and Proceedings Editors

Rıdvan Durak (Chair), Atatürk University, Turkey
Sebahattin Tüzemen (Co-Chair), Atatürk University, Turkey

### Scientific Committee

Semsettin Altındal,Gazi University, Turkey
Fikret Anlı, Sutcu Imam University, Turkey
Aytunc Ates, Yıldırım Beyazıt University, Turkey
Yıldırım Aydogdu, Gazi University, Turkey
Enise Ayyıldız, Erciyes University, Turkey
Laurent Jolissaint, Heig-Vd, Switzerland
Richard C. Jones, Texas Woman's University, USA
Mustafa Kaçal, Giresun University, Turkey
Abdulhalik Karabulut Atatürk University, Turkey
Bünyamin Karabulut, Ondokuz Mayıs University, Turkey

Journal of Physics: Conference Series **707** (2016) 011001 doi:10.1088/1742-6596/707/1/011001

Ferdi Akman, Bingöl University, Turkey
Emin Bacaksız, Karadeniz Technical University, Turkey
Mehmet Biber, Atatürk University, Turkey
Elif Boydaş, Atatürk University, Turkey
Guven Cankaya, Yıldırım Beyazıt University, Turkey
Lütfi Ozyüzer, Izmir Institute of Technology, Turkey
Seydi Doğan, Balıkesir University, Turkey
Oguz Dogan, Necmettin Erbakan University, Turkey
Songul Duman, Atatürk University, Turkey
Rıdvan Durak, Atatürk University, Turkey
Hasan Efeoğlu, Atatürk University, Turkey
Sezai Elagöz, Cumhuriyet University, Turkey
Kadir Esmer, Marmara University, Turkey
Fahrettin Goktaş, Yıldırım Beyazıt University, Turkey
Omer Güllü, Batman University, Turkey
Mohd Nizar Bin Hamidon, University Putra, Malaysia
İbrahim Han, İbrahim Çecen University, Turkey
Orhan İçelli, Yldız Technical University, Turkey
Rıdvan Karapınar, Yüzüncüyıl University, Turkey
Tahsin Kılıçoğlu, Batman University, Turkey
Atalay Küçükbursa, Dumlupınar University, Turkey
İlham Nasıroglu, Atatürk University, Turkey
Mohamad Missous, Manchester University, UK
Jean Michel Chauveau, University of Nice Sofia Antipolis, France
Reynier Peletier, University of Groningen, Holland
Recep Polat, Erzincan University, Turkey
Agnieszka Słowıkowska, Ins. of Astr. Univ. of Zielona Góra, Poland
Mustafa Sağlam, Atatürk University, Turkey
Önder Şimsek, Atatürk University, Turkey
Engin Tırasoglu, Karadeniz Technical University, Turkey
Sabahattin Tüzemen, Atatürk University, Turkey
Fatih Ucun Suleyman, Demirel University, Turkey
Nazım Uçar Suleyman, Demirel University, Turkey
Demet Yılmaz, Atatürk University, Turkey
Sinan Kaan Yerli, Ortadogu Technical University, Turkey
Yahya Kemal Yoğurtçu, İstanbul S. Zaim University, Turkey

## Local Organizing Committee

Mustafa Sağlam, Atatürk University, Turkey
Abdulhalik Karabulut, Atatürk University, Turkey
Mehmet Biber, Atatürk University, Turkey
Emre Gür, Atatürk University, Turkey
Neslihan Ekinci, Atatürk University, Turkey
Bahattin Abay, Atatürk University, Turkey
Yakup Kurucu, Atatürk University, Turkey
Sakir Aydoğan, Atatürk University, Turkey
Yüksel Ozdemir, Atatürk University, Turkey
Ali Gürol, Atatürk University, Turkey
Mutlu Kundakçı, Atatürk University, Turkey
Betul Güzeldır, Atatürk University, Turkey
Cahit Yeşilyaprak, Atatürk University, Turkey
Esra Kavaz, Atatürk University, Turkey
Esra Cinan, Atatürk University, Turkey
Erdem Şakar, Ataturk University, Turkey
Yavuz Güney, Ataturk University, Turkey

## Invited Speakers

Bruno Daudin, Institut NEEL / France
Bekir Aktaş, Gebze Technical University / Turkey
Mohd Nizar Bin Hamidon, Universiti Putra / Malaysia
Niyazi Meriç, Ankara University / Turkey
Alpan Bek, Middle East Technical University / Turkey
İlhami Gülçin, Member of TUBA, Atatürk University / Turkey
Coşkun Kocabaş, Bilkent University / Turkey
Sebahattin Tüzemen, Atatürk University / Turkey
Sait Eren San, Gebze Technical University / Turkey
Hüseyin Kurt, Istanbul Medeniyet University / Turkey

# Title and abstract of invited talks in order of presentation

**Mohd Nizar Bin Hamidon**, University Putra, Malaysia
**Title:** Future Trend in Sensor Technology

**Absrtact:** Smart' sensors are increasingly being employed, not only in industrial settings, but in every aspect of human life, where they are telling something about their environment by bridging them with the electronic world. Sensors are used to gather a wealth of information from the process that can improve operational efficiency and product quality which involving a mass of data that need to be analysis especially with the merging of the Internet of Things (IoT). They also included with new features such as communication capability and on-board diagnostics. Leading edge research in sensors has been propelled by the advancements made in fabrication, signal processing and material technology in the last decade. With these the scientific world is now on the verge of delivering sensors with radically new capabilities for the human societies. This talk deals with the future trend in sensor technology and several research work in Universiti Putra Malaysia related with them.

**Sait Eren San**, Gebze Technical University, Turkey
**Title:** Liquid Crystals and Conducting Polymers with Nanoparticles for Several Organic Electronics Applications

**Absrtact.** Liquid Crystals (LC) are highly nonlinear optical materials and they are sensitive against optical, electrical and magnetic fields. This sensitivity could be exploited for various device designs. In the scope of this presentation, I will firstly discuss the diffraction grating experiments, which are indeed base experiments for Holographic data storage and principally information could be stored at the order of Terabyte per cubic centimeter via this approach. Actually it was experimentally observed and shown that dye and Carbon nanoparticles doped LCs are promising candidates as storage mediums via Holography. Second case study is devoted to another critical experimental demonstration that proves the reorientation and conservation of Carbon Nanotubes (CNT) in the structure of nematic LC. Experimental concepts propose this CNT reorientation, which is of critical importance for potential CNT applications, via electrical, optical and magnetic fields. Third case study is a Hybrid Solar Cell, which is designed and proposed as a feasible and reasonable alternative, according to acquired efficiency with the employment of zinc oxide (ZnO) nanorods and ZnO thin films at the same time. Both of these ZnO structures were grown electrochemically and poly(3-hexylthiophene) : phenyl-C61-butyric acid methyl ester; (P3HT: PCBM) was used as an active polymer blend, which was found to be compatible to prepared Indium-Tin-Oxide (ITO) substrate base. This ITO base was introduced with mentioned ZnO structure in such a way that, the most efficient configuration was optimized to be ITO/ZnO film/ ZnO nanorod/ P3HT: PCBM/Ag. All ZnO works were carried out electrochemically, that is indeed for the first time and at relatively lower temperatures. Forth one is an Organic Field Effect Transistor (OFET) device, which was prepared by a side chain thiophene based-dielectric gel copolymer and it shows a higher field-effect mobility and lower threshold voltage. The device is a sandwich type and promising results could be attributed to the compatibility that comes from the similarity between the chemical structures of semiconductor and dielectric layers due to π-π interactions caused by intermolecular overlapping of p orbitals in aromatic thiophene groups existing in both layers. Actually this kind of devices could also be a good inspiration for artificial skin concepts, if pressure sensitivity reveals at the same time. 19 One of the proposed device without pressure deal is the one, prepared with poly (MMA-coMTM) has a field-effect mobility of 0.57 cm2 /Vs, a threshold voltage of -0.20 V, on/off current ratio of 103 while the device prepared with just PMMA has a field-effect mobility of 0.38 cm 2 /Vs, a threshold voltage of -0.26 V and on/off current ratio of 102. There are some positive qualitative results for pressure sensitivity of OFET concepts as well.

**A. Bek[1], M. E. Taşgın[2], H. E. Ünalan[1]**, [1]Middle East Technical University, Turkey, [2]Hacettepe University, Turkey
**Title:** Molecular Activation by Plasmon Enhanced Nonlinear Conversion

**Absrtact.** In this study we propose a method for deterministic activation of few molecules. We demonstrate effective background-free continuous wave nonlinear optical excitation of enhanced yellow fluorescent protein molecules that are sandwiched between asymmetrically constructed plasmonic gold nanoparticle clusters [1]. We observe that infrared photons are converted to visible photons through efficient plasmonic second harmonic generation. Our theoretical model and simulations demonstrate that nonlinear conversion of continuous wave light becomes possible by Fano resonance in the nonlinear response. We show that nonlinearity enhancement of plasmonic nanostructures via coupled quantum mechanical oscillators such as molecules can be several orders larger as compared to their classical counterparts [2-3]. Our numerical simulations demonstrate that observation of second harmonic generation with continuous wave laser becomes possible owing to the cooperative action of conversion enhancement through Fano resonance, hybridization in the plasmon absorption spectrum and the size asymmetry of nanoparticle dimers. We further investigate the Fano resonances in intrinsic nonlinear optical response of single or coupled metal nanoparticles. We show that Fano resonance conditions can be tuned to enhance nonlinear optical conversion. When a bare metal nanoparticle is coupled with a plasmonic nonlinear converter, second harmonic generation can be enhanced by several orders of magnitude. This phenomenon emerges due to path interference effects. It is shown that second harmonic generation enhancement in hybrid structures can be obtained even in the absence of coupled quantum emitters [4]. This is an important finding with the potential of facilitating use of purely metallic nanoparticles with appropriate plasmonic properties for efficient nonlinear conversion.

**İlhami Gülçin,** Atatürk University, Turkey
**Title:** Polyphenolic Antioxidant Compounds-Potent Carbonic Anhydrase Enzymes Inhibitors

**Abstract.** Polyphenols compounds are a structural class of mainly natural, but also synthetic or semisynthetic, organic chemicals characterized by the presence of large multiples of phenol structural units. On the other hand, an antioxidant molecule has been defined as any substance when found in low concentrations compared to that of an oxidizable substrate significantly delays or inhibits the oxidation. Antioxidant compounds can scavenge free radicals and increase shelf life by retarding the process of lipid peroxidation, which is one of the major reasons for deterioration of food, medicine and pharmaceutical products during processing and storage. The major antioxidant compounds are especially phenolics and flavonoids, which are responsible for their health benefits. Carbonic anhydrase (CA, EC 4.2.1.1) is a pH regulatory enzyme in all life kingdoms, being found in organisms all over the phylogenetic tree. It catalyzes the hydration of carbon dioxide ($CO_2$) and water ($H_2O$) to bicarbonate ion ($HCO_3^-$) and the corresponding dehydration of $HCO_3^-$ in acidic medium with regeneration of $CO_2$. CA isoforms are found in a variety of tissues. They participate in several crucial biological processes such as acid-base balance, respiration, $CO_2$ and ion transportation, bone resorption, ureagenesis, gluconeogenesis, lipogenesis and electrolyte secretion. On the other hand, the phenyl moiety of phenol was found to lay in the hydrophobic part of the CA active site, where $CO_2$, the physiologic substrate of the CAs, binds in the precatalytic complex, explaining thus the behavior of phenol as a unique $CO_2$ competitive inhibitor. This presentation consists of two main sections. The first section is devoted to main phenolic antioxidant compounds in the foodstuffs and beverages. The second general section is about some definitions of CA inhibitory effects of the main phenolic compounds used for antioxidant activity. The phenolic compounds and acids had marked especially CA I, and II inhibition effects and might be used as leads for generating CA isoenzyme inhibitors. This class of compounds may lead to isoform-selective inhibitors targeting just one or few of the medicinally relevant CAs. In addition, there are given some chemical and kinetic basis and technical details related to phenolic antioxidant compounds and CA isoenzymes.

**B. Aktaş, R. Topkaya,** Gebze Technical University, Turkey
**Title:** Spin Waves in Ultra-Thin Magnetic Multilayers

**Abstract.** Spin wave resonance (SWR) has been widely used by researchers to study magnetic materials since the basic theory was introduced first by Kittel, Herring, Amend and Rado [1-3]. The SWR modes can be excited even by a uniform rf field if certain boundary conditions are met. In this study, the fundamental aspects of the collective spin behaviour in magnetic thin films were investigated. A basic dynamic equation of motion was used together with suitable boundary conditions at the surface or interfaces of magnetic thin films or multilayers. Gilberttype damping parameter was assumed in order to simulate the SWR spectra. It has been realized that the theory developed for magnetic multilayered thin films is suitable to apply to also thick (bulk) plate-like samples in limiting case for a large number of layer. The spectra strictly depend on equilibrium direction of dc magnetization. Since the equilibrium direction continuously varies as the external field is swept, it has to be calculated for each value of the external field during the record of the SWR spectra. However, it is quite a lot of time consuming, that is, the computer speed may not be enough especially for multilayered structures. Therefore, we have developed a new technique and algorithm to speed up this process. It has been shown that the SWR technique is very suitable to study the spin dynamics and extract various relevant magnetic parameters.

**Bruno DAUDIN**, University Grenoble Alpes, France
**Title:** GaN Nanowires Grown by Molecular Beam Epitaxy: From Spontaneous Nucleation to Light Emitting Heterostructures

**Abstract.** Nitride-based light emitting diodes (LEDs) are now widely used for a wide range of applications, including domestic lightning. Present applications are based on the combination of blue LEDs with phosphors in order to produce white light. This is due to the limited efficiency of LEDs in the long wavelength range (green and red). After discussing the limitations of conventional InGaN/GaN LEDs, I will show how nanowire (NW) based LEDs could allow one to extend the present limitations and overcome the present limitations of conventional LEDs. In particular, I will report on the growth, structural and optical characterization of InGaN nanowire (NW) heterostructures grown by molecular beam epitaxy, in view of realizing green and red light emitting diodes (LEDs). As evidenced by a combination of Energy Dispersive X-rays Spectroscopy (EDS) and photoluminescence (PL) data, it will be shown that Indium content can be as large as 60%. The dominant role of carrier localization in InGaN NWs will be discussed in details as well as the potential of NW heterostructures for a new generation of LEDs.

**Coşkun Kocabaş**, Bilkent University, Turkey
**Title:** Graphene based optoelectronics from visible to microwave

**Abstract.** Graphene emerges as a viable material for optoelectronics because of its broad optical response and gate-tunable properties. Atomic thickness together with tunable high mobility charge carriers yields a unique platform for electrical control of light-matter interactions. For practical applications, however, single layer graphene has performance limits due to its small optical absorption defined by fundamental constants. In the first part of the talk will review the recent research on graphene based optoelectronic devices covering a broad spectrum from visible to microwave. In the second part I will summarize our work on application of graphene in adaptive camouflage systems.

**Hüseyin Kurt**, İstanbul Medeniyet University, Turkey
**Title:** Heusler Alloys: A Rich Class of Materials

**Abastract.** Materials with varying electronic and magnetic structures enable new functionalities. For example, a semiconductor can be used to turn on and off electric current, a superconductor enable dissipation-less electric current to flow, and a spin valve device can act as a valve for electric current controlled by external magnetic fields. Heusler alloys are usually ternary alloys that can be engineered to exhibit electronic and magnetic properties spanning a very wide class of materials. By engineering a well ordered face centered cubic structure novel band structures and magnetic properties can be achieved. During this talk I will give a recent overview of novel materials with novel functionalities obtained through judicious selection of elements and crystallizing them in the L21 or C1b Heusler crystal structures.

**Sebahattin Tüzemen,** Atatürk University, Turkey
**Title:** Approaching to Gravity?

**Abstract.** Following the discovery of Higgs Bosons (H) with a mass of around 126 GeV/c2, it is hypothesized in this study that the gauge bosons (theoretically predicted to be gravitons) in the mass interaction may be produced by H, just as photons are produced in the case of electromagnetic (EM) interaction. Although the hypothesis is extremely difficult to prove experimentally in gravitation mechanism, from the parallel estimation of —quantum efficienciesǁ of virtual gravitons produced by virtual H annihilation in gravitation and virtual photons produced by virtual electron-positron annihilation in EM interaction, due to Heisenberg's Uncertainty Principle (HUP), we evaluate that the relative ratio of photon to graviton intensities is in the order of 1036 which is in agreement with the relative strength between EM and gravitational interactions predicted in Quantum Field Theory (QFT).

**Niyazi Meriç,** Director Ankara University Institute of Nuclear Sciences, Turkey
**Title:** Education and Training in the Nuclear Field

**Abstract.** Bearing in mind the long lead times generally required for nuclear education and training, the establishment and preservation of an adequate nuclear workforce supply calls for systematic planning decades ahead. In this respect, contradictory energy policies can have grave effects. A deteriorated global context caused by the persistent financial crisis and the negative sentiments in the wake of the Fukushima accident heighten uncertainties and may exacerbate existing shortcomings. Indeed, shifting or deferred government decisions act as deterrent mechanisms in investment and employment, and have deleterious repercussions on the interest and engagement of younger people in the industry.

Journal of Physics: Conference Series **707** (2016) 011002 doi:10.1088/1742-6596/707/1/011002

# Some Pictures of the International Physics Conference at the Anatolian Peak 2016 (IPCAP2016), Erzurum-Turkey

**Picture 1**. Prof. Dr. Rıdvan DURAK (The Conference Chair ) giving a welcome remark at the opening of IPCAP 2016.

**Picture 2**. At the opening ceromony of IPCAP 2016.

Journal of Physics: Conference Series **707** (2016) 011002 doi:10.1088/1742-6596/707/1/011002

**Picture 3**. Some participants of IPCAP2016 in Nenehatun Cultural Center on the day of the conference.

**Picture 4**. Bruno DAUDIN Nanophysique et Semiconducteurs, Institut NEEL / FRANCE delivering a keynote talk.

**Picture 5**. Mohd Nizar Bin HAMIDON from Engineering Faculty, Universiti Putra / MALAYSIA delivering a keynote talk.

**Picture 6**.Registration, Exhibition.

**Picture 7**. IPCAP2016 chair Mr. Durak and invited speaker Mr. Hamidon interview with National TV TRT.

Journal of Physics: Conference Series 707 (2016) 011002 doi:10.1088/1742-6596/707/1/011002

**Picture 8.** Chat in a coffee break of the conference.

**Picture 9.** Poster session.

**Picture 10.** IPCAP2016 Chair Mr. Durak, invited speakers Mr. Daudin and Mr. Hamidon and IPCAP2016 Team.

**Picture 11**. Sait Eren San, Sebahattin Tüzemen, invited by Gebze Technical University and Atatürk University, Turkey.

**Picture 12**. Welcome reception.

**Picture 13**. Welcome reception.

Journal of Physics: Conference Series **707** (2016) 011002 doi:10.1088/1742-6596/707/1/011002

**Picture 14**. Invited speakers at the ipcap2016 gala dinner.

**Picture 15.** Lunch & Social Event in the Palandöken Ski Center.

**Picture 16.** Lunch & Social Event in the Palandöken Ski Center.

**Picture 17.** Now is the tme to dance the halay.

Journal of Physics: Conference Series **707** (2016) 011003 doi:10.1088/1742-6596/707/1/011003

# Peer review statement

All papers published in this volume of *Journal of Physics: Conference Series* have been peer reviewed through processes administered by the proceedings Editors. Reviews were conducted by expert referees to the professional and scientific standards expected of a proceedings journal published by IOP Publishing.

# Investigation of chemical effect on the absorption parameters for some selected indium complex at 59.54 keV photon energy

**F. Akman[1], F. Akman[2], M.R. Kaçal[3], M.F. Turhan[4], F. Akdemir[2], A. Araz[2] and R. Durak[2]**
[1]Bingöl University, 1200, Bingöl, Turkey
[2]Bitlis Eren University, 13000, Bitlis, Turkey,
[3]Giresun University, 28100, Giresun, Turkey,
[4]Afyonkocatepe University, 03200, Afyonkarahisar, Turkey.
[5]Atatürk University, 25240, Erzurum, Turkey,

E-mail: fakman@bingol.edu.tr

**Abstract.** In this work, the mass attenuation coefficients, molecular, atomic and electronic cross sections, effective atomic numbers and electron densities of some selected indium complexes such as $C_5H_{10}InNO_9$, $C_3H_6InNO_9$ and $C_9H_{10}InNO_9$, were determined with experimentally and theoretically using transmission geometry at 59.54 keV photon energy which emitted from $^{241}$Am annular source. The results were compared with the theoretical calculations which obtained from the WinXCOM program. Also, the results were interpreted based on some chemical parameters such as energy gap, ionization energy, electron affinity, hardness, chemical potential, electronegativity and global electrophilicity values of malonate, 2,2 dimethylmalonate and phenylmalonate which were calculated using the density functional theory (DFT/B3LYP).

## 1. Introduction

Indium is a soft and silvery metal. It is an important part of corrosion resistant mirror surface, solar panels, nuclear reactors and touch screens. The mass attenuation coefficient, molecular, atomic and electronic cross sections, effective atomic number and electron density are the basic quantities required in determining the penetration of X-ray or gamma photons in a material. These parameters are used in X-ray fluorescence surface chemical analysis, radiation physics, dosimetric computations for health physics, elemental analysis, basic studies of nuclear physics, etc. In composite materials such as soil, plastic, complex, alloy, the atomic number cannot be represented uniquely across the entire energy region, as in the case of elements, by a single number. This number is defined as effective atomic number in composite materials. The electron density is defined as the numbers of electrons per unit mass.

There are several studies in the literature about the mass attenuation coefficients, effective atomic numbers and electron densities [1-9]. İçelli et al [1] determined the molecular, atomic, electronic cross sections and effective atomic numbers for some boron compounds and the trommel sieve waste using an extremely narrow-collimated-beam transmission geometry in the energy range 15.74-40.93 keV. The mass attenuation coefficients, effective atomic numbers and electron densities for Cd, Se, Te in elemental state and semiconductor CdSe, CdTe were estimated using an Ultra-LEGe detector at different energies from 9.7 to 87.3 keV by Cevik et al. [2]. Kaewkhao et al. [3] obtained the mass attenuation coefficients, total interaction cross sections, effective atomic numbers, effective electron densities and photon mean free paths of the Cu/Zn alloy at 356, 511, 662, 835 and 1275 keV energies using a NaI(Tl) scintillation detector. The effective atomic numbers and electron densities were determined from the experimental values of mass attenuation coefficients for some natural minerals at 22.1, 25.0, 59.5 and 88.0 keV energies using a Si(Li) detector system by Han et al [4]. Sharma et al. [5] calculated the effective atomic numbers by two different methods such as ratio of atomic to electron cross section and logarithmic interpolation of molecular cross section for different chemical compositions of calcium-strontium-borate glasses in the energy range from 1 keV to 100 GeV. Sidhu et al. [6] obtained the total mass attenuation coefficients, total photon interaction cross sections, effective atomic numbers and electron densities for some dosimetric compounds using transmission geometry at 59.54 keV photon energy. To understand of interaction of Xray photons with bacteriorhodopsin, the mass attenuation coefficients, effective atomic numbers and electron densities of bacteriorhodopsin and its comprising amino acids for photon energies 1 keV to 100 GeV were calculated by Ahmadi et al. [7]. Akça and Erzeneoğlu [8] estimated the mass attenuation coefficients, molecular, atomic, and electronic cross sections, effective atomic numbers and electron densities for compounds of biomedically important some elements such as Na, Mg, Al, Ca and Fe at 59.5 keV energy using a Si(Li) detector and a $^{241}$Am radioactive source. Akman et al. [9] determined the effective atomic numbers and electron densities from the total mass attenuation coefficients for some selected samarium compounds in the energy range from 36.847 up to 57.142 keV using the transmission geometry. Akman et al. [2016] determined the K shell absorption jump ratios, jump factors, effective atomic numbers and electron densities for some selected gadolinium compounds and the results were interpreted according to some chemical parameters.

In this work, the mass attenuation coefficients, molecular, atomic and electronic cross sections, effective atomic numbers and electron densities of some selected indium complexes such as $C_5H_{10}InNO_9$, $C_3H_6InNO_9$ and $C_9H_{10}InNO_9$, were estimated with experimentally and theoretically using transmission geometry at 59.54 keV photon energy. The results were compared with the theoretical calculations which obtained from the WinXCOM program. Also, the results were interpreted based on some chemical parameters such as energy gap, ionization energy, electron affinity, hardness, chemical potential, electronegativity and global electrophilicity values of malonate, 2,2 dimethylmalonate and phenylmalonate which were calculated using the density functional theory (DFT/B3LYP).

## 2. Experimental process and data analysis

The experimental set-up used for determining the mass attenuation coefficients, atomic, and molecular cross sections effective atomic numbers and electron densities is shown in Fig. 1.

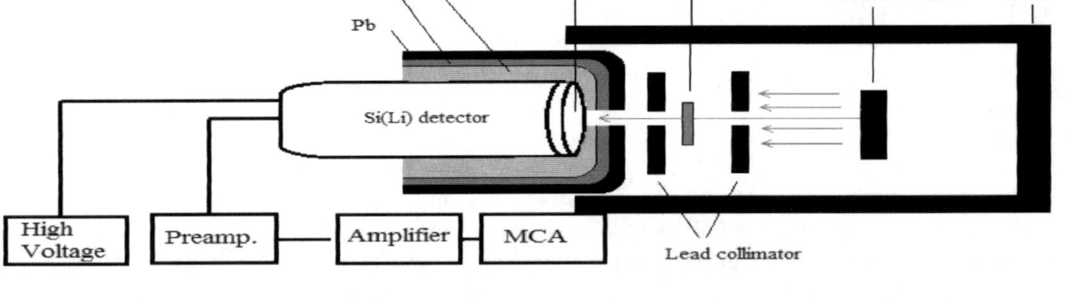


**Figure 1.** The experimental set-up

The samples were irradiated by 59.54 keV photon energy which emitted from 100 mCi annular $^{241}$Am radioactive source. The unattenuated ($I_0$) and attenuated (I) intensities counted with a Si(Li) detector having 160 eV resolution at 5.9 keV, active area 12.5 mm$^2$, sensitive depth 5 mm and Be window thickness 8 μm with coupled to 2048 multi-channel analyzer. The detector placed within the graded shield made from Pb, Fe and Al to filter of 26.4 keV energy coming from $^{241}$Am source and Np characteristic L X-rays. The detector energy calibration was done in the range 0 to 88 keV using standard test sources. The selected indium complexes such as $C_5H_{10}InNO_9$, $C_3H_6InNO_9$ and $C_9H_{10}InNO_9$ and pure In element were used as samples. The details of the preparation of samples were given in our earlier report [10]. The live time was selected as 1800 s for each sample. The areas under the attenuated and unattenuated photo-peak were estimated using Microcal Origin 7.5 demo software program. The net peak areas of each sample were determined after the background subtraction. The least square fit method with a Multi-Gaussian function was used to obtain the net peak area. A typical spectrum of $C_5H_{10}InNO_9$ sample is given in Fig. 2.

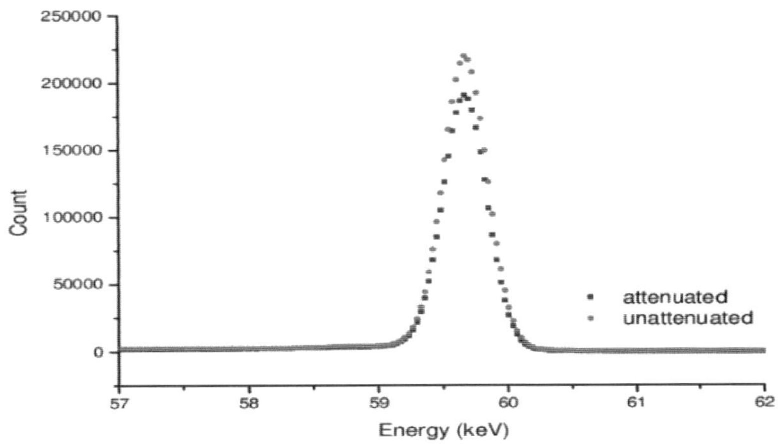


**Figure 2.** A typical spectrum of $C_5H_{10}InNO_9$

The energy gap, ionization energy, electron affinity, hardness, chemical potential, electronegativity and global electrophilicity parameters for malonate, 2,2 dimethylmalonate and phenylmalonate ions were calculated with DFT/B3LYP method using 631G+(d, p) basis set with the aid of Windows version of the GAUSSIAN 09 package program [11] and Gaussview molecular visualization program [12]. The calculated parameters for the present ions are listed in Table 1.

**Table 1.** Some parameters for ions obtained with DFT/B3LYP method using 631G+(d, p) basis set

| | DFT/B3LYP/631G+(d, p) | | |
|---|---|---|---|
| Parameters (eV) | Malonate | 2,2 dimethylmalonate | Phenylmalonate |
| Energy Gap (ΔE) | 4.67 | 3.88 | 2.89 |
| Ionization energy (I) | -2.66 | -2.39 | -1.94 |
| Electron affinity (A) | -7.34 | -6.28 | -4.83 |
| Hardness(η) | 2.34 | 1.94 | 1.44 |
| Chemical potential (μ) | 5.00 | 4.33 | 3.38 |
| Electronegativity (χ) | -5.00 | -4.33 | -3.38 |
| Global electrophilicity (ω) | 5.34 | 4.83 | 3.95 |

The frontier molecular orbitals and electrostatic potential surfaces (ESP) for malonate, 2,2 dimethylmalonate and phenylmalonate ions are given in Figs. 3-4, respectively.

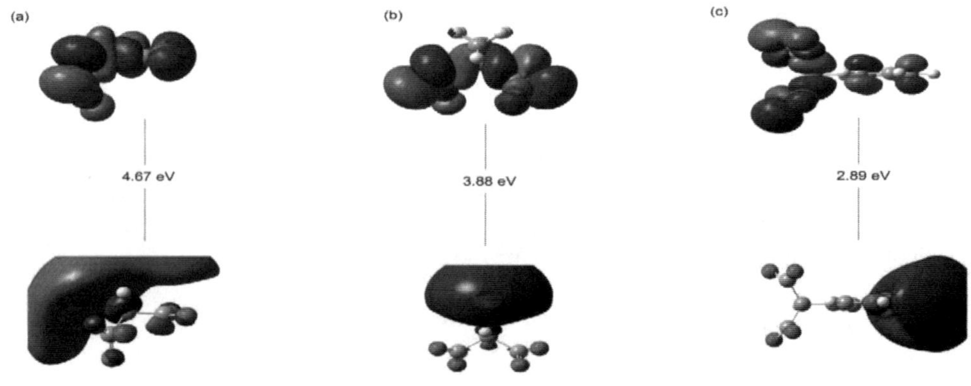


**Figure 3.** Frontier molecular orbitals for (a) malonate (b) 2,2 dimethylmalonate (c) phenylmalonate ions

The ESP was computed at B3LYP/631G+(d, p) by means of the optimized geometry for prediction of electron poor region (nucleophilic) and electron rich region (electrophilic). As seen from Fig. 4, the negative electrostatic potential is around the oxygen atoms which is seen as reddish blob, namely, the ion can be connected to the central atom from this points. The energy values of HOMO and LUMO and their energy gap reflect the chemical activity of the molecules. A small energy gap between HOMO and LUMO means low kinetic stability, more polarizable and high chemical reactivity of the molecule. The ionization energy and electron affinity can be expressed using HOMO and LUMO orbital energies (I=-EHOMO and A=-ELUMO). The global electrophilicity is the power of a ligand ($w=\mu 2/2\eta$), where $\eta$ is the stability ($\eta$=(-EHOMO+ELUMO)/2)) and $\mu$ is the chemical potential ($\mu$=1/2(EHOMO+ELUMO).

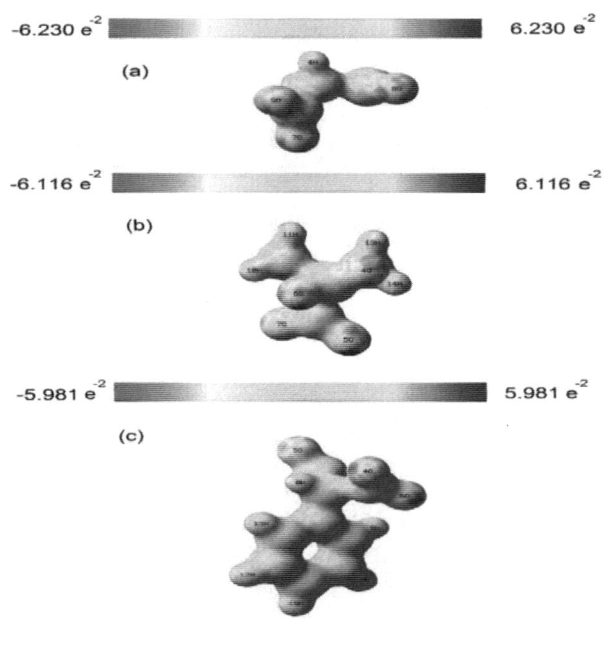


**Figure 4.** Electrostatic Potential (ESP) for (a) malonate (b) 2,2 dimethylmalonate (c) phenylmalonate ions

To obtain the mass attenuation coefficient for the present complexes, the following equation is used,

$$\mu/\rho = -\frac{1}{\rho x} ln\left(\frac{I}{I_0}\right) \tag{1}$$

where, $\mu/\rho$ is the mass attenuation coefficient, $I_0$ and I are the unattenuated and attenuated intensities and $\rho x$ is the mass per unit area of the complex and determined by,

$$\rho x = \frac{m}{\pi r^2} \tag{2}$$

here, r is the radius and m is the mass of the complex. For any compound, alloy, mixture or complex, the mass attenuation coefficient is determined by,

$$(\mu/\rho)_{complex} = \sum W_i(\mu/\rho)_i \tag{3}$$

in the equation, $(\mu/\rho)_{complex}$ is the mass attenuation coefficient of complex, $(\mu/\rho)_i$ is the mass attenuation coefficient of i.th constituent element in a complex and $W_i$ is the weight fraction and given by,

$$W_i = \frac{a_i A_i}{\sum_j a_j A_j} \tag{4}$$

where, $a_i$ is the number of atoms of i.th constituent element and $A_i$ is the atomic weight of i.th element in the complex. The total molecular cross section ($cm^2$/molecule) can be estimated using the following equation,

$$\sigma_{t,m} = \frac{1}{N}(\mu/\rho)\ complex \sum_i (n_i A_i) \tag{5}$$

here, $n_i$ and $A_i$ are the number of the atoms and the atomic weight of the i.th element in a complex and N is the Avogadro number. The total atomic cross section ($cm^2$/atom) can be obtained simply using the molecular cross section,

$$\sigma_{t,a} = \sigma_{t,m}\frac{1}{\sum_i n_i} \tag{6}$$

The total electronic cross section ($cm^2$/electron) is calculated theoretically for elements from the following equation,

$$\sigma_{t,e} = \frac{1}{N}\sum_i \frac{f_i A_i}{Z_i}(\mu/\rho)_i \tag{7}$$

in the equation, $f_i$ is the fractional abundance of the i.th constituent element with respect to total number of atoms and $Z_i$ is the atomic number.
From the values of atomic and electronic cross sections, the effective atomic number can be obtained semi-empirically using the following equation,

$$Z_{eff} = \frac{\sigma_{t,a}}{\sigma_{t,e}} \tag{8}$$

namely, the effective atomic number can be determined from the ratio of the atomic to electronic cross section. Lastly, the effective electron density (electrons/g) is estimated with the help of the effective atomic numbers,

$$N_E = \frac{Z_{eff}}{A_{tot}}(N n_{tot}) \tag{9}$$

here, $n_{tot}$ is the total number of atoms and $A_{tot}$ is the total atomic weight of complex.

## 3. Results and discussion

The mass attenuation coefficients ($\mu/\rho$), molecular ($\sigma_{t,m}$), atomic ($\sigma_{t,a}$) and electronic ($\sigma_{t,e}$) cross sections, effective atomic numbers ($Z_{eff}$) and electron densities ($N_E$) for some selected indium complexes at 59.54 keV are listed in Table 2 along with the theoretical calculated which is used WinXCOM program [13]. This program presents the attenuation coefficients of any substance as the sum of appropriately weighted contributions from the individual atoms. It is clearly seen from Table 2 that the $\mu/\rho$, $\sigma_{t,m}$, $\sigma_{t,a}$, $\sigma_{t,e}$, $Z_{eff}$ and $N_E$ values depend on the number of elements within complex. The maximum uncertainty in the determination of $\mu/\rho$, $\sigma_{t,m}$, $\sigma_{t,a}$, $Z_{eff}$ and $N_E$ values is estimated as %1.61. this uncertainty is attributed to systematic uncertainties, $I_0$ and I intensities uncertainties and mass per unit area measurements.

**Table 2.** The $\mu/\rho$, $\sigma_{t,m}$, $\sigma_{t,a}$, $\sigma_{t,e}$, $Z_{eff}$ and $N_E$ values at 59.54 keV

| | $\mu/\rho$ | | $\sigma_{t,m}$x$10^{23}$ | |
|---|---|---|---|---|
| Sample | Experimental | Theoretical | Experimental | Theoretical |
| In | 6.222±0.063 | 6.439 | 118.626±1.196 | 122.76 |
| $C_5H_{10}InNO_9$ | 2.366±0.032 | 2.284 | 134.741±1.821 | 130.06 |
| $C_3H_6InNO_9$ | 2.431±0.039 | 2.470 | 127.093±2.041 | 129.14 |
| $C_9H_{10}InNO_9$ | 1.948±0.021 | 2.025 | 126.482±1.369 | 131.46 |

**Table 2.** continued.

| $\sigma_{t,a}$x$10^{24}$ | | $\sigma_{t,e}$x$10^{25}$ | $Z_{eff}$ | | $N_E$x$10^{-23}$ | |
|---|---|---|---|---|---|---|
| Experimental | Theoretical | Theoretical | Experimental | Theoretical | Experimental | Theoretical |
| 1186.264±11.959 | 1227.59 | 250.53 | 47.350±0.477 | 49.00 | 5.819±0.059 | 6.02 |
| 51.824±0.701 | 50.02 | 15.30 | 33.882±0.458 | 32.71 | 74.720±1.010 | 72.13 |
| 63.546±1.020 | 64.57 | 18.21 | 34.903±0.560 | 35.47 | 59.208±0.951 | 60.16 |
| 42.161±0.456 | 43.82 | 14.03 | 30.040±0.325 | 31.22 | 76.439±0.827 | 79.45 |

According to the Table 2, the $\mu/\rho$, $\sigma_{t,m}$, $\sigma_{t,a}$, $Z_{eff}$ and $N_E$ values depend on the total atomic weight. The maximum differences between the measured and theoretical values of of $\mu/\rho$, $\sigma_{t,m}$, $\sigma_{t,a}$, $Z_{eff}$ and $N_E$ are found to be 3.80%. From these differences, we can sat taht the measured parameters are in good agreement with the theoretically calculated ones within the experimental uncertainties.

As seen from Table 1 and 2, the $\mu/\rho$, $\sigma_{t,m}$, $\sigma_{t,a}$, $Z_{eff}$ and $N_E$ values changed with energy gap, ionization energy, electron affinity, hardness, chemical potential, electronegativity and global electrophilicity. According to the Table 1 and 2, it is possible to say that the $\mu/\rho$, $\sigma_{t,a}$, and $Z_{eff}$ values increase with increasing energy gap, hardness, chemical potential and global electrophilicity values and decreasing ionization energy, electron affinity and electronegativity values. Also, the $N_E$ values increase with decreasing energy gap, hardness, chemical potential and global electrophilicity values and increasing ionization energy, electron affinity and electronegativity values. The same effects are shown in the theoretical values. To the best of the authors, the $\mu/\rho$, $\sigma_{t,m}$, $\sigma_{t,a}$, $Z_{eff}$ and $N_E$ values have been discussed based on these chemical parameters for the first time in the present work.

It can be concluded that the $\mu/\rho$, $\sigma_{t,m}$, $\sigma_{t,a}$, $Z_{eff}$ and $N_E$ values depend on the chemical environment of complex and these values depend on the total atomic weight and numbers of elements within complex. The measured values change with chemical parameters such as energy gap, ionization energy, electron affinity, hardness, chemical potential, electronegativity and global electrophilicity. It is believed that the measured $\mu/\rho$, $\sigma_{t,m}$, $\sigma_{t,a}$, $Z_{eff}$ and $N_E$ values are sufficiently reliable for In complexes. This method should be used for the other complexes or compounds at different energies.

**Acknowledgements**
The authors wish to thank Bitlis Eren University, Scientific and Technological Application and Research Centre for supporting the Gaussian 09 and Gaussian 05 view software.

## References

[1] İçelli O, Erzeneoğlu S and Bonçukcuoğlu R 2008 Nucl Instrum Meth B **266** 3226.
[2] Cevik U, Bacaksiz E, Damla N and Çelik A 2008 Radiat Meas **43** 1437.
[3] Kaewkhao J, Laopaiboon J and Chewpraditkul W 2008 J Quant Spectrosc RA **109** 1260.

[4] Han I, Demir L and Şahin M 2009 Radiat Phys Chem **78** 760.
[5] Sharma R, Sharma V, Singh PS and Singh T 2012 Ann Nucl Energy **45** 144.
[6] Sidhu BS, Dhaliwal AS, Mann KS and Kahlon KS 2012 Ann Nucl Energy **42** 153.
[7] Ahmadi M, Lunscher N and Yeow JTW 2013 Nucl Instrum Meth B **300** 30.
[8] Akça B and Erzeneoğlu SZ 2014 Sci Technol Nucl Instal **2014** 1.
[9] Akman F, Durak R, Turhan MF and Kaçal MR 2015 Appl Radiat Isotopes **101** 107.
[10] Akman F, Durak R, Kaçal MR and Bezgin F 2016 X-ray Spectrom **45** 103.
[11] Frisch MJ, Trucks AGW, Schlegel HB, Scuseria GE, Robb MA, Cheeseman JR, Scalmani G, Barone V, Mennucci B, Petersson GA, Nakatsuji H, Caricato M, Li X, Hratchian HP, Izmaylov AF, Bloino J, Zheng G, Sonnenberg JL, Hada M, Ehara M, Toyota K, Fukuda R, Hasegawa J, Ishida M, Nakajima T, Honda Y, Kitao O, Nakai H, Vreven T, Montgomery JA, Peralta JE, Ogliaro F, Bearpark M, Heyd JJ, Brothers E, Kudin KN, Staroverov VN, Kobayashi R, Normand J, Bakken K, Adamo C, Jaramillo J, Gomperts R, Stratmann RE, Yazyev O, Austin AJ, Cammi R, Pomelli C, Ochterski JW, Martin RL, Morokuma K, Zakrzewski VG, Voth GA, Salvador P, Dannenberg JJ, Dapprich S, Daniels AD, Farkas O, Foresman JB, Ortiz JV, Cioslowski J and Fox DJ 2010 Gaussian 09 *Gaussian Inc* Wallingford CT.
[12] Dennington R, Keith T and Millam J 2010 GaussView Version 5 Semichem Inc Shawnee Mission KS.
[13] Gerward L, Guilbert N, Jensen KB and Levring H 2001 Radiat Phys Chem **60** 23.

# Structure, Electronic and Nonlinear Optical Properties of Furyloxazoles and Thienyloxazoles

**Ozlem Dagli[1], Rabia Gok[2], Mehmet Bahat[2] and Akif Ozbay[2]**

[1] Gazi University, Faculty of Medicine, Ankara, Turkey
[2] Gazi University, Physics Department, Ankara, Turkey

bahat@gazi.edu.tr

**Abstract.** Geometry optimization, electronic and nonlinear optical properties of isomers of furyloxazole and thienyloxazole molecules are carried out at the B3LYP/6-311++G(2d,p) level. The conformational analysis of 12 compounds have been studied as a function of torsional angle between rings. Electronic and NLO properties such as dipole moment, energy gap, polarizability and first hyperpolarizability were also calculated.

## 1. Introduction

Thiophene, furan and oxazole are five membered heterocyclic aromatic molecules. They have been studied extensively in the literature because of their wide applications in various areas of science and industry [1-3]. Also, they are important building blocks for new materials possessing interesting electronic, mechanical and biological properties [2,3]. Discrete oligomers and their polymers such as polytiophenes, and polyfurans are widely studied by experimentally and computationally [4,5]. There have been few reports on mixed these oligomers and other five or six membered heterocycles. Quantum mechanical methods are able to describe the geometric and electronic structure of such kind of compounds in good agreement with experimental data. Theoretical investigations have been reported addressing the torsion potential of 2,2'-bithiophene, and bifuran, [6-7]. A literature review can be found in in these references. It becomes interesting to study composite molecules made of thiophene, furan and oxazoles. When two units connected together by a single bond they produce new composite molecules such as furyloxazoles and thienyloxazoles. Each of them have six isomers. It is important to understand their molecular structure and physical quantities to provide guidelines for the development of new materials.

Furyloxazoles and thienyloxazoles are composite molecules having subunits as furan and thiophene monomers connected to oxazole monomer by a single bond, respectively. Each oh them have six structural isomers. The subunit molecules can rotate around the single bond linking them together. The aim of present work is to obtain theoretical data of the structure, conformation, dipole moment, highest occupied molecular orbital energy (HOMO) and lowest unoccupied molecular orbital energy (LUMO) differences that is energy gap, dipole polarizability, and first hyperpolarizability of isomers of furyloxazole and thienyloxazole molecules.

## 2. Computations and Methodology

All the quantum chemical DFT calculations were performed with the Gaussian03 program [8]. The B3LYP hybrid functional of Becke [9,10] was employed here in combination with the 6-311++G(2d,p) basis set [11-12] for all calculations. The equilibrium geometries of 12 isomers of furyloxazole and thienyloxazole molecules have been obtained through geometry optimization. Vibrational frequencies were calculated to confirm the equilibrium geometries corresponding to the energy minima. The conformational analysis of all compounds have been studied as a function of inter-rings C-C dihedral angle varying between 0° and 180° with a step of 15°. Electronic and nonlinear optical (NLO) properties such as dipole moment, HOMO-LUMO energy gap, polarizability, anisotropic polarizability and first hyperpolarizability were also calculated. When a molecular system having a permanent dipole moment $\mu_0$ interact with a uniform external electric field $E$, molecule have an induced dipole moment $\mu_i$ as

$$ {}_i = {}_0 + \alpha_{ij}E_{ij} + \beta_{ijk}E_{ijk} + \cdots \qquad i,j,k = x,y,z \tag{1}$$

Where $\alpha_{ij}$ and $\beta_{ijk}$ are polarizability and first hyperpolarizability compenets, respectively. The averaged polarizabiliy $(\alpha)$, anisotropic polarizability $(\Delta\alpha)$ and total first hyperpolarizability $(\beta)$ were reported as

$$\alpha = \frac{1}{3}(\alpha_{xx} + \alpha_{yy} + \alpha_{zz})^{1/2} \tag{2}$$

$$\Delta\alpha = \left\{\frac{1}{2}\left[\left(\alpha_{xx} - \alpha_{yy}\right)^2 + (\alpha_{xx} - \alpha_{zz})^2 + \left(\alpha_{yy} - \alpha_{zz}\right)^2\right]\right\}^{1/2} \tag{3}$$

$$\beta = \left[\left(\beta_{xxx} + \beta_{xyy} + \beta_{xzz}\right)^2 + \left(\beta_{yyy} + \beta_{yzz} + \beta_{yxx}\right)^2 + \left(\beta_{zzz} + \beta_{zxx} + \beta_{zyy}\right)^2\right]^{1/2} \tag{4}$$

where x, y and z components of $\alpha$ and $\beta$ were obtained from Gaussian output.

## 3. Results and Discussion

The Figure 1 depicts the B3LYP/6-311++G(2d,p) optimized structures that completely planar. The changes in the electronic energy terms are caused by internal rotation around C-C bond between rings can be revealed by keeping all other geometry parameters fixed. In Figures 2 and 3 the relative torsion energy as a function of torsion angle are presented where the energy of global minimum structure is taken as a zero level.

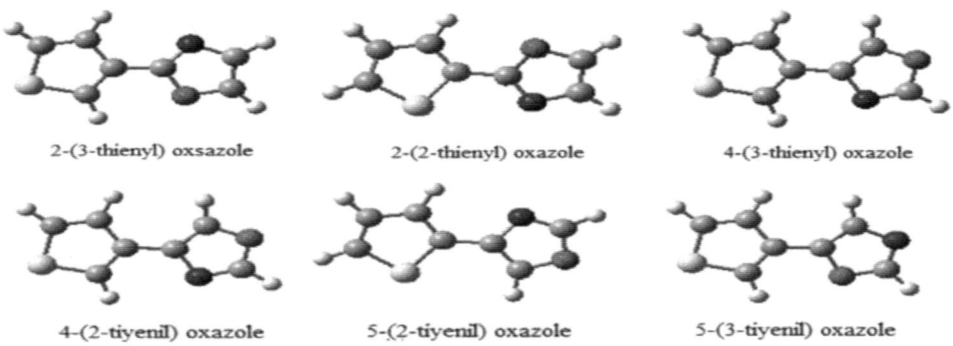


**Figure 1**. Molecular structures of thienyloxazoles (grey ball: C, white ball:H, blue ball:N, red ball:O, yellow: S. For furyloxazoles yellow:O).

Journal of Physics: Conference Series **707** (2016) 012002 doi:10.1088/1742-6596/707/1/012002

Figures 2 and 3 show that there are two minima on the potential energy surface. We take global minimum at 0°, and local minimum at 180°. The height of rotational barrier at 90° fall within the range between 3.8-6.4 kcal.mol$^{-1}$ and 3.9-6.0 kcal.mol$^{-1}$ for thienyloxazoles and furyloxazoles, respevtively.

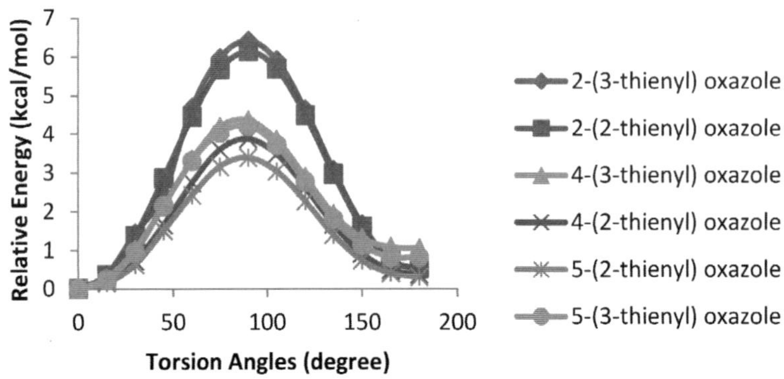


**Figure 2.** Relative torsion energies for thienyloxazoles.

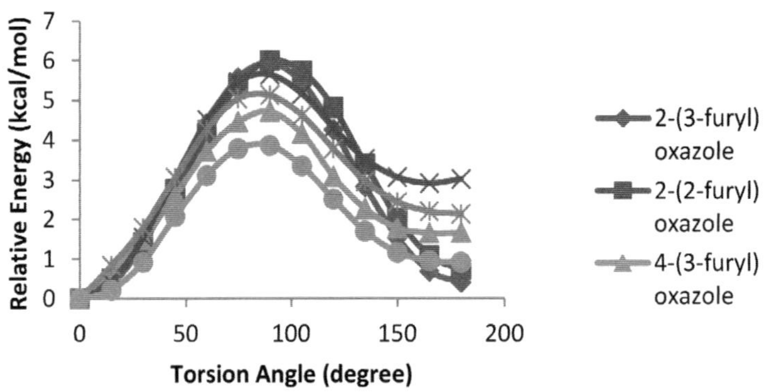


**Figure 3.** Relative torsion energies for thienyloxazoles.

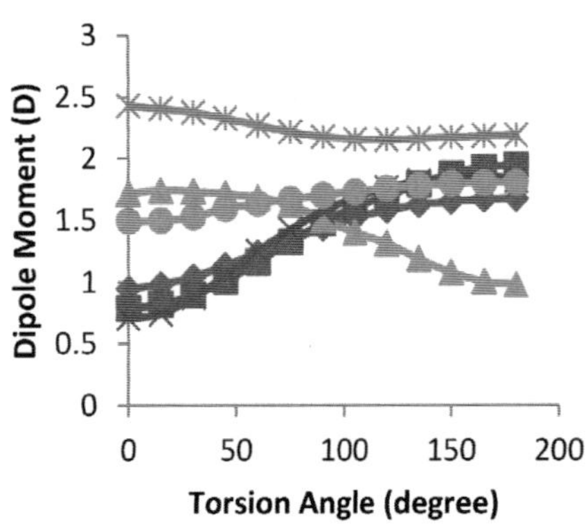


**Figure 4.** Dipole moments for thienyloxazoles.

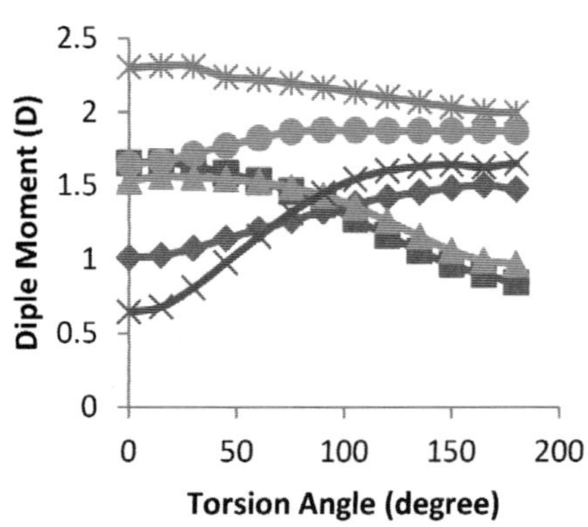


**Figure 5.** Dipole moments for thienyloxazoles.

Dipole moment is an important descriptor of charge distribution in a molecule. The dipole moments are given in Tables 1 and 2 and in Figures3 and 4 as a function of torsion angle. Figures show that dipole moment values of studied molecules are heavily depend on torsion angle with exceptions. The B3LYP/6-311++G(2d,p) calculated electric dipole moment, (H-L) energy gap, polarizability, anisotropic polarizability and first hyperpolarizability values are given in Table 1 and 2 for thienyloxazoles and furyloxazoles, respectively.

**Table 1.** Calculated electric dipole moment, (H-L) energy gap, polarizability, anisotropic polarizability and first hyperpolarizability of thienyloxazoles.

| Molecule | Dipole (Debye) | \|H-L\| eV | α (au) | Δα (au) | β (au) |
|---|---|---|---|---|---|
| 2-(3-thienyl) oxazole | 1.0118 | 4.88 | 112.9 | 57.6 | 155.3 |
| 2-(2-thienyl) oxazole | 1.6556 | 4.50 | 114.8 | 65.8 | 190.6 |
| 4-(3-thienyl) oxazole | 1.5288 | 5.27 | 91.1 | 39.1 | 48.5 |
| 4-(2-thienyl) oxazole | 0.6462 | 4.83 | 109.2 | 51.0 | 92.7 |
| 5-(2-thienyl) oxazole | 2.3022 | 5.25 | 114.6 | 54.8 | 51.5 |
| 5-(3-thienyl) oxazole | 1.6461 | 4.98 | 94.3 | 42.6 | 235.4 |

**Table 1.** Calculated electric dipole moment, (H-L) energy gap, polarizability, anisotropic polarizability and first hyperpolarizability of furyloxazoles.

| Molecule | Dipole (Debye) | \|H-L\| eV | α (au) | Δα (au) | β (au) |
|---|---|---|---|---|---|
| 2-(3-furyl) oxazole | 0.9459 | 5.15 | 94.5 | 35.0 | 100.1 |
| 2-(2-furyl) oxazole | 0.7889 | 4.67 | 100.1 | 40.8 | 45.1 |
| 4-(3-furyl) oxazole | 1.7181 | 5.50 | 91.6 | 30.6 | 117.1 |
| 4-(2-furyl) oxazole | 0.7071 | 4.96 | 95.5 | 40.0 | 25.1 |
| 5-(2-furyl) oxazole | 2.4286 | 4.66 | 99.4 | 37.5 | 185.4 |
| 5-(3-furyl) oxazole | 1.4911 | 5.15 | 94.1 | 32.2 | 169.1 |

The HOMO-LUMO energy gap is an important parameter determine the usefulness of molecules in technology. The calculated absolute HOMO-LUMO energy values ranges from 4.66 to 5.50 eV. The dipole polarizability values ranges from 91 to 115 au. The calculated hyperpolarizability values ranges from 45 to 235 au. that shows very low nonlinear optical properties.

## 4. References

[1] Torsi L, Malitesta C, Sabbatini L, Zambonin P, Dodabalapur A and Katz H E 1998 *Mat.Sci. Eng.* **5** 233-36
[2] Whalley A C, Steigerwald M L, Guo X and Nuckolls C 2007 *J. Am. Chem. Soc.* **129** 12590-91
[3] Shi M and Shoichet M S 2008 *J. Biomater. Sci. Polym.* **19** 1143-57
[4] Iraqi A, Barker G and Pickup D 2006 *React. Func. Poly.* **66** 195-200
[5] Moncada J L and Labbe A T 2006 *Chem Phys. Let.* **429** 161-165
[6] Raos G, Famulari A and Marcon V 2003 *Chem. Phys. Let.* **379** 364-72
[7] Blom J W and Wheler S E 2014 *J. Chem. Theory. Comp.* **10** 3647-55
[8] Frisch M J and et al., Gaussian 03W, Revision D.01, Gaussian, Inc: Pittsburgh PA (2003)
[9] Becke A D 1993 *J.Chem.Phys.* **98** 5648-60
[10] Lee C, Yang W and Parr R G 1988 *Phys. Rev. B* **37** 785-794
[11] Hehre W J, Ditchfield R and Pople J 1972 *J. Chem. Phys.* **56**, 2257-2261
[12] Hariharan P C, Pople J 1974 *J. Chem. Phys.* **27**, 209-214

# Approaching to Gravity?

**Sebahattin Tüzemen**

Department of Physics, Faculty of Science, Atatürk University, 25240 Erzurum, Turkey

E-mail: stuzemen@atauni.edu.tr

**Abstract.** It has been detected by the proton collision experiments in Large Hadron Collider of CERN that the Higgs Bosons (H) exists with a mass of around 126 GeV/$c^2$. It is predicted in this study that the mediating particles, which are theoretically invoked to be gravitons in the gravitational interaction, may be produced by H, just as photons are produced in the case of electromagnetic (EM) interaction. From the simultaneous evaluation of "quantum efficiencies" of virtual gravitons produced by virtual H annihilation in gravitation and virtual photons produced by virtual electron-positron annihilation in EM interaction, we predict that the relative ratio of photon to graviton intensities is in the order of estimated value in Quantum Field Theory (QFT).

## 1. Introduction

The results [1,2] provided by Large Hadron Collider of CERN proved that the Higgs Bosons (H) exist with a very high mass corresponding to Einstein's equivalent mass energy of around 126 GeV, although detection of these bosons is so difficult due to their very fast decay in apparently less than one septillionth of a second $(\tau_H \approx 10^{-24} s)$ [3]. Prediction of such high mass particle was a necessity to explain the symmetry breaking reality of universe between matter and anti-matter. This theoretical assumption eventually provided Higgs et al the 2013 Noble Prize in Physics, after 50 years.

The Standard Model (SM) of the universe predicts that the four kinds of mediating forces between the two physical objects or particles in space occurs when the two have the same kind of charge. A certain type of charge of a particle appears in the forms of electric charge, color charge, weak charge and mass charge that determine the role of participation in the interaction processes. These effects respectively correspond to electromagnetic, strong nuclear, weak nuclear and gravitational interactions. Elementary particles named leptons and quarks are the key particles that provide the three of the four kinds of charges except the mass charge, within any kind of particle or object. The fourth one entered the elementary particles' table as the Higgs Bosons, explaining the mass charge [1,2]. Mediating gauge bosons in any kind of interaction can be produced by the annihilation of the same kind of particle-antiparticle collisions of leptons and quarks. The SM establishes that gauge bosons; photons, W-Z bosons and gluons (g) perfectly explain the three of the four interaction forces; electromagnetic, weak and strong nuclear interaction except for gravitation. Gravitons (G) needed to be theoretically comprehended in Theory of Everything (TOE), fitting in the fourth interaction of the SM in Grand Unified Theory (GUT).

The TOE reduces to General Relativity of Einstein and the gravitation law of Newton in the classical and weak field limit as described by Feynman et al [4]. However, while the general theory of relativity is also macroscopic and the geometric theory of gravitation using the space-time curvature, present study proposes a microscopic model of mediating gravitons due to virtual Higgs boson annihilation, using similar predictions for the electric charge interaction in the SM, and evaluates the recently observed non-stability of these mass charge particles with extremely short lifetimes (see also Ref. [5] for extended discussion).

## 2. Fundamental principles

In the SM of particle physics uses the Heisenberg uncertainty principal (HUP),

$$\Delta m \approx \frac{\Delta E}{c^2} = \frac{\hbar}{c^2 \Delta t} \quad \text{or} \quad \Delta E \approx \frac{\hbar}{\Delta t} \qquad (1)$$

within the duration of any kind of fundamental interactions and, therefore the mediating particles known as virtual gauge bosons (photons, W-Z bosons, g and theoretically gravitons) are produced due to uncertainty in time, $\Delta t$. This means a particle with a mass of $\Delta m$ can exist if its duration is less than $\Delta t$. The production mechanisms are illustrated by the Feynman diagrams involving unstable particle-antiparticle annihilations. The more mass these virtual particles have the shorter the time they can exist, according to Eq.1. Because photons and gravitons are massless, they live forever and the electromagnetic and gravitational interaction can reach infinite distances with the speed of light, while the other two short distance interactions involving heavy W-Z bosons and gluons (g) occur only within the nuclei. Gluons cannot exhibit long distance effect due to g-g coupling, confining the particles within the nuclei [6].

Existence of carrier bosons that are needed for gravitational interaction is a mystery. One should think that gravitons can also be produced by the annihilation of these elementary particles. The question is could this be annihilation of H?

## 3. Results and discussion

Let us discuss the two long distance fundamental forces; the EM and the gravity. In the EM interaction, there appear force carrying virtual photons as mediating bosons due to HUP, and the most basic annihilation process is the electron-positron annihilation producing the two $\gamma$-photons. Similarly, in gravity, there appear mediating bosons, hypothetically named gravitons and let us assume that the utopic H annihilation continuously produce gravitons in vacuum as proposed by HUP before the actual H boson decays. This annihilation energies with an H mass of around $126\,GeV/c^2$ corresponds to virtual graviton frequencies in the order of $\nu \cong 10^{24}\,Hz$ which is nearly $10^4$ times greater than the most energetic $\hbar\nu$ $\gamma$-ray energy observed in the EM spectra. In the case of actual gravitons, such frequency corresponds to a wavelength of around 10 attometres (10-18 m) for an individual graviton and cannot have interaction by any kind of known material so that it cannot be sensed by any kind of presently designed detectors.

Gravitons should not be mixed up with the gravitational waves which were predicted by Einstein in 1916 on the basis of his general relativity theory, presuming that the space-time curvature fluctuates due to the rotation or any kinds of motion of the gravitating source until it was, 100 years after, directly detected by the LIGO Scientific Collaboration and Virgo Collaboration teams last week, observing gravitational waves from a pair of black holes by the Advanced LIGO detectors [7]. Period of these waves ranges from the age of universe to the orders of milliseconds, depending on the source whether it is initiated by some quantum fluctuations in the early universe or a rotating supernova. The wave and gravitons are supposed to propagate with the speed of light and this is also the expansion velocity of the gravitational effect unlike the case of Newtonian gravity that supposes infinite velocity. As has been discussed in detail [5], the angular momentum is conserved in the H annihilation by decaying into two gravitons with spins of 2 in opposite directions just as in the case of para-positronium decaying into two photons in its annihilation [8].

Let us now consider, simultaneously that an individual electron and an individual H boson as sources of respectively EM and gravitational interactions are separately propagating the particular force carrying bosons due to the requirement of HUP, having $\Delta x$ uncertainties in the order of the thermal de Broglie wavelengths of each individual particle. The ratio of the thermal de Broglie wavelengths is given by

$$\frac{\lambda_H}{\lambda_e} = \sqrt{\frac{m_e}{m_H}} \tag{2}$$

where $m_e$ and $m_H$ are respectively the electron and H masses. Intensities of propagated gauge bosons into empty space from the spherical surface areas of $4\pi r^2$ with diameters in the orders of the thermal de Broglie wavelengths $(\lambda = 2r)$ of each individual particle can be written as follows:

$$I_P = \frac{1}{(4\pi r^2)_e}(\eta_P \hbar\omega_P) = \frac{1}{\pi\lambda_e^2}(\eta_P \hbar\omega_P) \tag{3-a}$$

$$I_G = \frac{1}{(4\pi r^2)_H}(\eta_G \hbar\omega_G) = \frac{1}{\pi\lambda_H^2}(\eta_G \hbar\omega_G) \tag{3-b}$$

where $\eta_P$ is the photon emission rate and $\eta_G$ is the G emission rate, i.e. numbers of emitted force carrying bosons per second, and $\omega_P$ and $\omega_G$ are the gauge photon and gauge G angular frequencies, respectively. One should notice that Eqs. (3a and b) explain the inverse square laws.

Using Eq.(2) and considering photon and G energies ($\hbar\omega_P = m_e c^2$ and $\hbar\omega_G = m_H c^2$ respectively) correspond to Einstein's equivalent mass energy of annihilating particles, the ratio of photon to G intensities is given by the direct proportion of photon and G emission rates in respectively EM and gravitational interactions as follows:

$$\frac{I_P}{I_G} = \frac{\eta_P}{\eta_G} \tag{4}$$

It is basically considered [5] that the emission rates are basically given by the virtual annihilation rates around the actual particles and that depends on the availability of a certain particle defined by its lifetime. In the first case, electron-positron pair appears and annihilates and in the latter, H and a-H pair appears and annihilates in every short enough $\Delta t$ time duration given by HUP in Eq. (1). In other words the quantum efficiency $(Q)$ of mediating bosons in each interaction is proportional with the life times;

$$Q \propto \tau \tag{5}$$

where $\tau$ is the lifetime of each particle acting role in the interaction processes.

In the SM the mean lifetime of Higgs bosons is predicted to be in the order of $\tau_H = 1.56x10^{-22}$ $s$. Because of very limited detector resolutions, apparently this measurement can only be estimated within a factor of 1000 [9]. Therefore we take the lifetime for H as one septillionth of a second $(\tau_H \approx 10^{-24} s)$ [5], which fits in the width of the H lifetime within limiting factor.

According to HUP, average life of a virtual H pair around the incident H can only be;

$$\Delta t \approx \frac{\hbar}{\Delta E} = 5.2x10^{-27} s \tag{7}$$

where the reduced Planck constant, $\hbar \approx 6.6\times10^{-16} eV.s$ and $\Delta E \approx 126 \ \ GeV$ for H bosons. This means that only around a couple of hundred, $(\tau_H/_{\Delta t}) \approx 200$, virtual H annihilation is permitted producing virtual immortal G (since they are massless) before the actual unstable H boson decays within its lifetime which is assumed to be in the order of $\sim 10^{-24} s$ [5].

On the other hand, electrons, the basic source of EM interaction, are very stable and the electron's mean lifetime is given as $4.6x10^{26}$ years [10,11], which is much longer than the life of the universe. Therefore it has been considered [5] that an electron producing electric charge interaction is as old as the universe which allows electronic life time taken as $\tau_e = 14x10^9 \, years = 4.4x10^{17} s$ with a very high confidence and this means that the electron of incidence continuously produce virtual gauge photons by the virtual electron-positron annihilations since the beginning of the universe. Similar arguments in the above paragraph for the lifetime of virtual electron-positron pair around an actual electron having Einstein's equivalent mass energy of $0.511 \ \ MeV$ works out to be

$$\Delta t \approx \frac{\hbar}{\Delta E} = 1.3x10^{-21} s \tag{8}$$

Therefore, huge numbers of around $(\tau_e/_{\Delta t}) \approx 3.4x10^{38}$ virtual electron-positron annihilations have occurred around the actual electron since the beginning of universe, producing immortal gauge photons for the EM interaction.

This means that photon to graviton intensity ratio in Eq.(4) is in the order of $(3.4x10^{38}/200) \approx 10^{36}$ which is also equals to the ratio of photon to graviton emission rates produced in respectively EM and gravitational interactions. This should also correspond to the relative strength of the EM in comparison to the gravitation which is given as in the order of $10^{36}$ too, determined from "coupling constant" arising in the QFT.

## 4. Conclusion

It has been proposed that gauge bosons (gravitons) in the mass charge interaction may be produced by the annihilation of these mysterious H bosons with their anti-character just as photons are produced as gauge bosons in the case of electric charge interaction, due to basic principles of quantum electrodynamics. From the estimated "quantum efficiencies" of gravitons and photons emitted around an actual H and electron, respectively, it has been shown that relative strength of gravitational interaction initiated by an H is in the order of 10-36 of the EM interaction initiated by an electron and the ratio found by the predictions in this letter is quite consistent with the Quantum Field Theory. The huge difference between the EM and gravitational interaction comes from the fact that the EM is produced by a very stable source such as electrons with very long lifetimes while gravitation is produced by a very unstable source such as H bosons with lifetimes in the order of $10^{-24} s$.

**Acknowledgements**

I thank Research Associate Ali Baltakesmez of Çankırı Karatekin University for assisting in preparation of this manuscript.

**References**

[1] The CMS Collaboration 2012, Physics Letters B, 716, p.30–61.
[2] The ATLAS Collaboration 2012, Physics Letters B, 716, p.1-29.
[3] Williams B. (2014), An Investigation of Modern Physics.

[4] Feynman, R. P.; Morinigo, F. B.; Wagner, W. G.; Hatfield, B. 1995. Feynman Lectures on Gravitation. Addison-Wesley. ISBN 0-201-62734-5.
[5] Tüzemen, S 2016, Physical Science International Journal, 9(1):1-6.
[6] Ametller L I, Gava E, Paver N, and Treleani D 1985, Phys. Rev. D 32, p.1699
[7] LIGO Scientific Collaboration and Virgo Collaboration 2016, Physical Review Letters 116 (6).
[8] Karshenboim, Savely G. 2003, International Journal of Modern Physics A [Particles and Fields; Gravitation; Cosmology; Nuclear Physics] 19 (23): 3879–3896.
[9] Charley S. 2014, Measuring the Lifetime of the Higgs Boson, Symmetry. Available at www.symmetrymagazine.org.
[10] Particle Data Group 2012, Physical Review D 86 (1).
[11] Back, H. O.; et al. 2002, Physics Letters B 525(1-2), p.29–40.

# Calculation of K Shell Intensity Ratios and Line Widths of Ti and some of its compounds by means of 5,96 keV energy

***Oğuz Kağan Köksal[1], Gökhan Apaydın[1], Erhan Cengiz[1], Kazım Karabulut[1]**

Department of Physics, Faculty of Sciences, Karadeniz Technical University, 61080, Trabzon, Turkey
E-mail: okoksal@ktu.edu.tr

**Abstract.** K shell intensity ratios and Line Widths of pure Ti and some of its compounds have been determined experimentally using an Ultra-LEGe detector with resolution 140 eV at 5.9 keV. The samples were excited 5.96 keV photons emitted from a $^{55}$Fe radioisotope source with 50 mCi activity. The experimental values of the K shell intensity ratios have been compared with the experimental and theoretical values available in the literature for pure Ti and line widths have been only compared with a theoretical value for pure Ti.

## 1. Introduction

Information experimental result regarding the K X-ray intensity ratios for different elements is important because of its extensive use in atomic, molecular and nuclear physics, properties of superconductor, semiconductor thin films, etc., and non-destructive analysis of materials using energy dispersive X-ray fluorescence (ED-XRF) techniques [1].
It is well known that X-ray emission spectra are influenced by the chemical combination of X-ray emitting atoms with different ligands. The effects of the chemical structure are not large and a theoretical interpretation of these effects has not been established completely. Therefore, chemical effects have rarely been utilized in the characterization of materials. Furthermore; the process of particle size effects, vacancy production, data analysis procedures to acquire the areas under peaks, detector efficiency, and self- absorption in the specimen have to be well controlled. On the contrary, the measured X-ray fluorescence parameters can be erratic because of the effect of these conditions [2].

A number of researchers [1, 3-10] have worked on K-shell intensity ratios for various elements, compounds and alloys with different detector types and radioactive sources but not K line widths. So this paper is aimed to determine the K Shell X-ray the intensity ratios of Ti and some of compounds and K line widths. And this work is to interpret the measured results according to the chemical environment.

## 2. Experimental procedure

The geometry of the experimental set-up and the present experimental equipment have been described in previous study [10]. The purity of commercially obtained materials is better than 99%. Powder samples were got by purchasing from Sigma-Aldrich and Alfa Aesar company. In this experimental set-up, 5.9 keV X-ray photons emitted by an annular 50 mCi $^{55}$Fe radioactive sources were used to excite the samples. The fluorescence K X-rays from the sample were detected by a collimated Ultra-LEGe detector having an energy resolution of 140 eV at 5.96 keV. In the experimental determinations, spectral deconvolution is one of the main problems that arise when determining these parameters due to the strong peak overlapping in ED-XRF system. In the present work, a peak fitting program developed by Origin Company (Origin 7.0) was used to determine peak analysis. In the present study, Fig. 1 shows the typical K X-ray spectrum for Titanium. The experimental $K_i$ X-ray production cross-sections were acquired by the following relation:

$$\sigma_{X_i} = \frac{N_{Xi}}{I_0 G \varepsilon_{Xi} \beta_{Xi} m} [i = \alpha, \beta] \quad (1)$$

where $N_{Xi}$ is the measured intensity (area under the photo peak) corresponding to the $K_i$ group of X-rays, $I_0$ is the intensity of the incident radiation, G is the geometric factor, $\varepsilon_{Xi}$ the detection efficiency for the Ki group of X-rays and $\beta_{Xi}$ the self-absorption correction factor for the target material, which accounts for the absorption in the target of the incident photons and the emitted characteristic X-rays. mi is the thickness of the target in g/cm$^2$.

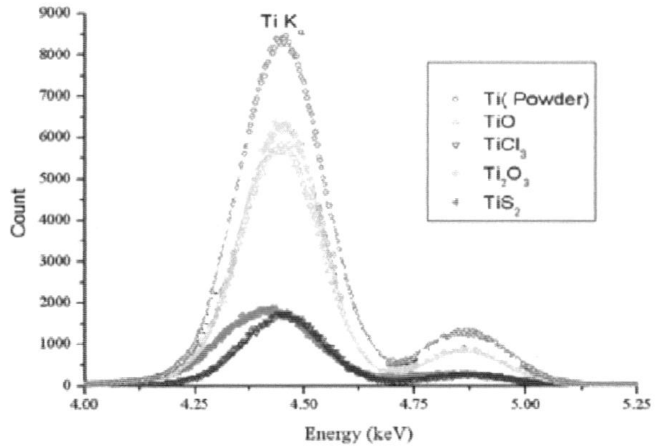


**Figure** 1. Typical K X-ray spectrum for Titanium and its compounds

where $N_{Xi}$ is the measured intensity (area under the photo peak) corresponding to the $K_i$ group of X-rays, $I_0$ is the intensity of the incident radiation, G is the geometric factor, $\varepsilon_{Xi}$ the detection efficiency for the Ki group of X-rays and $\beta_{Xi}$ the self-absorption correction factor for the target material, which accounts for the absorption in the target of the incident photons and the emitted characteristic X-rays. mi is the thickness of the target in g/cm$^2$.
The self-absorption correction was calculated using the equation[11]

$$\beta = \frac{1 - \exp\left[-\left(\frac{\mu_{inc}}{\cos\theta_1} + \frac{\mu_{emt}}{\cos\theta_2}\right)t\right]}{\left(\frac{\mu_{inc}}{\cos\theta_1} + \frac{\mu_{emt}}{\cos\theta_2}\right)t} \qquad (2)$$

where $\mu_{inc}$ and $\mu_{emt}$ are the mass attenuation coefficients [12] of incident photons and emitted characteristic X-rays, respectively; t is the thickness of the target in g cm$^{-2}$; the angles of the incident photons and the emitted X-rays with respect to the normal of surface, $\theta_1$ and $\theta_2$, were equal to 45$^0$ and 0$^0$ in the present experimental set-up, respectively.

The product $I_0G\varepsilon$, containing the terms related to the incident photon flux, geometrical factor and intrinsic absolute efficiency of the X-ray detector, was determined by collecting the $K_\alpha$ and $K_\beta$ X-ray spectra of samples of S, Cl, K, Ca and Ti for $^{55}$Fe in the same geometry using the equation

$$I_0G\varepsilon_{Ki} = \frac{N_{Ki}}{\sigma_{Ki}\beta_{Ki}m}[i = \alpha, \beta] \qquad (3)$$

where $N_{Ki}$ is the measured intensity (area under the photo peak) corresponding to the $K_i$ group of X-rays, $I_0$ is the intensity of the incident radiation, G is the geometric factor, $\varepsilon_{Ki}$ is the detection efficiency for the $K_i$ group of X-rays and $\beta_{Ki}$ is the self-absorption correction factor for the target material, which accounts for the absorption in the target of the incident photons and the emitted characteristic X-rays. m is the thickness of the target in g/cm$^2$.
The factor $I_0G\varepsilon_{Ki}$ was fitted as a function of energy using the following equation for $^{55}$Fe:

$$I_0G\varepsilon_{Ki} = A_0 + A_1E_x + A_2E_x^2 + A_3E_x^3 \qquad (4)$$

where $E_x$ is the X-ray energy and, $A_0$,$A_1$, $A_2$ and $A_3$ are constants evaluated from a fitting polynomial. The variations of the factors $I_0 G\varepsilon$ as a function of energy are shown in Fig. 2.

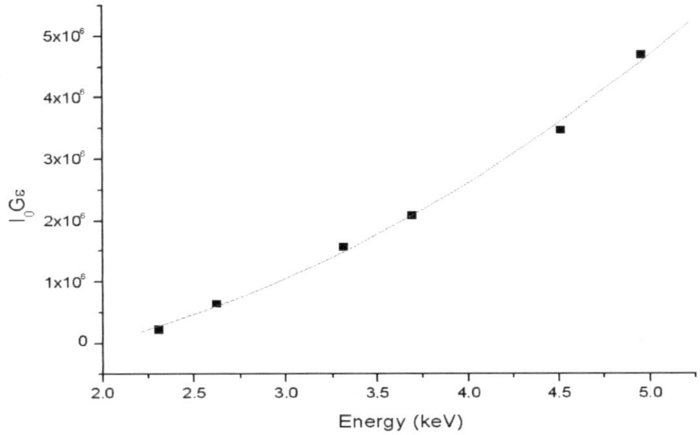


**Figure 2**. The variation in the factor $I_0$ Gε as a function of the mean K X-ray energy for $^{55}$Fe radioisotope

The theoretical K X-ray production cross sections were evaluated by using the relation;

$$\sigma_{Ki} = \sigma_K(E)\omega_K F_{Ki} \quad (5)$$

where $\sigma_K(E)$ is the K –shell photoionization cross-section of the given element for the excitation energy E [13], $\omega_K$ is the K-shell fluorescence yield [14], and $F_{Ki}$ is the emission rate of the fractional X-ray for $K_i$ X-rays [15].
The semi-empirical K shell fluorescence yields $\omega_K$ were measured according to the following equation:

$$\omega_K = \frac{\sum \sigma_{Ki}}{\sigma_K(E)} (i = \alpha, \beta) \quad (6)$$

where $\sum \sigma_{Ki}$ is the total K X-ray fluorescence cross-section obtained experimentally, $\sigma_K(E)$ is the theoretical K shell photoionization cross-section of a given element for the excitation energy E [13].

K shell line widths were calculated by dividing the fluorescence yield of the radiative transition rate [16] as using the formula shown below;

$$\Gamma_K = \frac{\Gamma_K(R)}{\omega_K} \quad (7)$$

Where $\Gamma_K$ is the K shell line width and $\omega_K$ is K shell fluorescence yield.
Intensity ratios of K shell were determined as ratio of cross-sections of $K_\beta$ and $K_\alpha$ with using the equation illustrated below;

$$K_\beta/K_\alpha = \frac{\sigma K_\beta}{\sigma K_\alpha} \quad (8)$$

Where $K_\beta/K_\alpha$ is K shell intensity ratio; $\sigma K_\alpha$ and $\sigma K_\beta$ are cross-sections belong with K shell.

## 3. Results and discussion

Experimental $K_\beta/K_\alpha$ x-ray intensity ratios values for Ti and its compounds, measured for incident photon energies 5,95 keV, are presented in Table 1. Experiental K X-ray intensity ratio values have been compared with theoretical estimates based on relativistic Hartree– Fock and Hartree–Slater theories calculated by [16,17], [18] and the other experimental values. The obtained $K_\beta/K_\alpha$ intensity ratios values are more compatible with relativistic Hartree-Fock theory's values rather than the other one.

**Table 1**. Intensity Ratios for Titanium and its compounds

| Element | Kβ/Kα | | | | | |
|---|---|---|---|---|---|---|
| | Experimental | | | Theoretical | | |
| | Measured | [19] | [10] | [17] | [16] | [18] |
| Ti Powder | 0,1049±0,0063 | 0,134 | 0,1282 | 0,1355 | 0,1137 | 0,1136 |
| $TiCl_3$ | 0,1139±0,0068 | --- | --- | --- | --- | --- |
| TiO | 0,1047±0,0063 | --- | --- | --- | --- | --- |
| $Ti_2O_3$ | 0,1141±0,0069 | --- | --- | --- | --- | --- |
| $TiS_2$ | 0,1039±0,0062 | --- | --- | --- | --- | --- |

The values are in the agreement between present results and theoretical predictions are within the range 15-24%, 0-9%, and 0-9% respectively. According to [19] and [10], the results of present study is different between 14-23% and 11-19%, respectively. Experimental K shell line width values for Ti and its compounds, measured for incident photon energies 5,95 keV, are presented in Table 2.

**Table 2**. Line Widths for Titanium and its compounds

| Samples | Kα Line Widths | Theoretical [14] |
|---|---|---|
| Ti Powder | 0,78±0,05 | 0,94 |
| TiO | 0,77±0,05 | --- |
| $TiCl_3$ | 0,87±0,05 | --- |
| $Ti_2O_3$ | 0,86±0,05 | --- |
| $TiS_2$ | 0,77±0,05 | --- |

The uncertainties in the K X-ray intensity ratios are estimated to be less than 6% and are found propagating the errors in various parameters used for the determination of the intensity ratios. The uncertainties in the parameters are listed in Table 3.

**Table 3**. Uncertainties in the quantities used to determine the parameters

| Quality | Nature of uncertainty | Uncertainty (%) |
|---|---|---|
| $N(K_i)$ ($i=\alpha, \beta$) | Counting statistic | $\leq 3$ |
| $I_0 G \varepsilon_{K_i}$ | Errors in different parameters used to evaluate factor | $\leq 2$ |
| β | Error in the absorption coefficients correction at incident and emitted photon energies | $\leq 3$ |
| t | Errors in the weight and thickness of the samples | $\leq 2$ |

In this study, $K_\beta/K_\alpha$ intensity ratios and K line widths for Ti compounds were experimentally investigated and the conclusions are summarized in four points.

First,$K_\beta/K_\alpha$ intensity ratios and K line widths vary with the change of the oxidation numbers. Second, it is known that different interatomic distances (lattice parameters) cause different interactions between the central atom and the ligand groups and these interactions can change the $K_\beta/K_\alpha$ intensity ratios K shell line widths. Third, K shell vacancies are filled by outer shell electrons and these electrons feel the effect of ligand atoms more strongly than the inner shell electrons. Fourth, the ligands bound to the central Ti atom have different electronegativity values and these differences can change $K_\beta/K_\alpha$ intensity ratios and K shell line widths. To acquire more absolute results about chemical effects on the cross sections and intensity ratios, it is planned to extend these measurements to various elements and compounds [20].

In this work, in order to reduce the absorption, thin samples were used as the target; furthermore, an absorption correction was also performed for each sample. In order to reduce the statistical error, the spectra were collected under the $K_\alpha$ and $K_\beta$ peaks. In the experimental determinations, spectral deconvolution is one of the main problems that arise when determining these parameters due to the strong peak overlapping in ED-XRF system. Good statistics is not enough for this purpose and a careful fitting methodology is required in order to obtain accurate values for the peak areas. Conventional fitting programs have not shown good performance, but the method of parameter optimization previously developed for X-ray fluorescence analysis. In the present work, a new peak fitting program developed by Origin Company (Origin 7.0) program was used to determine the accurate peak intensity.

The experimental values of K X-ray intensity ratios for titanium element and compounds have important role in the material, chemical sciences and industrial area.

## References

[1] Apaydın, G., Aylıkcı,V., Cengiz,E., Kaya,N., Kobya,Y., Tıraşoğlu, E. (2008). Determination of K Shell X-ray intensity ratios for some heavy elements. Radiation Physics and Chemistry 77, 923-927.

[2] Cengiz, E., Tıraşoğlu, E., Aylıkcı, V., Apaydın, G. (2010). The investigations on K and L X-ray fluorescence parameters of gold compounds. Radiation Physics and Chemistry 79, 809-815.

[3] McCrary, J.H., Singman, L.V., Ziegler, L.H., Looney, L.D., Edmonds, C.M.,Harris, C.E. (1971). K fluorescent X-ray relative intensity measurements. Physical Review A 4, 1745-1750.

[4] Campbell, J.L., Schulte, C.W. (1980). Energy-dispersive measurement of Ka X-ray line widths and relative intensities.Physical Review A 22,609-613.

[5] Bhuinya, C.R., Padhi, H.C. (1992). Alloying effect on the Kb/Ka intensity ratios of Ti, Cr and Ni. J. Phys. B 25, 5283–5287.

[6] Küçükönder, A., Sahin, Y., Büyükkasap, E., Kobya, A.I. (1993). Chemical effects on $K_\beta/K_\alpha$ X-ray intensity ratios in coordination compounds of some 3d elements. J. Phys. B: At. Mol. Opt. Phys. 26, 101-105.

[7] Be, M.M., Lepy, M.C., Plagnard, J., Duchemin, B., 1998. Measurement of relative X-ray intensity ratios for elements in the $22\leq Z\leq 29$ region. Appl. Radiat. Isot. 49, 1367-1372.

[8] Ertuğrul, M., Söğüt, Ö ., Şimşek, Ö ., Büyükkasap, E. (2001). Measurement of $K_\beta/K_\alpha$ intensity ratios for elements in the range $22\leq Z\leq 69$ at 59.5 keV. J. Phys. B: At. Mol. Opt. Phys. 34, 909–914.

[9] Mukoyama, T., Taniguchi, K., Adach, H. (2001). Single-atom approximation for K $\beta$ to- K $\alpha$ X-ray intensity ratios in chemical compounds of 3d elements. Phys. Rev. A 63, 042514 1-4.

[10] Ertuğral, B., Apaydın, G., Baltaş, H., Çevik, U., Ertuğrul, M., Kobya, A. İ. (2007). $K_\beta/K_\alpha$ X-ray intensity ratios for elements in the range $16\leq Z\leq 92$ excited by 5.9, 59.5, and 123.6 keV photons. Radiation Physics and Chemistry 76, 15–22.

[11] Hubbell J. H. Seltzer S. M. (1995). Natl. Inst. Stand. Phys. Lab. NISTIR. 5632.

[12] Web-1: /http://physics.nist.gov/S, consulted 22 March 2016.

[13] Scofield, J.H. (1973). Theoretical photoionization cross sections from 1 to 1500 keV. UCRL Report 51326. Lawrence Livermore Laboratory. Livermore. CA.

[14] Krause M. O., Oliver, J.H. (1979). J.Phys.Chem.Ref. Data,8,329.

[15] Broll, N. (1986). Quantitative X-ray fluorescence analysis X-Ray Spectrom. 15, 271–285.

[16] Scofield, J.H. (1974b). Relativistic Hartree-Slater values for K and L X-ray emission rates. At. Data Nucl. Data Tables, 14, 121-137.

[17] Scofield, J.H. (1974a). Exchange corretions of K X-ray emission rates. Phys. Rev. A ,9, 1041-1047.

[18] Manson, S.T., Kennedy, D.J. (1974). X-ray emission rates in the Hartree–Slater approximation. At. Data Nucl. Data Tables 14, 111–120.
[19] Salem, S.I., Panossian, S.L., Krause, R.A. (1974). Experimental K and L relative X-ray emission rates. At. Data Nucl. Data Tables 14, 91–109.
[20] Cengiz, E., Aylıkcı, V., Kaya, N., Apaydın, G., Tıraşoğlu, E. (2008). Chemical effects on K and L shell production cross sections and transfer probabilities in Nb compounds. Journal of Radıoanalytıcal and Nuclear Chemıstry 278,89-96.

# EPR Study of [Cu(Him)$_4$]·2Br Complex

**İ Yıldırım**[1] **and B Karabulut**[2]

[1]Department of Radiotherapy, Vocational School of Health Services, Biruni University, Topkapı 34010, Istanbul, Turkey
[2]Department of Computer Engineering, Faculty of Engineering, Ondokuz Mayıs University, Kurupelit 55139, Samsun, Turkey

iyildirim@biruni.edu.tr

**Abstract**. The single crystal of [Cu(Him)$_4$]·2Br (Him: $C_3H_4N_2$, imidazole) complex has been investigated at ambient temperature in three mutually perpendicular planes by EPR technique. The magnetic environments of [Cu(Him)$_4$]·2Br complex have been identified by EPR technique. The study reveals the existence of two magnetically inequivalent $Cu^{2+}$ sites. The principal values of g tensors were obtained. The EPR parameters show that the paramagnetic centers have rhombic symmetry.

**Keywords**. EPR, imidazole, $Cu^{2+}$, spin-spin interaction.

## 1. Introduction
The imidazole and imidazole derivatives ligands play an important role in the coordination chemistry [1]. Methyl imidazole and its derivative is extremely important biologically becomes one tooth ligand coordinated to the metal atom through the nitrogen atom. They are used in the in pharmacy, industry, electrochemical hybridization indicators and electroactive markers of DNA in many types of intensive imidazole derivatives [2-11]. Overall it is a very large work area of the imidazole complex. These are electronic structure solutions, analysis of spectroscopic and magnetic properties are most important. In this study, the EPR investigation of single crystal [Cu(Him)$_4$]·2Br complex has been performed g parameters have been calculated.

## 2. Experimental
EtOH solution of Him (0.6 g, 8.8 mmol) and EtOH solution of $CuBr_2$ (0.5 g, 2.2 mmol) was stirred at room temperature. After about six weeks, dark blue crystals of the [Cu(Him)$_4$]·2Br complex suitable for EPR analyses were collected. [Cu(Him)$_4$]·2Br complex has monoclinic space group $C2/c$ and the unit cell parameters are $a$=*14.739(4) Å*, $b$=*9.029(1) Å*, $c$=*14.269(4) Å*, $\beta$=*115.95(1)*° and $Z$=*4* [12].

EPR spectra were recorded on a Varian E-109C EPR spectrometer.

## 3. Results and discussion
[Cu(Him)$_4$]·2Br single crystal of suitable size are selected to record spectrum for every 30° angle between 0° and 180° by rotating the crystal in *a*c*, *ba** and *cb* planes at room temperature. The EPR spectra of [Cu(Him)$_4$]·2Br complex in Figure 1a when the magnetic field is in *a*b* plane 0° from *a**

axis, in Figure 1b when the magnetic field is in *cb* plane 30° from *b* axis. The angular variation of $g^2$ *a*c*, *ba** and *cb* planes are shown in Figure 2.

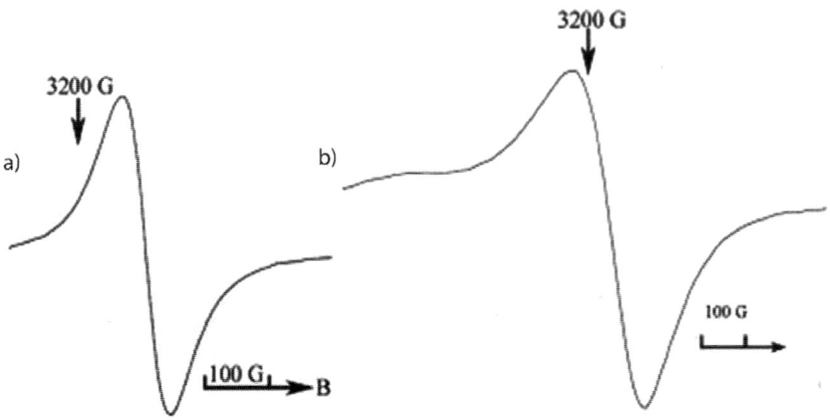


**Figure 1. a)** EPR spectra of $[Cu(Him)_4]\cdot 2Br$ single crystal when the magnetic field in the *a*c* plane 0° away from the *a**-axis. **b)** EPR spectra of $[Cu(Him)_4]\cdot 2Br$ single crystal when the magnetic field in the *cb* plane 30° away from the b-axis.

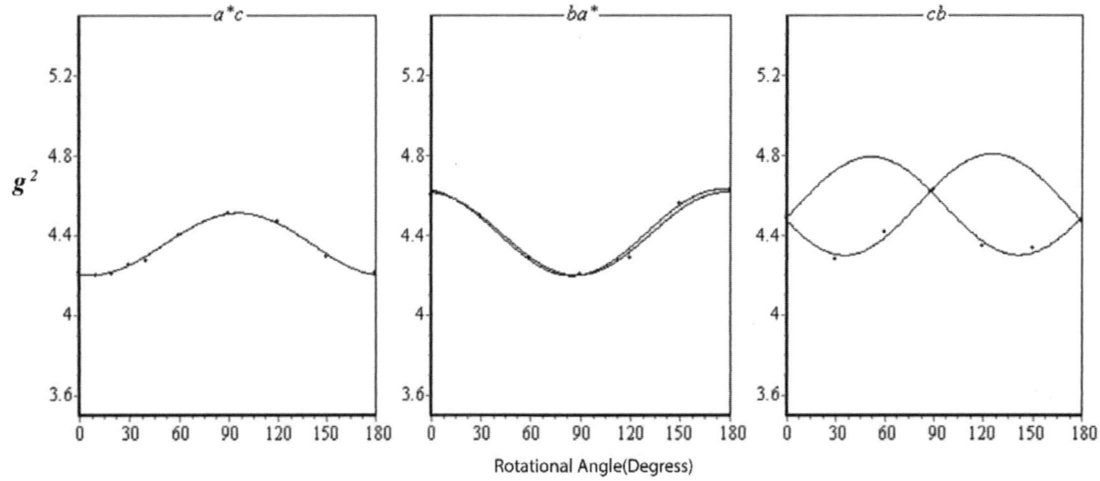


**Figure 2.** Angular variations of the EPR spectra of $[Cu(Him)_4]\cdot 2Br$ single crystal.

Since S=1/2 and I=3/2 for $Cu^{2+}$ ion, the spectrum consist of four lines. But, the EPR spectra of $[Cu(Him)_4]\cdot 2Br$ complex has one line in all orientation. It is considered that this is due to the spin-spin interaction and spin–lattice coupling. The angular variations of $g^2$ in three mutually perpendicular planes are shown in Figure 2. The angular variation of $g^2$ values for a specific line fits to the following the expression which is valid for each plane:

$$g_k^2(\theta) = g_{ii}^2 \cos^2\theta + g_{jj}^2 \sin^2\theta + 2g_{ij}^2 \sin\theta\cos\theta \qquad (1)$$

In this equation, i, j, k = x, y, z represent laboratory coordinates. θ is the rotation angle. $g_{ii}^2$, $g_{jj}^2$ and $g_{ij}^2$ correspond to tensor elements of g which can be found by a fitting procedure [13].

The principal values of g tensor for $[Cu(Him)_4]\cdot 2Br$ complex was evaluated by considering the following spin Hamiltonian,

$$\mathcal{H} = \beta\left(g_x B_x S_x + g_y B_y S_y + g_z B_z S_z\right) + A_x I_x S_x + A_y I_y S_y + A_z I_z S_z \quad (2)$$

The Hamiltonian includes electron Zeeman and hyperfine interactions. To find g value, we have used an iterative numerical technique [14]. g tensors are constructed and diagonalized to find their principal values and the results are found to indicate rhombic symmetry. The calculated g parameters are given in Table 1. From this table, we deduce that the $[Cu(Him)_4]\cdot 2Br$ complex has an octahedral environment with a rhombic distortion. This behaviour of the spectra corresponds to a paramagnetic sample with monoclinic crystal symmetry.

**Table 1.** Principal g values and their direction cosines of $[Cu(Him)_4]\cdot 2Br$ single crystal ($\Delta g=\pm 0.0005$).

| Site | g | Direction cosines | | |
|---|---|---|---|---|
| | | *a* | *b* | *c* |
| | $g_x$=2.050 | 0.997 | 0.054 | 0.061 |
| I | $g_y$=2.077 | -0.005 | 0.784 | -0.621 |
| | $g_z$=2.193 | -0.082 | 0.619 | 0.781 |
| | | *a* | *b* | *c* |
| | $g_x$=2.048 | 0.957 | 0.237 | 0.169 |
| II | $g_y$=2.079 | -0.291 | 0.762 | 0.578 |
| | $g_z$=2.193 | 0.008 | -0.602 | 0.798 |

**Acknowledgments**
This work was supported by the Ondokuz Mayıs University by project No PYO.FEN.1904.10.18.

**References**

[1] Navarro J A R and Lippert B 2001 *Coord. Chem. Rev.* **222** 219
[2] Jaouen G 2006 *Bioorganometallics. Biomolecules, Labeling, Medicine*, (Weinheim: Wiley–VCH Verlag GmbH & Co. KGaA)
[3] Prenesti E, Berto S and Daniele P G 2003 *Spectrochim. Acta A* **59** 201
[4] Kim E, Kamaraj K, Galliker B, Rubie N D, Moelnne-Loccoz P, Kaderli S, Zuberbulhler A D and Karlin K D 2005 *Inorg. Chem.* **44** 1238
[5] Zhang S S, Niu S Y, Qu B, Jie G F, Xu H and Ding C F 2005 *J. Inorg. Biochem.* **99** 2340
[6] Wang Q X, Jiao K, Sun W, Jian F F and Hu X 2006 *Eur. J. Inorg. Chem.* **9** 1838
[7] Wang Q X, Jiao K, Liu F Q, Yuan X L and Sun W 2007 *J. Biochem. Biophys. Methods* **70** 427
[8] Li F, Chen W, Tang Ch and Zhang S 2008 *Talanta* **77** 1
[9] Sava G, Clerici K, Capozzi I, Cocchietto M, Gagliardi R, Alessio E and Mestroni G 1999 *Anticancer Drugs* **10** 129

[10] Velders A H, Bergamo A, Alessio E, Zangrando E, Haasnoot J G, Casarsa C, Cocchietto M, Zorzet S and Sava G 2004 *J. Med. Chem.* **47** 1110
[11] Ravera M, Baracco S, Cassino C, Zanello P and Osella D 2004 *Dalton Trans.* **15** 2347.
[12] Parker O J and Breneman G L 1995 *Acta Crystallographica C* **51** 1097
[13] Misra S K and Sun J 1991 *Phys. Rev. B* **44** 10333
[14] Karabulut B and Tapramaz R 1999 *Radiat. Phys. Chem.* **55** 331

# Investigation of coherent to incoherent scattering cross section ratios of some foil metals depending on the temperature

**N. Çatal[1], M. Ertuğrul[2], Y. Özdemir[1]**

[1]Department of Physics, Faculty of Science, Ataturk University, Erzurum, Turkey
[2]Department of Electrical and Electronic, Faculty of Engineering, Ataturk University, Erzurum, Turkey

E-mail: nurselcatal@gmail.com

**Abstract.** In this study, it was aimed at examining the cross section ratios of coherent and incoherent scattering depending on the temperature for the elements Cd, In, Sn and Pb by 59.5 keV γ-rays from a 100 mCi $^{241}$Am radioisotope point source. The coherent and incoherent cross section of Cd, In, Sn and Pb have been measured by using a Si(Li) solid-state detector at temperature between 30-300 °C. Coherent to incoherent cross section ratios and FWHM (Full width at half maximum) of the elements have been calculated. Temperature-dependent changes of the parameters have been given in the graphical forms. Based on the results obtained, coherent to incoherent cross section ratios of the elements are dependent on the temperature. It is observed that coherent to incoherent cross section ratios of Cd, In, and Pb decrease with increasing temperature. For Sn, first of all coherent and incoherent intensity ratios decrease, then increase and decrease again respectively. To sum up, coherent to incoherent cross section ratios tend to decrease with increasing temperature.

## 1. Introduction

Coherent and incoherent scattering is a type of interaction of γ-rays and x-rays with matter. Coherent and Compton scattering are major processes for obtaining information about the tructural properties of matter. The bound electrons of an atom, which dominate coherent scattering for most of the X-ray and low-energy γ -ray regimes, make the contribution to coherent scattering. In condensed matter physics as well, coherent and Compton scattering have played a central role in the understanding of the excited states of many important systems.(İçelli and Erzenoğlu, 2002) The coherent scattering of photons by atoms includes Rayleigh scattering, nuclear Thomson scattering, Delbruck scattering and nuclear resonance scattering. Incoherent scattering is one of the major processes by which γ-rays interact with matter in the energy range from 0.1 to 5 MeV (Shivaramu et al., 1980; Kurucu et al., 1998). Coherent and incoherent scattering cross-sections are used in such diverse applications as medical X-ray technology, power reactor shielding, industrial radiation processing and analysis of nuclear physics experiments (Hubbell et al., 1975).

Differential cross section is used in the calculation of radiation attenuation, reactor shielding, industrial radiography, transport and energy deposition in medical physics and in a variety of other fields (Kurucu et al., 1998). Ertugrul (2001) investigated dependencies of the coherent to in- coherent intensity ratio on mean atomic number of the compounds. Icelli and Erzeneoglu (2002) experimentally investigated coherent to incoherent scattering differential cross section ratios of some elements at scattering angles of 55° and 115° at 59.54 keV. Sogut et al (2003) investigated atomic number of coherent to incoherent scattering intensity ratios in compounds depending on average atomic number. Simsek and Ertugrul (2004) measured the ratio of differential cross sections for coherent to Compton scattering of 59.54 keV at scattering angles of 40° and 135° for Zr, Nb and Mo targets. Singh et al. (2007) have measured effective atomic number of composite materials using coherent to incoherent scattering of 279 keV gamma rays. Singh et al. (2012) have measured coherent to incoherent cross section ratio of elements in the range $6 \leq Z \leq 82$ for 59.54 keV gamma photons. Singh et al. (2013) have measured the ratio of differential cross sections for coherent to incoherent scattering at 145 keV at scattering angles of 50°, 70° and 90° for C, Al, Fe, Mo, Sn, W, Au and Pb. Akkus et al (2015) have measured coherent to incoherent scattering differential cross-section ratios of Au at 59.54 keV depending on polar and azimuthal angular. As far as we know there is no study depending on the temperature even though several studies about coherent and incoherent scattering cross section ratios exist in the related literature. In the present work, coherent to incoherent differential cross-section ratios of Cd, In, Sn and Pb have been measured at temperature between 30°-300°C) at 59.54 keV photon energy by using Si (Li) solid-state detector.

## 2. Theory

The theory of scattering of X-rays from free electrons was studied by J.J. Thomson, and it is called as the Thomson scattering. The unpolarized Thomson scattering cross section contains Rayleigh's result (Strutt, 1871a,b). Differential Thomson scattering cross section per electron, is given as;

$$\frac{d\sigma^T}{d\Omega} = \frac{r_e^2}{2}\,(1 + cos^2\theta) \tag{1}$$

and the total, angle-integrated Thomson scattering cross section is;

$$\sigma^T = \frac{8\pi}{3} r_e^2 \tag{2}$$

where the classical electron radius is $r_e = e^2/m_e c^2$

The differential Rayleigh scattering cross section for elastic scattering of unpolarized photons through an angle y, and averaged over scattered-photon polarizations in form-factor approximation may be written

$$\frac{d\sigma}{d\Omega} = \frac{r_e^2}{2}\,(1 + cos^2\theta)[F(x,Z)\,]^2 \tag{3}$$

where $F\,(q,\,Z\,)$ is the atomic form factor, $r_e$ (2,8179 $x$ $10^{-15}$m) is the classical electron radius, $Z$ is the atomic number and $q$ is the momentum transfer parameter measured in units of 1/Å and defined as $q = \left(\sin{}^{\theta}/_{2}\right)\left({}^{1}/_{\lambda}\right)$ where $\lambda$ is the incident photon's wavelength and $\theta$ is the angle of scattering.

The incoherent scattering from free electrons is accurately described by Klein and Nishina (1929) theory. Klein and Nishina (1929) derived an expression applying relativistic quantum theory to the scattering from a free electron for the cross section singly differential in scattered photon angle. Theoretically, the differential incoherent scattering cross section per unit solid angle for elements is expressed in terms of the Klein–Nishina cross section as;

$$\frac{d\sigma_{(KN)}(\theta)}{d\Omega} = \frac{r_e^2}{2}[1 + k\,(1 - cos\ \theta\,]^{-2} \times [1 + k\left(1 + co\,s\,\theta + \frac{k^2(1 - co\,s\,\theta)}{1 + k(1 - co\,s\,\theta)}^2\right] \tag{4}$$

where $d\sigma_{(KN)}(\theta)/d\Omega$ is the differential cross section per electron for the number of photons scattered into the solid angle in the direction $\theta$, $r_e = e^2/m_e c^2$is the classical electron radius and $k = E(keV)/511.0034$ is the photon energy in units of the electron restmass energy. The Klein and Nishina (1929) theory is valid for the scattering of high-energy photons where the atomic electrons are relatively free. Departures from the Klein and Nishina (1929) theory occur in situations where the photon energies are comparable with the binding energies of the inner-shell electrons of the target. When the incident energy is comparable to the binding energy, the binding effects have to be taken into account. The electron binding effect is incorporated by incoherent scattering function S(x,Z). Many effects of the interaction of radiation with atoms depend on the incoherent scattering function S(x,Z). To obtain the bound electron differential incoherent scattering cross section, the free electron differential incoherent scattering cross section is then multiplied by the incoherent scattering function S(x,Z) as;

$$\frac{d\sigma_{inc}}{d\Omega} = S(x,Z)\frac{d\sigma_{KN}(\theta)}{d\Omega} \tag{5}$$

The coherent to Compton scattering cross section ratio becomes;

$$\frac{d\sigma_{coh}}{d\sigma_{inc}} \propto \frac{|F\,(q,Z)|^2}{S\,(q,Z)} \tag{6}$$

The theoretical values of $S\,(q, Z\,)$ and $F\,(q, Z\,)$ are tabulated (Hubbell et al.1975) on the basis of non-relativistic Hartree–Fock calculations.

## 3. Experimental procedure and calculations

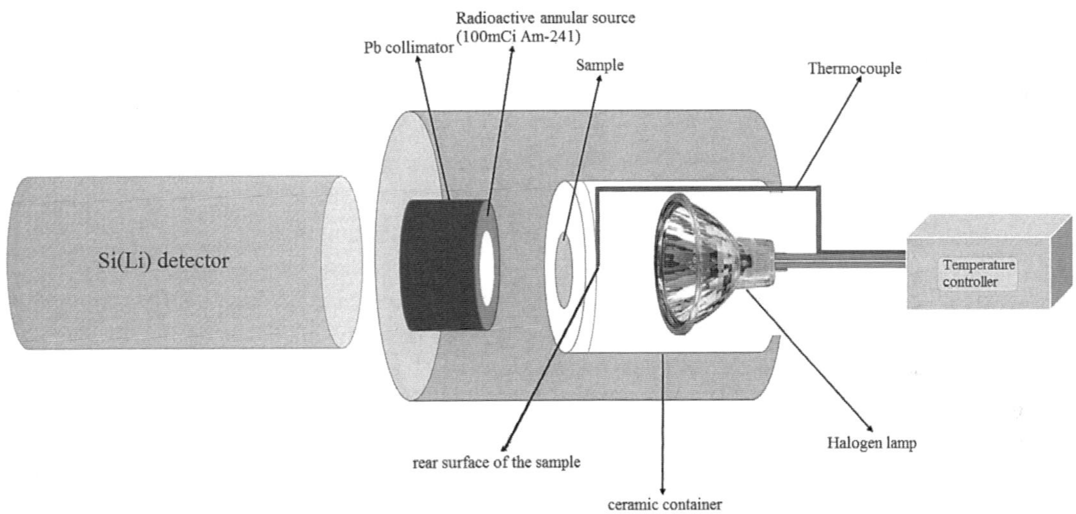


**Fig. 1.** Experimental set up

Pure elements which commercially obtained Cd (% 99.99), In (% 99.999), Sn % 99,9), and Pb (% 99,99) were taken from Alfa Aesar. The foil samples diameter were cut into 13 mm. High purity, thin uniform samples were excited using a radioactive annular source of Am-241 of strength 100 mCi and γ-photon energy 59.54 keV. A Si(Li) detector (FWHM=160 eV at 5.96 keV, active area 20 mm2 , sensitive depth 5 mm and Be window thickness 0.008 mm) with a multichannel analyzer was used to detect X-rays in the measurements.

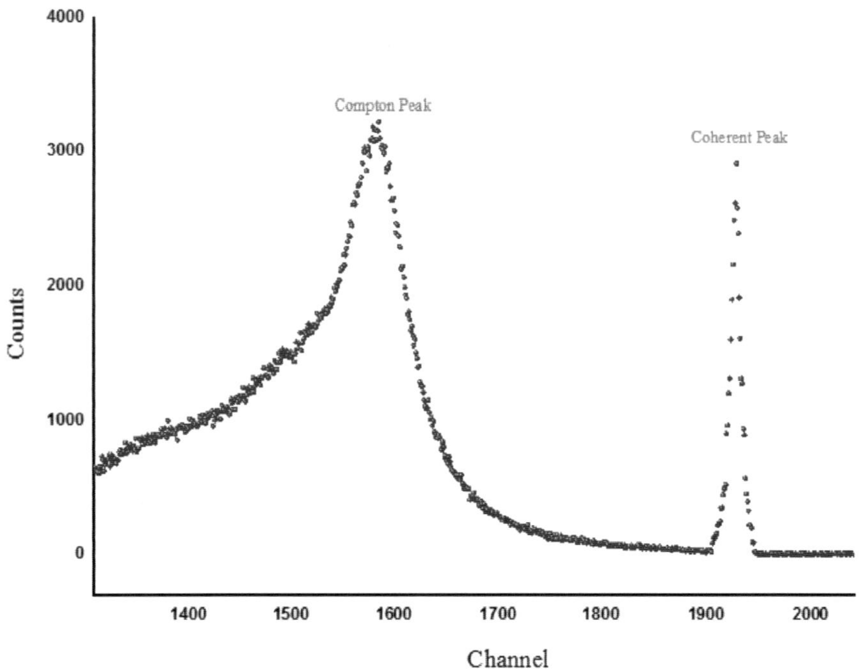


**Fig. 2.** A typical observed spectrum from In target at scattering of temperature 30 ˚C

The experimental setup consist of a Si (Li) detector, Am-241 radioactive annular source, ceramic and aluminum container, thermocouple and temperature controller as shown in Fig.1. During the study, the temperature was changed by 50 ˚C increments between 30-300 ˚C degrees. Coherent and Compton X-ray spectrums of the elements were measured at the given temperatures as in-situ. A typical Compton and coherent X-ray spectrum of In is shown in Fig. 2 as an example. All the X-ray spectra were carefully analyzed by means of the Origin 9.0 software program using a multi-Gaussian least-square fit method in order to determine the net peak. The coherent and incoherent scattering cross section ratios were determined from peak areas fitted to Gaussian function after applying necessary corrections to the data.
The intensities of coherent and Compton scattered peaks are corrected for photo peak efficiency, absorption in air between target and the detector and self-absorption in the target using the following equation:

$$N_{actual} = \frac{N_{obs}}{\varepsilon_\gamma \beta_{\gamma a} \beta_{\gamma t}} \tag{7}$$

where $N_{obs}$ is observed intensity under coherent or Compton peak, $\varepsilon_\gamma$is photo peak efficiency of gamma detector for coherent or Compton scattered photons, $\beta_{\gamma a}$ is correction factor for absorption of photons in air between target and detector and $\beta_{\gamma t}$ is self absorption correction factor for scattered photons in the target.

The ratio of coherent to Compton scattering cross-section based on the equation given above is given as follows:

$$\frac{d\sigma_{coh}}{d\sigma_{inc}} = \frac{N_{coh}}{N_{inc}} \frac{\beta_{\gamma t_{inc}}}{\beta_{\gamma t_{coh}}} \frac{\beta_{\gamma a_{inc}}}{\beta_{\gamma a_{coh}}} \frac{\varepsilon_{\gamma_{inc}}}{\varepsilon_{\gamma coh}} \tag{8}$$

where *Ncoh*/*Ninc* is the ratio of the number of counts under coherent and Compton scattered peaks, $\beta\gamma t$ (*inc*)/$\beta\gamma t$ (*coh*) is the ratio of self-absorption correction factors in the target for coherent and Compton scattered photons, $\beta\gamma a$ (*inc*)/$\beta\gamma a$ (*coh*) is the ratio of the air absorptions for Compton scattered and coherent scattered gamma rays in air and $\varepsilon\gamma$ (*inc*)/$\varepsilon\gamma$ (*coh*) is the ratio of photo-peak efficiencies of the SiLi detector for Compton and coherent scattered photons.

The self-absorption correction factor $\beta$ is given by;

$$\beta_{\gamma t} = \frac{1 - exp\left[(-1)\left(\frac{\mu_i}{\cos\theta_1} + \frac{\mu_s}{\cos\theta_2}\right)t_{\mathrm{t}}\right]}{\left(\frac{\mu_i}{\cos\theta_1} + \frac{\mu_s}{\cos\theta_2}\right)t_{\mathrm{t}}} \tag{9}$$

The air absorption correction is given by;

$$\beta_{\gamma \mathrm{a}} = \frac{1 - exp\left[(-1)\left(\frac{\mu_i}{\cos\theta_1} + \frac{\mu_s}{\cos\theta_2}\right)t_{\mathrm{a}}\right]}{\left(\frac{\mu_i}{\cos\theta_1} + \frac{\mu_s}{\cos\theta_2}\right)t_{\mathrm{t}}} \tag{10}$$

where $\mu_i$ and $\mu_s$ are the mass attenuation coefficients ($cm^2/g$) of incident photons and emitted characteristic X-rays respectively. $\theta_1$ and $\theta_2$ are the angles of incident photons and emitted X-rays with respect to the normal at the surface of the sample in the present setup and t is the mass thickness of the sample in $g/cm^2$. To estimate the self-absorption correction in the sample and the absorption correction in the air path we used the mass attenuation coefficients obtained by means of a computer program named WinXCom (Gerward et al., 2001, 2004) [initially developed as XCOM (Berger and Hubbell, 1995)]. This program provides total cross-sections and attenuation coefficients of elements, compounds or mixtures as well as partial cross-sections for incoherent and coherent scattering, photoelectric absorption and pair production both in the field of nucleus and electrons at energies from 1 keV to 100 GeV.

The photopeak efficiency curve for the SiLi detector was obtained experimentally in the range 26.345–661.657 keV photon energy by using $Am^{241}$, $Ba^{133}$, $Eu^{152}$ and $Cs^{137}$ radioactive calibration sources. Each of these radioactive sources of known activity was placed at position of the target and the energy spectra were recorded by the SiLİ detector. The equation used for the efficiency calculations is as follows (Demir and Tursucu, 2013)

The equation was calculated as;

$$\varepsilon_\gamma = -30{,}9037 + 15{,}51926E^1 - 2{,}95852E^2 + 0{,}16207E^3$$

The excitation energy can be supplied by raising the sample to a high temperature, by irradiating it with electromagnetic radiation, or by exposing it to an electrical arc or spark. If the sample is excited by radioactive sources as well as is exposed to different temperatures, how the compton and coherent scattering spectrum of the sample is affected? Compton and coherent scattering will be severely affected by fluctuations in temperature since signal is dependent on the number of atoms in the excited state. Temperatures can have a dramatic effect on the ratio of excited atomic particles and unexcited atomic particles. Compton and coherent scattering spectra where quantitative values are being measured would be temperature sensitive.

| Temperature (°C) | FWHM | | | | | | | |
|---|---|---|---|---|---|---|---|---|
| | Cd | | In | | Sn | | Pb | |
| | *Compton* | *Coherent* | *Compton* | *Coherent* | *Compton* | *Coherent* | *Compton* | *Coherent* |
| 30 | 15,6782 | 12,4088 | 16,845242 | 11,5405 | 15,40829 | 12,38435 | 17,45528 | 12,063242 |
| 100 | 17,1217 | 12,6414 | 17,139605 | 12,53922 | 15,49665 | 12,34261 | 18,23458 | 12,27721 |
| 150 | 18,4878 | 13,0887 | 18,818334 | 13,20909 | 18,27693 | 12,65987 | 20,25369 | 12,32614 |
| 200 | 20,2213 | 13,1546 | - | - | 19,17815 | 12,68604 | 20,40269 | 12,45229 |
| 250 | 20,1007 | 12,8393 | - | - | - | - | 20,72563 | 12,59632 |
| 300 | 21,9292 | 12,8116 | - | - | - | - | 20,80632 | 12,78905 |

## 4. Result and discussion

**Table 1.** Experimental values of differential cross section ratios of Cd, In, Sn and Pb at different temperatures.

| Temperature(°C) | $d\sigma_{coh}/ d\sigma_{inc}$ | | | |
|---|---|---|---|---|
| | *Cd* | *In* | *Sn* | *Pb* |
| 30 | 0,061694 | 0,050196 | 0,056002 | 0,759613 |
| 100 | 0,061611 | 0,048336 | 0,057275 | 0,742554 |
| 150 | 0,061137 | 0,046309 | 0,055654 | 0,729155 |
| 200 | 0,061091 | - | 0,053169 | 0,727328 |
| 250 | 0,060084 | - | - | 0,721082 |
| 300 | 0,058675 | - | - | 0,715505 |

**Table 2.** Experimental values of FWHM for Cd, In, Sn and Pb at different temperatures

Therefore in this study, effect of the temperature changes on the Compton and coherent scattering spectra of the some foil metals (Cd, In, Sn, Pb) have been investigated. Compton and coherent scattering spectrums of Mo, Nb, Zr and Y pure metals was determined using Si (Li) solid-state detector at between 30-300 ˚C degrees. The measured values of coherent to incoherent differential cross-section ratio of Cd, In, Sn and Pb for the different temperetures (30 °C-300 °C) are given in Table 1, and full width at half maximum (FWHM) values of Cd, In, Sn and Pb for the different temperetures (30 °C-300 °C) are given in Table 2. Figs. 3, 4, 5 and 6 show these changes at graphical form.There are no available data in literature for comparison with the present results.

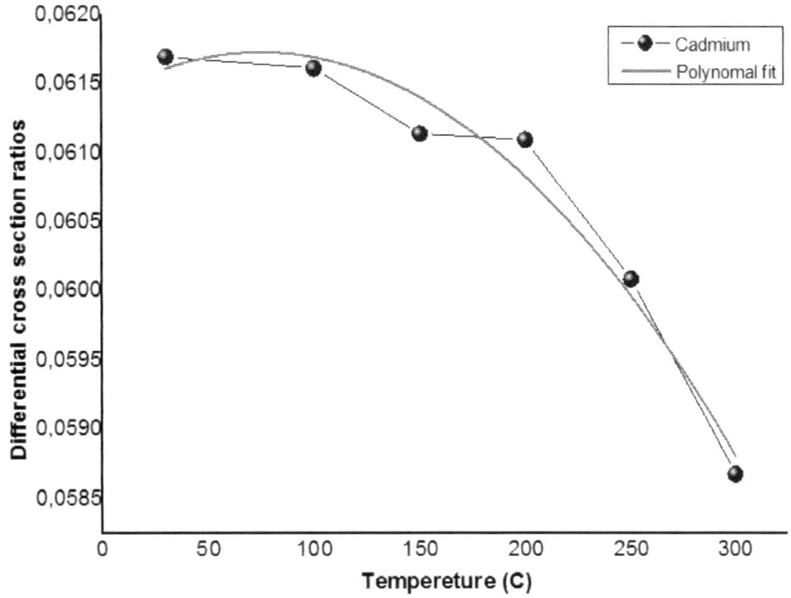


**Fig.3.** Coherent to incoherent cross section ratio as a function of temperature for Cd

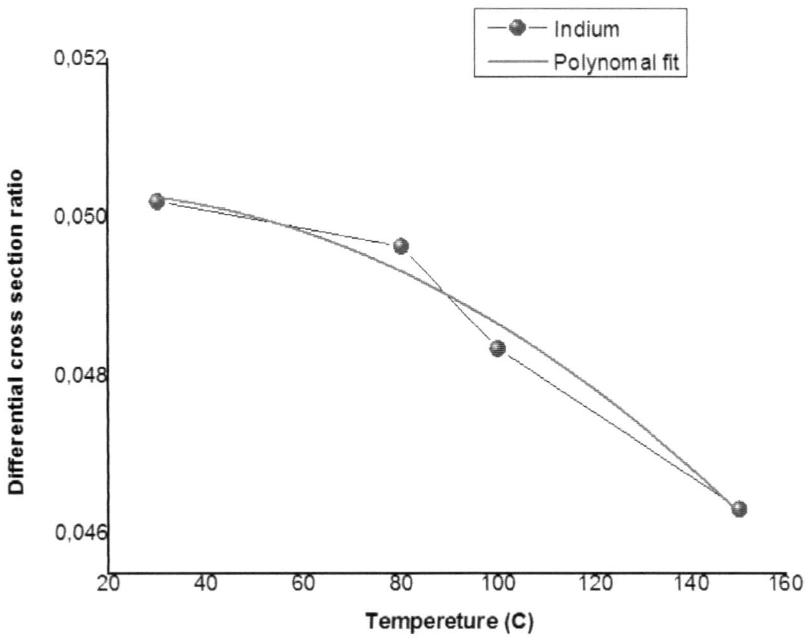


**Fig. 4.** Coherent to incoherent cross section ratio as a function of temperature for In

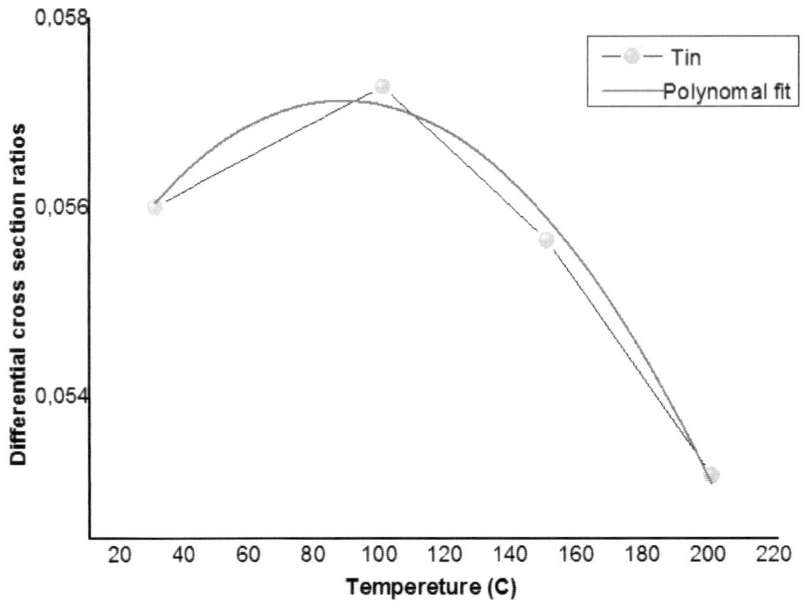


**Fig. 5.** Coherent to incoherent cross section ratio as a function of temperature for Sn

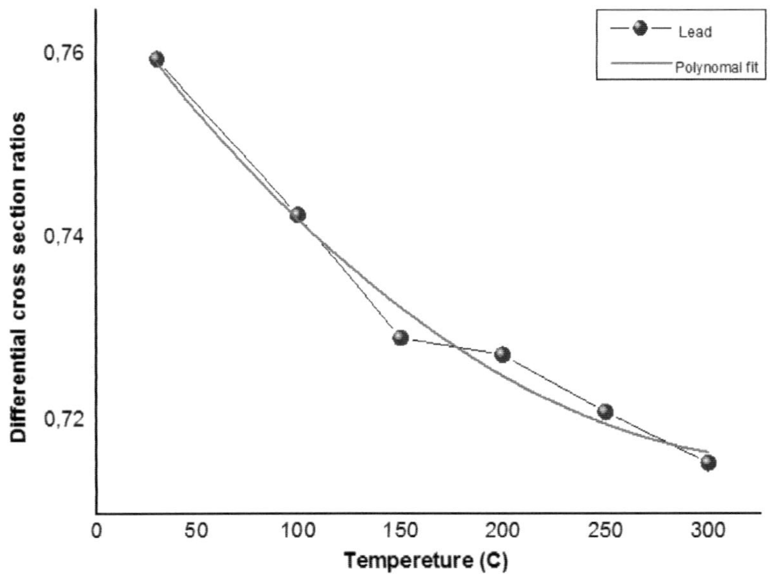


**Fig. 6.** Coherent to incoherent cross section ratio as a function of temperature for Sn

## References

[1] İcelli, O. ,Erzeneoglu, S. ,2002. Experimental study on ratios of coherent scattering to Compton scattering for elements with atomic numbers 26 Z 82 in 59.5ke V for 55° and 115°. Spectrochim. Acta Part B57, 1317–1323

[2] Kurucu, Y., Erzeneoglu, S., Durak, R., Sahin, Y., 1998. Measurement of Compton and coherent scattering differential cross sections. Turk. J. Phys. 22, 783–788.

[3] Hubbell, J.H., Veigele, Wm. J., Bringgs, E.A., Brown, R.T., Cromer, D.T., Hoverton, R.J., 1975. Atomic form factors, incoherent scattering functions, and photon scattering crosssections. J.Phys. Chem.Ref. Data 4,471.

[4] Ertugrul, M.,2001.Measurement of mean atomic number of compounds with scattered radiation. Instrum. Sci. Technol. 29(4), 309–315.

[5] Sogut et al., 2004. Dependence on average atomic number of coherent/incoherent scattering intensity ratios in compounds Journal of Quantitative Spectroscopy & Radiative Transfer 87, 15–23

[6] Simsek, O.,Ertugrul, M., 2004. Angular dependence ofc oherent to incoherent scattering differential cross-sectionratios for Zr, Nb and Mo elements. Rad. Meas. 38,271–276.

[7] Singh, M.P., Sandhu, B.S., Singh,B., 2007. Measurement of the effective atomic number of composite materials using Rayleigh to Compton scattering of 279 keV gamma rays. Phys.Scr. 76, 281–286.

[8] Singh, M.P., Sharma, A., Singh,B., Sandhu, B.S., 2013. An experimental study on cross-section ratio of coherent to in coherent scattering for 145 keV incident gamma photons. Rad. Meas.59, 30–36.

[9] Akkuş et al., 2015 Polar and azimuthal angular dependence of coherent to incoherent Scattering differential cross-sectionratios of Au at 59.54keV Radiation Physics and Chemistry 117, 167–171

[10] Klein, O., Nishina,Y., 1929. Uber die streuung von strahlung dureh freie elektronen nack der neuen relativistischen quanten dynamik Von Dirac .Z.Phys. 52, 583.

[11] Gerward, L., Guilbert, N., Jensen, B. K., Levring, H., 2001. X-ray absorption in matter reengineering XCOM. Rad. Phys. Chem. 60, 3–24.

[12] Demir, D., Tursucu, A., 2013. Measurement of effective atomic number of $Fe_xCr_{1-x}$ and $Fe_xNi_x$ alloys using scattering of gamma rays. J. Alloy. Compd. 581, 213–216.

# Graphene quantum dot with a hydrogenic impurity

**Neslihan Gökçek[1] and Bekir Sıtkı Kandemir[1]**

[1] Department of Physics, Faculty of Sciences, Ankara University, 06100 Tandoğan, Ankara, Turkey

E-mail: ngokcek@science.ankara.edu.tr

**Abstract.** We investigate the electronic energy levels of a circular graphene quantum dot with a hydrogenic impurity placed at the center of the dot. We have adopted a variational scheme in such a way that the trial wave functions take into account the carrier confinement due to the dot geometry as well as the effect of the magnetic field. We found that the presence of impurity strongly modifies the electronic spectrum of the graphene quantum dot. We also calculated the critical value of the impurity strength as a function of dot radius.

## 1. Introduction

Since the discovery of graphene [1, 2] impurity related effects in graphene have been extensively reported in the graphene literature [3–19]. The study of impurity states in graphene quantum dots (GQDs) is itself important, since it is expected that these effects will be more pronounced in such structures due to the reduction of the dimensionality. Since experiments with adatoms in monolayer graphene clearly adress the long range impurities [13], which are responsible for limiting the mobility of graphene [15], Coulomb like hydrogenic impurities are of particular interest in the relevant literature [7–10, 15]. Moreover, the strength of Coulomb interaction, which is proportional to dimensionless quantity $\overline{Z}$ ( $Z\ \alpha$ ) defined by the valency of impurity $Z$ and $\alpha = e^2/\epsilon\hbar v_F$, introduces a new kind of quantum electrodynamics regime (QED), which is strictly different from the usual perturbative one. Here, $\epsilon$ is the dielectric constant, and $v_F$ is the Fermi velocity for massless Dirac fermions which is of order of $10^6 ms^{-1}$. Since $\alpha$ is of order of unity for conventional $SiO_2$ substrate [7, 8], such a new regime needs nonperturbative methods to investigate long range impurity related effects.

On the one hand, due to their controllable sizes and shapes, GQDs are the center of both experimental and theoretical interest in graphene physics [20–63]. On the other hand, the formation of spin qubits in GQDs [46] with controllable valley splitting is another active research area in the field. The crucial requirement for this formation is to lift the valley degeneracy of the pristine graphene. It is shown that this is achievable in GQDs.

Since it is also possible to control where the adsorbate can be placed in graphene, controlling the impurity related effects are not only important for graphene, but also for graphene based nanostructures such as GQDs.

The continuum model of confined states of Dirac electrons in GQDs provides a simple physical insight into understanding the bound states in such structures. It is shown that [59, 60], it presents results in full agreement with those found from tight binding calculations (TBCs) for bound QD states in which their radii are much larger than the carbon-carbon distance. Even for small radii the agreement between continuum description and TBC is always reasonable for low-lying bound states with high values of the angular momentum. This establishes the use of continuum approach to study the confinement effects.

With these motivations, in the present work, we study the combined effects of quantum confinement due to GQD and hydrogenic donor impurity located at the center of the GQD. To do this, we take a circular confinement potential in order to model the GQD, and place the hydrogenic impurity at the center of dot. The calculation is performed within the framework of a variational approach based on using exact solutions of Dirac-Weyl equation in the absence of quantum dot potential as trial wave functions.

## 2. Theory

The Hamiltonian of a system consisting of an electron (hole) bound to a donor (acceptor) ion inside a parabolic graphene quantum dot , in the presence of a magnetic field perpendicular to GQD plane, is given by

$$\mathcal{H} = v_F\, \boldsymbol{\alpha}\cdot(\boldsymbol{p}+\frac{e}{c}\boldsymbol{A}) + \boldsymbol{\beta}\ \Delta_0 r^2 - \frac{Ze^2}{\epsilon r} \tag{1}$$

where $\boldsymbol{A}$ is the vector potential chosen at symmetrical gauge, i.e., $\boldsymbol{A} = B_0(-y,x,0)/2$ , $\Delta_0 = U_0/2R_o^2$ wherein $U_0$ and $R_0$ are the strength and the radius of GQD, respectively. In Eq.(1), $\boldsymbol{\alpha}$ and $\boldsymbol{\beta}$ are Dirac matrices, and the Fermi velocity $v_F = (3a/2)\,J_0$ is given in terms of $J_0$ and $a$, which are the resonance integral between nearest neighbour carbon atoms, and is of order of 2.7 eV, and the equilibrium bond length of the graphene, respectively. The solution of Eq.(1) in the absence of both GQD potential and the impurity is well established in Ref. 17. In order to study the combined effects of these two, we use the exact wave functions in their absence as trial wave functions corresponding to the Dirac-Weyl equation $\mathcal{H}\Psi_n = E_n\Psi_n$ in which four-component spinors are given by

$$\begin{aligned}
\Psi_n^{\mathbf{K}}(x,\varphi) &= R_n(x)e^{im\varphi}e^{-x^2/2}\begin{pmatrix}\mathbb{L}^0\\0\\0\\\mathbb{L}^{\mathbf{K}}\end{pmatrix}\\
\Psi_n^{\mathbf{K}'}(x,\varphi) &= R_n(x)e^{im\varphi}e^{-x^2/2}\begin{pmatrix}0\\\mathbb{L}^0\\\mathbb{L}^{\mathbf{K}'}\\0\end{pmatrix}
\end{aligned} \tag{2}$$

with $R_n(x) = [\nu!/2\pi(\nu+|m|)!]^{1/2}\gamma_0 x^{|m|}$. In Eq.(2) we have defined $\mathbb{L}^0$, $\mathbb{L}^{\mathbf{K}}$ and $\mathbb{L}^{\mathbf{K}'}$ in terms of Laguerre polynomials as $\mathbb{L}^0(x) = L_\nu^{|m|}(x^2)$,

$$\begin{aligned}
\mathbb{L}^{\mathbf{K}} &= \frac{ixe^{+i\varphi}}{\sqrt{n}}\begin{cases} L_\nu^{|m|+1}(x^2) & m \geqslant 0\\ -\frac{n}{x^2}L_{\nu+1}^{|m|-1}(x^2) & m<0\end{cases},\\
\mathbb{L}^{\mathbf{K}'} &= \frac{ixe^{-i\varphi}}{\sqrt{n}}\begin{cases} -\frac{n}{x^2}L_\nu^{|m|-1}(x^2) & m \geqslant 0\\ L_{\nu-1}^{|m|+1}(x^2) & m<0\end{cases},
\end{aligned} \tag{3}$$

respectively. The corresponding eigenvalues are given by $E_n = \lambda\epsilon_n$ wherein $\epsilon_n = \hbar v_F\sqrt{2n}/\ell_B$, $\lambda$ is the chirality index and takes $-1$ and $+1$ values, which correspond to valence and conduction bands of graphene, respectively. $\ell_B = (\hbar c/eB)^{1/2}$ is the magnetic confinement length, and $n = \nu + (|m|+m+\tau+1)/2$ where $\tau$ is considered as a pseudospin index whose $-1$ and $+1$ values specifies the two valleys, i.e., $\mathbf{K}$ and $\mathbf{K}'$, respectively. In the following, we will restrict ourselves to only conduction band. Extension of the results to the other band is straightforward. In Eq.(3), $x = \gamma_0 \rho$ is a dimensionless, and $\gamma_0$ is the inverse of characteristic length of the system. On the one hand, without spatial parabolic confinement, $\gamma_0$ is proportional to inverse of magnetic confinement length $\ell_B$, i.e., $\gamma_0 = \gamma = 1/\sqrt{2}\ell_B$. On the other hand, in the presence of spatial confinement, graphene system described by the Hamiltonian Eq.(1) suggest a natural unit of length $R = (\hbar v_F R_o^2/U_0)^{1/3}$ which characterises the size of a scalar parabolic quantum dot in graphene [20, 21]. Therefore, we use the ansatz of Ref. 21 to characterize the system by only one parameter, i.e., we write $\gamma_0^2 = \gamma^2 + R^{-2}$ and take $\gamma$ to be a variational parameter, so as to reduce to $1/\sqrt{2}\ell_B$ in the limit $R \to \infty$. Therefore, the variational eigenenergies of GQD is given by

$$E_n = \int d^2\boldsymbol{r} \Psi_n^\dagger \left[ \boldsymbol{\alpha} \cdot (-i\, \boldsymbol{\nabla} + \frac{e}{\hbar c}\mathbf{A}) + \boldsymbol{\beta}\, \Delta_0 r^2 - \frac{Ze^2}{\epsilon r} \right] \Psi_n. \tag{4}$$

Inserting Eqs. (2) and (3) into Eq.(4) we get a general expression for the variational energy as

$$\overline{E}_n = \left( \sqrt{n} - \frac{\bar{Z}}{2}\mathbb{M}_n^{\mathrm{I}} \right) \overline{\gamma}_0 + \frac{\sqrt{n}}{2\overline{l}_B^2}\frac{1}{\overline{\gamma}_0} + \frac{1}{4\overline{R}^3}\frac{1}{\overline{\gamma}_0^2}\mathbb{M}_n^{\mathrm{II}}. \tag{5}$$

In Eq.(5), the energy and all lengths are expressed in units of $J_0$ and $a$, respectively. The integrals $\mathbb{M}_n^{\mathrm{I}}$ and $\mathbb{M}_n^{\mathrm{II}}$ in Eq.(5) are defined in terms of Laguerre polynomials as

$$\mathbb{M}_n^{\mathrm{I}} = \widetilde{\mathbb{F}}_n + \frac{1}{n}\widetilde{\mathbb{G}}_n \tag{6}$$

with

$$\widetilde{\mathbb{F}}_n = \frac{\nu!}{(\nu + |m|)!} \int_0^\infty dr e^{-r} r^{|m|+\frac{1}{2}} \left| L_\nu^{|m|}(r) \right|^2$$

$$\widetilde{\mathbb{G}}_n = \frac{\nu!}{(\nu + |m|)!} \int_0^\infty dr e^{-r} r^{|m|-\frac{1}{2}} \mathbb{H}_{\nu,\alpha}^{|m|}(r).$$

and

$$\mathbb{M}_n^{\mathrm{II}} = \mathbb{F}_n - \frac{1}{n}\mathbb{G}_n$$

with

$$\mathbb{F}_n = \frac{\nu!}{(\nu + |m|)!} \int_0^\infty dr e^{-r} r^{|m|+1} \left| L_\nu^{|m|}(r) \right|^2$$

$$\mathbb{G}_n = \frac{\nu!}{(\nu + |m|)!} \int_0^\infty dr e^{-r} r^{|m|+1} \mathbb{H}_{\nu,\alpha}^{|m|}(r)$$

where

$$\mathbb{H}_{\nu,\alpha}^{|m|}(r) = \begin{cases} r \left| L_{\nu+\alpha}^{|m|+1}(r) \right|^2, & \alpha = \begin{cases} 0 & m \geqslant 0 \quad \tau = +1 \\ -1 & m < 0 \quad \tau = -1 \end{cases} \\ \frac{n^2}{r} \left| L_{\nu+\alpha}^{|m|-1}(r) \right|^2, & \alpha = \begin{cases} 0 & m \geqslant 0 \quad \tau = -1 \\ +1 & m < 0 \quad \tau = +1 \end{cases} \end{cases},$$

respectively. $\mathbb{M}_n^{\mathrm{II}}$ can be integrated analytically to yield $\mathbb{F}_n = 2\nu + |m| + 1$ and $\mathbb{G}_n = 2n - m$, respectively. Thus, we have exactly $\mathbb{M}_n^{\mathrm{II}} = -\tau$. For each $n$ values, $\mathbb{M}_0^{\mathrm{I}}$ can also be integrated analytically [64] so that, for instance, it yields $\mathbb{M}_0^{\mathrm{I}} = \Gamma(|m|+1/2)/\Gamma(|m|+1)$ for $n = 0$. Here, $\Gamma(x)$ is the Gamma function. To get an approximate analytical expression for the energy given by Eq.(5), we investigate some of its limiting cases. In the absence of impurity, since its minimization with respect to $\overline{\gamma}$ gives $\overline{\gamma} = 1/\sqrt{2}\overline{\ell}_B$, as $\overline{R}$ goes to infinity. By replacing this value back into the relevant variational energy, one gets $\overline{E}_n = \sqrt{2n}/\overline{\ell}_B$, which are the well-known Landau levels for massless graphene. Consequently, substituting this value back into Eq.(5) as a first approximation yields

$$\bar{E}_0 = -\frac{\bar{Z}}{2\bar{l}_0}\frac{\Gamma(|m|+1/2)}{\Gamma(|m|+1)} - \tau\frac{\bar{l}_0^2}{4\bar{R}^3} \tag{7}$$

where $1/\overline{\ell}_0^2 = 1/2\overline{\ell}_B^2 + 1/\overline{R}^2$ defines the effective confinement length. Eq.(7) may give an idea about the effect of hydrogenic donor impurity onto the energy spectrum of a parabolic GQD as well as the critical value of the impurity strength. In the absence of impurity and magnetic field, Eq.(7) works well, as reported in Ref. 65, for the graphene devices fabricated by Ponomarenko *et al* [66], wherein the GQDs have a gap value of $0.010 - 0.5eV$ for the GQDs with diameter of $D = 40 - 1.0nm$. Eq.(7) predicts their corresponding values as $0.014 - 0.57eV$. Even in the high magnetic field case it yields the well-known relativistic LLs, which are also confirmed experimentally [67]. From Eq. (7), we estimate the critical value of $\bar{Z}$ as

$$\bar{Z}_{crt} = \frac{\bar{l}_0^3}{2\bar{R}^3}\frac{\Gamma(|m|+1)}{\Gamma(|m|+1/2)}$$

This reduces to $\bar{Z}_{cr} = 1/2\sqrt{\pi}$ in the absence of magnetic field.

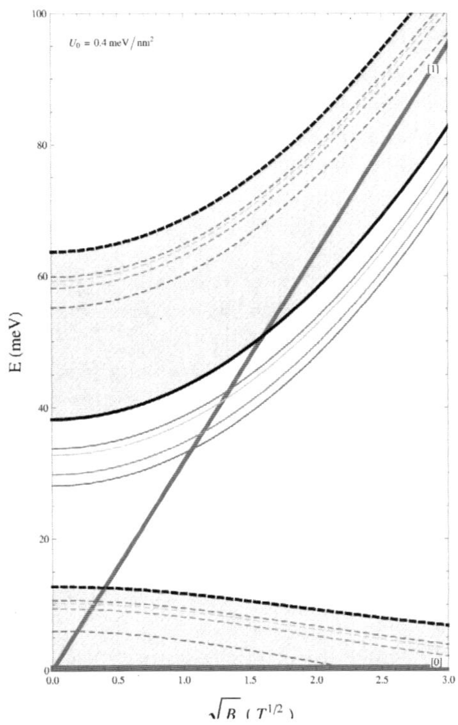


**Figure 1.** The magnetic field dependence of parabolic quantum dot energy spectra in the presence of hydrogenic donor impurity. While the bold gray lines (represented by labels $[n]$) correspond to the unperturbed LLs (in the limit $\overline{R} \to \infty$), the dashed (straight) curves represents $\mathbf{K}'(\mathbf{K})$ valleys. For all curves, from the bottom to the top, $m = 0, -1, -2$ and $-3$ levels are depicted by gray, red, green and blue colors.

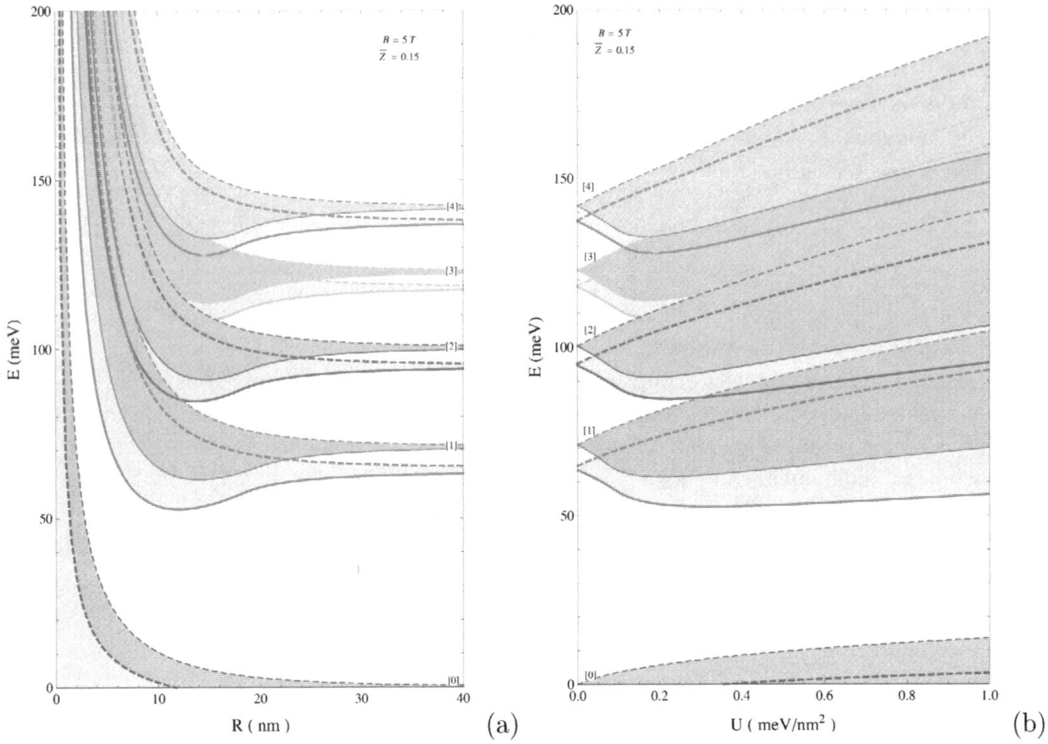


**Figure 2.** (Color online)(a) The lowest-lying energy eigenvalues of GQD as a function of effective dot radius $R = (\hbar v_F R_0^3/U_0)^{1/3}$ at a constant magnetic field and impurity strength, i.e., $B = 5$ T and $\overline{Z} = 0.15$, respectively. (b) The effective dot strength $U = U_0/R_o^2$ dependence of the same levels with $m = 0$.

FIG. 1 shows the ground- and the first excited-state energies of parabolic GQD as a function of $\sqrt{B}$ both in the presence ( thin dashed ( $\mathbf{K}'$) and thin straight ($\mathbf{K}$ ) lines) and absence ( bold dashed ( $\mathbf{K}'$) and bold straight ($\mathbf{K}$ ) lines) of hydrogenic donor impurity. We also included for comparison the first two LLs of the gapless graphene (thick straight gray lines). It is clearly seen that the impurity energy levels in GQD goes up to well known LLs with increasing B for a constant $U_0$. To understand the effect of confinement on impurity energy levels itself, for a fixed impurity strength, we plot the lowest-lying energies as a function of dot radius in FIG. 2. In both figure, to see better the valley splitting for the associated levels, space arising from this splitting is coloured by grey. It is seen that, since the narrower dot creates stronger confinement the magnitude of valley splitting, $\mathbf{K}(\mathbf{K}')$ becomes larger as the confinement becomes stronger. We found that, the magnitude of the valley degeneracy, with decreasing the strength of GQD, first drop slowly and then approach to its value in the gapless graphene, that is, zero. This is more clearly seen from the FIG. 2(b). However, introduction of donor impurity causes these levels to shift down to lower energies, and leads enhancement in the size of the valley splitting for constant $U$. On the other hand, inclusion of such a long range donor hydrogenic impurity leads to decrease in valley splitting.

In FIG. 3, the associated energy levels are plotted as functions of impurity strength $\bar{Z}$.

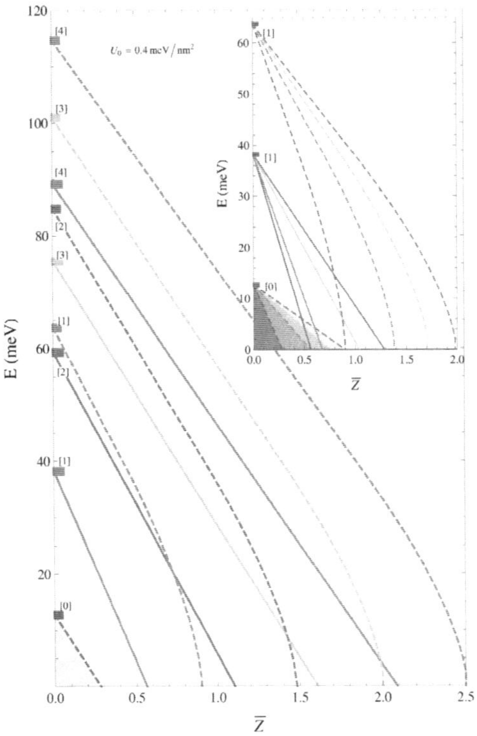


**Figure 3.** (Color online) GQD energies of a Dirac electron bound to a hydrogenic donor impurity as a function of impurity strength at a constant impurity strength with zero magnetic field. The solid curves represent the energy levels for $\tau = +1$, the dashed curves represent the those for $\tau = -1$. Inset shows the associated energy levels but with inclusion of the $m = -1$ (Red), $m = -2$ (Green) and $m = -3$ (Blue) as well as $m = 0$ (Gray).

## 3. Conclusion

In conclusion, we have investigated the effect of a hydrogenic donor impurity on the energy levels of a parabolic GQD. For this purpose, we use a variational scheme based on the choice of trial wave functions taking into account carrier confinement together with the effect of the magnetic field. On the basis of our results, we have found that the magnitude of valley splitting is strongly dependent on the impurity strength and the strength of the magnetic field, and concluded that the size of valley splitting can easily be controlled by the valency of the impurity. We hope that the presented results may be useful to gain insight how to control some electronic properties of GQDs with impurities.

## Acknowledgments

This work was supported by TÜBİTAK under Grant No. 113F103.

## References

[1] Novoselov K S, Geim A K, Morozov S V, Jiang D, Zhang Y, Dubonos S V, Grigorieva I V and Firsov A A *Science* 2004 **306** 666
[2] Novoselov K S, Jiang D, Schedin F, Booth T J, Khotkevich V V, Morozov S V and Geim A K 2005 USA *Proc. Natl. Acad. Sci.* **102** 10451
[3] Ostrovsky P M, Gornyi I V and Mirlin A D 2006 *Phys. Rev.* B **74** 235443
[4] Katsnelson M I 2006 *Phys. Rev.* B **74** 201401
[5] Pereira V M, Nilsson J and Castro Neto A H 2007 *Phys. Rev. Lett.* **99** 166802
[6] Fogler M M, Novikov D S and Shklovskii B I 2007 *Phys. Rev.* B **76** 233402
[7] Shytov A V, Katsnelson M I and Levitov L S 2007 *Phys. Rev. Lett.* **99** 236801
[8] Shytov A V, Katsnelson M I and Levitov L S 2007 *Phys. Rev. Lett.* **99** 246802
[9] Novikov D S 2007 *Appl. Phys. Lett.* **91** 102102
[10] Novikov D S 2007 *Phys. Rev.* B **76** 245435
[11] Terekhov I S, Milstein A I, Kotov V N and Sushkov O P 2008 *Phys. Rev. Lett.* **100** 076803
[12] Kotov V N, Pereira V M and Uchoa B 2008 *Phys. Rev.* B **78** 075433

[13] Chen J -H, Jang C, Adam S, Fuhrer M S, Williams E D and Ishigami M 2008 *Nature Phys.* **4** 377
[14] Pereira V M, Kotov V N and Castro Neto A H 2008 *Phys. Rev.* B **78** 085101
[15] Katsnelson M I, Geim A K 2008 *Phil. Trans. R. Soc.* A **366** 195
[16] Zhu W, Wang Z, Shi Q, Szeto K Y, Chen J and Hou J G 2009 *Phys. Rev.* B **79** 155430
[17] Kandemir B S and Mogulkoc A 2010 *Eur. Phys. J.* **74** 391
[18] Altanhan T and Kozal B 2012 *Eur. Phys. J.* **85** 222
[19] Jia-Lin Zhu and Songyang Sun 2012 *Phys. Rev.* B **85** 035429
[20] Silvestrov P G and Efetov K B 2007 *Phys. Rev. Lett.* **98** 016802
[21] Chen H, Apalkov V and Chakraborty T 2007 *Phys. Rev. Letts.* **98** 186803
[22] Rycerz A, Tworzydlo J and Beenakker C W J 2007 *Nat. Phys.* **3** 172
[23] De Martino A, Dell' Anna L and Egger R 2007 *Phys. Rev. Lett.* **98** 066802
[24] Sols F, Guinea F and Castro Neto A H 2007 *Phys. Rev. Letts.* **99** 166803
[25] Recher P, Trauzette B, Rycerz A, Blanter Ya. M, Beenakker C W J and Morpurgo A F 2007 *Phys. Rev.* B **76** 235404
[26] Trauzettel B, Bulaev D V, Loss D and Burkard G 2007 *Nat. Phys.* **3** 192
[27] De Martino A, Dell' Anna L and Egger R 2007 *Sol. State Commun.* **144** 547
[28] Wunsch B, Stauber T and Guinea F 2008 *Phys. Rev.* B **77** 035316
[29] Matulis A and Peeters F M 2008 *Phys. Rev.* B **77** 115423
[30] Katsnelson M I and Guinea F 2008 *Phys. Rev.* B **78** 075417
[31] Zhang Z Z, Chang K and Peeters F M 2008 *Phys. Rev.* B **77** 235411
[32] Hewageegana P and Apalkov V 2008 *Phys. Rev.* B **77** 245426
[33] Schnez S, Ensslin K, Sigrist M and Ihn T 2008 *Phys.Rev.* B **78** 195427
[34] Recher P, Nilsson J, Burkard G and Trauzettel B 2009 *Phys. Rev.* B **79** 085407
[35] Romanovsky I, Yannouleas C and Landman U 2009 *Phys. Rev.* B **79** 075311
[36] Stander N, Huard B and Goldhaber-Gordon D 2009 *Phys. Rev. Lett.* **102** 026807
[37] Young A. F. and Kim P 2009 *Nat. Phys.* **5** 222
[38] Wurm J, Rycerz A, Adagideli İ, Wimmer M, Richter K and Baranger H U 2009 *Phys. Rev. Lett.* **102** 056806
[39] Masir M R, Matulis A and Peeters F M 2009 *Phys. Rev.* B **79** 155451
[40] Wang D and Jin G 2009 *Europhys. Lett* **88** 17011
[41] Giavaras G, Maksym P A and Roy M 2009 *J. Phys.: Condens. Matter* **21** 102201
[42] Peres N M R, Rodrigues J N B, Stauber T and dos Santos L J M B 2009 *J. Phys.: Condens. Matter* **21** 344202
[43] Wang D and Jin G 2009 *Phys. Lett.* A **373** 4082
[44] Bardarson J H, Titov M and Brouwer P W 2009 *Phys. Rev. Lett.* **102** 226803
[45] Giavaras G and Nori F 2010 *Applied Phys. Lett.* **97** 243106
[46] Recher P and Trauzettel B 2010 *Nanotechnology* **21** 302001
[47] Giavaras G, Maksyma P A and Roy M 2010 *Physica E* **42** 715
[48] Yan C-H and Wei L-F 2010 *J. Phys.: Condens. Matter* **22** 295503
[49] Horing N J M and Liu S Y 2010 *J. Phys.: Condens. Matter* **22** 025502
[50] Horing N J M and Liu S Y 2010 *J. Phys.: Condens. Matter* **22** 159801
[51] González J W, Pacheco M, Rosales L and Orellana P A 2010 *Europhys. Lett.* **91** 66001
[52] Kim S C, Park P S and Yang S -R E 2010 *Phys. Rev.* B **81** 085432
[53] Libich F, Rotter S, Güttinger J, Stampfer and Burgdörfer C 2010 *J. Phys. Rev.* B **81** 245411
[54] Wimmer M, Akhmerov A R and Guinea F 2010 *Phys. Rev.* B **82** 045409
[55] Giavaras G and Nori F 2011 *Phys. Rev.* B **83** 165427
[56] Masir M R, Vasilopoulos P and Peeters F M 2011 *J. Phys.: Condens. Matter* **23** 315301
[57] Molitor F, Güttinger J , Stampfer C, Dröscher S, Jacobsen A, Ihn T and Ensslin K 2011 *J. Phys.: Condens. Matter* **23** 243201
[58] Rozhkova A V, Giavaras G, Bliokh Y P , Freilikher V and Nori F 2011 *Physics Reports* **503** 77
[59] Pal G, Apel W and Schweitzer L 2011 *Phys. Rev.* B **84** 075446
[60] Zarenia M, Chaves A, Farias G A and Peeters F M 2011 *Phys. Rev.* B **84** 245403
[61] Tahir M and Sabeeh K 2012 *J. Phys.: Condens. Matter* **24** 135005
[62] Kim S C, Lee J W and Yang S-R E 2012 *J. Phys.: Condens. Matter* **24** 495302
[63] Grujić M , Zarenia M, Tadić M and Peeters F M 2012 *Phys. Scr. T* **149** 014056
[64] Gradshteyn I S and Rzyhik I M 2000 *Table of Integrals, Series and Products* 6th edn. ( New York: Academic Press)
[65] Kandemir B S and Omer G 2013 *Eur. Phys. J.* B **86** 299
[66] Ponomarenko L A, Schedin F, Katsnelson M I, Yang R, Hill E W, Novoselov K S and Geim A K 2008

*Science* **320** 356
[67] Molitor F, Jacobsen A, Stampfer C, Güttinger J, Ihn T and Ensslin K 2009 *Phys. Rev.* B **79** 075426

Journal of Physics: Conference Series **707** (2016) 012009 doi:10.1088/1742-6596/707/1/012009

# Oxygen partial pressure effects on the magnetron sputtered $WO_3$ films

**G Merhan Muğlu [1], E Gür[1]**

[1]Department of Physics, Faculty of Science, Atatürk University, 25240, Erzurum, Turkey.

E-mail: *emregur@atauni.edu.tr*

**Abstract.** Electrochromism is changing color of a substance in response to the applied an external electric field and the phenomenon is reversible. $WO_3$ is very attractive material due to its electrochromic properties as well as it is also attractive for many different applications such as gas sensors, phosphorous screen, textile, glass industry. In this study, it is aimed to provide optimization of the optical and structural characteristics of $WO_3$ by changing the growth parameters mainly the oxygen partial pressure. The partial pressure of oxygen was changed with increments of 0.7 mTorr. For the analysis, X-ray Diffraction (XRD), absorption, Raman spectroscopy measurements were used. When $O_2$ gas increased, peaks belong to the $WO_3$ was observed in XRD patterns at the 2 theta angles of 23.0, 11.0, 23.5 and 28.5 angles corresponding to the (002), (020) and (220) planes, respectively. This shows that there is a significant effect of increasing $O_2$ partial pressure in the formation of $WO_3$ films. The bandgap energy of the $WO_3$ thin films are found to be around 3.0 eV. Raman measurements showed vibrational modes of W-O-W stretching and bending modes which shows small shifts depending on the partial pressures of the $O_2$. Obtained results indicated that better crystal structure is obtained with higher $O_2$ gas partial pressure.

## 1. Introduction

Tungsten trioxide thin films have been extensively studied because of their potential application in electrochromic devices. The optical properties of these films can be changed in a reversible and persistent way under the influence of an applied voltage [1]. Electrochromism refers to the reversible change of color of thin films due to a small change in the voltage. This is important for smart windows and display applications [2-4]. It is widely accepted that the electrochromic process involves the simultaneous injection (or extraction) of electrons and charge compensating ions into interstitial sites of the $WO_3$ matrix [5]. For the polycrystalline films, the injected electrons behave as free-carriers leading to an increase in the near infrared reflectivity [6]. Nevertheless, scattering of the excess free electrons by extended defects, leads to lower reflectivity in this region [7].

Various methods have been used to prepare $WO_3$ thin films, including thermal evaporation [8,10,13,14], chemical vapor deposition (CVD) [12,15], sputtering [9,11,16,17,18,19] and pulsed laser deposition (PLD) [20,21]. In this study $WO_3$ thin film were deposited on the glass substrate by R.F magnetron sputtering a 99.9% pure tungsten target.

## 2. Experimental

RF magnetron sputtering has advantageous of having high deposition rates, sputtering of insulating targets, low power and low pressure operation conditions. $WO_3$ thin films were deposited on the Si (111), Si(100), ITO substrates by R.F. magnetron sputtering using a 99.9% pure 2'' diameter W target reactively. The vacuum chamber was evacuated to $5 \times 10^{-7}$ Torr base pressure before the deposition. Both argon and oxygen flows were controlled by mass flow controllers. The total pressure in the deposition chamber was kept constant at 12-15 mTorr. It was aimed to provide optimization of the optical and structural characteristics of $WO_3$ by changing the growth parameters mainly the oxygen partial pressure. The partial pressure of oxygen was changed with increments of 0.7 mTorr. While the applied power to the target was kept constant at 100 W and the substrate temperature was maintained at 350K.

Initially the deposition parameters were optimized to obtained $WO_3$ films with high transparency. Later, all the parameters except the deposition time were kept constant. The deposition parameters are listed in Table 1.

**Table 1.** R.F. sputtered parameters used for deposition of $WO_3$ thin films.

| Growth material | Base Pressure (torr) | Growth Pressure (mTorr) | Temperature (Celcius) | Gas1 Ar (sccm) | Gas2 $O_2$ (sccm) | Growth Rate (A°/s) | Growth Time | Substrate |
|---|---|---|---|---|---|---|---|---|
| $WO_3$ | $1,5 \times 10^{-6}$ | 15,70 | 330 | 12.5 | **0.5** | 1,5 | 130 | Si (111)<br>Si (100)<br>İTO glass |
| $WO_3$ | $6 \times 10^{-7}$ | 18,35 | 311 | 12.5 | **1.5** | 0,2 | - | Si (111)<br>Si (100)<br>İTO glass |
| $WO_3$ | $1,3 \times 10^{-6}$ | 18,88 | 300 | 12.5 | **1** | 0,6 | 54 | Si (111)<br>Si (100)<br>İTO glass |
| $WO_3$ | $7,8 \times 10^{-7}$ | 20,3 | 300 | 12.5 | **2** | 0,1 | 60 | Si (111)<br>Si (100)<br>İTO glass |

## 3. Results and Discussion

Increasing $O_2$ partial pressure led to the formation of XRD peaks for $WO_3$ films grown on ITO and the p-Si substrate. No XRD peak was observed for the thin films grown at the lowest oxygen partial pressure. XRD Peaks belong to the $WO_3$ films was observed at 23.0, 11.0, 23.5 and 28.5 angles corresponding to the (002), (020) and (220) planes, respectively. The peak intensity ratio of the of (002) plane to the other planes in films grown on the ITO substrate significantly has increased with increasing $O_2$ partial pressure. This shows that there is a significant effect of increasing $O_2$ partial pressure in the formation of $WO_3$ films. XRD results suggests that better crystal structure is possible by increasing $O_2$ partial pressure in $WO_3$ thin films.

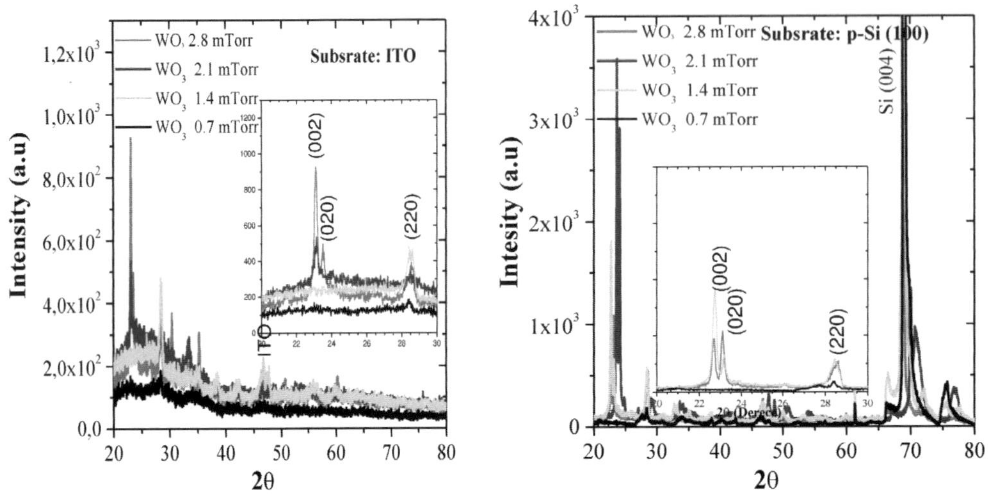


**Figure 1.** XRD pattern of $WO_3$ thin films grown as depending on to $O_2$ partial pressure.

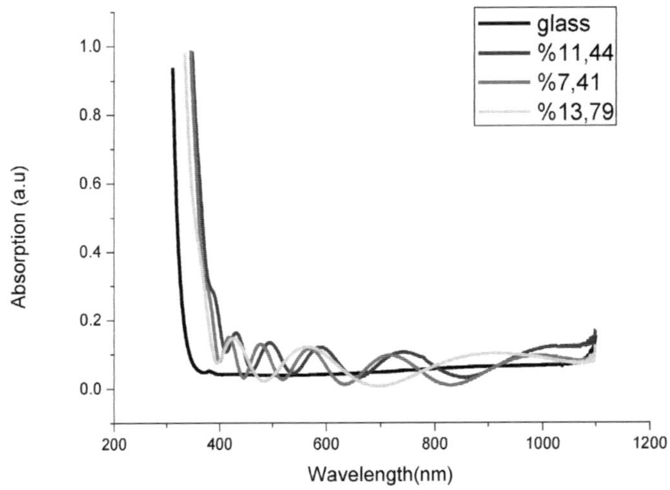


**Figure 2.** Absorption pattern of $WO_3$ thin films grown depending on to $O_2$ partial pressure.

Absorption plot of films that is grown at the highest $O_2$ partial pressure from the grown $WO_3$ thin films is shown in Fig.2. Starting around 400 nm, absorption spectra increases corresponding to the near band edge absorption of the grown $WO_3$ thin films. In the figure above, absorption spectrum of the pure glass substrates has shown in the plot which is used as the reference [20]. Very sharp absorption band edges of the sputtered $WO_3$ thin films suggest the successful high quality of the grown thin films.

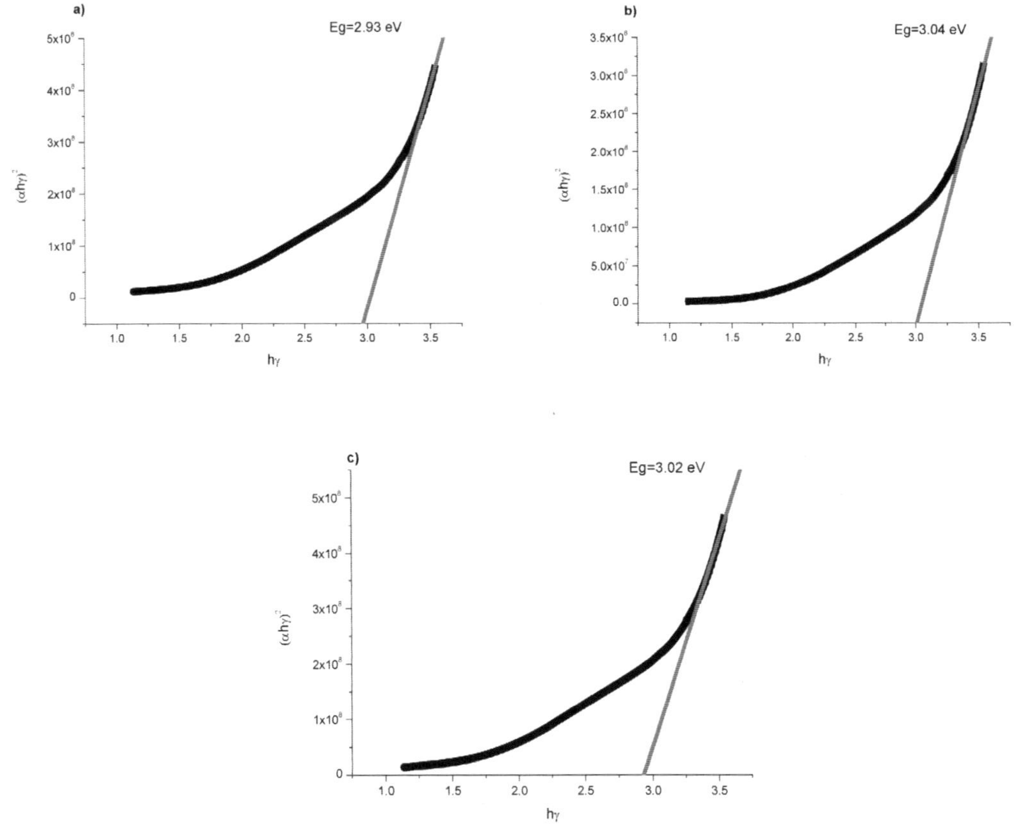


**Figure 3**.The absorbance spectra were used to determine optical band gap energies of $WO_3$ on glass substrate.

Bandgap of the $WO_3$ thin films is shown in Fig.3 by using the following relation for near band edge optical absorption of semiconductors,

$$\alpha = \left(\frac{K}{h\gamma}\right)(h\gamma - E_g)^{2} \tag{1}$$

where $\boldsymbol{\alpha}$ is absorption co-efficient, hv is the photon energy, K is a constant, $E_g$ is the optical band gap and n is a constant (n equals 1/2 for allowed direct transitions) [25-29]. Fig. 3 shows the plot of $(\alpha h\nu)^2$ versus hv for 4.45% $O_2$ partial pressure, 11.44% $O_2$ partial pressure, 13.79% O2 partial pressure on glass substrate thin films; the linear portion of the curve is extrapolated to the energy (hv) axis and the intercept gives the value of optical band gap $E_g$ of the film [30]. The optical band gap values for 4.45% $O_2$ on glass substrate, 11.44% $O_2$, 13.79 % $O_2$ thin films, determined from Fig. 3(a)-(c) as the 2.93 eV, 3.04 eV, 3.02 eV respectively.

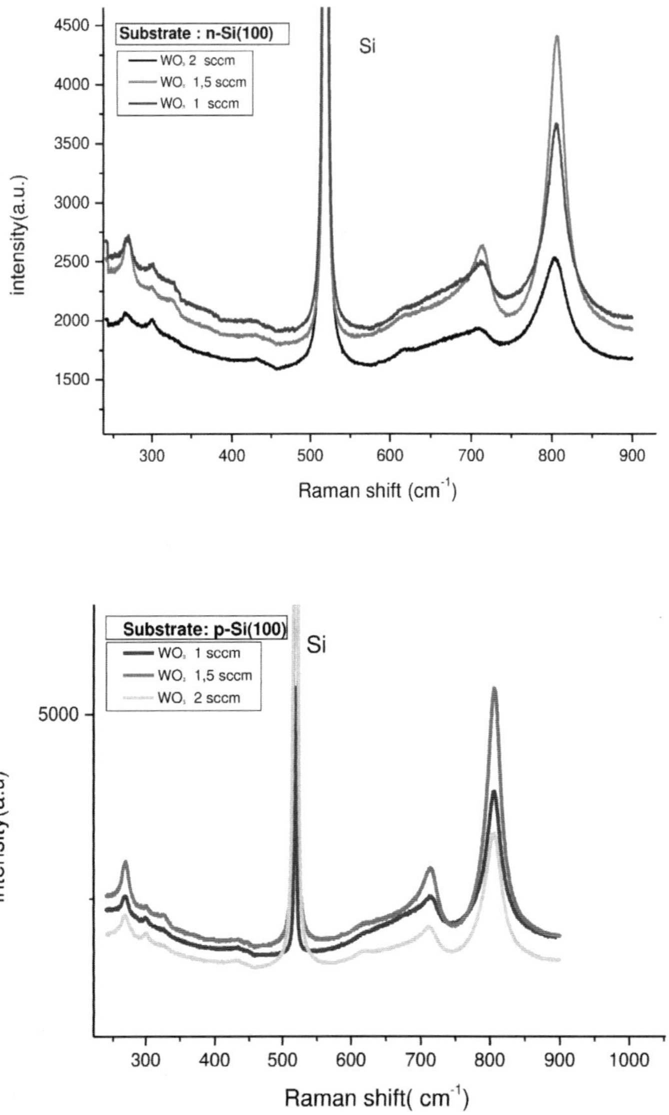


**Figure 4.** Raman spectra of $WO_3$ thin films grown depending on to $O_2$ partial pressure.

The Raman spectroscopy is effective in revealing structural features and structural transitions in $WO_3$ owing to its high sensitivity to the small changes in the lattice positions of the tungsten atoms. The $WO_3$ exhibits the following sequence of crystal phases with increasing temperature: triclinic , monoclinic , orthorombic , tetragonal [21]; in particular, bulk $WO_3$ is monoclinic-γ between 20°C and 330°C and it is orthorombic-β between 330°C and 740°C [22]. When Raman spectrum of the as prepared $WO_3$ thin films are examined, it is seen four well defined peaks located at 807, 710, 327, and 267 $cm^{-1}$. The high-frequency peaks at 807 and 710 $cm^{-1}$ were attributed to the W–O–W stretching vibration mode, while the low frequency ones (i.e., 327 and 267 $cm^{-1}$) were attributed to the W–O–W bending mode. The peaks shown in the present study similar to those for monoclinic $WO_3$ thin films [23,24]. $WO_3$ thin films were not shown for 0.7 mTorr. The details of the Raman peaks, wavenumbers and Full Width at Half Maximum (FWHM) obtained is given in table 1 and 2, respectively.

**Table 2.** Peak analysis of $WO_3$ thin films in Raman Spectroscopy

| **Si n-tipi** | **% 4.45 $O_2$ (0.5 sccm)** | **% 7.41 $O_2$ (1 sccm)** | **% 11.44 $O_2$ (1.5 sccm)** | **% 13.79 $O_2$ (2 sccm)** |
|---|---|---|---|---|
| 1) (w-O-W) stretching mode | _ | 804,965 | 805,594 | 804,126 |
| 2) (w-o-W) stretching mode | _ | 711,399 | 714,545 | 708,881 |
| 3)(W-O-W) bending mode | _ | 300,42 | 300 | 300,629 |
| 4)(W-O-W) Bending mode | _ | 268,741 | 268,741 | 265,385 |
| **Si p-tipi** | **% 4.45 $O_2$ (0.5 sccm)** | **% 7.41 $O_2$ (1 sccm)** | **% 11.44 $O_2$ (1.5 sccm)** | **% 13.79 $O_2$ (2 sccm)** |
| 1) (W-O-W) Stretching mode | _ | 805,385 | 805,594 | 805,385 |
| 2) (W-O-W) Stretching mode | _ | 714,755 | 714,126 | 712,238 |
| 3) (W-O-W) bending mode | _ | 299,79 | 300,839 | 301,259 |
| 4) (W-O-W) bending mode | _ | 267,902 | 268,112 | 267,692 |

**Table 3.** FWHM of $WO_3$ thin films in Raman Spectroscopy

| **Si n-tipi** | **% 4.45 $O_2$ (0.5 sccm)** | **% 7.41 $O_2$ (1 sccm)** | **% 11.44 $O_2$ (1.5 sccm)** | **% 13.79 $O_2$ (2 sccm)** |
|---|---|---|---|---|
| (W-O-W) stretching mode | _ | 24,71823 | 23,5692 | 30,94485 |
| (W-O-W) stretching mode | _ | 29,81282 | 24,4564 | 31,11964 |
| (W-O-W) bending mode | _ | 4,80163 | 10,48127 | 7,70834 |
| (W-O-W) Bending mode | _ | 9,73598 | 12,22645 | 17,08973 |
| **Si p-tipi** | **% 4.45 $O_2$ (0.5 sccm)** | **% 7.41 $O_2$ (1 sccm)** | **% 11.44 $O_2$ (1.5 sccm)** | **% 13.79 $O_2$ (2 sccm)** |
| (W-O-W) stretching mode | _ | 24,70637 | 23,60933 | 26,19571 |
| (W-O-W) stretching mode | _ | 17,0363 | 21,83434 | 19,42854 |
| (W-O-W) Bending mode | _ | 5,78893 | 7,22312 | 9,74574 |
| (W-O-W) Bending mode | _ | 14,24921 | 9,9938 | 13,97798 |

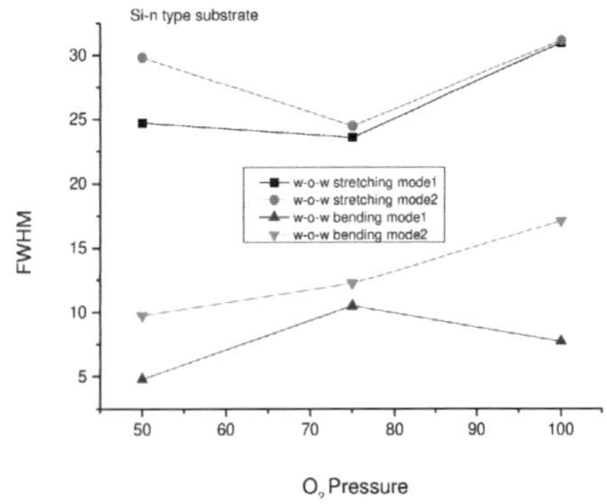


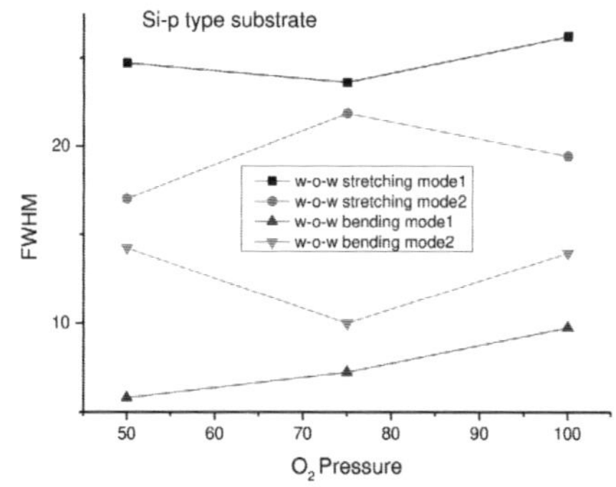


**Figure 5.** FWHM of WO3 thin films in Raman Spectroscopy for n and p type Si

## 4. Conclusion

To summarize, in the present study, the structural, optical properties and the electrochromic characteristics of oxygen sputtered $WO_3$ films have been investigated. $WO_3$ thin films were not shown 4.45% $O_2$ peaks whereas the best peaks were shown 11.44% $O_2$. Tungsten oxide is one of the most promising inorganic materials which exhibit excellent electrochromic, photochromic and gasochromic properties and it has been widely investigated to be used in electrochromic, gasochromic, solar energy, optical modulation, writing–reading–erasing optical devices.

## References

[1] Regragui M, Jousseaume V, Addou M, Outzourhit A, Bernede J C and El-Idrissi B 2001 *Thin Solid Films* **397** 238–243
[2] Roth S, Ignatowitz M, Muller P, Monch W and Oesterschulze E, 2011 *Microelectronics J.* **88** 2349–2351
[3] Granqvist C G 2012 *Sol. Energy Mater. Sol. Cells* **99** 1–13.
[4] Mihelcˇicˇ M, Jerman I and Orel B 2013 *Prog. Org. Coat.* **76** 1752–1755.
[5] Dautremont-Smith W C 1982 *Displays* **3** pp 67-80.
[6] Goldner R B, Chapman R L, Foley G, Goldner E L, Haas T, Norton P, Seward G and Wong K K 1986 *Sol. Energy Mater.* **14** 195.
[7] Goldner R B, Brofos A, Foley G, Goldner E L, Haas T E, Henderson W, Norton P, Ratnam B A, Weis N and Wong K K 1984 *SPIE Proc.* **502** 54.
[8] Brezesinski T, Fattakhova-Rohlfing D, Sallard S, Antonietti M, and Smarsly B M 2006 *Small* **2(10)**, 1203
[9] Berveglieri G, Depero L, Groppelli S, and Nelli P 1995 *Sens. Actuators,* B 1 **26(1–3),** 89.
[10] Cantalini C, Sun H T, Faccio M, Pelino M, Santucci S, Lozzi L, and Passacantando M 1996 *Sens. Actuators,* B **31(1–2)**, 81
[11] Di Giulio M, Manno D, Micocci G, Serra A, and Tepore A 1997 *J. Phys. D: Appl. Phys.* **30(23),** 3211
[12] Tong M, Dai G, Wu Y, He X, Gao D and Mater J 2001 *Sci.* **36(10)**, 2535
[13] Agrawal A and Habibi H 1989, *Thin Solid Films* **169(2),** 257.
[12] LeGore L J Greenwood, O D Paulus, Frankel D J, Lad R J, Vac J 1997 *Sci. Technol,* A **15(3)**, 1223.
[14] Tagtstrom P and Jansson U 1999 *Thin Solid Films* **352(1–2),** 107.
[15] Akram H, Tatsuoka H, Kitao M, and Yamada S 1987 *J. Appl. Phys.* **62(5)**, 2039.
[16] Kaneko H, Nishimoto S, Miyake S K, and Suedomi N 1986 *J. Appl. Phys.* **59(7)**, 2526.

[17] Kobayashi Y, Terada S, and Kubota K 1989 *Thin Solid Films* **168(1),** 133.
[18] Li M, Altman E I,Posadas A, and Ahn C H 2004 *Thin Solid Films* **446(2)**, 238.
[19] Garg A, Leake J A and Barbe Z H 2000 J.Phys. D: *Appl. Phys.* **33(9)**, 1048.
[20] Wang C, Sahu D R, Wang S, Chung-Kwei C and Huang J 2012 *J. Phys. D: Appl. Phys.* **45**, 225-303.
[21] Lu D Y, Chen J, Chen H J, Gong L, Deng S Z, Xu N S 2007 *Appl. Phys. Lett,* **90**, 041919
[22] Cazzanelli E, Vinegoni C, Mariotto G, Kuzmin A, Purans J 1999 *Sol. St. Ionics*, **123**, 67
[23] Zhou L, Ren Q J, Zhou X F, Tang J W, Chen Z H and Yu C Z 2008 *Microporous Mesoporous Mater* **109**, 248–257.
[24] Siciliano T, Tepore A, Micocci G, Serra A, Manno D and Filippo E 2008 Sens. *Actuators,* B **133**, 321–326.
[25] Hamberg L 1984 Indium tin oxide thin films: basic optical properties and applications to energy efficient windows, *Ph.D. Thesis, Chalmers University of Technology*, Gothenburg, Sweden.
[26] Born M, Wolf E 1983 *Principles of Optics* (Oxford: Pergamon Press)
[27] Heavens O S, 1970 *Optical Properties of Thin Films*, Dover, New York; Heavens O.S. 1965 *Thin Film Physics,* Methuen, London.
[28] Pankove J I, 1971 Optical Processes in Semiconductors, *Prentice- Hall, Englewood Cliffs,* N.J.
[29] Ozer N, Tepehan F, Tepehan G 1997 *Proc. SPIE—Int. Soc. Opt. Eng. (USA)* **3138**, 9–31.
[30] Miller E L, Paluselli D, Marsen B, Rocheleau R E 2004 *Thin solid Films* **466**, 307.

# Optoelectronics Properties Tunability by Controlled Deformation

**F.H. Alharbi[1,2], P. Serra[1,3], M.A. Carignano[1], and S. Kais[1,2,4]**

[1] Qatar Environment & Energy Research Institute, Qatar Foundation, Doha, Qatar
[2] College of Science & Eningeering (QEERI), Hamad Bin Khalifah University, Doha, Qatar
[3] Facultad de Matemática, Astronomía y Física, Universidad Nacional de Córdoba and IFEG-CONICET, Ciudad Universitaria, X5016LAE Córdoba, Argentina.
[4] Department of Chemistry, Physics, and Birck Nanotechnology Center, Purdue University, West Lafayette, Indiana 47907, USA.

E-mail: `falharbi@qf.org.qa`

**Abstract.** Manipulating energy levels while controlling the electron localization is an essential step for many applications of confined systems. In this paper we demonstrate how to achieve electron localization and induce energy level oscillation in one-dimensional quantum systems by externally controlling the deformation of the system. From a practical point of view, the one-dimensional potentials can be realized using layered structures. In the analysis, we considered three different examples. The first one is a graded quantum well between confining infinite walls where the deformation is modeled by varying slightly the graded well. The second systems is a symmetric multiple quantum well between infinite walls under the effect of biasing voltage. The third system is a layered 2D hybrid perovskites where pressure is used to induce deformation. The calculations are conducted both numerically and analytically using the perturbation theory. It is shown that the obtained oscillations are associated with level avoided crossings and that the deformation results in changing the spatial localization of the electrons.

## 1. Introduction

Over the past decade, manipulating electrons in quantum-confined system has progressed remarkably [1, 2] and thus allows developing devices for many applications like quantum computing [3, 4], spectroscopy [5, 6], ultra-fast and sensitive optoelectronics [7, 8], optical tunability [9, 10], and negative heat capacity [11]. In such devices, it is essential to have well controlled manipulation means. Different manipulation mechanisms are used currently such as superconductor circuits [12, 13] and spins in solids [14] and molecules [15–17]. However, the controllability of confined quantum systems in general and in particular for selected energy levels is still a major challenge [18, 19] that has stimulating the exploration of many new concepts [20–24]. One of the largely investigated control techniques relies on the utilization of Stückelberg oscillations. By passing an avoided crossing twice, the interference due to the dynamical phase between transitions can be either constructive or destructive. This is known as Landau-Zener-Stückelberg (LZS) interference [13, 21, 25]. Usually, the levels' oscillation occurs in the time domain. However, Nori and co-workers showed that this is achievable as well in the space domain using spatially inhomogeneous magnetic field [26]. Furthermore, similar oscillatory behaviors originated from different external influences and not intended for LZS interferometry, have been reported in other types of systems. For example, using self assembled quantum dots [27] under a strain field, quantum dots in micro cavities under terahertz laser excitation [28] or coupled semiconductor nano rings with varying inter-ring distance [29].

In this paper we present practical methods to manipulate electron localization and to cause energy levels' oscillation in quantum confined systems using controlled deformations that can, in principle, be applied at room temperature. These methods can be used to realized LZS interferometry where the oscillation is caused by various means. By deforming the quantum system, the system potential energy is slightly altered and hence the states and their energies are changed accordingly. The practical application of these ideas depends on the magnitude of the level oscillation and displacement of the localization centers, and how these are coupled to the externally controlled deformations. We found that the control of the state localization is easily achievable with deformations. The control of the energy levels requires careful engineering addressing the binding strength of the electron to particular sites that can be obtained using, for example, quantum dots. The assumed controlled deformations are applied either by voltage biasing or pressure across 2D layered systems such as graded semiconductor quantum wells [30–32] or stacks of 2D hybrid perovskites [9, 33]. Thus, the problem becomes a piece-wise constant potential and hence it has an analytical solution [34, 35] where the eigenfunctions, which are exponential and trigonometric functions with continuous logarithmic derivative, and the eigenenergies are obtained as solutions of transcendental algebraic equations. Also, the problem can be solved numerically using various methods like finite difference (FDM) [36, 37], finite element (FEM) [38, 39], and spectral methods [40–44]. In this work, we use a flexible FDM [37] to calculate the eigen-pairs.

The numerical calculations show that energy level oscillations are achievable and manipulatable in the studied systems by controlled deformations. Furthermore, the obtained oscillations are associated with avoided crossings. Thus, it can be used to realize LZS interferometry using alternative means other than the time-varying field. It is shown also that beside the oscillations, the deformation results in changing the spatial localization of the electrons. This should provide means to manipulate the electrons and hence can be used for other applications of quantum-confined systems. The next section presents three different scenarios of controlled deformations, followed by a final section with our general conclusions.

## 2. Controlled deformations

In order to explore possible scenarios in which localization/delocalization transitions occur via controlled deformations and how are they linked to energy level oscillations we studied three different confined systems. The first one is a graded quantum well between infinite walls where the deformation is modeled by varying spatially the graded potential. The second studied systems is a symmetric multiple quantum well between infinite walls under the effect of biasing voltage. The third system is a layered 2D hybrid perovskites where the pressure is used to induce deformation. Tuning the properties through pressure has been applied for purely inorganic systems [32,45]. However, the softer character of organic layers should lead to more pronounced effects in 2D hybrid materials [9, 10]. For the calculations, we use a flexible finite FDM [37] for the calculations as aforementioned and atomic units are assumed throughout the paper.

### 2.1. *Graded quantum well between infinite walls*

The system analyzed in this subsection is a square well between two infinite walls separated by a distance $L$. So, the potential (Figure-1) is simply

$$V(x) = \begin{cases} 0 & \text{if} \quad 0 < x < d \text{ or } d + a < x < L \\ -U & \text{if} \quad d < x < d + a \\ \infty & \text{otherwise} \end{cases} , \tag{1}$$

where $L$, $d$, $a$ and $U$ are all positive constants and $d + a < L$. The potential deformation is modeled by varying $d$, which is the distance between the left infinite wall and the edge of the well. Namely, the deformation represents to displace the position of the potential well with respect to the confining walls. As an example of the effects we show some of the results obtained with $L = 15$, $a = 2$, and $U = 3$. In this case, we obtain two states with energy levels below zero (shown in lower left of Figure-1) that are bounded stated to the small well. The energies above zero as a function of $d$ are shown in the right panel of Figure-1, up to the 20th level. It is clear that these states are oscillating with $d$ and that the amplitude of the oscillations depends not in a monotonic way with the order of the level.

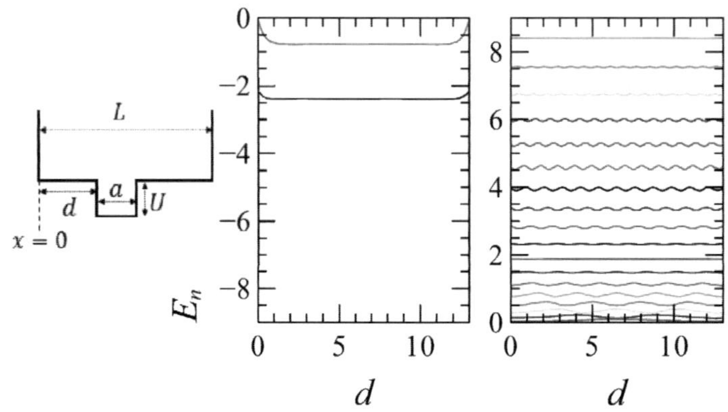


**Figure 1.** (left) The studied single-well structure. (middle and right panels) Numerical calculation of the first 20 eigenenergies for the example system ($L = 15$, $a = 2$, $U = 3$). The eigenenergies less than zero are shown in the bottom left panel while those greater than zero are shown in the right panel.

In order to have an analytical perspective that could shed light into the different contributions that result on the observed oscillatory behavior, the non degenerate eigenenergies (above zero in the used model) are calculated using perturbation theory (PT). In this case, the unperturbed structure is assumed to be an infinite walls potential and the perturbation is due to the small finite well. So, the Hamiltonian can be rewritten as $H = T + V_0 + V_1$, where

$$V_0(r) = \begin{cases} 0 & \text{if} \quad 0 < x < L \\ \infty & \text{otherwise} \end{cases} \tag{2}$$

and

$$V_1(r) = \begin{cases} -U & \text{if} \quad d < x < d + a \\ 0 & \text{otherwise} \end{cases} \tag{3}$$

The well-known eigenvalues of $H_0$ $(T + V_0)$ are $E_m^{(0)} = \frac{1}{2}k_m^2 = \frac{1}{2}\left(\frac{m\pi}{L}\right)^2$, where $m = 1, 2, \ldots,$ is the sum of peaks and valleys of the corresponding eigenfunctions, which are

$$\psi_m(x) = \sqrt{\frac{2}{L}} \sin(k_m x) \quad . \tag{4}$$

Here, $m$ is used in this subsection as the counter for the states with energies above zero and should not be confused with $n$ that counts all the states. With the exception of this subsection,

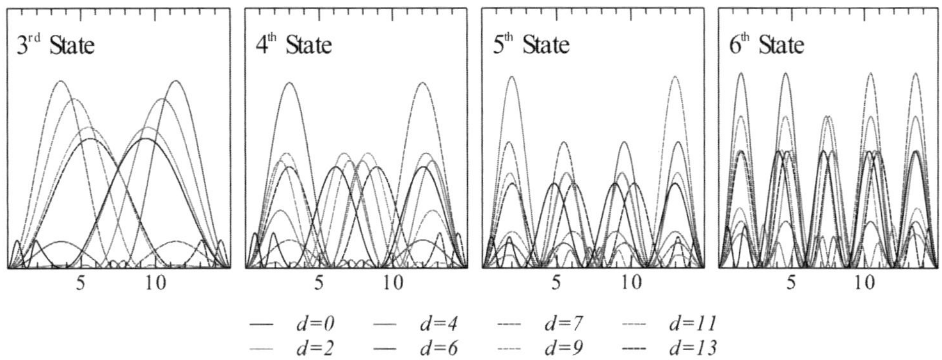


**Figure 2.** The densities for different states as indicated and selected values if $d$. As the small well is displaced from left to right there is a concurrent displacement of the density that can be concentrated on either side depending on the state level being observed.

$n$ is used to describe the energy levels throughout this paper. The first order correction to the energies is given by

$$\begin{aligned}\Delta_m^{(1)} &= \langle m\,|V_1|\,m\rangle = -U\int_d^{d+a}\psi_m^2(x)\,dx \\ &= -U\frac{a}{L} + U\frac{1}{m\pi}\sin\left(m\pi\frac{a}{L}\right)\cos\left(m\pi\frac{(2d+a)}{L}\right).\end{aligned} \tag{5}$$

It is clear from the above equation that the corrections are oscillating with $d$ at a frequency of $2m\pi/L$, which matches the oscillations obtained by the numerical calculations especially for higher levels. Eq.-5 also implies that the oscillation amplitude is proportional to $U$ and is sinusoidal with $m$ and $a$, providing additional means to design and control the oscillations. In particular, the amplitude dependency on $m$ explains the behavior observed on Figure 1 showing stronger oscillations at intermediate values of $m$.

The localization of the electron, as described by the density $\rho(x) = |\psi(x)|^2$, is also affected by the deformation. In Figure 2 we show the resulted densities for the first four positive eigenenergies, $n$ =3, 4, 5, and 6, for a selection of different $d$ values. There is a clear shifting from one side of the system to the other on the most probable position of the particle as a function of the system state. For example, for the third state, the particle is localized in the right side of the well for $d < (L-a)/2$. The localization is shifted toward the left side for $d > (L-a)/2$. This shifting effect persists up to higher states, although the side chosen by the system depends, among other factor, on the odd/even character of the quantum number.

So far we have shown that energy levels' oscillations (for energies above zero) are directly related to the system's deformation and therefore providing a way to externally control the particular values of the energy spectrum of a system. As aforementioned in the introduction section, many similar oscillations, due to other causes, were reported and in many of them the oscillations are associated with crossing avoidance. The oscillations observed in this example are also associated with density swapping between states. In Figure 3, the energies and densities of the third and forth eigenstates are shown. In the left panel, it is clear that there is a crossing avoidance at $d = 6.5$. This is associated with a density swapping where for $d = 6$, the third state is localized in the left of the well while the fourth state is localized in the right. This swapping happens in the reverse order for $d = 7$.

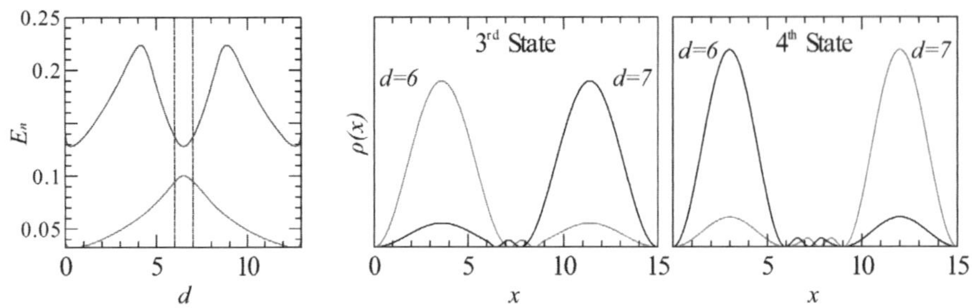


**Figure 3.** (color online) (left) The energies for the third and forth eignestates vs. $d$. (middle) The density for the third eigenstate for $d = 6$ and $d = 7$. (right) The density for the firth eigenstate for $d = 6$ and $d = 7$.

*2.2. A symmetric multiple quantum well between infinite walls deformed by biasing*

In this subsection and the next we consider two examples that could be practically achieved with currently known materials and technologies, allowing the experimental exploration of controlled energy level oscillations and the associated localization/delocalization of the particles. In this case, we start with a symmetric double-well system confined between two infinite walls. Then, a biasing voltage is applied to deform the original system into a tilted one. The unbiased potential, represented in the sketch of Figure 4, is mathematically described by the following potential:

$$V(x) = \begin{cases} 0 & \text{if} \quad |x| < d \text{ or } d + a \leq |x| < L \\ -U & \text{if} \quad d < |x| < d + a \\ \infty & \text{otherwise} \end{cases}, \tag{6}$$

where $L$, $d$, $a$ and $U$ are positive constants and $d + a < L$. Equivalent systems can be made although with finite potential walls. For example, using semiconductor multiple quantum wells [46, 47] or a double quantum dot system [48, 49] can be used to realize a device having the qualitative features of Eq. (6). Having this examples in mind, the parameters used in this analysis are within the ranges of practical values. We chose $L = 200$, $a = 40$ and $U = 0.02$ that represent 10 nm, 2 nm and 0.54 eV, respectively. By applying a biasing voltage $V_b$, the potential is tilted by an additional term of $-V_b x$ providing the externally controlled deformation. In our example, the voltage is varied between -0.15 and 0.15 mV. The resulting energies are shown in Figure 4 as a function of $V_b$, where it can be seen the emergence of an oscillating pattern having similar characteristics to the example of the previous section. For example, the amplitude of the oscillation is a function of the energy level, as it was found by the PT analysis presented above.

Analogously to the case of Subsection A, in this example the energy oscillation has an associated crossing avoidance and localization transition. In Figure 5 we show the energies and densities corresponding to the $11^{\text{th}}$ and $12^{\text{th}}$ eigenstates. In the left panel, it is clear that there is a crossing avoidance around $V_b = 0$ mV. The concurrent density swapping happens upon a very small change in the biasing: for $V_b = -0.01$ mV, the $11^{\text{th}}$ state is localized in the left of the well while the $12^{\text{th}}$ state is localized in the right. The picture is completely reversed for $V_b = 0.01$ mV.

*2.3. A layered 2D hybrid perovskites deformed by applying pressure*

The second possible practical deformation method that we consider relies on the application of an external pressure. To realize this system we propose a layered 2D hybrid perovskites. These structures are composed of alternating layers of hybrid 2D framework of octahedra inorganic and organic cations providing the wells. Moreover, the system could be intercalated between

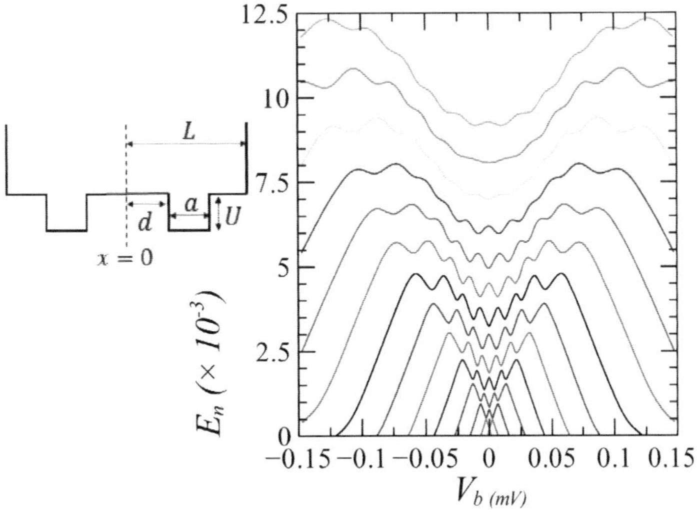


**Figure 4.** The sketch shows the double-well symmetric structure. The left figure displays the energy levels as a function of the biasing voltage $V_b$ and the induced oscillations. The different colors represent the different energy levels.

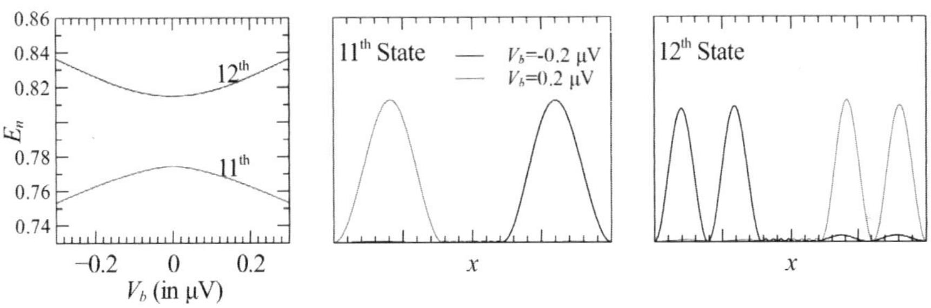


**Figure 5.** (left) The energies for the 11$^{th}$ and 12$^{th}$ eignestates vs. $V_b$. (middle) The density for the 11$^{th}$ eigenstate for $V_b = -0.2$ $\mu$V and $V_b = 0.2$ $\mu$V. (right) The density for the 12$^{th}$ eigenstate for $V_b = -0.2$ $\mu$V and $V_b = 0.2$ $\mu$V.

softer polymeric layers allowing large deformations and increasing the degree of external control by the application of moderate pressure as compared with purely inorganic crystals [50–52]. These materials can be analyzed as superlattices, as shown by Even et al. [9] and the references therein.

The most studied 2D hybrid materials are $[RNH_3]_2(CH_3NH_3)_{m-1}Pb_mI_{3m+1}$, with R representing an organic group. In these materials, $Pb_mI_{3m+1}$ forms the octahedra that include the $(CH_3NH_3)$ cations forming the wells. On the other hand, $[RNH_3]_2$ constitutes the barriers and by changing the organic group (R), the equilibrium separation between the wells can be tailored [9]. The depth of the well is estimated to be $U_1 = 0.9$ eV [53]. As for the well width, it depends on the number of the octahedra layers. For $m$ =1, 2, 3, and 4, the thicknesses $a$ are 0.65, 1.25, 1.85, and 2.5 nm respectively [53].

In this subsection, we study the effect of the pressure on the energy levels on the $[RNH_3]_2(CH_3NH_3)_{m-1}Pb_mI_{3m+1}$ quantum wells with 1.25 nm barrier. A finite number of wells

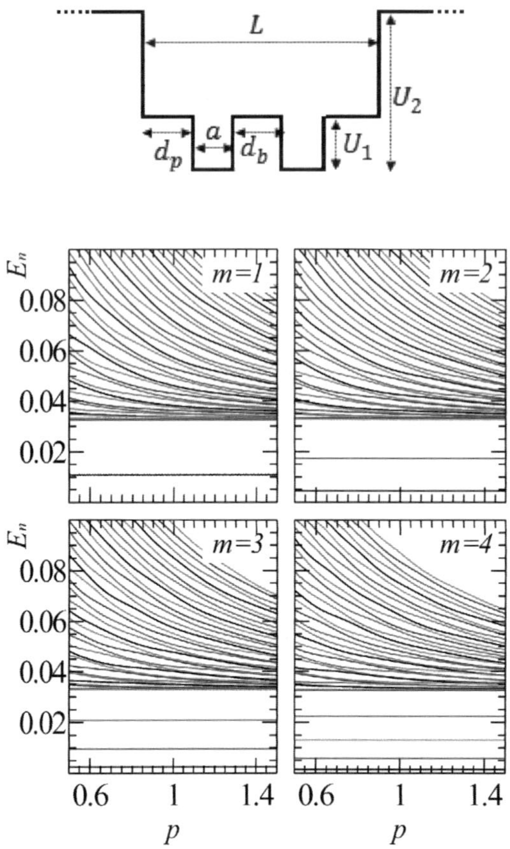


**Figure 6.** (top) Model potential to represent 2D layer perovskites confined between insulating walls, in this case for two layers creating two confining wells. The deformation affects the external confining layers of width $d_p$. (bottom) Numerical calculation of the first 40 eigenenergies for a two wells system with $m$ =1, 2, 3, and 4 as a function of the deformation factor.

will be used where the whole system is then placed between two insulators with a barrier height of $U_2$ = 2.7 eV, as described in Figure 6 for the case of two wells [9]. In our approximation, the external pressure only affects the confining external layers by changing their thickness by a factor $p$. The eigenvalues show the avoidance crossing characteristics as in the previous cases, although now as a function of the deformation parameter $p$. This is explicitly displayed in Figure 7, in which we display two states corpsonding to the case of $m = 3$. As the system is deformed, the 14$^{\text{th}}$ energy level has the particle shifting its preferential position from the central region between the wells for $p = 0.6$, to the outermost layers for $p = 1.4$. The energy level immediately above, the situation is exactly the opposite.

## 3. Conclusion

In this paper, we have analyzed the interplay between the detailed dimensions of a confining potential, energy level oscillations and particle's localization. It was shown that the obtained oscillations are associated with level avoided crossings and that the deformation induces a change in the preferential spatial localization of the electrons. This effect is quite general, as shown by

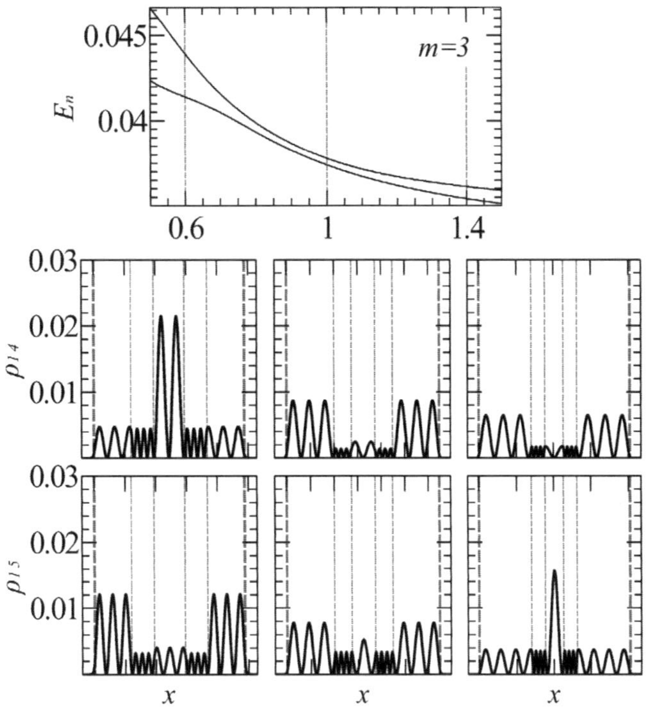


**Figure 7.** (top) Eigenvalues corresponding the $14^{\text{th}}$ and $15^{\text{th}}$ energy levels for the case of two wells and $m = 3$ as function of the deformation factor (the pressure $p$). The red vertical dashed lines correspond to the the selected deformations for which the densities are plotted for the two states as function of distance. (bottom) The densities of the two states as function of distance where the dashed green lines are corresponding to the edges of the insulating walls while the dashed red lines are corresponding to the small wells edges.

the the three described examples, and therefore it can be realized in different ways. The external controlling parameter could be an applied voltage or pressure, and other procedures are certainly possible. Obtaining a practical way to experimentally control the electron localization would enable the manipulation of charges in quantum confined systems and provide guidelines for the design of devices. The approach that we have used to describe the is based on simple 1D problems, but the consistent presence of this effect on multiple scenarios suggests that this phenomena is general and could be realized in 3D systems at the nanometer scale.

## References

[1] Petta J, Johnson A, Taylor J, Laird E, Yacoby A, Lukin M, Marcus C, Hanson M and Gossard A 2005 *Science* **309** 2180–2184
[2] Taylor J, Petta J, Johnson A, Yacoby A, Marcus C and Lukin M 2007 *Physical Review B* **76** 035315
[3] Schoelkopf R and Girvin S 2008 *Nature* **451** 664–669
[4] Vion D, Aassime A, Cottet A, Joyez P, Pothier H, Urbina C, Esteve D and Devoret M H 2002 *Science* **296** 886–889
[5] Astafiev O, Zagoskin A M, Abdumalikov A, Pashkin Y A, Yamamoto T, Inomata K, Nakamura Y and Tsai J 2010 *Science* **327** 840–843
[6] Berns D M, Rudner M S, Valenzuela S O, Berggren K K, Oliver W D, Levitov L S and Orlando T P 2008 *Nature* **455** 51–57

[7] Niemczyk T, Deppe F, Huebl H, Menzel E, Hocke F, Schwarz M, Garcia-Ripoll J, Zueco D, Hümmer T, Solano E *et al.* 2010 *Nature Physics* **6** 772–776
[8] Zecherle M, Ruppert C, Clark E, Abstreiter G, Finley J and Betz M 2010 *Physical Review B* **82** 125314
[9] Even J, Pedesseau L and Katan C 2014 *ChemPhysChem* **15** 3733–3741
[10] Lanty G, Jemli K, Wei Y, Leymarie J, Even J, Lauret J S and Deleporte E 2014 *The Journal of Physical Chemistry Letters* **5** 3958–3963
[11] Serra P, Carignano M A, Alharbi F H and Kais S 2013 *Europhysics Letters* **104** 16004
[12] Oliver W D, Yu Y, Lee J C, Berggren K K, Levitov L S and Orlando T P 2005 *Science* **310** 1653–1657
[13] Sillanpää M, Lehtinen T, Paila A, Makhlin Y and Hakonen P 2006 *Physical review letters* **96** 187002
[14] Hanson R and Awschalom D D 2008 *Nature* **453** 1043–1049
[15] Vandersypen L M and Chuang I L 2005 *Reviews of modern physics* **76** 1037
[16] Criger B, Passante G, Park D and Laflamme R 2012 *Philosophical Transactions of the Royal Society A: Mathematical, Physical and Engineering Sciences* **370** 4620–4635
[17] Oh S, Huang Z, Peskin U and Kais S 2008 *Physical Review A* **78** 062106
[18] Schirmer S G, Pullen I C and Solomon A I 2005 *Journal of Optics B: Quantum and Semiclassical Optics* **7** S293
[19] Ferrón A, Serra P and Osenda O 2013 *Journal of Applied Physics* **113** 134304
[20] Awschalom D D, Bassett L C, Dzurak A S, Hu E L and Petta J R 2013 *Science* **339** 1174–1179
[21] Shevchenko S, Ashhab S and Nori F 2010 *Physics Reports* **492** 1–30
[22] Yao N Y, Jiang L, Gorshkov A V, Maurer P C, Giedke G, Cirac J I and Lukin M D 2012 *Nature Communications* **3** 800
[23] Zhang J, Yung M H, Laflamme R, Aspuru-Guzik A and Baugh J 2012 *Nature Communications* **3** 880
[24] Zhao Y J, Fang X M, Zhou F and Song K H 2012 *Physical Review A* **86** 052325
[25] Shevchenko S, Ashhab S and Nori F 2012 *Physical Review B* **85** 094502
[26] Zhang J N, Sun C P, Yi S and Nori F 2011 *Physical Review A* **83** 033614
[27] Sheng W and Leburton J P 2002 *Physical review letters* **88** 167401
[28] Sherwin M S, Imamoglu A and Montroy T 1999 *Physical Review A* **60** 3508
[29] Chwiej T and Szafran B 2008 *Physical Review B* **78** 245306
[30] Lenchyshyn L, Liu H, Buchanan M and Wasilewski Z 1996 *Journal of applied physics* **79** 8091–8097
[31] Högele A, Seidl S, Kroner M, Karrai K, Warburton R J, Gerardot B D and Petroff P M 2004 *Physical review letters* **93** 217401
[32] Bardyszewski W and Łepkowski S 2012 *Physical Review B* **85** 035318
[33] Mitzi D B, Chondroudis K and Kagan C R 2001 *IBM journal of research and development* **45** 29–45
[34] Flügge S 1994 *Practical quantum mechanics* (Springer Verlag)
[35] Serra P and Kais S 2012 *Journal of Physics B: Atomic, Molecular and Optical Physics* **45** 235003
[36] Harrison P 2000 *Quantum Wells, Wires and Dots: Theoretical and Computational Physics* 1st ed (Chichester: John Wiley & Sons)
[37] Alharbi F 2008 *Optical and quantum electronics* **40** 551–559
[38] Nakamura K, Shimizu A, Koshiba M and Hayata K 1989 *Quantum Electronics, IEEE Journal of* **25** 889–895
[39] Le K Q 2009 *Microwave and Optical Technology Letters* **51** 1–5
[40] Alharbi F 2010 *Physics Letters A* **374** 2501–2505
[41] Alharbi F 2009 *Optical and quantum electronics* **41** 751–760
[42] Alharbi F and Scott J C 2009 *Optical and quantum electronics* **41** 583–597
[43] Alharbi F H and Kais S 2013 *Physical Review E* **87** 043308
[44] Lin W, Kovvali N and Carin L 2006 *Computer physics communications* **175** 78–85
[45] Meyer R, Dahl M, Schaack G, Waag A and Boehler R 1995 *Solid state communications* **96** 271–278
[46] Lumb M P, Yakes M K, González M, Vurgaftman I, Bailey C G, Hoheisel R and Walters R J 2012 *Applied Physics Letters* **100** 213907
[47] Christmann G, Askitopoulos A, Deligeorgis G, Hatzopoulos Z, Tsintzos S I, Savvidis P G and Baumberg J J 2011 *Applied Physics Letters* **98** 081111
[48] Stafford C A and Wingreen N S 1996 *Physical review letters* **76** 1916
[49] Hu X and Sarma S D 2000 *Physical Review A* **61** 062301
[50] Frost J M, Butler K T, Brivio F, Hendon C H, van Schilfgaarde M and Walsh A 2014 *Nano Letters* **14** 2584–2590
[51] Angel R, Zhao J and Ross N 2005 *Physical review letters* **95** 025503
[52] Schaak R E and Mallouk T E 2002 *Chemistry of Materials* **14** 1455–1471
[53] Tanaka K and Kondo T 2003 *Science and Technology of Advanced Materials* **4** 599–604

# Convergent sum of gradient expansion of the kinetic-energy density functional up to the sixth order term using Padé approximant

**A. Sergeev[1], F.H. Alharbi[1,2], R. Jovanovic[1], and S. Kais[1,2,3]**

[1] Qatar Environment & Energy Research Institute, Qatar Foundation, Doha, Qatar.
[2] College of Science & Eningeering (QEERI), Hamad Bin Khalifah University, Doha, Qatar.
[3] Department of Chemistry, Physics, and Birck Nanotechnology Center, Purdue University, West Lafayette, Indiana 47907, USA.
Hi

E-mail: `falharbi@qf.org.qa`

**Abstract.** The gradient expansion of the kinetic energy density functional, when applied to atoms or finite systems, usually grossly overestimates the energy in the fourth order and generally diverges in the sixth order. We avoid the divergence of the integral by replacing the asymptotic series including the sixth order term in the integrand by a rational function. Padé approximants show moderate improvements in accuracy in comparison with partial sums of the series. The results are discussed for atoms and Hooke's law model for two-electron atoms.

## 1. Introduction

For all electronic systems, Hohenberg and Kohn [1] proved the existence of a "universal functional" that accounts for the total energy except those contributions due to an external potential (mainly due to the electron-nuclei Coulombic interaction). The main challenge is that the universal functional is not known and thus Kohn and Sham [2] devised an alternative approach to approximate the "universal functional". In this approach, the kinetic term for non-interacting electrons was treated exactly but the exchange-correlation term are left unknown. Furthermore, in their initial work, Hohenberg and Kohn [1] suggested that the components of the "universal functional"; namely, the kinetic, exchange, and correlation contributions, can be expanded using gradient based expansions.

Since then, many approximations have been developed. For kinetic energy, the starting point has been mostly Thomas–Fermi functional. In the limit of infinitely extended uniform density, the kinetic energy $T_{\rm s}[\rho]$ is simply Thomas–Fermi kinetic energy $T_{\rm TF}[\rho]$ (in density functional form) [3, 4, 5]. For slowly varying densities, corrections to $T_{\rm TF}$ can be derived using gradient expansions. Different approaches have been used and they agrees in most cases. The most frequently used procedures are Kirzhnits expansion [6, 7, 8, 9], partition-function approach [10, 7], and Grammaticos-Voros algebraic method [11, 12]. As a result, the kinetic energy density functional (KEDF) can be arranged in the following series:

$$T_{\rm s}[\rho(\mathbf{r})] = T_0 + T_2 + T_4 + T_6 + \ldots, \qquad (1)$$

where $T_0 = T_{\mathrm{TF}}$, which will be presented in the next section, $T_n = \int \tau_n d\mathbf{r}$, and the kinetic energy densities $\tau_n$ collect terms that have gradients of the total order $n$. With the exception of solid state systems [13], the sixth-order term of the gradient expansion generally diverges for a finite system.

To eliminate the divergence, many additional approximations have been suggested. Pearson and Gordon [14] suggested local truncation of the gradient expansion with additional corrections to avoid vanishing densities. Also, many alternative approximate re-summation methods [15, 16, 17, 18] including Padé approximants [19, 20, 21] were suggested. The suggested Padé approximants are mostly based on $\tau_2/\tau_0$ expansions to resemble different asymptotic behaviors [19, 20, 21]. Obviously, each approach has advantages and disadvantages and each has a range of applicability.

In this paper, we present a systematic approach that allows to avoid regions where higher order terms are relatively large and are no longer "small corrections" while taking advantage of these terms in areas where they are small and improve the accuracy. So, we use a Padé approximant to have a convergent sum of gradient expansion approximation up to the sixth-order term. In many cases, adding of the correction $T_4[\rho]$ usually makes the result less accurate, because this correction is valid only asymptotically for heavy atoms. Moreover, since $T_6[\rho]$ diverges for finite systems, the knowledge of $\tau_6$ is assumingly -from the first glance- useless. By using the proper Padé approximant, it is possible to "sum" approximately the gradient expansion of kinetic energy so that the approximant has some asymptotic behavior, but suppresses divergent terms in a region away from asymptotics.

The proposed Padé approximant improves the accuracy when there exists a significant region of $\mathbf{r}$ where

$$|\tau_6(r)| < |\tau_4(r)|. \tag{2}$$

In case of a finite system, $\rho(r)$ decreases exponentially for large distance, and all gradients of $\rho$ are roughly proportional to $\rho$. Therefore, taking into account Equations (5) and (6), for large $r$ $\tau_4(r) \sim \rho^{1/3}(r)$ and $\tau_6(r) \sim \rho^{-1/3}(r)$, and for sufficiently large $r$, the inequality (2) is not satisfied. We guess that for small $r$, the behavior of $\tau_4$ and $\tau_6$ may depend on the presence of Coulomb singularity in the potential. In case of atoms, we have unfavorable inequality $|\tau_6(r)| > |\tau_4(r)|$ both for small $r$ and for large $r$, with a narrow intermediate region where the inequality (2) holds. We expect this approximation to work very well for confined harmonic potentials such the case of treating quantum dots [22, 23].

## 2. The method

As aforementioned, in the limit of uniform density [3, 4], the kinetic energy $T_s[\rho]$ is

$$T_{\mathrm{TF}} = C_{\mathrm{TF}} \int \rho^{5/3}\, d\mathbf{r},\; C_{\mathrm{TF}} = \frac{3}{10}\left(3\pi^2\right)^{2/3} \tag{3}$$

Any effect due to slowly varying densities can be accounted for approximately as corrections to Eq. 3 using gradient expansions. So, the kinetic energy density functional can be organized in a series as in Eq. 1. Using Kirzhnits expansion approach, Kirzhnits [6], Hodges [7], and Murphy [8] have derived the second-, forth-, and sixth-order terms respectively. These terms are

$$\tau_2 = \frac{1}{72}\frac{\left(\nabla\rho(\vec{r})\right)^2}{\rho(\vec{r})}, \tag{4}$$

$$\tau_4 = \frac{\left(3\pi^2\right)^{-2/3}}{540}\rho^{1/3}\left[\left(\frac{\nabla^2\rho}{\rho}\right)^2 - \frac{9}{8}\left(\frac{\nabla^2\rho}{\rho}\right)\left(\frac{\nabla\rho}{\rho}\right)^2 + \frac{1}{3}\left(\frac{\nabla\rho}{\rho}\right)^4\right]. \tag{5}$$

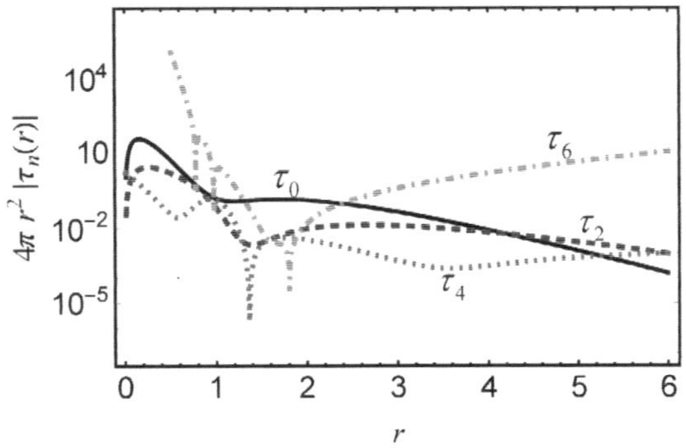


**Figure 1.** Gradient expansion corrections for the density of kinetic energy of beryllium atom, in logarithmic scale. The black, blue, red, and green lines are for $\tau_0$, $\tau_2$, $\tau_4$, and $\tau_6$ respectively. Obviously, the correction $\tau_6$ diverges as $r \to \infty$, while $\tau_4$ converges, but very slowly.

and

$$\tau_6 = \frac{(3\pi^2)^{-4/3}}{45\,360}\rho^{-1/3}\left[13\left(\frac{\nabla\nabla^2\rho}{\rho}\right)^2 + \frac{2575}{144}\left(\frac{\nabla^2\rho}{\rho}\right)^3 + \frac{249}{16}\left(\frac{\nabla\rho}{\rho}\right)^2\left(\frac{\nabla^4\rho}{\rho}\right)\right.$$
$$+\frac{1499}{18}\left(\frac{\nabla\rho}{\rho}\right)^2\left(\frac{\nabla^2\rho}{\rho}\right)^2 - \frac{1307}{36}\left(\frac{\nabla\rho}{\rho}\right)^2\left(\frac{\nabla\rho\cdot\nabla\nabla^2\rho}{\rho^2}\right) + \frac{343}{18}\left(\frac{\nabla\rho\cdot\nabla\nabla\rho}{\rho^2}\right)^2$$
$$\left. + \frac{8341}{72}\left(\frac{\nabla^2\rho}{\rho}\right)\left(\frac{\nabla\rho}{\rho}\right)^4 - \frac{1\,600\,495}{2592}\left(\frac{\nabla\rho}{\rho}\right)^6\right]. \quad (6)$$

Generally, the sixth-order term diverges for atoms and molecules. Actually, it diverges when the density vanishes at any position. A study of the gradient expansion for atomic isoelectronic series [24] shows that that $T_0 + T_2$ is more accurate approximation than $T_0 + T_2 + T_4$ in almost all cases, except for highly ionized Ar-like ions. A similar study for all neutral atoms [25] shows that $T_0 + T_2$ is more accurate than $T_0 + T_2 + T_4$ for all atoms up to potassium. Only for heavier atoms, the situation reverses. It rises the question of whether the forth order term $T_4$ and the diverging sixth-order density $\tau_6$ could be of any use for the case of finite systems.

A typical behaviour of the gradient correction for an atomic system is illustrated here for the case of a beryllium atom. We took the electron density from the Hartree–Fock data of Clementi tables [26]. The components of kinetic energy gradient expansion up to the 6-th order term are shown on Figure 1. It is clear that in a narrow region of $1.6 < r < 1.9$, the gradient expansion seems convergent, i.e. $|\tau_6| < |\tau_4| < |\tau_2| < |\tau_0|$. However, everywhere else $\tau_6$ is no longer a small correction to the lower order terms. For sufficiently large $r > 6$, we have an opposite inequality $|\tau_6| > |\tau_4| > |\tau_2| > |\tau_0|$, that means that all corrections become erroneous.

So, the objective is to "sum" the gradient expansion at each point $r$ by replacing it by another function that has some asymptotic behavior, but suppresses the divergent terms in a region away from asymptotics. Let's consider a family of densities parameterized by a dummy variable $g > 0$,

$$\rho_g(\vec{r}) = g\rho(r), \quad (7)$$

so that the original density $\rho$ can be recovered from this family of functions by setting $g = 1$. Then, the gradient expansion for the density can be organized as in the following series

$$\tau_g(\vec{r}) = g^{5/3}\left(\tau_0(\vec{r}) + g^{-2/3}\tau_2(\vec{r}) + g^{-4/3}\tau_4(\vec{r}) + g^{-6/3}\tau_6(\vec{r}) + \ldots\right). \quad (8)$$

The expression in parenthesis in Eq. 8 is obviously a power series in

$$x = g^{-2/3}. \tag{9}$$

So, asymptotically for small $x$ we have

$$f(x) = \tau_0 + \tau_2 x + \tau_4 x^2 + \tau_6 x^3 + \ldots, \tag{10}$$

where

$$f(x) = g^{-5/3}\tau_g(\vec{r}). \tag{11}$$

We would like to replace $f(x)$ by an approximation $\tilde{f}(x)$ having the same asymptotic expansion as $f(x)$ given by Eq. 10. We seek to have $\tilde{f}(1) \approx \tau_0 + \tau_2 + \tau_4 + \tau_6$ if $|\tau_6| < |\tau_4| < |\tau_2| < |\tau_0|$ and $\tilde{f}(1) \approx \tau_0 + \tau_2$ if $|\tau_6|$ is large.

The Padé approximant [2/1] to the series Eq. (10) is defined as a ratio of two polynomials of degree 2 and 1 in $x$,

$$f^{[2/1]}(x) = \frac{a_0 + a_1 x + a_2 x^2}{b_0 + b_1 x} \tag{12}$$

with $b_0 \neq 0$ that has the same expansion in $x$ up to the third order,

$$f^{[2/1]}(x) = f(x) + o(x^3). \tag{13}$$

Explicitly

$$f^{[2/1]}(x) = \tau_0 + \tau_2 x + \frac{\tau_4^2 x^2}{\tau_4 - \tau_6 x}. \tag{14}$$

Notice that if $\tau_6 \to \infty$, then $f^{[2/1]}(1) = \tau_0 + \tau_2$.

We define $\tau^{[2/1]} = f^{[2/1]}(1) = \tau_0 + \tau_2 + \frac{\tau_4^2}{\tau_4 - \tau_6}$ and $T^{[2/1]} = 4\pi \int_0^\infty r^2 \tau^{[2/1]}\, dr$ where the later equation is written in case of spherical symmetry. If the integrand has poles, i.e. $\tau_4 = \tau_6$ for some $r$, then the integral can be defined through Cauchy principal value.

In similar way, we consider here the Padé approximant $\tau^{[1/1]} = \tau_0 + \frac{\tau_2^2}{\tau_2 - \tau_4}$ along with the partial sum $\tau_0 + \tau_2 + \tau_4$. Even if the integral of this partial sum converges, the convergence is slow because of behavior $\sim \rho^{1/3}$ for large $r$. For the Padé approximant $\tau^{[1/1]}$, much faster convergence $\sim \rho^{7/3}$ is expected as obtained and presented in the next section.

## 3. Results and discussions

### *3.1. Hooke's law model for two-electron atoms*

The Hamiltonian for the Hooke's law model for two-electron atoms is given by

$$H = -\frac{1}{2}(\nabla_1^2 + \nabla_2^2) + \frac{\omega^2}{2}(r_1^2 + r_2^2) + \frac{1}{r_{12}}. \tag{15}$$

The ground state wavefunction is known analytically for some values of $\omega$. For $\omega = 1/2$, it is [27, 28]

$$\Psi(r_1, r_2) = N_0 \left(1 + \frac{1}{2} r_{12}\right) e^{-(1/4)(r_1^2 + r_2^2)}, \tag{16}$$

where

$$N_0 = \left[4\pi^{5/2}\left(8 + 5\pi^{1/2}\right)\right]^{-1/2}. \tag{17}$$

and the density is (corrected Eq. (7) from [27])

$$\rho(r) = N_0^2 e^{-\frac{1}{2}r^2} \left\{ \left(\frac{\pi}{2}\right)^{1/2} \left[\frac{7}{4} + \frac{1}{4}r^2 + \left(r + \frac{1}{r}\right) \mathrm{erf}(2^{-1/2} r)\right] + e^{-\frac{1}{2}r^2} \right\}. \tag{18}$$

If $\omega \neq 1/2$, then we used a variational approach with the trial wavefunction in the form similar to Hylleraas variational function for helium [29, 30, 31]

$$\Psi(\mathbf{r}_1, \mathbf{r}_2) = \sum_{i,j,k \geq 0,\, i+j+k=N} C_{ijk} \left( r_1^i r_2^j e^{-\alpha_1 r_1 - \alpha_2 r_2} + r_2^i r_1^j e^{-\alpha_1 r_2 - \alpha_2 r_1} \right) r_{12}^k \tag{19}$$

with $\alpha_1 = 2\sqrt{\omega}$, $\alpha_2 = 3.2\sqrt{\omega}$ (ideally, the values of $\alpha_1$ and $\alpha_2$ should be chosen by minimizing the energy, but since it involves non-linear equations, we did it numerically for $N = 8$, and then tried to approximate the near optimal values by a simple analytic formula). Calculations were done for increasing set of basis functions with $N = 10$, 12, and 16 for $\omega = 1/2$, and the results agree with the results determined from the exact solution for the density, 18. For several other values of $\omega$, calculations were done for $N = 10$. The results are shown in Table 3.1 and for the case of $\omega = 1/2$, the gradient expansion corrections and their sums up to $n$-th order are shown on Fig. 2 and Fig. 3 respectively. For $\omega < 1$, it appears that the Padé approximant [2/1] is the most accurate, and for $\omega \geq 1$ the partial sum $T_0 + T_2$ is the best.

**Table 1.** Comparison of kinetic energy obtained from density $\rho$ by summation of gradient expansion using different methods, for Hooke's law model. (The results are presented as percent error.)

| | | Percent error ($E_T = 100\,|T_s - T|/T_s$) | | | | |
|---|---|---|---|---|---|---|
| $\omega$ | $T_s$ | $E_{T_0}$ | $E_{T_0+T_2}$ | $E_{T_0+T_2+T_4}$ | $E_{T[1/1]}$ | $E_{T[2/1]}$ |
| 1/4 | 0.30036 | -12.7 | -1.67 | 15.6 | 0.48 | -1.15 |
| 1/2 | 0.63525 | -11.9 | -0.78 | 16.5 | 1.27 | -0.26 |
| 1 | 1.32757 | -11.3 | -0.19 | 15.4 | 1.81 | 0.33 |
| 4 | 5.62884 | -10.7 | 0.45 | 15.1 | 2.4 | 0.98 |

Practically, for the system of two non-interacting particles with the same density $\rho$, $\psi(\vec{r}_1, \vec{r}_2) = \phi(\vec{r}_1)\phi(\vec{r}_2)$ where $\phi_1^2 = \phi_2^2 = \frac{1}{2}\rho$. In the ground state, $\phi_1 = \phi_2 = \sqrt{\rho}$. So, we obtain

$$\begin{aligned} T_s &= -\int d\mathbf{r}_1\, d\mathbf{r}_2\, \psi(\mathbf{r}_1, \mathbf{r}_2) \frac{1}{2} (\nabla_1^2 + \nabla_2^2) \psi(\mathbf{r}_1, \mathbf{r}_2) \\ &= -\int d\mathbf{r}\, \phi(\mathbf{r}) \nabla^2 \phi(\mathbf{r}) = \int \tau_K(\rho)\, d\mathbf{r} \end{aligned} \tag{20}$$

where

$$\tau_K(\rho) = \frac{1}{8} \frac{(\nabla\rho)^2}{\rho} - \frac{1}{4} \nabla^2 \rho. \tag{21}$$

In radial coordinates,

$$T_s = 4\pi \int r^2\, \tau_K(\rho)\, dr, \tag{22}$$

where

$$\tau_K(\rho) = \frac{1}{8} \frac{(\rho')^2}{\rho} - \frac{1}{4} \rho'' - \frac{1}{2} \frac{\rho'}{r}. \tag{23}$$

Because of spherical symmetry, the integration reduces to integration over radius, $\tilde{T} = 4\pi \int_o^\infty r^2\, \tilde{\tau}\, dr$, where $\tilde{\tau}$ is an approximation to the density of kinetic energy, and $\tilde{T}$ is corresponding approximation to the total kinetic energy.

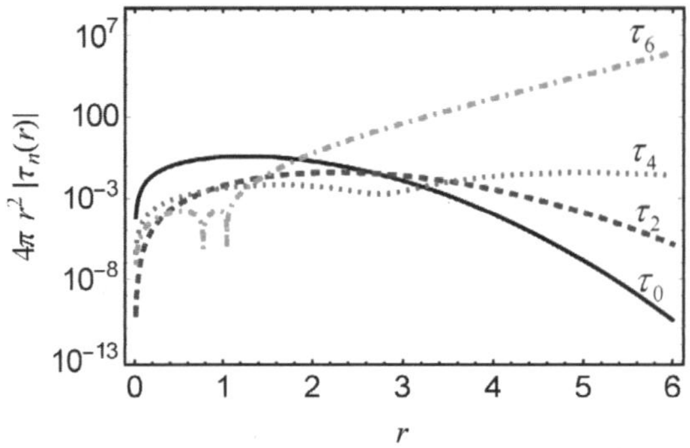


**Figure 2.** Gradient expansion corrections for the density of kinetic energy, in logarithmic scale. The black, blue, red, and green lines are for $\tau_0$, $\tau_2$, $\tau_4$, and $\tau_6$ respectively. Obviously, the correction $\tau_6$ diverges as $r \to \infty$, while $\tau_4$ converges, but very slowly.

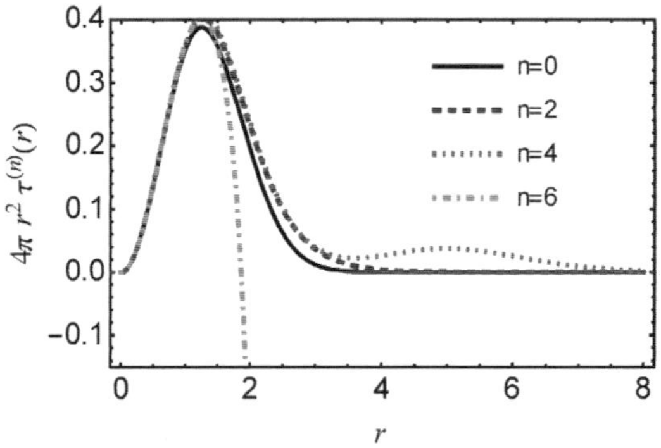


**Figure 3.** Sum of several first terms of gradient expansion, $\tau^{(0)} = \tau_0$ (black), $\tau^{(2)} = \tau_0 + \tau_2$ (blue), $\tau^{(4)} = \tau_0 + \tau_2 + \tau_4$ (red), and $\tau^{(6)} = \tau_0 + \tau_2 + \tau_4 + \tau_6$ (green) for Hooke's law model. Clearly, the 6-th order sum diverges when $r > 1$. It appears that the most accurate partial sum is $\tau^{(2)}$, and the curve for Padé approximant $\tau^{[2/1]}$ coincides with $\tau^{(2)}$ within accuracy of the plotting.

**Table 2.** Comparison of kinetic energy obtained from density $\rho$ by summation of gradient expansion using different methods, for atoms. (The results are presented as percent error.)

| | | Percent error ($E_T = 100\,|T_s - T|/T_s$) | | | | |
|---|---|---|---|---|---|---|
| Atom | $T_{\mathrm{HF}}$ (a.u.) | $E_{T_0}$ | $E_{T_0+T_2}$ | $E_{T_0+T_2+T_4}$ | $E_{T^{[1/1]}}$ | $E_{T^{[2/1]}}$ |
| He | 2.8617 | -10.5 | 0.59 | 3.57 | 2.01 | 0.53 |
| Li | 7.4328 | -10.1 | 0.62 | 3.09 | 2.00 | 0.61 |
| Be | 14.573 | -9.9 | 0.50 | 2.9 | 1.86 | 0.53 |
| Ne | 128.55 | -8.4 | -0.55 | 0.95 | 0.50 | -0.51 |
| Ar | 526.81 | -7.0 | -0.49 | 0.69 | 0.32 | -0.43 |
| Kr | 2752.05 | -5.9 | -0.69 | 0.18 | -0.09 | -0.63 |
| Xe | 7232.13 | -5.2 | -0.68 | 0.07 | -0.17 | -0.62 |

### *3.2. Atomic densities*

For the atoms, we calculated the densities using Hartree–Fock wavefunctions from Clementi Atomic Data Tables [26]. The results of summation of the gradient expansion are given in Table

3.2. For He, we found that the gradient expansion and its summations are similar to the case of Hooke's law model for two-electron atoms, but the sixth-order term diverges even faster. As a result, the Padé approximant improves the partial sum $\tau_0 + \tau_2$ only in a narrow range of $r \sim 0.6$, where $\tau_6$ is comparable with $\tau_0$. Outside of this range, $\tau^{[2/1]} \approx \tau^{(2)}$. Therefore, results of integration of these functions are very close, but still the Padé approximant gives some small improvement in comparison with the partial sum.

While for Li and Be, we found that the gradient expansion and its summation is similar to the case of helium atom, but the sixth-order term diverges much faster. As a result, the Padé approximant does not improve accuracy in comparison with the partial sum.

For heavier noble-gas atoms (Ne, Ar, Kr, and Xe), we found that the Padé approximant [2/1] is still slightly more accurate than the sum of two terms, but the Padé approximant [1/1] is even more accurate. Calculations show that in these cases $\tau_4 < \tau_2$ in the most important area of integration when $\tau(r) > 0.01$. It explains the good accuracy of the Padé approximant [1/1] which is based on fourth order expansion. However, we found that $\tau_6(r) > \tau_4(r)$ for almost all $r$, which is probably the reason of relatively poor performance of the Padé approximant [2/1] that is based on sixth order expansion.

## 4. Conclusion

Here, we take advantage of the known higher order corrections to the density of the kinetic energy, $\tau_4$ and $\tau_6$, to improve the accuracy of the gradient expansion. Simple adding of the correction $T_4 = \int \tau_4(\mathbf{r})\, d\mathbf{r}$ usually makes the result less accurate, because this correction is valid only asymptotically for heavy atoms. Moreover, since $T_6 = \int \tau_6(\mathbf{r})\, d\mathbf{r}$ is given by a divergent integral for finite systems, the knowledge of $\tau_6$ is assumingly useless, from the first glance. In this paper, we present a systematic approach that allows to avoid regions where higher order terms are relatively large and are no longer "small corrections" while taking advantage of these terms in areas where they are small and improve the accuracy.

We expect that this approximation can improve the accuracy in case when there exists a significant region of $\vec{r}$ where $|\tau_6(\vec{r})| < |\tau_4(\vec{r})|$. We found that for atoms, this region is quite narrow, presumably because of the presence of Coulomb singularity. In this case the method has only small advantage over the method of plane summation. Our second example is Hooke's law model for two-electron atoms, where the Coulomb attraction $-Z/r$ is replaced by a harmonic binding $\frac{\omega^2}{2}r^2$. In the latter case, we found that the method works only in case of sufficiently weak binding, $\omega \leq 1/2$. So, we expect this approximation to work very well for confined harmonic potentials such the case of treating quantum dots.

## References

[1] Hohenberg P and Kohn W 1964 *Physical review* **136** B864
[2] Kohn W and Sham L J 1965 *Physical Review* **140** A1133
[3] Lieb E H and Simon B 1973 *Physical Review Letters* **31** 681–683
[4] Lieb E H and Simon B 1977 *Advances in Mathematics* **23** 22–116
[5] Sergeev A, Jovanovic R, Kais S and Alharbi F H 2015 *Physica Scripta* **90** 125401
[6] Kirzhnits D 1957 *Soviet Phys. JETP* **5**
[7] Hodges C 1973 *Canadian Journal of Physics* **51** 1428–1437
[8] Murphy D 1981 *Physical Review A* **24** 1682
[9] Salasnich L 2007 *Journal of Physics A: Mathematical and Theoretical* **40** 9987
[10] Jennings B, Bhaduri R and Brack M 1975 *Physical Review Letters* **34** 228
[11] Grammaticos B and Voros A 1979 *Annals of physics* **123** 359–380
[12] Grammaticos B and Voros A 1980 *Annals of Physics* **129** 153–171
[13] Yan Z, Perdew J P, Korhonen T and Ziesche P 1997 *Physical Review A* **55** 4601
[14] Pearson E W and Gordon R G 1985 *The Journal of Chemical Physics* **82** 881–889
[15] Laricchia S, Fabiano E, Constantin L and Della Sala F 2011 *Journal of Chemical Theory and Computation* **7** 2439–2451

[16] Deb B and Ghosh S 1983 *International Journal of Quantum Chemistry* **23** 1–26
[17] Gázquez J and Robles J 1982 *The Journal of Chemical Physics* **76** 1467–1472
[18] Acharya P K, Bartolotti L J, Sears S B and Parr R G 1980 *Proceedings of the National Academy of Sciences* **77** 6978–6982
[19] DePristo A E and Kress J D 1987 *Physical Review A* **35** 438
[20] Vitos L, Skriver H L and Kollár J 1998 *Physical Review B* **57** 12611
[21] Cedillo A, Robles J and Gázquez J L 1988 *Physical Review A* **38** 1697
[22] Gülveren B 2012 *Solid State Sciences* **14** 94–99
[23] Smith R and Wilkin N 2000 *Physical Review A* **62** 061602
[24] Wang W P 1980 *The Journal of Chemical Physics* **73** 416–418
[25] Murphy D R and Wang W P 1980 *The Journal of Chemical Physics* **72** 429–433
[26] Clementi E and Roetti C 1974 *Atomic data and nuclear data tables* **14** 177–478
[27] Kais S, Herschbach D, Handy N, Murray C and Laming G 1993 *The Journal of chemical physics* **99** 417–425
[28] Kais S, Herschbach D and Levine R 1989 *The Journal of Chemical Physics* **91** 7791–7796
[29] Hylleraas E 1929 *Z. Physik* **54** 347–366
[30] Sergeev A and Kais S 1999 *Journal of Physics A: Mathematical and General* **32** 6891
[31] Sergeev A V and Kais S 2001 *International Journal of Quantum Chemistry* **82** 255–261

# Electrical and photoelectrical characteristic investigation of a new generation photodiode based on bromothymol blue dye

**A Gencer Imer[1], A Tombak[2] and A Korkut[1]**

[1]Department of Physics, Yuzuncu Yil University, 65080, Van, Turkey
[2]Department of Physics, Batman, University, Batman, Turkey

E mail: gencerimerarife@gmail.com

**Abstract.** Bromothymol blue (BTB) with the molecular formula of $C_{27}H_{28}Br_2O_5S$ was grown onto p-Si substrate to fabricate heterojunction by spin coating technique. The current voltage (*I-V*) measurements of diode were carried out in dark and under different illumination intensity at room temperature. The photoelectrical properties of heterojunction based on BTB were investigated using the illumination intensity dependent *I-V* data. The results showed that photo current of diode increases with the increase in light intensity. Also, the electrical parameters of device were determined via *I-V*, and capacitance- voltage (*C-V*), conductance-voltage (*G-V*) measurements at different frequencies. It is observed that the excess capacitance is created at low frequencies due to the contribution of interface states charge which can follow the alternative current signal to capacitance. It is stated that, both the electrical & photoelectrical parameters of diode can be changed, and also the performance of the device could be affected by the organic thin film interlayer.

## 1. Introduction

The organics or molecular materials have a wide variety of applications in modern electronic technology owing to structural flexibility, lower production cost, easy of synthesis and the possibility of using for large area device production [1,5]. Also, the organic thin films can be easily deposited onto different kind of substrates via deep coating [6], spray pyrolysis [7], spin coating [8-13], which are very common and cost effective growing methods. Namely, electronic and optoelectronic properties of organic/inorganic semiconductor devices have been studied as Schottky didoes [5-7], photosensor [13], photodiodes [6,12], and solar cells [10-15] by a great number of authors. Kılıcoglu et.al [10] fabricated the *TFF/p-InP* heterojunction, and investigated the electrical and photoelectrical properties of that device. It was found that this device can be used in the photovoltaic applications, because of the strong light response. Ocak [16], and Güllü [17,18] have reported the value of barrier height for the organic inorganic heterojunction devices can be modified by the organic interlayer at the MS interface owing to the tunneling effect through the organic interlayer.

In the present study, we have tried to study rectifier and photoresponce behavior of the bromothymol blue/p-Si (organic/inorganic semicondutor) contacts. The aim of this study to fabricate an *Al/BTB/p-Si* structure and to investigate the electrical and photoelectrical properties of device via current voltage (*I-V*) measurement in dark and under different illumination intensities; and capacitance-voltage (*C-V*), conductance-voltage (*G-V*) measurements at different frequencies. Both the fundamental photoelectrical parameters and electrical parameters such as ideality factor ($n$), the barrier height ($\Phi_b$) and series resistance ($R_s$) are evaluated by different methods.

## 2. Experimental procedure

For the fabrication of *Al/BTB/p-Si* heterojunction, p-Si substrate (with (100) orientation and 1-10 Ω-cm resistivity) was chemically cleaned via RCA method. For the formation of ohmic contact, the aluminum metal was evaporated to substrate; the obtained structure was annealed at 570 °C for 3 min in $N_2$ atmosphere. For the synthesis of BTB thin films, bromothymol blue was dissolved in methanol, and the precursor solution of $1x10^{-2}$ $molL^{-1}$ BTB was prepared at fist. The precursor solution was coated onto the front side of silicon substrate with a rotating speed of 2000 rpm for 1 min by spin coater. Later, top contacts were formed onto the obtained BTB/p-Si structure using through a shadow mask with the shape of circular dots of 1.5 mm in diameter. The contacts having were labeled as D1, D2, D3. The *I-V* measurements of diode were done at room temperature using Keithley 2400 sourcemeter in dark and under a Newport 96000 solar simulator with AM 1.5 global filter with different illumination intensities. The *C-V, G-V* measurements at different frequencies were carried out by means of HP Agilent 4294A impedance analyzer.

## 3. Results and discussions

Current-voltage (I-V) measurements of the *BTB*/p-Si contacts were carried out in dark at room temperature to investigate the electrical properties of them. Semi-logarithmic *I-V* plots of *Al/BTB/p-Si* heterostructure for each diode have been represented in Figure 1. The structure has a good rectifying behaviour.

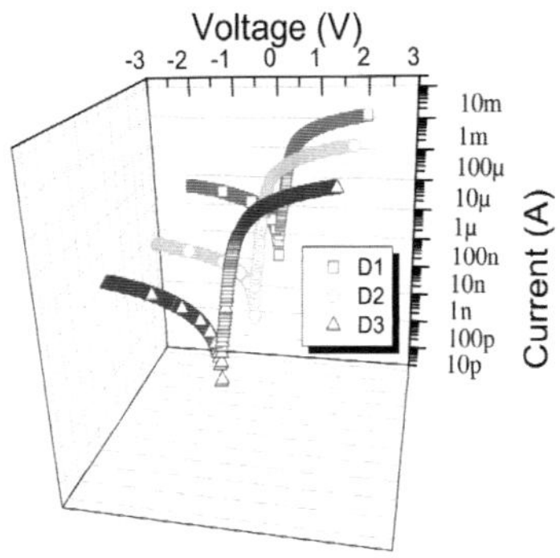


**Figure 2.** Experimental forward and reverse bias current versus voltage characteristics of the *Al/BTB/p-Si* structure for different dots

Thus, it can be said that organic-inorganic heterojunction reveals the property of Schottky like MS junction, and thermionic emission (TE) theory can be used to extract electrical parameters. According to the theory, the net current with series resistance ($R_s$) can be expressed as [19],

$$I = I_0 \exp\left[\left(\frac{q(V - IR_s)}{nkT}\right) - 1\right] \tag{1}$$

where $q$ is the electronic charge, $V$ is the applied voltage, $k$ is the Boltzmann constant, $T$ is the ambient temperature in Kelvin, and $I_0$ is the saturation current obtained by extrapolation of the linear portion of semi-log *I-V* plot to zero applied voltage and is defined as;

$$I_0 = AA^*T^2 \exp\left(-\frac{q\phi_b}{kT}\right) \tag{2}$$

here, $\phi_b$ is the zero bias barrier height of device, $A$ is the effective contact area (0.018 $cm^{-2}$), $A^*$ is the Richardson constant and is equals to 32 $Acm^{-2}K^{-2}$ for p-Si [20]. $n$ is an ideality factor which is a dimensionless quantity. It can be obtained from the slope of the linear portion of the forward bias semi-log *I-V* plot through the relation:

$$n = \frac{q}{kT}\frac{dV}{d\ln(I)} \tag{3}$$

Furthermore, the value of the barrier height of the device can be obtained as,

$$\phi_b = \frac{kT}{q}\ln\left(\frac{AA^*T^2}{I_0}\right) \tag{4}$$

The values of the barrier height and ideality factor of the fabricated structure were determined and given in Table 1.

The mean value of barrier height and ideality factor was evaluated as 0.729 eV and 1.41 using Eqs. (3) and (4), respectively. $\phi_b$ value of heterojunction is 0.729 eV, which is 0.149 eV higher than that of conventional Al/p-Si MS contact (0.58 eV) [21]. This enhancement may be derived from the saturation of dangling bonds at the surface of the inorganic semiconductor by the bromothymol blue thin film [22]. The modification in that of the barrier height by the organic interfacial layer has been studied in the many of research effort [7-13, 21-28]. For instance, the quantity of increment of 0.19 eV in the value of energy barrier was reported by Ugur et al [12] for the Al/p-Si heterojunction with quinoline yellow interfacial layer.

In addition, the mean ideality factor of heterostructure was found to be as 1.41, (n>1). So, the diode deviates from ideality because of organic interface layer at the interface. The higher ideality factor value of diode may be originated from the interface state density at the interface, existence of series resistance and presence of native oxide at interface [21-27]. Deviation from the linearity at high forward bias voltages in semi-log *I-V* plot is giving rise to the effect of series resistance, $R_s$. In order to determine of the $R_s$, modified Norde' function [28] can be used as follows:

$$F(V) = \frac{V}{\gamma} - \frac{kT}{q}\ln\left(\frac{I(V)}{AA^*T^2}\right) \tag{5}$$

where $\gamma$ is the first integer greater than $n$, ( $\gamma = 2$ *herein*). *F(V)* vs. *V* curve of the Al/BTB/p-Si device is shown in Fig.3. The barrier height values of heterojunction can be obtained by using the equation,

$$\phi_b = F(V_{min}) + \frac{V_{min}}{\gamma} - \frac{kT}{q} \tag{6}$$

where $F(V_{min})$ is the minimum value of $F(V)$ and $V_{min}$ is the corresponding voltage value. The values of $F(V_{min})$ and $V_{min}$ are determined, and listed in the Table 1 for different contacts. In this method, the series resistance $R_s$ can be calculated through the relation,

$$R_s = \frac{kT(\gamma - n)}{qI_{\min}} \tag{7}$$

where $I_{min}$ is the corresponding current value at the $V_{min}$. The values of the barrier height and series resistance of each diodes for *Al/BTB/p-Si* heterojunction were calculated, and also listed in Table 1, the average value of series resistance for that heterojunction was extracted as as 0.771 eV and 980 Ω by using Eqs (6) and (7), respectively. The value of barrier height obtained from Norde' function is very close to that of obtained forward bias semi-log *I-V*.

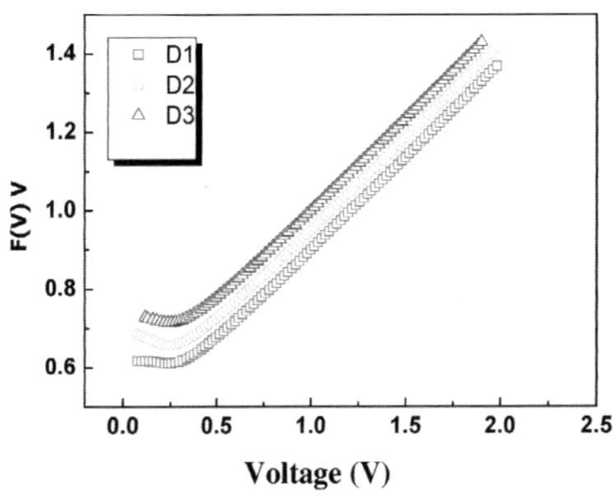


**Figure 3.** Experimental *F(V)* vs. *V* plots for different dots of *Al/BTB/p-Si* structure at the dark

**Table 1.** Electrical characteristic parameters for different contacts of *Al/BTB/p-Si* structure

| ID | From I-V plot | | From Norde plot | | | |
|---|---|---|---|---|---|---|
| | n | $\Phi_b$ (eV) | $F(V)_m$ (V) | $V_{min}$ (V) | $R_s$ (kΩ) | $\Phi_b$ (eV) |
| D1 | 1.3 | 0.75 | 0.662 | 0.276 | 0.74 | 0.77 |
| D2 | 1.4 | 0.72 | 0.658 | 0.236 | 1.25 | 0.75 |
| D3 | 1. | 0.72 | 0.671 | 0.306 | 0.95 | 0.79 |

The effect of illumination on the *I-V* characteristic of the heterojunction was carried out and seen in Figure 4. The reverse bias current increases extremely with the increasing light intensity, while forward bias current of that is almost constant. This indicates the generated electron-hole pairs in organic-semiconductor junction by incident photons. Photocurrent is produced by photo generation of the free carriers contributing to the reverse current after the process of absorption of photon energy at the junction. Furthermore, photocurrent and the photovoltage are increased with the increasing light intensity as seen from this figure. The characteristic photovoltaic parameters such as short circuit current $I_{SC}$, open circuit voltage $V_{OC}$ were defined as 54.68 μA and 406 mV under 100 mW/cm$^2$ illumination, respectively. These characteristic parameters of *Al/BTB/p-Si* structure under different illumination intensities were listed in the Table 2.

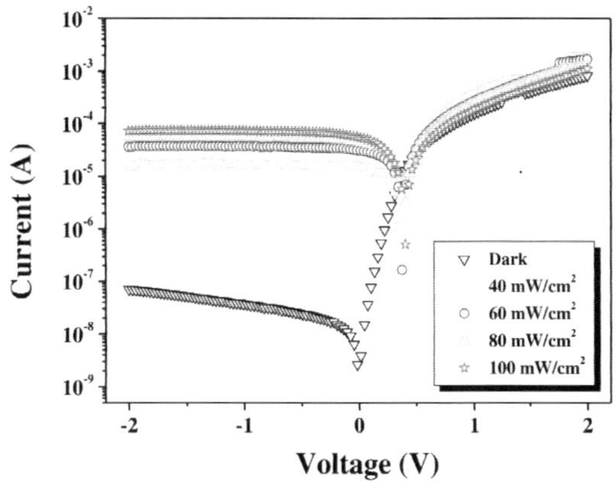


**Figure 4.** Semilogaritmic *I-V* plots of *Al/BTB/p-Si* at the dark and under different illumination intensities

The power conversion efficiency (PCE) was evaluated, and given in the Table 2. The obtained characteristic parameters of studied device are poor for the solar cell application, but suitable for photodiode, photosensor application [30,6].

In order to determine the phoconductive responsivity of the diode to the different illumination intensity, the following equation can be used [20,31];

$$R = I_{PH}/L.A \tag{8}$$

where, *A* is the area, which is exposed to illumination. The calculated values of *R* were given in Table 2. It is understood from table that these values exponentially increase with incident light intensity. This means that the responsivity to the light for this diode is dependent to the illumination power which is resulted in the generated electron-hole pairs by photons.

**Table 2.** The values of $V_{OC}$, $I_{SC}$, *PCE, R* at different light intensity

| Light Intensity (mW/cm$^2$) | $V_{OC}$ (mV) | $I_{SC}$ (μA) | *PCE* | *Responsivity, R* (A/W) (x10$^{-4}$) |
|---|---|---|---|---|
| 40 | 347 | 14.69 | 0.38 | 3.96 |
| 60 | 375 | 30.69 | 0.46 | 5.69 |
| 80 | 395 | 44.28 | 0.49 | 6.36 |
| 100 | 407 | 54.78 | 0.43 | 6.79 |

*C-V* measurements of *Al/BTB/p-Si* heterojunction were carried out in the range from 100 to 500 kHz frequencies. The Figure 5 presents the relation of capacitance with the applied voltage at different frequency for the studied photodiode. It is seen from the figure that the capacitance of diode is almost constant up to the voltage of ~ 1 V, reaches to the maximum value, and forms a peak owing to the interface states at the interface of the device. The maximum value of *C-V* plots increases at low frequency regime, as a result of that the interface states charges could follow the alternating signal (a.c.) [10,15].

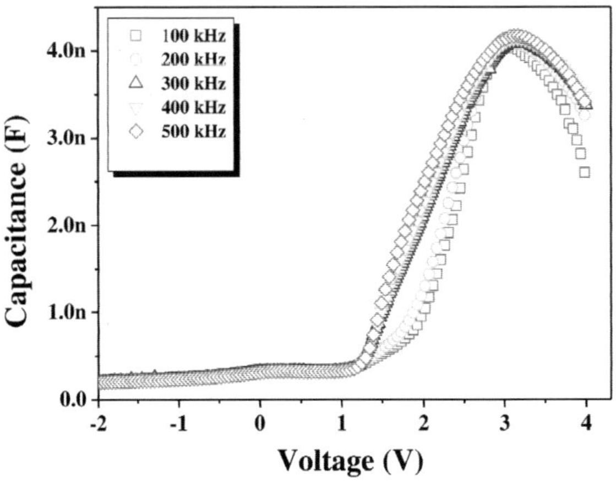


**Figure 5.** Capacitance vs. voltage plots of *Al/BTB/p-Si* structure under different frequencies.

In addition, the capacitive property of the diode can be analyzed as a function of reverse bias voltage. Thus, the depletion region capacitance of device can be described as [20],

$$\frac{1}{C^2} = \frac{2(V_{bi} + V)}{q\varepsilon_s\varepsilon_0 A^2 N_a} \tag{9}$$

where $V_{bi}$ the built in potential, which is estimated from the intersection of linear portion of the reverse bias $C^{-2}$-$V$ plot, $\varepsilon_S$ is the dielectric constant of semiconductor (=11.7 for Si), $\varepsilon_0 = 8.85 \times 10^{-14}$ F/m, $A$ is the effective diode area. $N_a$ is the density of ionized acceptors, which is found from the slope of linear portion of the same plot. The Figure 6 shows the $C^{-2}$ -$V$ plot of *Al/BTB/p-Si* structure for 500 kHz, as a representative example. The values of $V_{bi}$ and $N_a$ were calculated as 0.829 V and $4.81 \times 10^{15}$ cm$^{-3}$ to the $V$ axis, respectively. The barrier height value can be obtained from the relation;

$$\phi_{b(C-V)} = eV_{bi} + kT\ln(\frac{N_v}{N_a}) \tag{10}$$

The second term in the Eq. (10) is known as the potential difference between the top of the valance band in the neutral region of p-Si, in which $N_v = 1.04 \times 10^{19}$ cm$^{-3}$. Therefore, the $\phi_b$ value was calculated as 1.028 eV using Eq. (10) from reverse bias $C^{-2}$-$V$ characteristic of the heterostructure.

This barrier height value is higher than that of obtained from the *I-V* data. The reason of this difference can be the presence of trap states and interfacial layer, interfacial layer, and oxide layer at interface of the device [31,32].

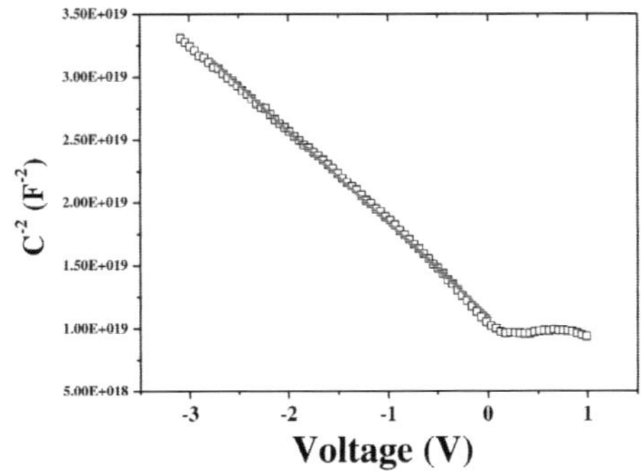


**Figure 6.** $C^{-2}$ -*V* plot of *Al/BTB/p-Si* structure for 500 kHz

The conductance (*G*) was measured, and the Figure 7 exhibits (*G-V*) characteristics of *Al/BTB/p-Si* heterojunction at different frequencies. It is seen from the figure the value of conductance for the studied diode increases as the frequency increases for the forward bias voltage, while that remains almost constant for the reverse bias voltage. Such behaviour in the *G-V* characteristics of specimen may be ascribed to the presence of series resistance, and very thin insulator layer on the semiconductor [6,10,11,30].

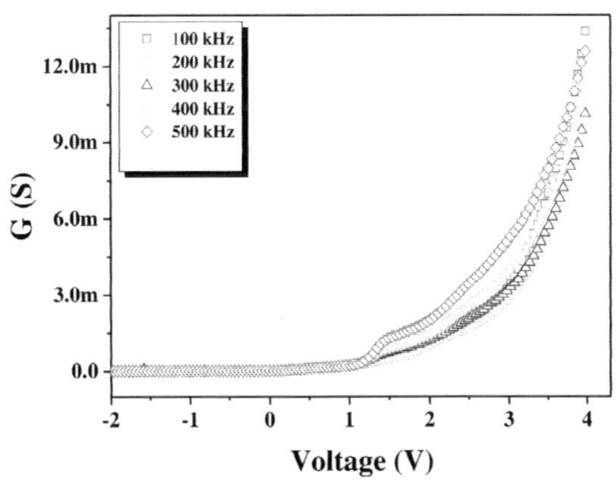


**Figure 7.** Conductance vs. voltage plots of *Al/BTB/p-Si* structure at different frequency

In order to evaluate the density of interface states ($D_{it}$) profile as a function of frequency, the reliable method is developed by Hill-Coleman [33], according to this method, $D_{it}$ can be estimated by the following relation,

$$D_{it} = \frac{2}{qA} \frac{G_m/\omega}{\left[\left(\frac{G_m}{\omega C_{ox}}\right)^2 + \left(1 - \frac{C_m}{C_{ox}}\right)^2\right]} \tag{11}$$

where $G_m$ is the maximum conductance obtained from the *G-V* plot, ω is the angular frequency, $C_m$ is the capacitance value corresponded to the $G_m$, $C_{ox}$ is the capacitance of the interface layer in accumulation regime. The value of $D_{it}$ diminishes with the increasing frequency, and varies from ~3.5 x $10^{11}$ to 5.9 x $10^{10}$ $eV^{-1}cm^{-2}$ as seen in the Figure 8 for the studied heterojunction. This value for the studied hetero structure is lower than both the that of conventional Al/p-Si diode (~$10^{13}$ $eV^{-1}cm^{-2}$ at room temperature) [34], and that of MIS structures reported in the literature [7,11,24]. Thus, both the electrical & photoelectrical parameters of diode can be modified, and also the performance of the device can be improved by the organic thin film interlayer.

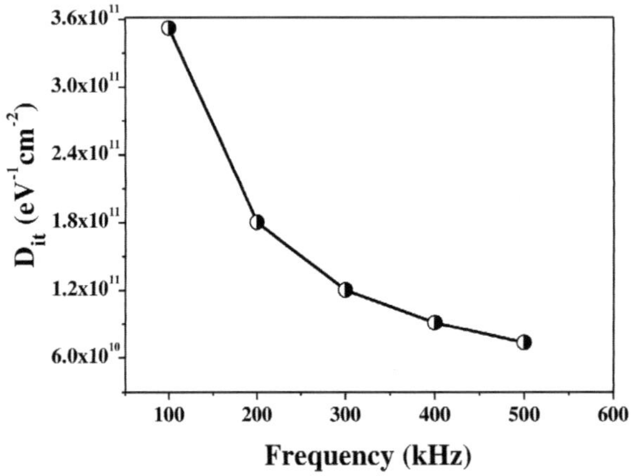


**Figure 8.** Density of interface states vs. frequency plot of *Al/BTB/p-Si* structure

## 4.Conclusion

In this work, *Al/BTB/p-Si* heterojunction was fabricated by forming the *BTB* organic thin film on the p- type silicon crystal using spin coating technique. The photoelectrical characteristics of heterojunction were investigated under different illumination intensities. The photocurrent and photovoltage of the diode increase with the illumination intensity. Also, photoconductive responsivity increases with incident light power. The value of *s* is found as 0.614 for the diode, which suggest continuous distribution of localised interface states. The characteristic photovoltaic parameters such as $I_{sc}$, $V_{oc}$ were determined for different illumination. The obtained results conclude that the forming of the *BTB* organic thin interlayer promotes the photoconductivity due to generation of the electron-hole pairs. It is good candidate for the photo sensor and photo diode applications. Furthermore, the main electrical characteristic parameters for different contacts of *Al/BTB/p-Si* diode, were determined using *I-V*, and frequency dependent *C-V*, and *G-V* measurements at room temperature. $D_{it}$ value of diode is significantly lower in comparison with that of commercial Al/p-Si diode. The obtained results confirm that the electrical properties of the heterojunction can be improved using the interface layer based on organic dye.

**Acknowledgements:** This work has been partially supported by the Yuzuncu Yil Scientific Research management (YYU-BAPB) projects under the contract 2012-FED-B007, 2015-FBE-YL007 and 2015-FBE-YL347. We would like to also thank the Science and Technology Application and Research Center in Dicle University (DUBTAM) for their support.

**References**

[1] Temirci C, Gülcan M, Goksen K and Sonmez M 2011 *Microelectron. Eng.* **88** 41
[2] Gencer Imer A, Temirci C, Gülcan M and Sonmez M 2014 *Mater. Sci. Semic. Proc.* **28** 31
[3] Chang H, Chen T L, Huang K D, Chien S H, Hung K C 2010 *J. Alloy Compound.* **504** 435
[4] Baviskar P, Zhang J, Gupta V, Chand S, Sankapal B 2012 *J. Alloy Compd.*, **510** 33
[5] Temirci C, Gülcan M, Goksen K and Sonmez M 2010 *Microelectron. Eng.* **87** 2282
[6] Phan D T et al 2012 *Solar Energy* **86** 2961
[7] Yakuphanoglu F, Ocak Y S, Kılıcoglu T and Farooq W A 2011 *Microelectron. Eng.* **88** 2951
[8] Tombak A, Ocak Y S, Asubay S, Kilicoglu T and Kahraman F 2014 *Mater. Sci. Semi. Proc.* **24** 187
[9] Gümüş A and Altındal S 2014 *Mater. Sci. Semi. Proc.* **28** 66
[10] Kilicoglu T, Tombak A, Ocak Y.S, and Aydemir M 2014 *Microelectron. Eng.* **129** 91
[11] Soylu M et al 2013 *Microelectron. Relia.* **53** 1901
[12] Ugur A, Gencer Imer A, and Ocak Y S 2015 *Mater. Sci. Semi. Proc.* **39** 569
[13] El-Menyawy E M, and Zedan I T 2015 *Spectrochimica Acta Part A: Molecular and Biomolecular Spectroscopy* **137** 810
[14] Yahia I S, Shakra A M, Fadel M, H.S. Hafez H, Micheal S, and Yakuphanoglu F 2014 *Sci. Semi. Proc.* ***28*** 77
[15] Tascıoglu I, Farooq W A, Turan R, Altındal S, Yakuphanoglu F 2014 *J. Alloy Compd.,* **590** 157
[16] Ocak Y S, Kulakci M , Kilicoglu T, Turan R, and Akkılıç K 2009 *Synth. Met.* **159** 1603
[17] Gullu O 2010 *Microelectron. Eng.,* **87** 648
[18] Gullu O, Asubay S, Biber M, Kilicoglu T and. Turut A 2010, *Eur. Phys. J. Appl. Phys.* ***50*** *10401*
[19] Rhoderick E H 1978 *Metal-Semiconductor Contacts* Oxford University Press, p. 121
[20] Sze S M 1978 *Physics of Semiconductor Devices,* 2nd ed. John Wiley& Sons p. 256
[21] Ocak Y S, Guven R G, Tombak A, Kilicoglu T, Guven K, and Dogru 2013 *Philos. Mag.* **93** 2172
[22] Çaldıran Z, Deniz A, Aydoğan Ş, Yesildag A, and Ekinci D 2012 *Superlattice Microst,* **56** 45
[23] Akkılıç K, Ocak Y S, Kilicoglu T and İlhan S, Temel H 2010 *Curr. Appl. Phys.* **10** 337
[24] Kilicoglu T and Ocak Y S 2011 *Microelectron. Eng.,* **88** 150
[25] Gullu O, and Turut A 2010 *Microelectron. Eng.,* **87** 2482
[26] Çaldıran Z, Deniz A R, Mehmet Coşkun F, Aydoğan S, Yesildag A, and Ekinci D 2014 *Journal of Alloys and Compounds* **584** 652
[27] Çaldıran Z, Aydoğan S, Yesildag A, Ekinci D, Kurudirek S V, and Türüt A 2015 *Mater. Sci. Semic. Proc.* **34** 58
[28] Norde H 1979 *J. Appl. Phys.*, **50** 5052
[29] Kazim S, Alia V, Zulfequar M, Hak M M, and Husain M 2013 *Physica B* **393** 310
[30] Soylu M et al 2014 *Sol. Ener. Mater. Sol. Cells* **124** 180
[31] Tecimer H, Türüt A, Uslu H, Altındal S, Uslu I 2013 *Sens. Actuators A* **199** 194
[32] Gullu O, Kilicoglu T, and Turut A 2010 *J. Phys. Chem. Solids* **71** 351
[33] Hill W A and Coleman C C 1980 *Solid State Electron.* **23** 987

# Investigation of the Electrical Characteristics of Al/p-Si/ Al Schottky Diode

Elvan Şenarslan, Betül Güzeldir and Mustafa Sağlam

Atatürk University, 25240, Erzurum, Turkey,

E-mail: elvan@atauni.edu.tr

**Abstract:**In this study, p-type Si semiconductor wafer with (100) orientation, 400 μm thickness and 1-10 Ω cm resistivity was used. The Si wafer before making contacts were chemically cleaned with the Si cleaning procedure which for remove organic contaminations were ultrasonically cleaned at acetone and methanol for 10 min respectively and then rinsed in deionized water of 18 MΩ and dried with high purity $N_2$. Then respectively RCA1(i.e., boiling in $NH_3$+$H_2O_2$+6$H_2O$ for 10 min at 60°C ), RCA2 (i.e., boiling in HCl+$H_2O_2$+6$H_2O$ for 10 min at 60°C ) cleaning procedures were applied and rinsed in deionized water followed by drying with a stream of $N_2$. After the cleaning process, the wafer is immediately inserted in to the coating unit. Ohmic contact was made by evaporating of Al on the non-polished side of the p-Si wafer pieces under ~ 4,2 $10^{-6}$ Torr pressure. After process evaporation, p-Si with omic contac thermally annealed 580 °C for 3 min in a quartz tube furnace in $N_2$. Then, the rectifier contact is made by evaporation Al metal diameter of about 1.0 mm on the polished surface of p-Si in turbo molecular pump at about ~ 1 $10^{-6}$ Torr. Consequently, Al/ p-Si /Al Schottky diode was obtained. The *I–V* measurements of this diode performed by the use of a KEITLEY 487 Picoammeter/Voltage Source and the *C–V* measurements were performed with HP 4192A (50–13 MHz) LF Impedance Analyzer at room temperature and in dark.

## 1. Introduction

The metal–semiconductor (M–S) contact has a very important role in all solid state devices. The reliability and thermal stability of the M-S contacts is of great practical importance in device technology [1-2]. The contact can be rectifying or ohmic contact [3]. Unlike rectifying, ohmic contac has a negligible resistance regardless of the polarity of the applied voltage. Schottky diodes have a great importance technologically. Therefore, M-S contact is one of the most widely used rectifying contacts in the electronics industry [4]. The performance of Schottky barrier diodes is drastically determined by the interface quality between deposited Schottky metal and semiconductor surface. Also, the reliability of Schottky barrier diodes depends significantly on the interdiffusion and interaction among those materials. Electronic properties of a Schottky diode are characterized by its barrier height and ideality factor parameters. The interface states play an important role on determination of Schottky barrier height and other characteristics parameters and these can affect device performance, stability and reliability [5–6]. This study reports current–voltage (*I–V*) and capacitance–voltage (*C–V*) characteristics of Al/p-Si/Al diode.

## 2. Experimental

p-type Si wafer pieces of (1 0 0) orientation and one side polished were used in this study. For this, the wafer cleaned with acetone and methanol ultrasonically for 10 min respectively and then rinsed in deionized water of 18 MΩ and dried with high purity $N_2$. Then, the wafer was chemically cleaned with using the RCA1 cleaning procedure followed RCA2. Again, the wafer was rinsed in de-ionized water of 18 MΩ and dried with high purity $N_2$. After the cleaning process, the wafer is immediately inserted in to the coating unit. Ohmic contact was made by evaporating of Al on the non-polished side of the p-Si wafer pieces under ~ 4,2 $10^{-6}$ Torr pressure. After process evaporation, p-Si with omic contac thermally annealed 580 °C for 3 min in a quartz tube furnace in $N_2$. Then, the rectifier contact is made by evaporation Al metal diameter of about 1.0 mm on the polished surface of p-Si in turbo molecular pump at about ~ 1 $10^{-6}$ Torr. Consequently, Al/ p-Si /Al Schottky diode was obtained. The *I–V* measurements of this diode performed by the use of a KEITLEY 487 Picoammeter/Voltage Source and the *C–V* measurements were performed with HP 4192A (50–13 MHz) LF Impedance Analyzer at room temperature and in dark.

## 3. Results and discussion

When the Schottky barrier diodes are considered, it is assumed that the forward-bias current the device is due to the thermoionic emission current and it can be expressed as [2, 6]

$$I = A.J = I_0\left[\exp\left(\frac{q(V - IR_s)}{nkT}\right) - 1\right] \tag{1}$$

where $I_0$ is the saturation current derived from the straight line intercept of ln I at $V$=0 and is given by

$$I_0 = \left[AA^*T^2\exp\left(\frac{-q\Phi_b}{kT}\right)\right] \tag{2}$$

where $\Phi_b$ is the effective barrier height at zero bias, *R*s is the series resistance of the substrate and dominates in the higher bias region and IRs is the voltage drop across the series resistance, $A^*$ is the effective Richardson constant and equals 32 A $cm^{-2}$ $K^{-2}$ for p-type Si, *A* is the diode area, *n* is an ideality factor and is a measure of convenience of the diode to thermionic emission. It is obtained from the slope of the straight line region of the semi-log forward bias *I–V* characteristics and it can be written as:

$$n = \frac{kT}{e}\frac{dV}{\ln I} \quad\text{.......(4)}$$

In addition, the barrier height can be obtained from the equation,

$$e\Phi_b = kT\,ln\left(AA^*T^2 / I_0\right) \quad\text{.......(5)}$$

Figure 1 shows the experimental semilog forward and reverse bias *I–V* characteristics of the Al/p-Si/Al structure. From fig 1, an experimental barrier height value of 0.697 eV for the Al/p-Si Schottky diode was calculated with the help of equation (5), and an experimental value of 1.02 for *n* using equation (4). The ideality factors slightly greater than unity are attributed to secondary mechanisms at the interface [6].

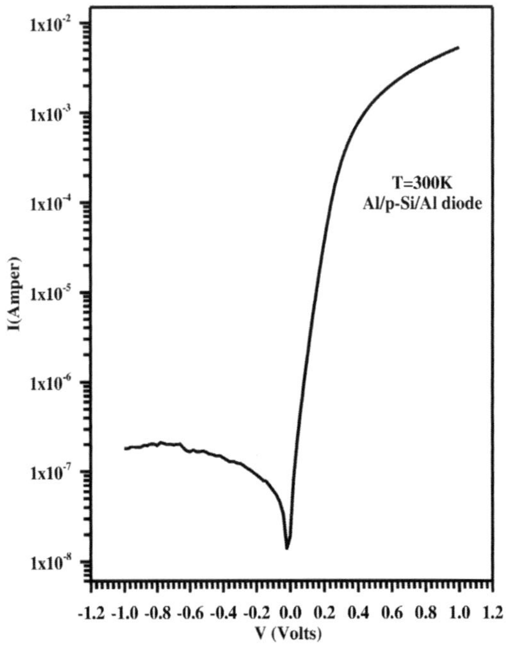


**Fig. 1.** The semi-log forward and reverse bias current–voltage characteristics of Al/p-Si/Al diode at room temperature.

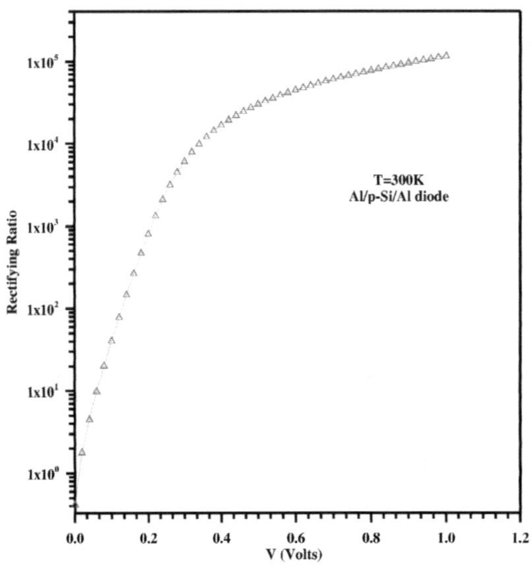


**Fig. 2.** The rectifying ratio versus applied voltage plot of the Al/p-Si/Al diode at room temperature.

As seen in Fig. 2, the rectification ratio increases with increase in bias voltage. It is clear from this figure that the device has a good rectifying property.

It is well known that the downward concave curvature of the forward bias current–voltage plots at sufficiently large voltages is caused by the presence of series resistance, in addition to the interface states, which are in equilibrium with the semiconductor [7–8]. The $R_s$ values have been calculated by using a method developed by Norde [6]. The following function has been defined in the modified Norde's method [6]:

$$F(V)=\frac{V}{\gamma}-\frac{1}{\beta}ln\left(\frac{I(V)}{AA^{*}T^{2}}\right) \quad (6)$$

where $\gamma$ is an arbitrary constant greater than the ideality factor, $I(V)$ is current obtained from the $I$–$V$ curve and the other parameters are described above. Thus, the effective barrier height and series resistance can be determined by

$$\Phi_b = F_m + \left[\frac{(\gamma-n)}{n}\right]\left[\frac{V_m}{\gamma}-\frac{kT}{q}\right] \quad (7)$$

$$R_s = (\gamma-n)\frac{kT}{qI_m} \quad (8)$$

Once the minimum of the F(V)—V plot is determined, the barrier height can be obtained from here. Where $F_m$ is the minimum point of F(V) curve, and $V_m$ is the corresponding voltage, $I_m$ is the current corresponds to the minimum $V_m$. A plot of F(V) versus V for the Al/ p-Si/Al structure is shown in Fig 3.

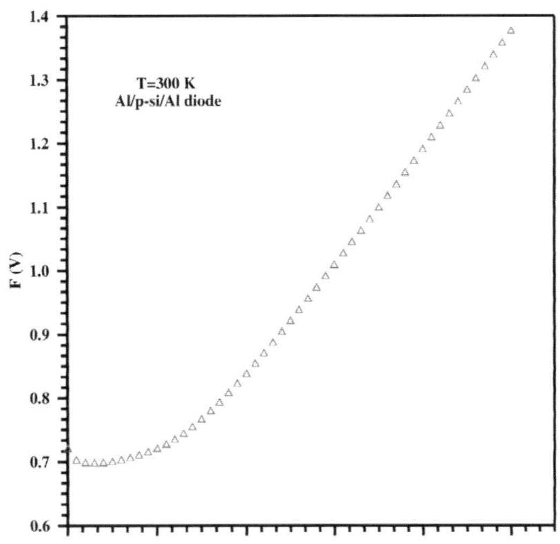


**Fig. 3.** F(V) versus V plot of the Al/ p-Si/Al structure at room temparature.

From the F–V plot by using F($V_m$)= 0.696 V and $V_m$= 0.06 V values, the values of $\Phi_b$ and $R_s$ of the Al/ p-Si/Al diode have been determined as 0.697 eV and 0.578 kΩ, respectively. There is a equal the $\Phi_b$ values obtained from the forward bias ln $I$–$V$ and Norde functions.

In metal–semiconductor contacts, the depletion layer capacitance, C can be expressed as [9]:

$$C^{-2} = \frac{2(V_d + V)}{\varepsilon_s \varepsilon_0 q A^2 N_v} \qquad (9)$$

where A is the area of the contact, $V_d$ is the diffusion potential at zero bias and determined from the extrapolation of the linear $C^{-2}$–$V$ plot. The value of the barrier height can be calculated from Eq. $\Phi_b = E_f + V_d$, where $E_f$ is the potential difference between the Fermi energy level and the top of the valence band and can be calculated from the following relation:

$$E_f = kT \ln\left(\frac{N_a}{N_v}\right) \qquad (10)$$

$\varepsilon_s$ is the dielectric constant of the semiconductor, and $N_a$ is the acceptor concentration of p-type semiconductor substrate. Measurement of the depletion region capacitance under forward bias is difficult. Because the diode is conducting and the capacitance is shunted by a large conductance. But, the capacitance can be easily measured as a function of the reverse bias.

The forward and reverse bias $C$–$V$ and reverse bias $C^{-2}$ –$V$ characteristics characteristics of the Al/ p-Si /Al diode with measured at different frequencies are shown respectively in Fig. 4 and Fig. 5.

As can be seen from the Figure 4 and Figure 5, the peak values of the forward biased capacitance are increased as the operating frequency is decreased. The dependence of the capacitance of such a junction upon frequency can also arise due to the presence of deep lying impurities in the depletion region. Presence of deep traps in the depletion region of the junction makes the junction capacitance a complicated function of the bias voltage and the measuring frequency. The capacitance is nearly constant at the reverse bias region. In the ideal case, the $C$–$V$ characteristics of metal–semiconductor or metal–insulator–semiconductor structures show an increase in capacitance with the increasing forward bias voltage the capacitance decreases with the applied bias, which reveals that a depletion region exists at the Al/p-Si junction, and the depletion region width increases with the reverse bias. For the forward biased capacitance, the peak capacitance values of the Al/ p-Si/Al diode are 218, 121, 75, 61, 54, 50, and 42pF for the 50, 100, 200, 300, 400, 500 and 1000 kHz, respectively. The peak values of the forward biased capacitance are increased as the operating frequency is decreased.

**Table 1**. The experimentally parameters obtained from reverse bias $C^{-2}$-$V$ characteristics as a function of frequency for Al/ p-Si/Al diode at room temperature.

| **f (kHz)** | $V_d(eV)$ | $N_a(cm^{-3})$ | $E_f(eV)$ | $\Phi_b(eV)$ |
|---|---|---|---|---|
| 50 | 0.660 | 2.104 E+14 | 0.280 | 0.940 |
| 100 | 0.661 | 2.066 E+14 | 0.280 | 0.941 |
| 200 | 0.661 | 2.053 E+14 | 0.281 | 0.941 |
| 300 | 0.661 | 2.043 E+14 | 0.281 | 0.941 |
| 400 | 0.662 | 2.034 E+14 | 0.281 | 0.943 |
| 500 | 0.662 | 2.035 E+14 | 0.281 | 0.943 |
| 1000 | 0.662 | 2.039 E+14 | 0.281 | 0.943 |

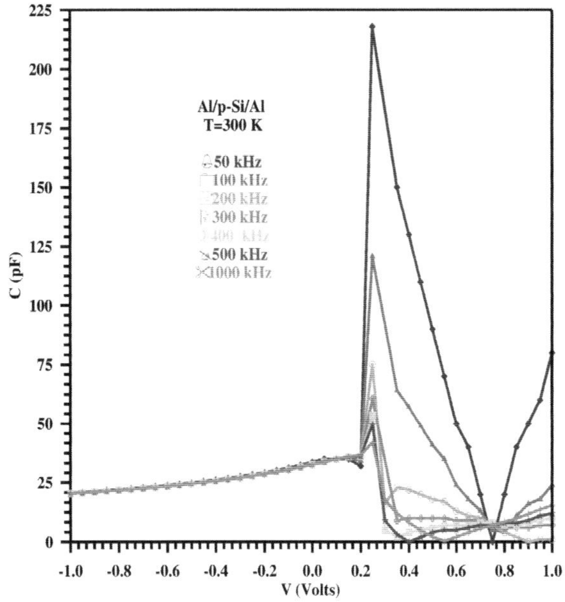


**Fig. 4**. The *C–V* characteristics of the Al/ p-Si /Al diode at room temperature.

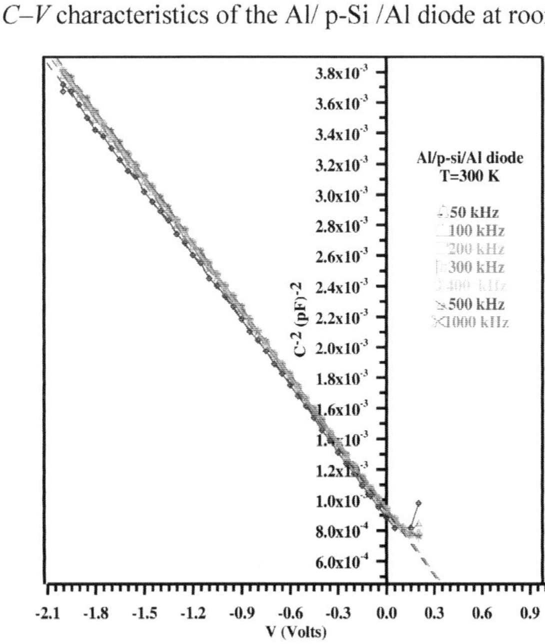


**Fig. 5**. $C^{-2}$ *–V* characteristics of the Al/ p-Si /Al diode at room temperature.

As shown from table 1; barrier heights, obtained from *I–V* measurements are different than those obtained from *C–V* measurements. Some general causes for these differences in barrier height have been mentioned in the literature, such as surface contamination at the interface, deep impurity levels, an intervening insulating layer, quantum mechanical tunneling, image force lowering and edge leakage currents [10, 11, 12]

## 4. Conclusions

In this study, we prepared Al/ p-Si/Al Schottky diode and investigated electrical characteristics of this structure with *I-V* and *C-V* measurement at room temperature. There is a equal barrier height values obtained from the forward bias ln *I–V* and Norde functions for this diode. This indicates that the diode to be compatible with the Norde functions. There is a difference barrier height values obtained from *I-V* and *C-V* measurement. As known, this difference is due to different nature of the *I-V* and *C-V* measurement techniques [1, 4].

## References

[1] E. H. Rhoderick and R. H Williams 1988 Metal-Semiconductor Contacts 2nd edn (Oxford:Clarendon)
[2] C Nuhoğlu, E Ayyıldız, M Saglam, A Turut 1998 Applied Surface Science 135, 350-356.
[3] Pierret R F 1996 Semiconductor Device Fundementals, Purdue University Pres.
[4] G Y Robinson, C W Wilmsen 1985 Physics and Chemistry of III–V Compound Semiconductor Interfaces, Plenum, New York.
[5] P L Hanselaer, W H Lafle´re, R L Van Meirhaeghe, F Cardon 1984 J. Appl. Phys. 56, 2309.
[6] H Norde (1979). J. Appl. Phys., 50, 5052
[7] Gupta R K, R A Singh 2004 Mater. Chem. Phys. 86 279.
[8] S Aydogan, M Saglam, A Turut 2008 Microelectron Eng. 85, 278.
[9] A V Ziel 1968 Solid State Physical Electron, second ed., Englewood Cliffs, Prentice-Hall, NJ, pp 380-381.
[10] E H Rhoderick 1998 Metal–Semiconductor Contacts, Clarendon, Oxford.
[11] F E Cimilli, M Saglam, H Efeoglu, A Turut 2009 PhysicaB, 404, 1558–1562.
[12] F E Cakıcı, M Saglam, B Guzeldır 2014 Material Science in Semiconductor Processing, 28, 121–126.

# Graphene quantum dot with a hydrogenic impurity

**Neslihan Gökçek[1] and Bekir Sıtkı Kandemir[1]**

[1] Department of Physics, Faculty of Sciences, Ankara University, 06100 Tandoğan, Ankara, Turkey

E-mail: ngokcek@science.ankara.edu.tr

**Abstract.** We investigate the electronic energy levels of a circular graphene quantum dot with a hydrogenic impurity placed at the center of the dot. We have adopted a variational scheme in such a way that the trial wave functions take into account the carrier confinement due to the dot geometry as well as the effect of the magnetic field. We found that the presence of impurity strongly modifies the electronic spectrum of the graphene quantum dot. We also calculated the critical value of the impurity strength as a function of dot radius.

## 1. Introduction

Since the discovery of graphene [1, 2] impurity related effects in graphene have been extensively reported in the graphene literature [3–19]. The study of impurity states in graphene quantum dots (GQDs) is itself important, since it is expected that these effects will be more pronounced in such structures due to the reduction of the dimensionality. Since experiments with adatoms in monolayer graphene clearly adress the long range impurities [13], which are responsible for limiting the mobility of graphene [15], Coulomb like hydrogenic impurities are of particular interest in the relevant literature [7–10, 15]. Moreover, the strength of Coulomb interaction, which is proportional to dimensionless quantity $\overline{Z}$ ( $Z\ \alpha$ ) defined by the valency of impurity $Z$ and $\alpha = e^2/\epsilon\hbar v_F$, introduces a new kind of quantum electrodynamics regime (QED), which is strictly different from the usual perturbative one. Here, $\epsilon$ is the dielectric constant, and $v_F$ is the Fermi velocity for massless Dirac fermions which is of order of $10^6 ms^{-1}$. Since $\alpha$ is of order of unity for conventional $SiO_2$ substrate [7, 8], such a new regime needs nonperturbative methods to investigate long range impurity related effects.

On the one hand, due to their controllable sizes and shapes, GQDs are the center of both experimental and theoretical interest in graphene physics [20–63]. On the other hand, the formation of spin qubits in GQDs [46] with controllable valley splitting is another active research area in the field. The crucial requirement for this formation is to lift the valley degeneracy of the pristine graphene. It is shown that this is achievable in GQDs.

Since it is also possible to control where the adsorbate can be placed in graphene, controlling the impurity related effects are not only important for graphene, but also for graphene based nanostructures such as GQDs.

The continuum model of confined states of Dirac electrons in GQDs provides a simple physical insight into understanding the bound states in such structures. It is shown that [59, 60], it presents results in full agreement with those found from tight binding calculations (TBCs) for bound QD states in which their radii are much larger than the carbon-carbon distance. Even for small radii the agreement between continuum description and TBC is always reasonable for low-lying bound states with high values of the angular momentum. This establishes the use of continuum approach to study the confinement effects.

With these motivations, in the present work, we study the combined effects of quantum confinement due to GQD and hydrogenic donor impurity located at the center of the GQD. To do this, we take a circular confinement potential in order to model the GQD, and place the hydrogenic impurity at the center of dot. The calculation is performed within the framework of a variational approach based on using exact solutions of Dirac-Weyl equation in the absence of quantum dot potential as trial wave functions.

## 2. Theory

The Hamiltonian of a system consisting of an electron (hole) bound to a donor (acceptor) ion inside a parabolic graphene quantum dot , in the presence of a magnetic field perpendicular to GQD plane, is given by

$$\mathcal{H} = v_F\, \boldsymbol{\alpha} \cdot (\boldsymbol{p} + \frac{e}{c}\boldsymbol{A}) + \boldsymbol{\beta}\, \Delta_0 r^2 - \frac{Ze^2}{\epsilon r} \tag{1}$$

where $\boldsymbol{A}$ is the vector potential chosen at symmetrical gauge, i.e., $\boldsymbol{A} = B_0(-y, x, 0)/2$ , $\Delta_0 = U_0/2R_o^2$ wherein $U_0$ and $R_0$ are the strength and the radius of GQD, respectively. In Eq.(1), $\boldsymbol{\alpha}$ and $\boldsymbol{\beta}$ are Dirac matrices, and the Fermi velocity $v_F = (3a/2)\, J_0$ is given in terms of $J_0$ and $a$, which are the resonance integral between nearest neighbour carbon atoms, and is of order of 2.7 eV, and the equilibrium bond length of the graphene, respectively. The solution of Eq.(1) in the absence of both GQD potential and the impurity is well established in Ref. 17. In order to study the combined effects of these two, we use the exact wave functions in their absence as trial wave functions corresponding to the Dirac-Weyl equation $\mathcal{H}\Psi_n = E_n\Psi_n$ in which four-component spinors are given by

$$\begin{aligned}
\Psi_n^{\mathbf{K}}(x,\varphi) &= R_n(x)e^{im\varphi}e^{-x^2/2}\begin{pmatrix} \mathbb{L}^0 \\ 0 \\ 0 \\ \mathbb{L}^{\mathbf{K}} \end{pmatrix} \\
\Psi_n^{\mathbf{K}'}(x,\varphi) &= R_n(x)e^{im\varphi}e^{-x^2/2}\begin{pmatrix} 0 \\ \mathbb{L}^0 \\ \mathbb{L}^{\mathbf{K}'} \\ 0 \end{pmatrix}
\end{aligned} \tag{2}$$

with $R_n(x) = [\nu!/2\pi(\nu + |m|)!]^{1/2}\gamma_0 x^{|m|}$. In Eq.(2) we have defined $\mathbb{L}^0$, $\mathbb{L}^{\mathbf{K}}$ and $\mathbb{L}^{\mathbf{K}'}$ in terms of Laguerre polynomials as $\mathbb{L}^0(x) = L_\nu^{|m|}(x^2)$,

$$\begin{aligned}
\mathbb{L}^{\mathbf{K}} &= \frac{ixe^{+i\varphi}}{\sqrt{n}}\begin{cases} L_\nu^{|m|+1}(x^2) & m \geqslant 0 \\ -\frac{n}{x^2}L_{\nu+1}^{|m|-1}(x^2) & m < 0 \end{cases}, \\
\mathbb{L}^{\mathbf{K}'} &= \frac{ixe^{-i\varphi}}{\sqrt{n}}\begin{cases} -\frac{n}{x^2}L_\nu^{|m|-1}(x^2) & m \geqslant 0 \\ L_{\nu-1}^{|m|+1}(x^2) & m < 0 \end{cases},
\end{aligned} \tag{3}$$

respectively. The corresponding eigenvalues are given by $E_n = \lambda\epsilon_n$ wherein $\epsilon_n = \hbar v_F\sqrt{2n}/\ell_B$, $\lambda$ is the chirality index and takes $-1$ and $+1$ values, which correspond to valence and conduction bands of graphene, respectively. $\ell_B = (\hbar c/eB)^{1/2}$ is the magnetic confinement length, and $n = \nu + (|m| + m + \tau + 1)/2$ where $\tau$ is considered as a pseudospin index whose $-1$ and $+1$ values specifies the two valleys, i.e., $\mathbf{K}$ and $\mathbf{K}'$, respectively. In the following, we will restrict ourselves to only conduction band. Extension of the results to the other band is straightforward. In Eq.(3), $x = \gamma_0 \rho$ is a dimensionless, and $\gamma_0$ is the inverse of characteristic length of the system. On the one hand, without spatial parabolic confinement, $\gamma_0$ is proportional to inverse of magnetic confinement length $\ell_B$, i.e., $\gamma_0 = \gamma = 1/\sqrt{2}\ell_B$. On the other hand, in the presence of spatial confinement, graphene system described by the Hamiltonian Eq.(1) suggest a natural unit of length $R = (\hbar v_F R_o^2/U_0)^{1/3}$ which characterises the size of a scalar parabolic quantum dot in graphene [20, 21]. Therefore, we use the ansatz of Ref. 21 to characterize the system by only one parameter, i.e., we write $\gamma_0^2 = \gamma^2 + R^{-2}$ and take $\gamma$ to be a variational parameter, so as to reduce to $1/\sqrt{2}\ell_B$ in the limit $R \to \infty$. Therefore, the variational eigenenergies of GQD is given by

$$E_n = \int d^2\boldsymbol{r}\Psi_n^\dagger \left[\boldsymbol{\alpha}\cdot(-i\,\boldsymbol{\nabla} + \frac{e}{\hbar c}\mathbf{A}) + \boldsymbol{\beta}\,\Delta_0 r^2 - \frac{Ze^2}{\epsilon r}\right]\Psi_n. \tag{4}$$

Inserting Eqs. (2) and (3) into Eq.(4) we get a general expression for the variational energy as

$$\overline{E}_n = \left(\sqrt{n} - \frac{\bar{Z}}{2}\mathbb{M}_n^{\rm I}\right)\overline{\gamma}_0 + \frac{\sqrt{n}}{2\bar{l}_B^2}\frac{1}{\overline{\gamma}_0} + \frac{1}{4\overline{R}^3}\frac{1}{\overline{\gamma}_0^2}\mathbb{M}_n^{\rm II}. \tag{5}$$

In Eq.(5), the energy and all lengths are expressed in units of $J_0$ and $a$, respectively. The integrals $\mathbb{M}_n^{\rm I}$ and $\mathbb{M}_n^{\rm II}$ in Eq.(5) are defined in terms of Laguerre polynomials as

$$\mathbb{M}_n^{\rm I} = \widetilde{\mathbb{F}}_n + \frac{1}{n}\widetilde{\mathbb{G}}_n \tag{6}$$

with

$$\widetilde{\mathbb{F}}_n = \frac{\nu!}{(\nu+|m|)!}\int_0^\infty dr e^{-r} r^{|m|+\frac{1}{2}}\left|L_\nu^{|m|}(r)\right|^2$$

$$\widetilde{\mathbb{G}}_n = \frac{\nu!}{(\nu+|m|)!}\int_0^\infty dr e^{-r} r^{|m|-\frac{1}{2}}\mathbb{H}_{\nu,\alpha}^{|m|}(r).$$

and

$$\mathbb{M}_n^{\rm II} = \mathbb{F}_n - \frac{1}{n}\mathbb{G}_n$$

with

$$\mathbb{F}_n = \frac{\nu!}{(\nu+|m|)!}\int_0^\infty dr e^{-r} r^{|m|+1}\left|L_\nu^{|m|}(r)\right|^2$$

$$\mathbb{G}_n = \frac{\nu!}{(\nu+|m|)!}\int_0^\infty dr e^{-r} r^{|m|+1}\mathbb{H}_{\nu,\alpha}^{|m|}(r)$$

where

$$\mathbb{H}_{\nu,\alpha}^{|m|}(r) = \begin{cases} r\left|L_{\nu+\alpha}^{|m|+1}(r)\right|^2, & \alpha = \begin{cases} 0 & m \geqslant 0 \quad \tau = +1 \\ -1 & m < 0 \quad \tau = -1 \end{cases} \\ \frac{n^2}{r}\left|L_{\nu+\alpha}^{|m|-1}(r)\right|^2, & \alpha = \begin{cases} 0 & m \geqslant 0 \quad \tau = -1 \\ +1 & m < 0 \quad \tau = +1 \end{cases} \end{cases},$$

respectively. $\mathbb{M}_n^{\rm II}$ can be integrated analytically to yield $\mathbb{F}_n = 2\nu + |m| + 1$ and $\mathbb{G}_n = 2n - m$, respectively. Thus, we have exactly $\mathbb{M}_n^{\rm II} = -\tau$. For each $n$ values, $\mathbb{M}_0^{\rm I}$ can also be integrated analytically [64] so that, for instance, it yields $\mathbb{M}_0^{\mathrm{I}} = \Gamma(|m| + 1/2)/\Gamma(|m| + 1)$ for $n = 0$. Here, $\Gamma(x)$ is the Gamma function. To get an approximate analytical expression for the energy given by Eq.(5), we investigate some of its limiting cases. In the absence of impurity, since its minimization with respect to $\overline{\gamma}$ gives $\overline{\gamma} = 1/\sqrt{2}\ell_B$, as $\overline{R}$ goes to infinity. By replacing this value back into the relevant variational energy, one gets $\overline{E}_n = \sqrt{2n}/\overline{\ell}_B$, which are the well-known Landau levels for massless graphene. Consequently, substituting this value back into Eq.(5) as a first approximation yields

$$\bar{E}_0 = -\frac{\bar{Z}}{2\bar{l}_0}\frac{\Gamma(|m|+1/2)}{\Gamma(|m|+1)} - \tau\frac{\bar{l}_0^2}{4\bar{R}^3} \tag{7}$$

where $1/\overline{\ell}_0^2 = 1/2\overline{\ell}_B^2 + 1/\overline{R}^2$ defines the effective confinement length. Eq.(7) may give an idea about the effect of hydrogenic donor impurity onto the energy spectrum of a parabolic GQD as well as the critical value of the impurity strength. In the absence of impurity and magnetic field, Eq.(7) works well, as reported in Ref. 65, for the graphene devices fabricated by Ponomarenko *et al* [66], wherein the GQDs have a gap value of $0.010 - 0.5eV$ for the GQDs with diameter of $D = 40 - 1.0nm$. Eq.(7) predicts their corresponding values as $0.014 - 0.57eV$. Even in the high magnetic field case it yields the well-known relativistic LLs, which are also confirmed experimentally [67]. From Eq. (7), we estimate the critical value of $\bar{Z}$ as

$$\bar{Z}_{crt} = \frac{\bar{l}_0^3}{2\bar{R}^3}\frac{\Gamma(|m|+1)}{\Gamma(|m|+1/2)}$$

This reduces to $\bar{Z}_{cr} = 1/2\sqrt{\pi}$ in the absence of magnetic field.

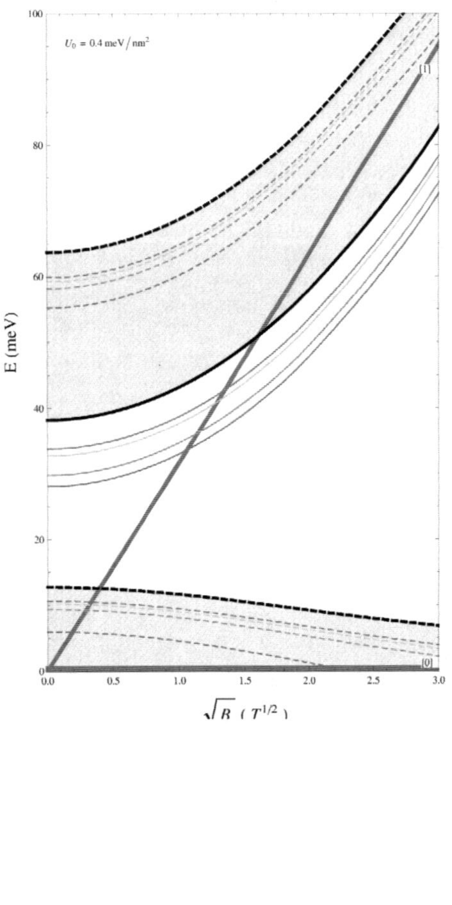


**Figure 1.** The magnetic field dependence of parabolic quantum dot energy spectra in the presence of hydrogenic donor impurity. While the bold gray lines (represented by labels $[n]$) correspond to the unperturbed LLs (in the limit $\overline{R} \to \infty$), the dashed (straight) curves represents $\mathbf{K}'(\mathbf{K})$ valleys. For all curves, from the bottom to the top, $m = 0,-1,-2$ and $-3$ levels are depicted by gray, red, green and blue colors.

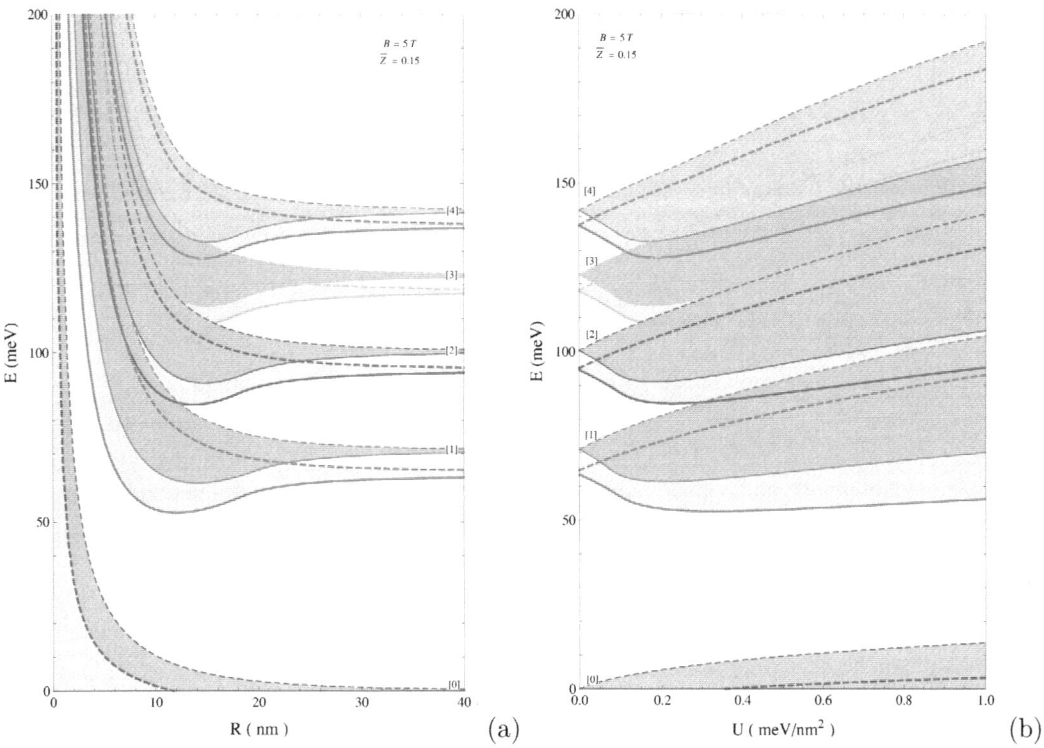


**Figure 2.** (Color online)(a) The lowest-lying energy eigenvalues of GQD as a function of effective dot radius $R = (\hbar v_F R_0^3/U_0)^{1/3}$ at a constant magnetic field and impurity strength, i.e., $B = 5$ T and $\overline{Z} = 0.15$, respectively. (b) The effective dot strength $U = U_0/R_o^2$ dependence of the same levels with $m = 0$.

FIG. 1 shows the ground- and the first excited-state energies of parabolic GQD as a function of $\sqrt{B}$ both in the presence ( thin dashed ( $\mathbf{K}'$) and thin straight ($\mathbf{K}$ ) lines) and absence ( bold dashed ( $\mathbf{K}'$) and bold straight ($\mathbf{K}$ ) lines) of hydrogenic donor impurity. We also included for comparison the first two LLs of the gapless graphene (thick straight gray lines). It is clearly seen that the impurity energy levels in GQD goes up to well known LLs with increasing B for a constant $U_0$. To understand the effect of confinement on impurity energy levels itself, for a fixed impurity strength, we plot the lowest-lying energies as a function of dot radius in FIG. 2. In both figure, to see better the valley splitting for the associated levels, space arising from this splitting is coloured by grey. It is seen that, since the narrower dot creates stronger confinement the magnitude of valley splitting, $\mathbf{K}(\mathbf{K}')$ becomes larger as the confinement becomes stronger. We found that, the magnitude of the valley degeneracy, with decreasing the strength of GQD, first drop slowly and then approach to its value in the gapless graphene, that is, zero. This is more clearly seen from the FIG. 2(b). However, introduction of donor impurity causes these levels to shift down to lower energies, and leads enhancement in the size of the valley splitting for constant $U$. On the other hand, inclusion of such a long range donor hydrogenic impurity leads to decrease in valley splitting.

In FIG. 3, the associated energy levels are plotted as functions of impurity strength $\bar{Z}$.

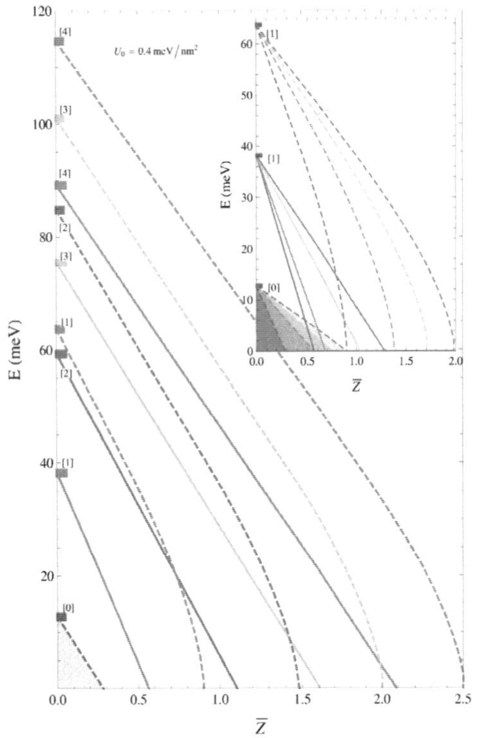


**Figure 3.** (Color online) GQD energies of a Dirac electron bound to a hydrogenic donor impurity as a function of impurity strength at a constant impurity strength with zero magnetic field. The solid curves represent the energy levels for $\tau = +1$, the dashed curves represent the those for $\tau = -1$. Inset shows the associated energy levels but with inclusion of the $m = -1$ (Red), $m = -2$ (Green) and $m = -3$ (Blue) as well as $m = 0$ (Gray).

## 3. Conclusion

In conclusion, we have investigated the effect of a hydrogenic donor impurity on the energy levels of a parabolic GQD. For this purpose, we use a variational scheme based on the choice of trial wave functions taking into account carrier confinement together with the effect of the magnetic field. On the basis of our results, we have found that the magnitude of valley splitting is strongly dependent on the impurity strength and the strength of the magnetic field, and concluded that the size of valley splitting can easily be controlled by the valency of the impurity. We hope that the presented results may be useful to gain insight how to control some electronic properties of GQDs with impurities.

## Acknowledgments

This work was supported by TÜBİTAK under Grant No. 113F103.

## References

[1] Novoselov K S, Geim A K, Morozov S V, Jiang D, Zhang Y, Dubonos S V, Grigorieva I V and Firsov A A *Science* 2004 **306** 666
[2] Novoselov K S, Jiang D, Schedin F, Booth T J, Khotkevich V V, Morozov S V and Geim A K 2005 USA *Proc. Natl. Acad. Sci.* **102** 10451
[3] Ostrovsky P M, Gornyi I V and Mirlin A D 2006 *Phys. Rev.* B **74** 235443
[4] Katsnelson M I 2006 *Phys. Rev.* B **74** 201401
[5] Pereira V M, Nilsson J and Castro Neto A H 2007 *Phys. Rev. Lett.* **99** 166802
[6] Fogler M M, Novikov D S and Shklovskii B I 2007 *Phys. Rev.* B **76** 233402
[7] Shytov A V, Katsnelson M I and Levitov L S 2007 *Phys. Rev. Lett.* **99** 236801
[8] Shytov A V, Katsnelson M I and Levitov L S 2007 *Phys. Rev. Lett.* **99** 246802
[9] Novikov D S 2007 *Appl. Phys. Lett.* **91** 102102
[10] Novikov D S 2007 *Phys. Rev.* B **76** 245435
[11] Terekhov I S, Milstein A I, Kotov V N and Sushkov O P 2008 *Phys. Rev. Lett.* **100** 076803
[12] Kotov V N, Pereira V M and Uchoa B 2008 *Phys. Rev.* B **78** 075433

[13] Chen J -H, Jang C, Adam S, Fuhrer M S, Williams E D and Ishigami M 2008 *Nature Phys.* **4** 377
[14] Pereira V M, Kotov V N and Castro Neto A H 2008 *Phys. Rev.* B **78** 085101
[15] Katsnelson M I, Geim A K 2008 *Phil. Trans. R. Soc.* A **366** 195
[16] Zhu W, Wang Z, Shi Q, Szeto K Y, Chen J and Hou J G 2009 *Phys. Rev.* B **79** 155430
[17] Kandemir B S and Mogulkoc A 2010 *Eur. Phys. J.* **74** 391
[18] Altanhan T and Kozal B 2012 *Eur. Phys. J.* **85** 222
[19] Jia-Lin Zhu and Songyang Sun 2012 *Phys. Rev.* B **85** 035429
[20] Silvestrov P G and Efetov K B 2007 *Phys. Rev. Lett.* **98** 016802
[21] Chen H, Apalkov V and Chakraborty T 2007 *Phys. Rev. Letts.* **98** 186803
[22] Rycerz A, Tworzydlo J and Beenakker C W J 2007 *Nat. Phys.* **3** 172
[23] De Martino A, Dell' Anna L and Egger R 2007 *Phys. Rev. Lett.* **98** 066802
[24] Sols F, Guinea F and Castro Neto A H 2007 *Phys. Rev. Letts.* **99** 166803
[25] Recher P, Trauzette B, Rycerz A, Blanter Ya. M, Beenakker C W J and Morpurgo A F 2007 *Phys. Rev.* B **76** 235404
[26] Trauzettel B, Bulaev D V, Loss D and Burkard G 2007 *Nat. Phys.* **3** 192
[27] De Martino A, Dell' Anna L and Egger R 2007 *Sol. State Commun.* **144** 547
[28] Wunsch B, Stauber T and Guinea F 2008 *Phys. Rev.* B **77** 035316
[29] Matulis A and Peeters F M 2008 *Phys. Rev.* B **77** 115423
[30] Katsnelson M I and Guinea F 2008 *Phys. Rev.* B **78** 075417
[31] Zhang Z Z, Chang K and Peeters F M 2008 *Phys. Rev.* B **77** 235411
[32] Hewageegana P and Apalkov V 2008 *Phys. Rev.* B **77** 245426
[33] Schnez S, Ensslin K, Sigrist M and Ihn T 2008 *Phys.Rev.* B **78** 195427
[34] Recher P, Nilsson J, Burkard G and Trauzettel B 2009 *Phys. Rev.* B **79** 085407
[35] Romanovsky I, Yannouleas C and Landman U 2009 *Phys. Rev.* B **79** 075311
[36] Stander N, Huard B and Goldhaber-Gordon D 2009 *Phys. Rev. Lett.* **102** 026807
[37] Young A. F. and Kim P 2009 *Nat. Phys.* **5** 222
[38] Wurm J, Rycerz A, Adagideli İ, Wimmer M, Richter K and Baranger H U 2009 *Phys. Rev. Lett.* **102** 056806
[39] Masir M R, Matulis A and Peeters F M 2009 *Phys. Rev.* B **79** 155451
[40] Wang D and Jin G 2009 *Europhys. Lett* **88** 17011
[41] Giavaras G, Maksym P A and Roy M 2009 *J. Phys.: Condens. Matter* **21** 102201
[42] Peres N M R, Rodrigues J N B, Stauber T and dos Santos L J M B 2009 *J. Phys.: Condens. Matter* **21** 344202
[43] Wang D and Jin G 2009 *Phys. Lett.* A **373** 4082
[44] Bardarson J H, Titov M and Brouwer P W 2009 *Phys. Rev. Lett.* **102** 226803
[45] Giavaras G and Nori F 2010 *Applied Phys. Lett.* **97** 243106
[46] Recher P and Trauzettel B 2010 *Nanotechnology* **21** 302001
[47] Giavaras G, Maksyma P A and Roy M 2010 *Physica E* **42** 715
[48] Yan C-H and Wei L-F 2010 *J. Phys.: Condens. Matter* **22** 295503
[49] Horing N J M and Liu S Y 2010 *J. Phys.: Condens. Matter* **22** 025502
[50] Horing N J M and Liu S Y 2010 *J. Phys.: Condens. Matter* **22** 159801
[51] González J W, Pacheco M, Rosales L and Orellana P A 2010 *Europhys. Lett.* **91** 66001
[52] Kim S C, Park P S and Yang S -R E 2010 *Phys. Rev.* B **81** 085432
[53] Libich F, Rotter S, Güttinger J, Stampfer and Burgdörfer C 2010 *J. Phys. Rev.* B **81** 245411
[54] Wimmer M, Akhmerov A R and Guinea F 2010 *Phys. Rev.* B **82** 045409
[55] Giavaras G and Nori F 2011 *Phys. Rev.* B **83** 165427
[56] Masir M R, Vasilopoulos P and Peeters F M 2011 *J. Phys.: Condens. Matter* **23** 315301
[57] Molitor F, Güttinger J , Stampfer C, Dröscher S, Jacobsen A, Ihn T and Ensslin K 2011 *J. Phys.: Condens. Matter* **23** 243201
[58] Rozhkova A V, Giavaras G, Bliokh Y P , Freilikher V and Nori F 2011 *Physics Reports* **503** 77
[59] Pal G, Apel W and Schweitzer L 2011 *Phys. Rev.* B **84** 075446
[60] Zarenia M, Chaves A, Farias G A and Peeters F M 2011 *Phys. Rev.* B **84** 245403
[61] Tahir M and Sabeeh K 2012 *J. Phys.: Condens. Matter* **24** 135005
[62] Kim S C, Lee J W and Yang S-R E 2012 *J. Phys.: Condens. Matter* **24** 495302
[63] Grujić M , Zarenia M, Tadić M and Peeters F M 2012 *Phys. Scr. T* **149** 014056
[64] Gradshteyn I S and Rzyhik I M 2000 *Table of Integrals, Series and Products* 6th edn. ( New York: Academic Press)
[65] Kandemir B S and Omer G 2013 *Eur. Phys. J.* B **86** 299
[66] Ponomarenko L A, Schedin F, Katsnelson M I, Yang R, Hill E W, Novoselov K S and Geim A K 2008

*Science* **320** 356
[67] Molitor F, Jacobsen A, Stampfer C, Güttinger J, Ihn T and Ensslin K 2009 *Phys. Rev.* B **79** 075426

# Effects of Post Annealing on I-V-T Characteristics of (Ni/Au)/$Al_{0.09}Ga_{0.91}N$ Schottky Barrier Diodes

**Abdullah Akkaya[1] and Enise Ayyıldız[2]**

[1] Ahi Evran University, Mucur Technical Vocational Schools, Tech. Prog. Dept., 40500 Kırşehir, Turkey

[2] Erciyes University, Science Faculty, Physics Dept., Talas Yolu 38039 Kayseri, Turkey

E-mail: abdullah.akkaya@ahievran.edu.tr

**Abstract.** Post annealing is a simple, effective and suitable method for improving the diode parameters, especially when the used chemically stable substrates like Si, III-N and ternary alloys. In our work, we were applied this method to (Ni/Au)/$Al_{0.09}Ga_{0.91}N$ Schottky Barrier Diodes (SBDs) and investigated by temperature-dependent current-voltage (I-V-T) characteristics at optimum conditions. Optimum annealing temperature was 600°C, which it's determined with respect to have a highest barrier height value.

The temperature-dependent electrical characteristics of the annealed at 600°C (Ni/Au)/$Al_{0.09}Ga_{0.91}N$ SBDs were investigated in the wide temperature range of 95-315K. The diode parameters such as ideality factor (n) and Schottky barrier height ($\Phi_{b0}$) were obtained to be strongly temperature dependent. The observed variation in $\Phi_{b0}$ and n can be attributed to the spatial barrier inhomogeneities in Schottky barrier height by assuming a triple Gaussian distribution (TGD) of barrier heights (BHs) at 95-145K, 145-230K and 230-315K. The modified Richardson plots and $T_0$ analysis was performed to provide an experimental Richardson constants and bias coefficients of the mean barrier height. Furthermore, the chemical composition of the contacts was examined by the XPS depth profile analysis.

## 1. Introduction

GaN and its ternary alloys, especially AlGaN, are of interest in the fabrication of high-power, high-temperature and high-frequency electronic devices, as well as, transistors, light emitting and detecting devices. Devices based on AlGaN layers or bulk includes bipolar junction transistors [1], solar blind photodiodes [2], light emitting diodes and high electron mobility transistors [3]. These exciting applications brings numerous challenges, such as making reliable, high quality, efficient, thermally stable metal [4-7] contacts to nitride semiconductor, which is crucial for device performance. With respect to electronic devices such as metal semiconductor field effect transistor (MESFETs), metal oxide semiconductor field effect transistors (MOSFETs), modulation doped field effect transistors (MODFETs) and high power switching devices, main process improvements are needed in the areas of stable and reproducible Schottky and ohmic contacts to GaN based materials [8]. Also, for advancement of these devices, electrical properties and effect of post annealing have to be fully investigated. Especially, more detailed characterizations of metal/AlGaN interfaces seem to be necessary. Furthermore, the chemical structure of as-deposited and annealed metal/AlGaN interfaces also have to be investigated to give clear sight about properties of Schottky barrier diodes (SBDs).

In this work we focused on thermal annealing effects on the electrical characteristics of (Ni/Au)/$Al_{0.09}Ga_{0.91}N$ SBDs. Annealing treatment was performed at a temperature ranging from 100 to 600°C in increments of 100°C for 3 min in high purity argon ambient. The thermal stability of the Ni/Au Schottky contacts was evaluated by considering the change of the Schottky barrier height (SBH) with the annealing temperature. Also, the temperature-dependent electrical characteristics of the as-deposited and annealed at 600°C (Ni/Au)/$Al_{0.09}Ga_{0.91}N$ SBDs were investigated in the wide temperature range of 95-315K. Furthermore, the chemical composition of the contacts was examined by the XPS depth profile analysis.

## 2. Experimental

In this study, unintentionally doped (uid) n-type $Al_{0.09}Ga_{0.91}N$ epitaxial layers grown by MOCVD on a HPSI-SiC substrate were used. The epistructure of the wafer consists of 2 nm thin layer of GaN cap layer for protection purposes, $Al_{0.09}Ga_{0.91}N$ layer with a thickness of 21 nm, AlN layer with a thickness of 3 nm, Fe doped GaN buffer layer with a thickness of 1.8 μm and a 4H-SiC high-purity semi-insulating substrate. The AlGaN substrates were cleaned consecutively acetone, methanol, trichloroethylene, deionized water (18 MΩ) 5 min. using ultrasonic agitation in each step. The substrates were then dried with high-purity nitrogen. After cleaning organic residuals, the substrates were dipped in aqua regia (3:1 HCl:$HNO_3$) to remove the native oxide from the front surface of the substrate and boiling KOH (1 M) to reduce the surface roughness, respectively. The Ti/Al (25 nm/105 nm) Ohmic metallization was deposited using magnetron dc sputtering for Ti and thermal evaporation for Al and a standard lift-off process was used to pattern the contacts. The contacts were annealed at 850°C for 1 min in flowing high purity (6N) argon gas in a quartz tube furnace. The Ni/Au (30 nm/50 nm) Schottky metallization was then deposited using magnetron dc sputtering for Ni and thermal evaporation for Au and a standard lift-off process was used to pattern to form Schottky contacts with a diameter of 0.4 mm.

I-V-T measurements of the as-deposited and annealed (Ni/Au)/$Al_{0.09}Ga_{0.91}N$ SBDs were accomplished by employing a computer-controlled HP 4140B picoamperemeter with a $LN_2$ cooled cryostat. Annealing was performed in the quartz tube furnace with flow (~10 ccm/s) of high purity Ar gas ambient. Annealing time was 3 minutes.

## 3. Results and discussion

### *3.1. I-V measurements and effect of annealing on diode parameters*

Fig. 1 shows the forward and reverse I-V characteristics of the as-deposited and annealed (Ni/Au)/$Al_{0.09}Ga_{0.91}N$ SBDs. For the as-deposited Ni/Au Schottky contact, the leakage current at -3 V is $7.18x10^{-7}$ A. For the diode annealed at temperature 400°C, 500°C and 600°C, the leakage currents are $2.06x10^{-7}$ A, $1.61x10^{-7}$ A and $1.21x10^{-7}$ A at -3 V, respectively. It can be seen from Fig. 1 that the thermal annealing effect on reverse bias current was small under the annealing temperature of 400°C.

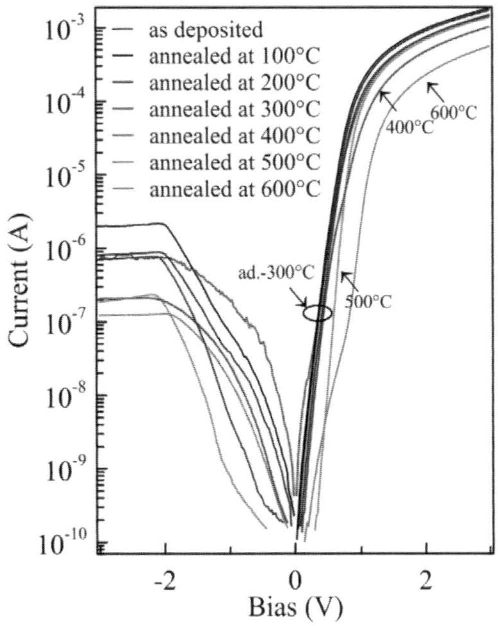


**Fig. 1.** I-V characteristics of the as-deposited and annealed (Ni/Au)/$Al_{0.09}Ga_{0.91}N$ SBD.

However, the ratio of forward bias current/reverse bias current at given bias (rectification factor) was increased approximately 4 times after the annealing at 500°C (Fig 2.). Therefore, the annealing temperature of 500°C was crucial for the studied SBDs.

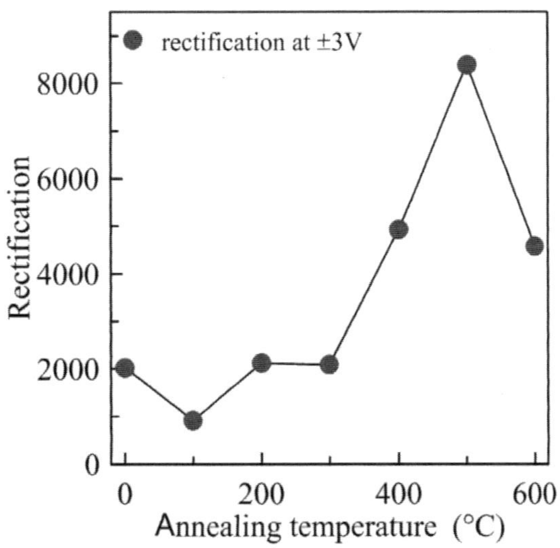


**Fig 2.** The variation of rectification factor of the as-deposited and annealed at ±3V.

I-V curves were analyzed using the thermionic emission theory (TET). According to the TET, the current flowing through the metal–semiconductor contact under forward bias V is given by (for V > 3kT/q) [9]

$$I = I_0 exp\left[{}^{qV}/_{nkT}\right] \tag{1}$$

where $I_0$ the saturation current is defined by

$$I_0 = AA^*T^2 exp\left[{}^{-q\Phi_{bo}}/_{kT}\right]. \tag{2}$$

The quantities $V$, $n$, $T$, $A$, $A^*$, $q$, $k$, and $\Phi_{bo}$ are the applied bias voltage, the ideality factor, the temperature in Kelvin, the effective diode area, the effective Richardson constant of 29.3 $AK^{-2}cm^{-2}$ for $Al_{0.09}Ga_{0.91}N$ [10], the electron charge, the Boltzmann constant, and the apparent barrier height at zero bias determined from the I-V data, respectively.

The experimental values of the barrier height $\Phi_{bo}$ and the ideality factor $n$ after each annealing step were calculated from the intercepts and slopes of the straight-line portions of the semilog forward bias I-V characteristics using Eqs. (1) and (2), respectively. The values of $n$ and $\Phi_{bo}$ obtained depending on the annealing temperature for the diodes are given in Table 1. The variation of SBH of the Ni/Au Schottky contacts as a function of annealing temperature is shown in Fig. 3.

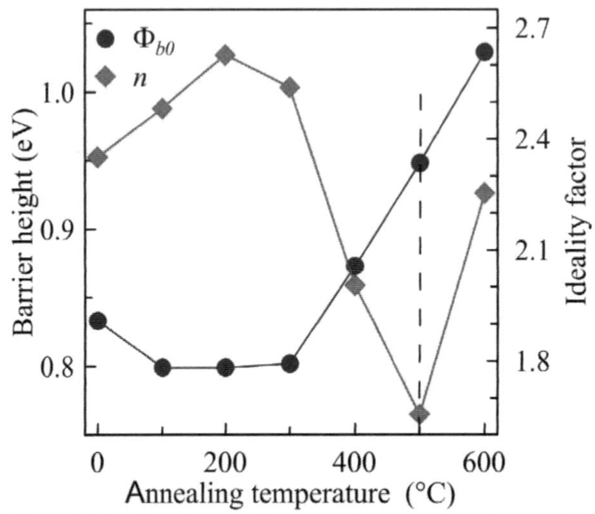


**Fig 3.** The variation of SBH and ideality factor of the as-deposited and annealed (Ni/Au)/$Al_{0.09}Ga_{0.91}N$ SBD.

Table 1 show that the Schottky barrier height determined of the as-deposited (Ni/Au)/$Al_{0.09}Ga_{0.91}N$ SBD was 0.833 eV which agrees with the literature [11]. The SBH is lower than the theoretical value which can be explained by the existence of the interfacial layer, structural defects such as the stacking faults, and dislocations on the AlGaN [11, 12]. The highest SBH was obtained to be 1.029°eV after annealing at 600°C for Ni/Au Schottky contact. Moreover, the rectification feature of the diode after annealing at 700°C was deteriorated (not shown here). The results in Table 1 show that the values of ideality factor are higher than unity for all contacts. For an ideal diode, the value of ideality factor should be nearly equal to one. A departure from the ideality could be caused by the image force lowering, generation and recombination of carriers in the space charge region, interface states, interfacial layer, and thermionic field emission [13-15]. These effects invariably increase the ideality factor [11].

The above observations reveal that before the values of the ideality factor fluctuate around 2 by annealing temperature up to 400°C and then decreases at the above annealing temperatures of 500°C. This situation encountered may be due to the change of the interface state distribution and/or chemical composition formed between the metal and the semiconductor at the interface [15-17]. This is confirmed by XPS analysis, as will be seen later. Therefore, it can be said that the performance and reliability of Schottky contacts depend on the nature of interfacial layer at the MS contacts which affected by annealing temperature. The nature of interfacial reactions between the metal and semiconductor plays an important role in determining the quality of Schottky barrier in AlGaN.

The SBHs and serial resistance ($Rs$) of the Ni/Au/n-GaN Schottky diodes are determined using the Norde method [18]. The plots of $F(V)$ versus $V$ for the SBD using data in Fig. 1 at the different annealing temperatures are shown in Fig. 4. The values of obtained $Rs$ of the SBDs are also given in

Table 1. As can be seen from Table 1, the obtained *Rs* values of the diodes are nearly constant up to 500°C.

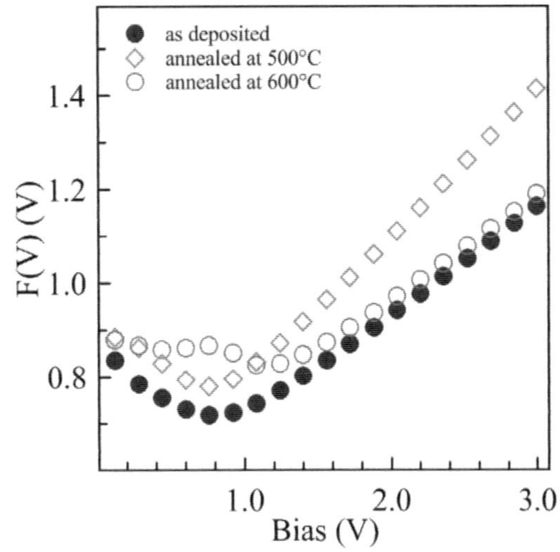


**Fig. 4.** The plot of *F(V)* versus *V* for the as-deposited and annealed (Ni/Au)/$Al_{0.09}Ga_{0.91}N$ SBDs.

Another way to determine diode parameters is to use Cheung's functions [19]. In this method, the slope and y-axis intercept of a plot of *dV/d(lnI)* vs. *I* give *Rs* and *nq/kT* and a plot of *H(I)* vs. I gives also a straight line with the y-axis intercept equal to $n\Phi_b$ (Fig. 5b) The slope of this plots also provides a second determination of *Rs*, which can be used to check the consistency of Cheung's approach [19]. The values of obtained *Rs*, *n* and $\Phi_b$ of the SBDs are given in Table 1.

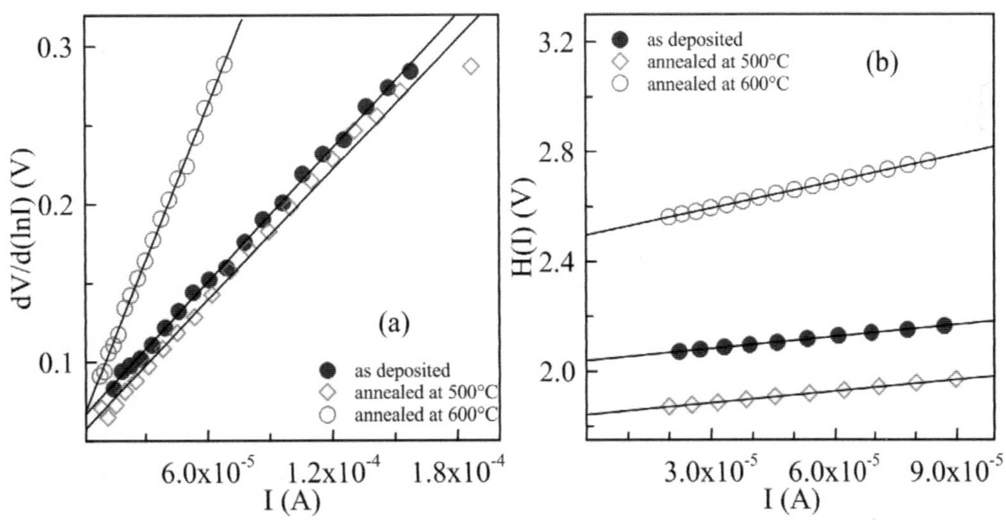


**Fig. 5.** The experimental *dV/d(ln I)* versus *I* (a) and *H(I)* versus *I* plots (b) for one of the as-deposited and annealed (Ni/Au)/$Al_{0.09}Ga_{0.91}N$ SBD.

**Table 1.** The obtained characteristics diode parameters of the (Ni/Au)/$Al_{0.09}Ga_{0.91}N$ SBDs before and after thermal annealing.

| AT (°C) | n | $\Phi_{b0}$ (eV) | $I_0$ (A) | Cheung | | | | Norde | |
|---|---|---|---|---|---|---|---|---|---|
| | | | | n | Rs (kΩ) | $\Phi_{b0}$ (eV) | Rs (kΩ) | $\Phi_{b0}$ (eV) | Rs (kΩ) |
| - | 2.351 | 0.833 | $1.39x10^{-10}$ | 2.472 | 1.445 | 0.825 | 1.442 | 0.892 | 1.558 |
| 100 | 2.484 | 0.799 | $5.07x10^{-10}$ | 2.335 | 1.142 | 0.819 | 1.138 | 0.865 | 1.102 |
| 200 | 2.628 | 0.799 | $5.23x10^{-10}$ | 2.294 | 1.207 | 0.836 | 1.2018 | 0.877 | 1.095 |
| 300 | 2.540 | 0.802 | $4.51x10^{-10}$ | 3.391 | 1.289 | 0.747 | 1.286 | 0.900 | 1.290 |
| 400 | 2.004 | 0.873 | $2.94x10^{-11}$ | 3.459 | 1.877 | 0.778 | 1.868 | 0.970 | 1.781 |
| 500 | 1.655 | 0.948 | $1.64x10^{-12}$ | 2.114 | 1.416 | 0.871 | 1.422 | 1.014 | 1.371 |
| 600 | 2.254 | 1.029 | $7.16x10^{-14}$ | 2.589 | 3.214 | 0.964 | 3.228 | 1.088 | 3.166 |

Kim et al. [20] have investigated the Schottky behavior of Ni/AlGaN/GaN (x=0.29) Schottky diodes under time dependent and fixed annealing temperature conditions by I-V measurements. They have reported that the barrier height change from 0.74°eV (as-deposited) to 0.82°eV under 170°min annealing at 400°C in $N_2$ ambient. They have explained the variation of barrier height upon annealing by the post annealing reduced the density of the electrically active states at the Schottky metal/AlGaN surface, leading to decrease of reverse leakage current, n, and saturation current and increase of SBH.

Akkaya et al. [21] have reported that the value of SBH of the as-deposited Ni/Au Schottky contact on n-GaN substrate was 0.560°eV (obtained from I-V) and 0.622°eV (obtained from C-V) and SBH increases with increase in the annealing temperature. After the annealing at 700°C, the value of the SBH of Ni/Au Schottky contact was found to be 0.911°eV (I-V) and 1.765°eV (C-V). In this study, increment in barrier height by as much as % 62.5 was successfully recorded by thermal annealing at 700°C. Furthermore, the variation of Schottky barrier heights and ideality factors with annealing temperature was attributed to the interfacial reactions of Ni/Au with GaN layer.

### *3.2. I-V-T Measurement results*

The semi-logarithmic forward I-V characteristics of the as-deposited and annealed at 600°C (Ni/Au)/$Al_{0.09}Ga_{0.91}N$ SBDs in the temperature range of 95–305°K by the step of 10°K are shown in Fig. 6. The experimental values of the apparent SBH and the ideality factor were determined from intercepts and slopes of the linear regions of the forward-bias *lnI* versus *V* plots according to the TET at each temperature, respectively. Least square fittings and other calculations are performed via computer program SeCLaS [22, 23]. The values of the SBH and the ideality factor of the as-deposited Ni/$Al_{0.09}Ga_{0.91}N$ SBD varied from 0.268 eV and 7.727 at 95 K to 1.873°eV and 0.913 at 315°K, respectively. The values of the SBH and the ideality factor of the annealed at 600°C (Ni/Au)/$Al_{0.09}Ga_{0.91}N$ SBD varied from 0.256°eV and 5.386 at 95 K to 0.772°eV and 2.000 at 315°K, respectively.

Fig. 7 shows the variation of the SBH and the ideality factor as a function of temperature for as-deposited and annealed SBD. As can be seen from Fig. 7, the values of ideality factor decrease while the values of SBH increase with increasing temperature. Such a temperature dependent behavior of the SBH and the ideality factor is commonly observed in real SBD and attributed to the lateral inhomogeneity of the SBH at the MS interface [24-26]. Similar trends have been reported for diodes on AlGaN [17, 21, 23, 27-31] as well as any other semiconductors [24, 32, 33].

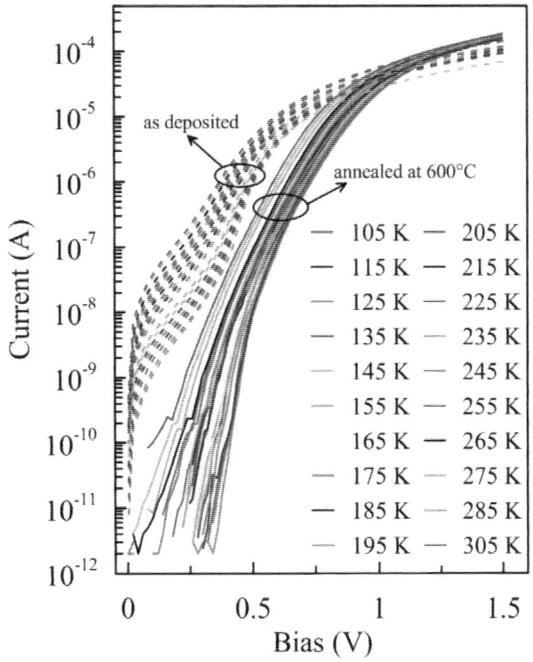


**Fig. 6.** The semi-logarithmic forward bias I-V characteristics for the as-deposited and annealed (Ni/Au)/$Al_{0.09}Ga_{0.91}N$ SBDs in the temperature range of 105–305°K.

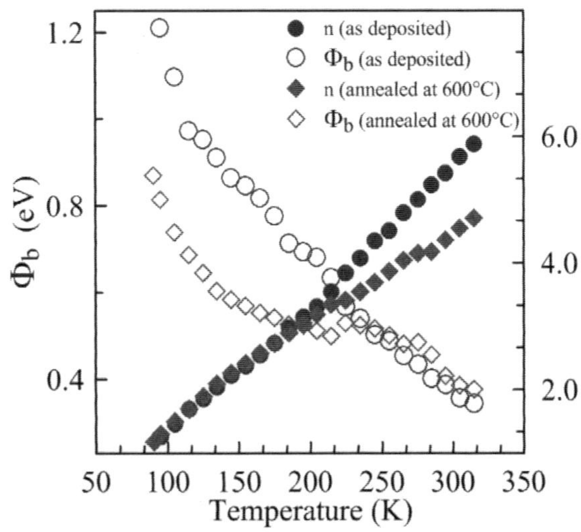


**Fig. 7.** The SBH and the ideality factor of the as-deposited and annealed (Ni/Au)/$Al_{0.09}Ga_{0.91}N$ SBDs as a function of temperature.

As explained in [34, 35], since the current transport across the metal-semiconductor (MS) interface is a temperature-activated process, electrons at low temperatures are able to surmount the lower barriers. Therefore, the current transport will be dominated by the current flowing through the patches of lower SBH, leading to a larger ideality factor. As the temperature increases, more and more electrons have sufficient energy to surmount the higher barriers. As a result, both SBH and n are strongly dependent on temperature. The nature and origin of these anomalies in some studies have been explained on the basis of the TET which takes into account the lateral barrier distribution of the SBHs due to inhomogeneities prevailing at the MS interface [23].

### *3.3. Triple Gaussian distribution of barrier heights*

There are two main method used to explain the Schottky barrier inhomogeneity. One of is Tung's pinch-off model [25, 36] and other approach is the parallel conduction model based on different patches having different barrier height which do not affect each other. According to the latter model, the total current through the diode is the sum of the currents flowing all of these patches in the present of the whole contact area and at thermodynamic equilibrium (V=0), the temperature dependence barrier height and variation of the ideality factor with temperature can be described by the following equations [37] ·

$$\Phi_b = \overline{\Phi_b} - {q\sigma_{s0}^2}/{2kT} \tag{3}$$

$$1/n - 1 = -\rho_2 + {a\rho_3}/{2kT}. \tag{4}$$

$\sigma_{s0}$, $\rho_2$ and $\rho_3$ can be found by using a changing in the $\Phi_b$ and *1/n-1* with *1/2kT*, respectively (Fig. 8 and Table 2). In our case, these graphics indicates SBH can be modeled with three set Gaussian distributions.

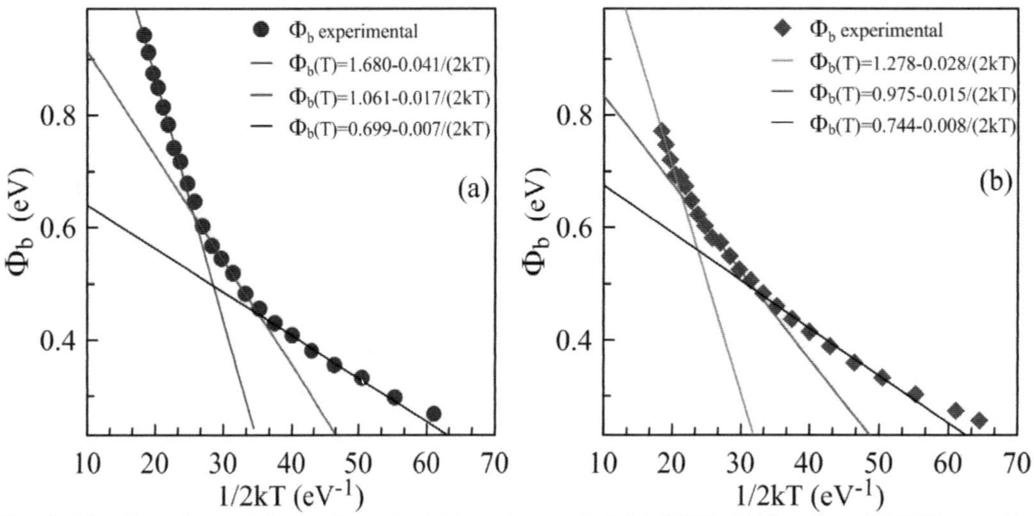


**Fig. 8.** The $\Phi_b$ values of the as-deposited (a) and annealed (b) (Ni/Au)/$Al_{0.09}Ga_{0.91}N$ SBD as a function of *1/2kT*.

**Table 2.** Three set Gaussian distribution parameters of the as-deposited and annealed (Ni/Au)/$Al_{0.09}Ga_{0.91}N$ SBDs.

| As deposited | | | |
|---|---|---|---|
| | $\Phi_b^1$ (eV) | $\Phi_b^2$ (eV) | $\Phi_b^3$ (eV) |
| | 1.680 | 1.061 | 0.699 |
| $\sigma_{si}$ | 0.202 | 0.132 | 0.086 |
| Annealed at 600°C | | | |
| | $\Phi_b^1$ (eV) | $\Phi_b^2$ (eV) | $\Phi_b^3$ (eV) |
| | 1.278 | 0.975 | 0.744 |
| $\sigma_{si}$ | 0.168 | 0.123 | 0.090 |

Richardson plots ($ln(I_0/T^2)$ *vs.* $1000/T$) have bowing of the curve. That is impossible to linear fit our data in the whole temperature range. However, for T>210 K, the experimental data lie on a straight line. The value of $A^*$ obtained from the intercept of the straight portion at the ordinate is equal to 1.859 $Acm^{-2}K^{-2}$ for as-deposited SBD and 1.951 $Acm^{-2}K^{-2}$ for annealed SBD, which is much smaller than the theoretical value of 29.3 $Acm^{-2}K^{-2}$ for electrons in undoped $Al_{0.09}Ga_{0.91}N$.

As above mentioned, the conventional Richardson plots have showed nonlinearity at low temperatures. To explain these discrepancies in Richardson plot can be rewritten as [38]

$$ln\left(^{I_0}/_{T^2}\right) - \left(^{q^2\sigma_{s0}^2}/_{2k^2T^2}\right) = ln(AA^*) - \left(^{q\overline{\Phi_b}}/_{2kT}\right). \tag{5}$$

Using the experimentally obtained $I_0$ data, the $ln(I_0/T^2)-q^2\sigma_{s0}^2/2k^2T^2$ values were calculated for both three values of $\sigma_{s0}$ associated with the Gaussian distributions of SBHs. The best linear fitting to these modified experimental data are exhibited by the solid curves in Fig. 9 which represent the true activation energy plots in the respective temperature regimes. The obtained Richardson constant values are very close to the theoretical value for n-type $Al_{0.09}Ga_{0.91}N$.

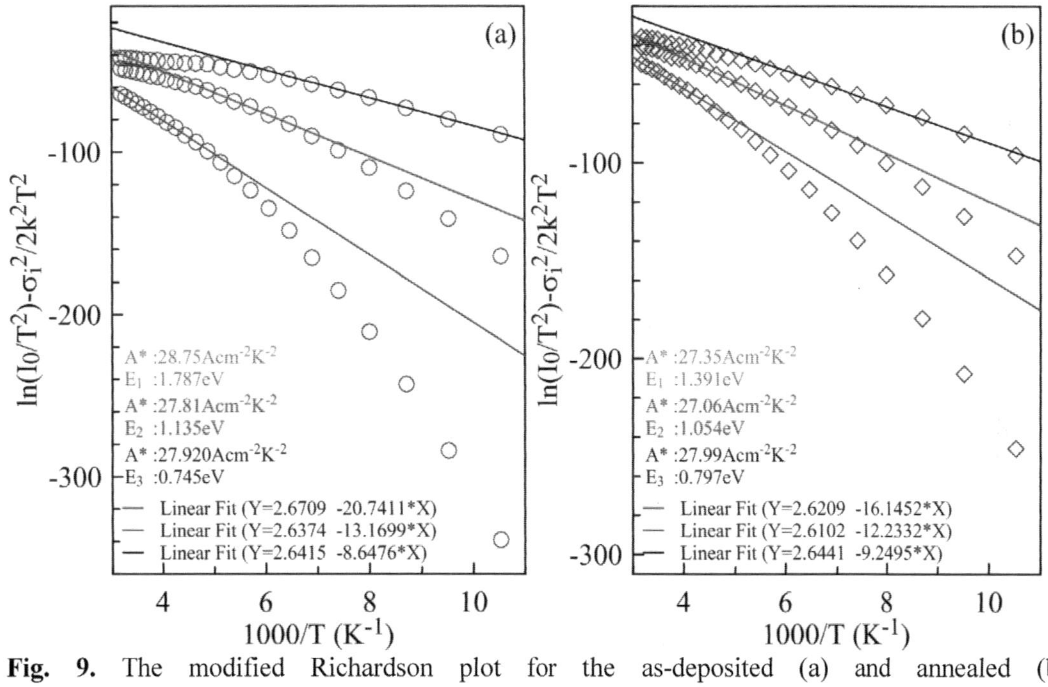


**Fig. 9.** The modified Richardson plot for the as-deposited (a) and annealed (b) (Ni/Au)/$Al_{0.09}Ga_{0.91}N$ SBDs according to the three set Gaussian distribution of SBHs. The solid lines show the best fitting of the data in the corresponding temperature regime.

### *3.4. XPS measurements results*

The XPS depth profile was employed in order to investigate the interfacial reaction between Ni/Au metal layers and AlGaN. Fig. 10 shows the XPS depth profiles of (Ni/Au)/$Al_{0.09}Ga_{0.91}N$ SBD as-deposited and after annealing at 600°C. The as-deposited layer reveals a relatively sharp interface, which indicates that there is no significant inter-diffusion between the metal layers and AlGaN. The depth profile of the contact annealed at 600°C is shown in Fig. 10b. It should be noted that some amount of Ga out-diffused into the metal layers. This is indicative of possible reaction between the Ni/Au layers and the AlGaN, resulting in the formation of gallide phases (GaNi) at the interface.

When the XPS spectra evaluated one by one for Al, Ga, N, Au and Ni peak shifting point out a intermetallic compounds (such as gallide phases and NiAu alloys) exist in interface even the contact was not anneal. Furthermore these findings indicate a few clues about interface. One of its Au was spread to almost all contact and small shift in binding energy could be caused from reaction between the Au and excess Ga and Ni. Shift in binding energy of Ni was verified by this situation. Secondly, when the diode annealed, significantly amount of Ga out diffused from substrate to the contact [21].

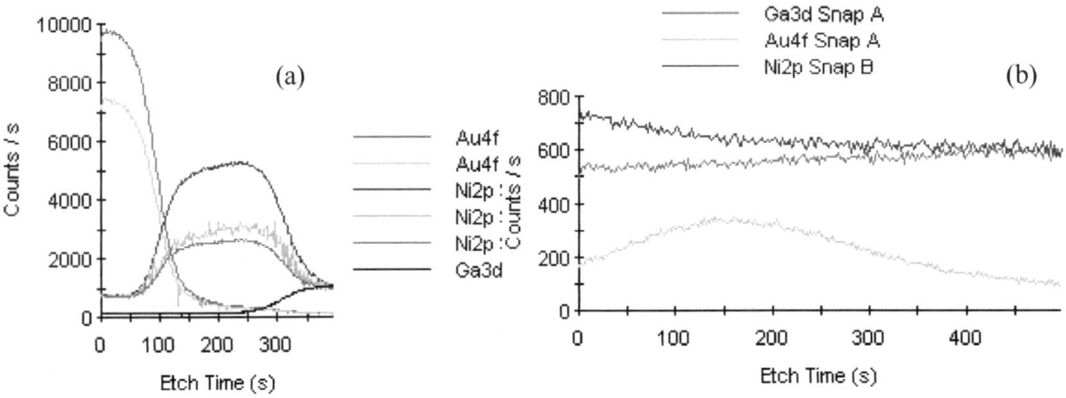


**Fig. 10.** XPS depth profiles of (Ni/Au)/$Al_{0.09}Ga_{0.91}N$ SBD (a) as-deposited sample and (b) 600°C annealed sample.

## 4. Conclusion

The effect of thermal annealing on electrical and structural properties of the (Ni/Au)/$Al_{0.09}Ga_{0.91}N$ SBD were investigated using I-V and XPS measurements. The value of SBH of the as-deposited (Ni/Au)/$Al_{0.09}Ga_{0.91}N$ SBD was found to be 0.833°eV. Furthermore, it is shown that the value of SBH increases with increase in the annealing temperature. After the annealing at 600°C, the value of the SBH was found to be 1.029°eV. The obtained results are shown that the optimum annealing temperature for the Ni/Au Schottky contact is 500-600°C. In other word, the partial stability observed up to the 400°C annealing temperature. Gaussian distribution parameters of as-deposited and annealed (Ni/Au)/$Al_{0.09}Ga_{0.91}N$ SBD indicates that annealed SBD become a more homogeneous. It was concluded that the Ni/Au Schottky contact is a suitable choice for the fabrication of AlGaN based high power device applications. Also, the analysis of XPS depth profile revealed that the formation of the gallide and intermetallic phases exist at metal/GaN/AlGaN interface. The variation of SBH and ideality factors with annealing temperatures may be attributed to the interfacial reactions of Ni/Au with GaN/AlGaN layer.

**Acknowledgements**: "This work was supported by the Scientific and Technological Research Council of Turkey (TÜBİTAK) under the Grant of 110T369 and Scientific Research Projects Unit of Erciyes University under the Grants of FDD-2012-4124 EÜBAP. The authors are grateful to TÜBİTAK and Erciyes University for their financial support"

**References**

[1] Kumakura K and Makimoto T 2008 High performance pnp AlGaN/GaN heterojunction bipolar transistors on GaN substrates *Appl Phys Lett* **92**

[2] Tut T, Yelboga T, Ulker E and Özbay E 2008 Solar-blind AlGaN-based p-i-n photodetectors with high breakdown voltage and detectivity *Appl Phys Lett* **92**

[3] Wang C, Zhang K, He Y L, Zheng X F, Ma X H, Zhang J C and Hao Y 2014 Effects of Annealing on Schottky Characteristics in AlGaN/GaN HEMT with Transparent Gate Electrode *Chinese Phys Lett* **31**

[4] Ao J P, Naoi Y and Ohno Y 2013 Thermally stable TiN Schottky contact on AlGaN/GaN heterostructure *Vacuum* **87** 150-4

[5] Readinger E D, Robinson J A, Mohney S E and Therrien R 2005 Thermal stability of metallizations on GaN/AlxGa1-xN/GaN heterostructures *Semicond Sci Tech* **20** 389-97

[6] Kaminska E, Piotrowska A, Golaszewska K, Lukasiewicz R, Szczesny A, Kowalczyk E, Jagodzinski P, Guziewicz M, Kudla A, Barcz A and Jakiela R 2005 Thermally stable transparent Ru-Si-O Schottky contacts for n-type GaN and AlGaN *Mater Res Soc Symp P* **831** 251-6

[7] Jeon C M, Jang H W and Lee J L 2003 Thermally stable Ir Schottky contact on AlGaN/GaN heterostructure *Appl Phys Lett* **82** 391-3

[8] Arulkumaran S, Egawa T, Ishikawa H, Umeno M and Jimbo T 2001 Effects of annealing on Ti, Pd, and Ni/n-A10.11Ga0.89N Schottky diodes *Ieee T Electron Dev* **48** 573-80

[9] Rhoderick E H and Williams R H 1988 *Metal-Semiconductor Contacts* (Oxford: Clarendon Press )

[10] Ambacher O 1998 Growth and applications of Group III nitrides *J Phys D Appl Phys* **31** 2653-710

[11] Motayed A, Sharma A, Jones K A, Derenge M A, Iliadis A A and Mohammad S N 2004 Electrical characteristics of AlxGa1-xN Schottky diodes prepared by a two-step surface treatment *J Appl Phys* **96** 3286-95

[12] Morkoç H 2009 *Handbook of Nitride Semiconductors and Devices, Materials Properties, Physics and Growth* vol 1 (Weinheim: Wiley-VCH)

[13] Card H C and Rhoderick E H 1971 Studies of tunnel MOS diodes I. Interface effects in silicon Schottky diodes *Journal of Physics D: Applied Physics* **4** 1589

[14] Padovani F A and Stratton R 1966 Field and thermionic-field emission in Schottky barriers *Solid State Electron* **9** 695-707

[15] Motayed A and Mohammad S N 2005 Chemicophysical surface treatment and the experimental demonstration of Schottky-Mott rules for metal/semiconductor heterostructure interfaces *J Chem Phys* **123**

[16] Wilmsen C 1985 Physics and Chemistry of Compound III–V Semiconductor Interfaces. (New York: Plenum Press) p 465

[17] Yıldırım N, Ejderha K and Türüt A 2010 On temperature-dependent experimental I-V and C-V data of Ni/n-GaN Schottky contacts *J Appl Phys* **108** 114506

[18] Norde H 1979 A modified forward I-V plot for Schottky diodes with high series resistance *J Appl Phys* **50** 5052-3

[19] Cheung S K and Cheung N W 1986 Extraction of Schottky Diode Parameters from Forward Current-Voltage Characteristics *Appl Phys Lett* **49** 85-7

[20] Kim H, Schuette M, Jung H C, Song J H, Lee J, Lu W and Mabon J C 2006 Passivation effects in Ni/AlGaN/GaN Schottky diodes by annealing *Appl Phys Lett* **89** 053516

[21] Akkaya A, Esmer L, Kantar B B, Cetin H and Ayyildiz E 2014 Effect of thermal annealing on electrical and structural properties of Ni/Au/n-GaN Schottky contacts *Microelectron Eng* **130** 62-8

[22] Akkaya A and Ayyildiz E 2012 Programming with VEE Pro™ for the Capacitance-Voltage (C-V), Capacitance-frequency (C-f) Measurements with Agilent E4980a LCR Meter and

Basic Diode Parameter Calculations Software (SecLaS). In: *29th International Physics Congress,* (Bodrum-Turkey

[23] Akkaya A, Karaaslan T, Dede M, Çetin H and Ayyıldız E 2014 Investigation of temperature dependent electrical properties of Ni/Al0.26Ga0.74N Schottky barrier diodes *Thin Solid Films* **564** 367-74

[24] Song Y P, Vanmeirhaeghe R L, Laflere W H and Cardon F 1986 On the Difference in Apparent Barrier Height as Obtained from Capacitance-Voltage and Current-Voltage-Temperature Measurements on Al/P-Inp Schottky Barriers *Solid State Electron* **29** 633-8

[25] Werner J H and Guttler H H 1991 Barrier Inhomogeneities at Schottky Contacts *J Appl Phys* **69** 1522-33

[26] Sullivan J P, Tung R T, Pinto M R and Graham W R 1991 Electron-Transport of Inhomogeneous Schottky Barriers - a Numerical Study *J Appl Phys* **70** 7403-24

[27] Tekeli Z, Altındal S, Cakmak M, Özçelik S, Caliskan D and Özbay E 2007 The behavior of the I-V-T characteristics of inhomogeneous (Ni/Au)-Al0.3Ga0.7N/AlN/GaN heterostructures at high temperatures *J Appl Phys* **102**

[28] Iucolano F, Roccaforte F, Giannazzo F and Raineri V 2007 Barrier inhomogeneity and electrical properties of Pt/GaN Schottky contacts *J Appl Phys* **102** 113701

[29] Doğan S, Duman S, Gürbulak B, Tüzemen S and Morkoç H 2009 Temperature variation of current–voltage characteristics of Au/Ni/n-GaN Schottky diodes *Physica E* **41** 646-51

[30] Lim W, Jeong J H, Lee J H, Hur S B, Ryu J K, Kim K S, Kim T H, Song S Y, Yang J I and Pearton S J 2010 Temperature dependence of current-voltage characteristics of Ni-AlGaN/GaN Schottky diodes *Appl Phys Lett* **97** 242103

[31] Akkaya A, Esmer L, Karaaslan T, Cetin H and Ayyildiz E 2014 Electrical characterization of Ni/Al0.09Ga0.91N Schottky barrier diodes as a function of temperature *Mat Sci Semicon Proc* **28** 127-34

[32] Anilturk O S and Turan R 2000 Electrical transport at a non-ideal CrSi2-Si junction *Solid State Electron* **44** 41-8

[33] Cetin H and Ayyildiz E 2005 Temperature dependence of electrical parameters of the Au/n-InP Schottky barrier diodes *Semicond Sci Tech* **20** 625-31

[34] Chand S 2004 On the intersecting behaviour of current-voltage characteristics of inhomogeneous Schottky diodes at low temperatures *Semicond Sci Tech* **19** 82-6

[35] Tung R T 1992 Electron-Transport at Metal-Semiconductor Interfaces - General-Theory *Phys Rev B* **45** 13509-23

[36] Tung R T 1991 Electron-Transport of Inhomogeneous Schottky Barriers *Appl Phys Lett* **58** 2821-3

[37] Chand S and Kumar J 1996 On the existence of a distribution of barrier heights in Pd2Si/Si Schottky diodes *J Appl Phys* **80** 288-94

[38] Zhu S Y, Van Meirhaeghe R L, Detavernier C, Cardon F, Ru G P, Qu X P and Li B Z 2000 Barrier height inhomogeneities of epitaxial CoSi2 Schottky contacts on n-Si (100) and (111) *Solid State Electron* **44** 663-71

# Effects of ageing on the electrical characteristics of Zn/ZnS/n-GaAs/In structure

**B. Güzeldir and M. Sağlam**
Atatürk University, 25240, Erzurum,
Turkey,

E-mail: bguzeldir@atauni.edu.tr

**Abstract**. Zn/ZnS/n-GaAs/In structure has been fabricated by the Successive Ionic Layer Adsorption and Reaction (SILAR) method and the influence of the time dependent or ageing on the characteristic parameters are examined. The current-voltage (I-V) of the structure have been measured immediately, 1, 3, 5, 15, 30, 45, 60, 75, 90, 105, 120, 135, 150 and 165 days after fabrication of this structure. The characteristics parameters of this structure such as barrier height, ideality factor, series resistance are calculated from the I-V measurements. It has been seen that the changes of characteristic parameters such as barrier height, ideality factor and series resistance of Zn/ZnS/n-GaAs/In structure have lightly changed with increasing ageing time.

## 1. Introduction

Metal-semiconductor contacts (MS) have been extensively studied on GaAs which is already largely used in industry and laboratories [1-3]. MS contacts are the basis of large number of compound semiconductor electronic devices, including microwave diodes, field-effect transistors (FETs), solar cells, photo detectors. In addition, the thermal stability of metal-semiconductor contacts is of great importance in device technology because they usually have to be subjected to elevated temperatures at some stage of manufacture of the device [3-6].

The SILAR method [7,8] is based on successive adsorption and reaction of species on the substrate surface from aqueous solutions. In the SILAR method, the substrate is immersed into separate cation and anion precursor solutions and rinsed with puried water after each immersion. The SILAR method is suitable for growing thin multilayer structure due to low deposition temperature since diffusion of ions in the thin film is low. Also among various deposition techniques, SILAR method is simple, less expensive and less time consuming for preparation of thin semiconductor chalcogenide films.

In this study Zn/ZnS/n-GaAs/In structure has been fabricated directly on n-GaAs semiconductor by the SILAR method. After the device fabrication, its performance and stability depending on time are important matters in the device manufacturing. Therefore, the purpose of this paper is to characterize how the I-V characteristics parameters such as barrier height, ideality factor, series resistance and so on of Zn/ZnS/n-GaAs/In structure have changed with increasing ageing time. The I-V measurements of the structure have been carried out under laboratory conditions in dark at room temperature.

## 2. Experimental process and data analysis

The Zn/ZnS/n-GaAs/In structure used in this study were fabricated using n-type single crystal n-GaAs wafer with (100) surface orientation and 2.5x1017 cm-3 carrier concentration. n-GaAs wafer was sequentially cleaned with trichloroethylene (CHClCCl2), acetone (CH3COCH3) and methyl alcohol (CH3OH), etched in a sequence of sulfuric acid (H2SO4) and hydrogenperoxide (H2O2) and finally de-ionized water of resistivity of 18MΩ cm for a prolonged time, before the Zn/ZnS/n-GaAs/In structure fabrication process. During to each cleaning step, the wafer was rinsed thoroughly in de-ionized water. After surface cleaning of n-GaAs, high purity ın was coated with at a pressure about10-5 Torr in high vacuum system. To perform the ohmic contact, n-GaAs wafer was annealed at 425 0C for 3min. After ohmic contact was made, the ohmic contact side and the edges of the n-GaAs semiconductor substrate was covered by wax so that the polished and cleaned front side of the semiconductor sample was exposed to the cationic precursor solution employed for SILAR method. To deposit ZnS one SILAR growth cycle involves the following four steps: A well-cleaned glass substrate is immersed in the first reaction vessel containing aqueous cation precursor 0.1M ZnCl2 solution at pH 5.5. After the cation immersion, the substrate is moved to the rinsing vessel where it is washed with purified water. The sulfide ions were adsorbed from an aqueous 0.05 M Na2S solution with pH 12. After anion immersion the substrate was washed as described above, thus the first SILAR growth cycle is finished. The above cycle was repeated and the optimized dipping was 45 dipping times to get enough film thickness. Zn dots with diameter of about 1.0 mm were evaporated on the ZnS thin film in vacuum coating unit at about 10-5 torr. In this way, Zn/ZnS/n-GaAs/In sandwich structure was obtained.

The I-V characteristics of this structure were measured immediately using a HP 4140B picoampermeter and a HP model 4192A LF impedance analyser, respectively, at room temperature in the dark. The measurements were also repeated 1, 3, 5, 15, 30, 45, 60, 75, 90, 105, 120, 135, 150 and 165 days after fabrication of the Zn/ZnS/n-GaAs/In structure in order to observe the effect of the ageing.

## 3. Results and discussion

Figure 1 shows the forward and reverse bias I-V curves measured immediately, 1, 3, 5, 15, 30, 45, 60, 75, 90, 105, 120, 135, 150 and 165 days after fabrication of the Zn/ZnS/n-GaAs/In structure. A standard Schottky barrier theory can explain with I-V curve, where the forward current is due to the injection of majority carriers from the semiconductor into the metal, while the reverse current is due to the thermal emission of electrons over the barrier $\Phi_B$ from metal to the conduction band of the semiconductor and can be given by a simple equation [1]

$$I = I_s \exp\left[ (\frac{qV}{nkT}) - 1 \right] \tag{1}$$

where $I$ is the current density, $V$ the voltage, $q$ the electronic charge, $k$ the Boltzmann constant, n is the ideality factor and $I_s$ is the saturation current. In Schottky theory, $I_s(T)$ is mainly determined by the barrier height $\Phi_b$ and the effective Richardson constant $A^*$:

$$I_s = AA^*T^2 \exp\left( \frac{-q\Phi_b}{kT} \right) \tag{2}$$

and hence

$$\Phi_b = \frac{kT}{q} \ln\left( \frac{AA^*T^2}{I_s} \right) \tag{3}$$

The slope of linear portion of the ln(I-V) curve gives the ideality factor and can be express as below [1]

$$n = \frac{q}{kT} \ln\left( \frac{dV}{d\ln I} \right) \tag{4}$$

The barrier height value of Zn/ZnS/n-GaAs/In structure was calculated with the help of Eq. (2) from the y-axis intercepts of the semilog-forward bias I–V plots, and the value of the ideality factor n was obtained using Eq. (3) from the linear region of these plots.

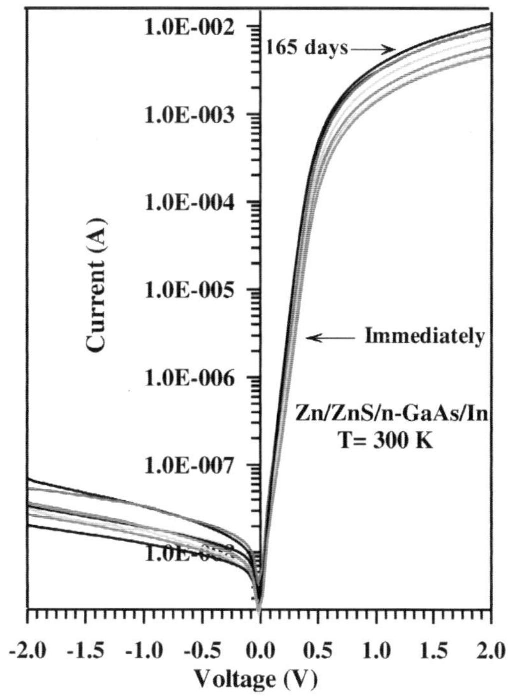


**Figure 1.** The forward and reverse bias current–voltage characteristics of Zn/ZnS/n-GaAs/In structure as a function of ageing time.

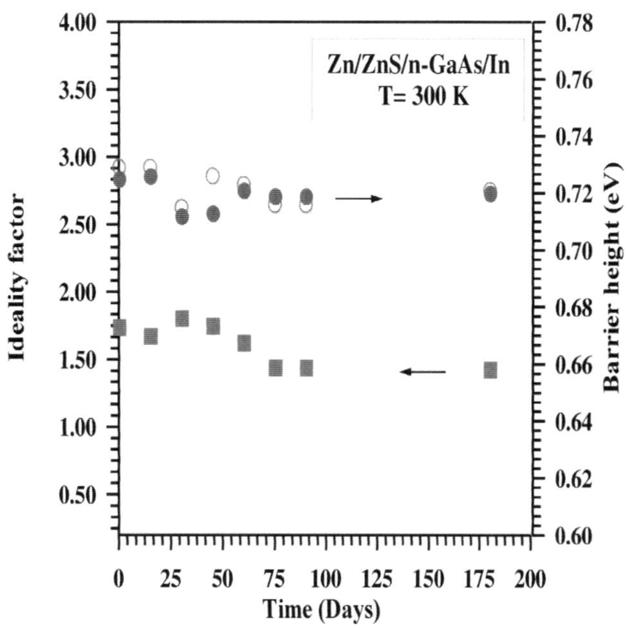


**Figure 2.** The ideality factor and barrier height versus the increasing ageing time of Zn/ZnS/n-GaAs/In structure.

The values of n and $\Phi_b$ parameters versus ageing time are shown in Figure 2. While n decreases with an increase in ageing time, $\Phi_b$ increases, as seen from Figure 2. The ideality factor, *n*, is introduced to take into account the deviation of the experimental I-V data from the ideal TE model. This can be attributed to the interfacial layer of ZnS layer on the n-GaAs surface. For MS contacts it is known that the contact characteristics are controlled by Fermi level pinning due to the interface states. That is, it can be concluded that the barrier height determined from the I-V characteristics is controlled by the interface states in equilibrium with the semiconductor [9].
Norde proposed a method to determine value of the series resistance. The following function has been defined in the modified Norde's method [10]:

$$F(V) = \frac{V}{\gamma} - \frac{kT}{e} In\left(\frac{I(V)}{AA^*T^2}\right) \tag{5}$$

where $\gamma$ is the first integer (dimensionless) greater than *n*. factor, I(V) is current obtained from the I–V curve and the other parameters are described above. From Norde's functions, the effective barrier height and $R_s$ value can be determined as

$$\Phi_b = Fm + \left[\frac{(\gamma - n)}{n}\right]\left[\frac{V_m}{\gamma} - \frac{kT}{e}\right] \tag{6}$$

$$R_s = (\gamma - n)(\frac{kT}{eI_m}) \tag{7}$$

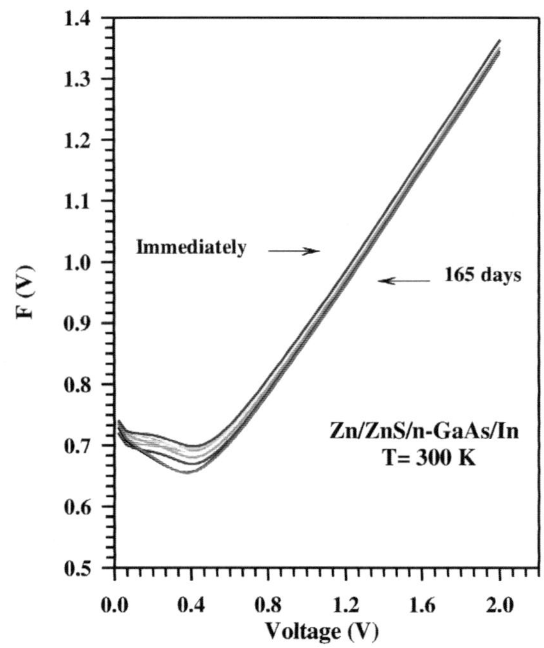


**Figure 3.** Experimental F(V)–V curves of Zn/ZnS/n-GaAs/In sandwich structure with increasing ageing time

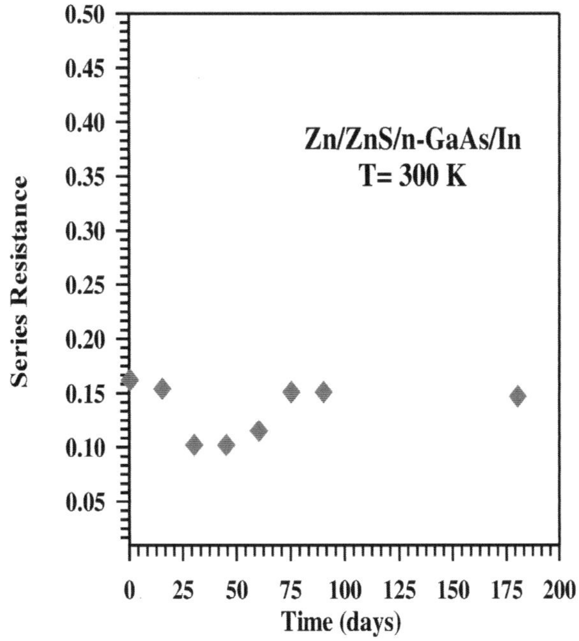


**Figure 4.** Series resistance versus the increasing ageing time of Zn/ZnS/n-GaAs/In structure.

Figure 3 shows the F(V)–V plots of the Zn/ZnS/n-GaAs/In structure as a function of ageing time. From the F(V)–V plots, the some parameters of the Zn/ZnS/n-GaAs/In sandwich structure, ( $\Phi_b$ , Rs) have been determined. It is seen that the values of the series resistance have increased with ageing time in Figure 4 The reason of this increase is that the doping agent in ZnS is gradually replaced by oxygen, which leads to a slow decrease of its conductivity. In addition to the ZnS layer, the time scale of the ageing may partly suggest an ionic activity within the thin film on n-GaAs semiconductor surface and it can be attributed to the migration of charged ions through the thin layer [9].

In this study, Zn/ZnS/n-GaAs/In structure has been fabricated by the Successive Ionic Layer Adsorption and Reaction (SILAR) method. The I-V characteristics of Zn/ZnS/n-GaAs/In structure have been investigated as a function of the ageing time. The characteristic parameters such as barrier height, ideality factor, series resistance of structure are calculated from the I-V measurement. The ageing effect has been attributed to anionic activity within the thin film on n-GaAs.

## References

[1] Rhoderick E H and Williams R H 1988 Metal-Semiconductor
Contacts 2nd edn (Oxford: Clarendon)
[2] Türüt A, Sağlam M 1992 Physica B **179** 285.
[3] Sağlam M, Ates A, Yıldırım M A, Güzeldir B 2010 Current Appl Phys **10** 513.
[4] Tung R T 1992 Phys Rev B 45 13509
[5] Osvald J, Dobrocka E 1996 Semicond Sci Technol **11** 1198.
[6] Osvald J, Horvath Zs J 2004 Appl Surf Sci **234** 349.
[7] Nicolau Y F 1985 Appl Surf Sci 1**061** 22
[8] Sankapal B R, Mane R S, Lokhande C D 2000 Mater Res Bull **35** 177.
[9] Sze S M 1981 Physics of Semiconductor Device second ed Wiley New York p. 245.
[10] Norde H 1979 J Appl Phys **50** 5052.

# Electronic, Optic and Dynamic Properties of Rhombohedral BiTeBr

**M N Secuk[1] and H Akkus[2]**
[1]Yuzuncu Yil University,Van, TURKEY
[2]Yuzuncu Yil University,Van, TURKEY

E-mail: [1]m.nurullahsecuk@yyu.edu.tr
[2]harunakkus@yyu.edu.tr

**Abstract.** Electronic, optic and thermodynamic properties of rhombohedral BiTeBr crystal were investigated under the local density approximation (LDA) using the density functional theory. No spin-orbit interaction (SOI) were taken into account during calculations and band gap of 0.946 eV was found as which is different from experimental results but in good agreement with previous theoretical studies without SOI. For this new type ferroelectric BiTeBr from ground states features lattice parameters, electronic total density of states (TDOS), partial density of states (PDOS) and electronic band structure, from optical properties dielectric function, refraction index, extinction, reflection and absorption coefficients, number of effective valance electrons, volume and surface loss functions, among thermodynamic properties change in Helmholtz free energy, internal energy, entropy and constant volume specific heat as a function of temperature were studied.

## 1. Introduction

There exist many studies on structural, electronic, elastic, dynamic, thermodynamic and optic properties of crystalline structures [1-9]. Ferroelectric crystals are being used in industrial fields extensively due to their dielectric, piezoelectric and pyroelectric properties and especially semiconductors which inherently owns ferroelectric features are being utilized in memory materials, sound sonar detectors, convertors, etc [10, 17]. Among ferroelectric materials $BaTiO_3$ is the first invented one, then SbSI group of $A^{V}B^{VI}C^{VII}$ semiconductors which are extremely sensitive in photoconductivty were found and studied [18]. The need for multi-functional, cheaper materials of high efficiency, caused an increase in interest for new types of ferroelectrics [19-26]. Among the type new ferroelectric semiconducting materials of $A^{V}B^{VI}C^{VII}$ bismuth tellurohalids (BiTeCl, BiTeBr, BiTeI) are of high importance because they own potential of being a part of spintronic studies because of giant Rashba split, as memory materials and they are topological insulators (insulating in the bulk and conducting at the surface), surfaces can be of n or p type semiconductor depending on the outermost layer of surface after the cut[27, 28].

Thus we investigated structural properties of rhombohedral BiTeBr crystal using density functional theory under the local density approximation.

Shevelkov et al [29] found crystalline structure of bismuth tellurobromide as rhombohedral of point group -3m and space group 156 (P-3m1) and with lattice parameters of a= 8.062 and c=12.259 Bohr and unit cell volume of $V_0$= 689.996 $(Bohr)^3$ given Figure 1. We firstly optimized structure of BiTeBr by total energy optimization with respect to cutoff kinetic energy of plane waves ($E_{cut}$) and number grids for k points (ngkpt), then by atomic optimization for lattice parameters, unit cell volume and reduced coordinates and finally by a two steps volume optimization for lattice parameters and unit cell volume with and without optimized reduced coordinates. We found 40 Ha for $E_{cut}$, 12x21x12 for ngkpt, 186 k points and best results we found in volume optimization with optimized reduced coordinates as 634.223 $(Bohr)^3$. Nearest calculated value of volume to the experimental one is of the step of volume optimization by reduced coordinates found in atomic optimization as 634.223 $(Bohr)^3$. So, remaining calculations were done by using results of this step where we optimized lattice parameters as 7.809 and 12.006 Bohr.

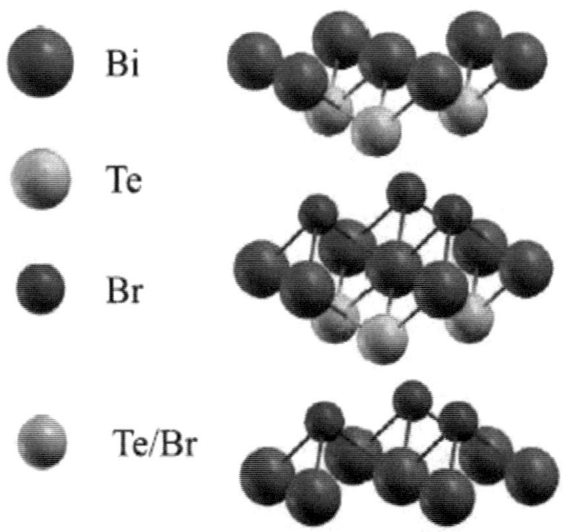


**Figure 1.** Crystal structure of double layer bismuth tellurochloride [43].

## 2. Computational Method

Electronic, optic and thermodynamic properties of rhombohedral BiTeBr crystal were investigated via ABINIT [31] under density functional theory. Calculations were performed within LDA by the FHI98PP self-consistent pseudopotentials [32] with the Ceperley-Alder-Perdew-Wang scheme which considers the exchange-correlation effects [33, 34].

Conjugate gradient minimization method [35] was utilized to solve Kohn-Sham equations [36] and plane augmented waves were used as basis set for electronic wave functions. True valance electrons were chosen as $6s^2 6p^3$ for Bi, $5s^2 5p^4$ for Te and $4s^2 4p^5$ for Br. Structural optimizations were done to a good convergence at 40 Hartree of cutoff energy and 186 k points using 12x12x12 Monkhorst-Pack mesh grid [37] in BiTeBr crystal, but for better results higher values of were used in electronical and optical calculations.

## 3. Electronic Properties

This step was performed without SOI by pseudo potential method within LDA based on density functional theory (DFT). A grid of 24x24x24 was chosen and producing 1872 k points were found for best convergence. BiTeBr crystal has 9 valence bands and additional 9 conduction bands were used for calculations of electronic band structure where Fermi level was set to zero and as it is seen in figure 2 BiTeBr crystal has a direct band gap at high symmetry point of A with a value of 0.946 eV which is near to and better than those calculations done without SOI (1.1 eV) [38] but deviates from experiment because of not considering SOI and Rashba split.

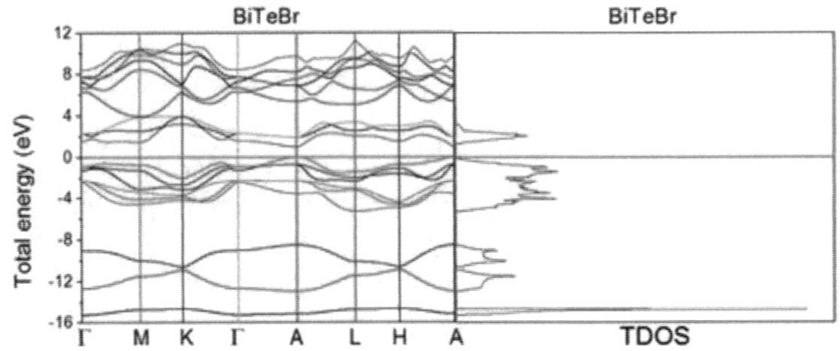


**Figure 2.** Electronic band structure and DOS for BiTeBr with $E_F$ adjusted to 0 eV.

Some of existing experimental and theoretical results on band gap of BiTeBr are given in table 3. The difference between experimental results and calculated ones originates from two main factors as utilizing pseudo-potential method and ignoring SOI. Pseudo-potential method and inherent intractability of density functional theory can cause different estimations of band gaps up to 50 % error, mostly underestimated [39], but sometimes overestimated band gaps can be gathered in calculations [38]. Overestimation in band gap is attributed to the nature of pseudo-potentials used and large spin orbit coupling (SOC) effect of Bi atom [41], which causes Rashba splitting in real crystal and band gap decreases.

**Table 1.** Theoretical and experimental results on band gap of BiTeBr in eV.

| Ref. [30] | Ref. [38] | Present work |
|---|---|---|
| (Exp.) | (Teo.) | (Teo) |
| 0.62 | 1.1 | 0.946 |

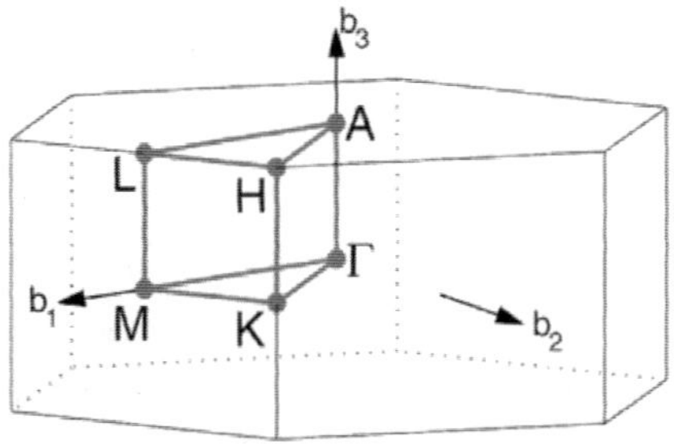


**Figure 3.** High symmetry points and paths in first Brillouin zone in reciprocal space [42].

As it is seen in figure 2 the paths of Γ-M-K-Γ and L-H-A are non-degenerate but the Γ-A path is highly degenerate.

## 4. Dynamic Properties

In bismuth tellurobromide crystal, unit cell contains 3 atoms at zero pressure. In the end of crystallization of BiTeI at zero pressure, Bi atoms are located at points of (0, 0, 0), Te atoms are located at (2/3, 1/3, 0.6928) while Br atoms are at (1/**3, 2/3, 0.2510)** in the reduced coordinates, inside the unit cell.

Figures below show the phonon band structure and phonon density of states for BiTeBr.

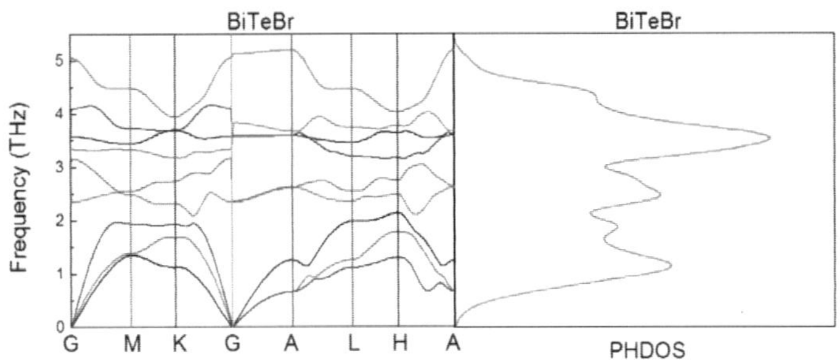


Fig. Phonon band structure and PHDOS

There are 3 atoms per unit cell the crystal, 9 phonon branches totally appear. While three of them are the acoustic branches, the remaining six are optical ones. Degeneracy exists especally between high symmetry points of G-A in Brillouin zone of BiTeBr crystal. As expected, acoustic phonon dispersion curves are linear as a function of *k* for small values of *k*.

Calculated frequency values of LO modes at the center of BZ are 2.37, 3.17 and 3.33 THz, while those of TO modes are at 3.60, 4.08 and 5.05 THz for BiTeBr.

There is a gap between the acoustic and the optic phonon branches. So, BiTeBr is a phononic crystal.

## 5. Optic Properties

20x20x20 M.P. mesh grid with 770 k points for optical calculations of BiTeBr crystal which is an optic crystal exhibiting symmetry of the point group -3m. Therefore the linear frequency dependent dielectric tensor for BiTeBr crystal has two non-zero and independent components in X and Z axes and calculated imaginary ($\varepsilon_1$) and real ($\varepsilon_2$) parts of these components are given in figures 5a and 5b for two axis separately.

(a)

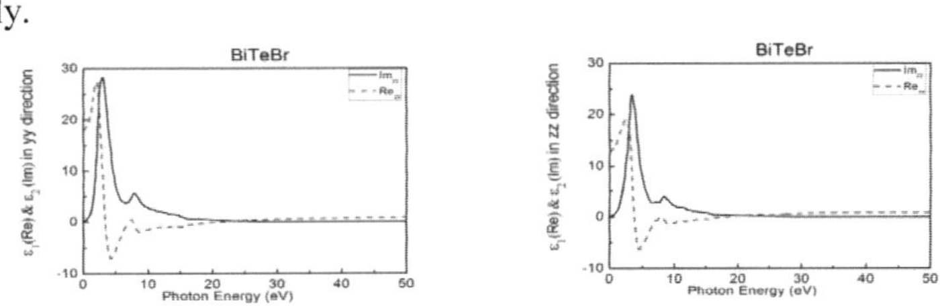


**Figure5.** Imaginary and real parts of dielectric functions of BiTeBr in a) X and b) Z directions.

As seen from these figures, real parts of linear dielectric functions, $\varepsilon_2$, reaches maximum values at 2.07 and 2.25 eV of photon energies for 11 (X) and 33 (Z) directions and the static dielectric constants for X and Z directions are 10.76 and 12.86, respectively.

The main peak values in the imaginary parts of linear dielectric functions, $\varepsilon_1$, (calculated in present work) are at 2.98 and 7.82 eV for X, and 3.34, 7.21 and 8.28 eV for Z directions, respectively. All of these values are at violet and near ultraviolet regions of spectrum.

For X direction, regions of 0-2 eV, 4.30-7.45 eV and above 8.60 eV, for Z direction 0-2.27 eV, 4.32-8.03 and above 9.23 eV are of normal dispersion where $\varepsilon_2$ increases as photon energy increses. Regions of transmition are 0-1.24 eV and above 17 eV, while 1.36-5.60 eV region is where reflection is strong and 5.60-17 eV is the region of absorption for Z axis while same regions for X axis are 0-1.60 eV and above 12 eV, 1.60-6.44 eV and 6.44-12 eV, respectively.

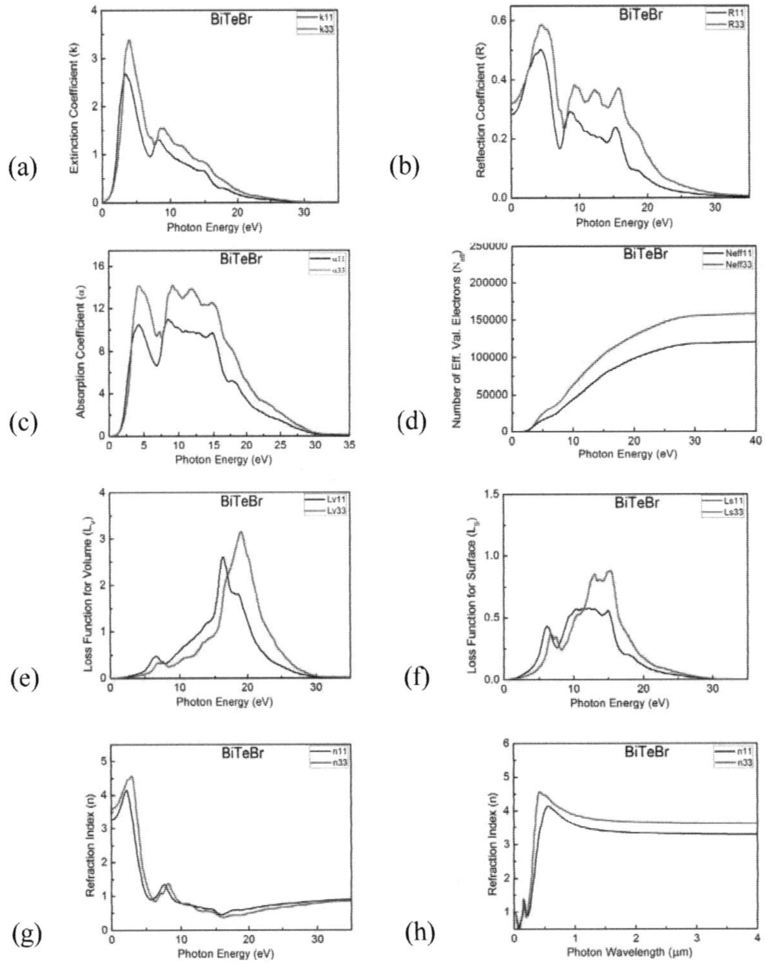


**Figure 6.** The calculated extinction coefficient (k), reflection (R) and absorption coefficients (α), number of effective valance electrons ($N_{eff}$), energy loss functions for both volume and surface, and refractive index (n) functions for BiTeBr in 11 (X) and in 33 (Z) directions.

Variations of some of optical properties as a function of energy (a-g) and of wavelength (h) are given above in figure 6. There are some different results in literature compared to ours which can be due to ignoring SOI and discrepancy of DFT [40].

## 6. Conclusions

Electronic, optic ad dynamic properties of BiTeBr were investigated by using ABINIT [31] code within the local density approximation in the density functional theory. Calculated lattice parameters and experimental ones are in agreement and calculated thermodynamic features are found as expected. Phonon band graph is very similar to the study of Sklyadneva [44] et al.

## Acknowledgments

The calculations reported in this paper were partially done at TUBITAK ULAKBIM, High Performance and Grid Computing Center (TRUBA resources) and Theoretical Physics Lab. of Physics Department at Yuzuncu Yil University.

## References

[1] Alward J F, Fong C Y, El-Batanouny M and Wooten F 1978 Electronic and optical properties of SbSBr, SbSI and SbSeI Solid *State Com.* **25** 307
[2] Edwardson P J, Boyer L L, Newman R L, Fox D H, Hardy J R, Flocken J W, Guenther R A and Mei W 1989 Ferroelectricity in perovskitelike $NaCaF_3$ predicted ab initio *Phys. Rev.* B **39** 9738
[3] Akkus H and Erdinc B 2009 First-principles study of $NaCdF_3$ *Phys. Status Solidi* B **246** 1334
[4] Flocken J W, Guenther R A, Hardyand J R and Boyer L L 1985 First-principles study of structural instabilities in halide-based perovskites: Competition between ferroelectricity and ferroelasticity *Phys. Rev.* B **31** 7252
[5] Bingol S, Erdinc B and Akkus H 2015 Electronic band structure, optical, dynamical and thermodynamic properties of cesium chloride (CsCl) from first-principles *International Journal for Simulation and Multidisciplinary Design Optimization (IJSMDO)* **6**:A7
[6] Gulebaglan S E 2012 The bowing parameters of $Ca_{\chi}Mg_{1-\chi}O$ ternary alloys *Modern Physics Letters* B **26** 1250199–8
[7] Erdinc B, Soyalp F and Akkus H 2011 First-principles investigation of structural, electronic, optical and dynamical properties in CsAu *Cent. Eur. J. Phys.* 9 **(5)** 1315-20
[8] Erdinc B, Akkus H and Goksen K 2010 Electronic and optical properties of GaS: A firstprinciples study *Gazi University Journal of Science (GU J Sci)* **23(4):**413–22
[9] Erdinc B, Aycibin M, Secuk M N, Gulebaglan S E, Doğan E K and Akkus H 2014 Theoretical Study of Rhombohedral $NaCaF_3$ Crystal in the predicted ferroelectric phase *Gazi University Journal of Science GU J Sci)* **27(4):**1093–97
[10] Akkus H and Mamedov A 2006 Ab-initio calculation of band structure and linear optical properties of SbSI in para- and ferroelectric phases *Open Physics* **5(1)** 25–34
[11] Akkus H, Kazempour A, Akbarzadeh H and Mamedov A M 2007 Band structure and optical properties of SbSeI: density-functional calculation *Physica Status Solidi* B **244 (10)** 3673–83
[12] Akkus H, Mamedov A, Kazempour A and Akbarzadeh H 2008 Band structure and optical properties of antimony-sulfobromide: density functional calculation *Open Physics* **6(1)** 64–75
[13] Koc H, Akkus H and Mamedov A M 2012 Band structure and optical properties of BiOCl: Density functional calculation *Gazi University Journal of Science (GU J Sci)* **25(1)** 9–17
[14] Akkus H, Cabuk S and Mamedov A M 2010 Linear and nonlinear optical susceptibilities in some ferroelectrics: Ab-initio calculation *Int. J. Nanoelectronics and Materials* **3** 53–67
[15] Akkus H 2009 Density functional calculation of the electronic structures of some $A^5B^6C^7$-type crystals *International Journal of Modern Physics* B **23** (1) 97–104

[16] Scott J F, Araujo C A and Mcmillan L D 1989 Ferroelectric Memory Applications *IEEE 1989 Ultrasonics Symposium: Proceedings* **Vols 1 and 2** 299–308
[17] Uchino K 2000 *Ferroelectirc Devices (*New York: Marcel Dekker Inc.)
[18] Fatuzzo E, Harbeke G, Merz W J, Nitsche R, Roetschi H and W. Ruppel 1962 Ferroelectricity in SbSI *Phys. Rev.* **127** 2036
[19] Duran D, Erdinc B, Aycibin M and Akkus H 2015 Linear optical properties of ferroelectric semiconductor $Bi_2NbO_5F$ crystal *Ferroelectrics* **486** 25–32
[20] Ota Y, Abe T and Inushima T 2011 Phonon frequencies of ferroelectric semiconductor SbSBr by first principle calculation *Ferroelectrics* **414** 113
[21] Audzijonis A, Zigas L and Kvedaravicius A 2012 The nature of anharmonicity and anomalous piezoelectric properties of ferroelectric SbSBr crystal *Physica* B **407** 774
[22] Akkus H and Mamedov A M 2007 Linear and nonlinear optical properties of SbSI: First-principle calculation *Ferroelectrics* **352** 148
[23] Li J F, Viehland D, Bhalla A and Cross L E 1992 Pyro-optic studies for infrared imaging *J. App. Phys.* **71** 2106
[24] Akkus H and Mamedov A M 2007 Ab initio calculations of the electronic structure and linear optical properties, including self-energy effects, for paraelectric SbSI J. *Phys. Condens. Matter* **19** 116207
[25] Audzijonis A and Sereika R 2010 The thermodynamic functions of SbSBr crystal *Phase Trans.* **83** 389
[26] Dogan E K, Aycibin M, Gulebaglan S E, Secuk M N, Erdinc B and Akkus H 2013 Structural, thermodynamic and phonon properties of SbSI and SbSBr single crystals *Journal of the Korean Physical Society* **63** 2133–37
[27] Sakano M *et al* 2013 Strongly spin-orbit coupled two-dimensional electron gas emerging near the surface of polar semiconductors *Phys. Rev. Lett.* **110** 107204
[28] Zhiyong Z, Cheng Y and Schwingenschlög U 2013 Orbital-dependent Rashba coupling in bulk BiTeCl and BiTeI *New Journal of Physics* **15** 023010
[29] Shevelkov A V, Dikarev E V, Shpanchenko R V and Popovkin B A 1995 Crystal structures of bismuth tellurohalides BiTeX (X = Cl, Br, I) from X-Ray Powder Diffraction Data *J. of Sol. Stat.Chem.* **114** 379–84
[30] Akrap A, Teyssier J, Magrez A, Bugnon P, Berger H, Kuzmenko A B, Marel van der D 2014 Optical properties of BiTeBr and BiTeCl *Phys. Rev.* B **90** 035201
[31] Gonze X *et al* 2002 First-principles computation of material properties: the ABINIT software project *Comput. Mat. Sci.* **25** 478
[32] Fuch M and Scheffler M 1999 Ab initio pseudopotentials for electronic structure calculations of poly-atomic systems using density functional theory *Comput. Phys. Commun.* **119** 67-98
[33] Perdew J P and Wang Y 1992 Accurate and simple analytic representation of the electron-gas correlation energy *Phys. Rev.* B **45** 13244
[34] Perdew J P, Burke K and Ernzerhof M 1996 Generalized gradient approximation made simple *Phys. Rev. Lett.* **77** 3865
[35] Payne M C, Teter M P, Allan D C, Arias T A and Joannopoulos J D 1992 Iterative minimization techniques for ab initio total-energy calculations: molecular dynamics and conjugate gradients *Rev. Mod. Phys.* **64** 1045
[36] Kohn W and Sham L J, 1965 Self-consistent equations including exchange and correlation effects *Phys. Rev.* **140** A1133
[37] Monkhorst H J and Pack J D 1976 Special points for Brillouin-zone integrations *Phys. Rev.* B **13** 5188
[38] Zhou S, Long J and Huang W 2014 Theoretical prediction of the fundamental properties of ternary bismuth tellurohalides *Materials Science in Semiconductor Processing* **27** 605–10
[39] Wang Z M 2014 *$MoS_2$: Materials, Physics, and Devices, Lecture Notes in Nanoscale and Technology* **21** 58

[40] Jacimovic J *et al* 2014 Enhanced low-temperature thermoelectrical properties of BiTeCl grown by topotactic method *Scripta Materialia* **76** 69

[41] Zhiyong Z, Yingchun C and Schwingenschlögl U 2013 Orbital-dependent Rashba coupling in bulk BiTeCl and BiTeI *New Journal of Physics* **15** 023010

[42] http://exciting-code.org/forum/t-222178/k-space-high-symmetry-points

[43] Eremeev S V, Rusinov I P, Nechaev I A and Chulkov E V 2013 Rashba split surface states in BiTeBr *New Journal of Physics* **15** (2013) 075015

[44] Sklyadneva I Y, Heid R, Bohnen K P, Chis V, Volodin V A, Kokh K A, Tereshchenko O E, Echenique P M and Chulkov E V 2012 Lattice dynamics of bismuth tellurohalides *Physical Review* B **86**, 094302

# Effects Of the γ- radiation on the electrical characteristics of the Au/n-Si/Au-Sb Schottky diode

**M. Abdolahpour Salari[1], E. Şenarslan[1], B. Güzeldir[1] and M. Sağlam[1]**

[1]Department of Physics, Faculty of Science, Atatürk University, 25240, Erzurum, Turkey.

E-mail: maryam_salari@yahoo.com

**Abstract.** In this study is investigated of effects of the γ-radiation on current-voltage *(I-V)* and capacitance-voltage *(C-V)* characteristics of Au/n-Si/Au-Sb Schottky diode at room temperature. Initially, the ohmic contact has been made on n-Si crystal with Au-Sb alloy. Then, the rectifier contact is made by evaporation Au metal diameter of about 1.0 mm to the other surface of n-Si in turbo molecular pump at about $10^{-7}$ Torr. The *I–V* measurements of this diode performed by the use of a KEITLEY 487 Picoammeter/Voltage Source and the *C–V* measurements were performed with HP 4192A (50–13 MHz) LF Impedance Analyzer at room temperature and in dark. Then, this diode was subjected to γ-radiation, and I-V and C-V measurements were taken again. Consequently, examines the difference between these two measurements.

## 1. Introduction

The metal–semiconductor (MS) contact in the semiconductor device technology is still investigated and has attracted much attention during recent years [1–3]. The performance and stability of MS is of great importance to the electronic devices. MS contacts have an important role in the development of semiconductor devices due to their applications in various electronic and optoelectronic devices [4]. The parameters which characterize such a contact depend on the method used during the fabrication [5]. The semiconductor crystal surfaces are usually covered with thin layer of native oxide and organic contaminants in the laboratory environment [6,7]. In many cases, the Schottky barrier height on chemically prepared semiconductor surfaces indicates the presence of interfacial layer whose thickness is dependent on exposure time of the semiconductor surface in the chemically cleaned substrates [2,8]. Recently, the studies of the radiation damage of the semiconductor-based devices have increasingly demanded the attention of researchers. Although the impact of the radiation on the performance of the devices has clearly been known as various deformations, there is yet no convenient explanation to all types of crystals or devices. There are two types of basic radiation-damage mechanisms for devices: ionization damage and displacement lattice damage. Ionizations of the material create free charge, which can move in the material. This damage is generally harmless for the device operation. But displacement lattice damage in semiconductor based devices can have a significant impact on their electrical properties, through the creation of stable radiation defects, which have one and/or more levels in the bandgap [9]. Well-known examples for silicon are the creation of group V donor–vacancy pairs or the creation of interstitial [10–13].

In many cases, the charge of the shallow dopants is compensated by the radiation levels, resulting in a lowering of the free carrier density and an increase in resistivity [14]. Therefore, it is important to determine how the Schottky contact parameters give a response to the γ-irradiation, since all semiconductor-based devices communicate one another over the MS contacts in an electronic circuit. The Effects Of the γ- radiation on the I–V and C–V characteristics of Au/n-Si/Au-Sb Schottky contact has been studied and reported in this article.

## 2. Experimental

n- Si wafer pieces of (100) orientation and one side polished were used in this study. Firstly, the wafer was sequentially cleaned with trichloroethylene, acetone and methanol ultrasonically for 3 min respectively and then rinsed in deionized water of 18 MΩ and dried with high purity $N_2$. The Si wafer were chemically cleaned using the RCA cleaning procedure (i.e., 10 min boiling in $NH_3+H_2O_2+6H_2O$ followed by 10 minutes in $HCl+H_2O_2+6H_2O$ at 60°C) before making contacts. Preceding each cleaning step, the samples were rinsed thoroughly in deionized water with ultrasonic vibration for 5 min and were finally dried by high purity nitrogen ($N_2$) atmosphere. After the cleaning process, the wafer is immediately inserted in to the vacuum room. Extensive alloys such as Au–Sb are used for n-Si ohmic contact. We have used Au–Sb alloy for ohmic contacts are made by evaporating on the non-polished side of the n-Si wafer pieces when vacuum was decreased by $10^{-6}$ Torr and then by thermal annealing at 300 °C for 3 min inflowing nitrogen in a quartz tube furnace. One of them was immediately inserted into the evaporation chamber to form the Au Schottky contacts. The metal gate, Au, was then deposited through a mask by thermal evaporation in dots of the shape of approximately 1mm diameter. Schottky contacts were formed when vacuum was decreased by $10^{-6}$ Torr. In this way, the Au/p–Si/Au-Sb Schottky diode was obtained. The I–V and C–V characteristics of the diode were measured using a Keithley 487 Pico-amperometer/voltage source and HP model 4192 ALF impedance analyzer under dark conditions. And then, this sample was subjected to γ-radiation, and then I-V and C-V measurements were taken again.

## 3. Results and discussion

Fig. 1 shows the forward and reverse bias I–V characteristics of the diode before and after irradiation. The effect of γ-radiation is clearly shown. It is seen that after γ-radiation in the diode, leakage current has been decreased. The other effect of irradiation is a reduction in a forward bias current.

According to Thermionic emission theory, forward bias current of a Schottky diode depending on the applied potential given by equations (1). The values of ideality factor (n) and barrier height ($e\Phi_b$) are obtained from these equalities (equations (2) and (3)).

$$I= I_0[exp(eV/nkT)-1] \quad (1)$$

$$n= e/kT\, dV/(d(lnI)) \quad (2)$$

$$e\Phi_b= kTln((AA^*T^2/I_0) \quad (3)$$

Where

$$I_0 = AA^*T^2 exp(-q\Phi_b/kT) \quad (4)$$

is the saturation current, A is the diode area, $A^*$ is the Richardson constant and equals 112 $A/cm^2\,K^2$ for n type Si, T is the temperature in Kelvin, $\Phi_b$ is the effective barrier height at zero bias, where q is the electron charge, V is the applied voltage, k is the Boltzmann constant, n is the ideality factor, and it is determined from the slope of the linear region of the forward bias lnI–V characteristic through the relation (2), where n equals unity for an ideal diode. However, n has usually a value greater than unity [9].

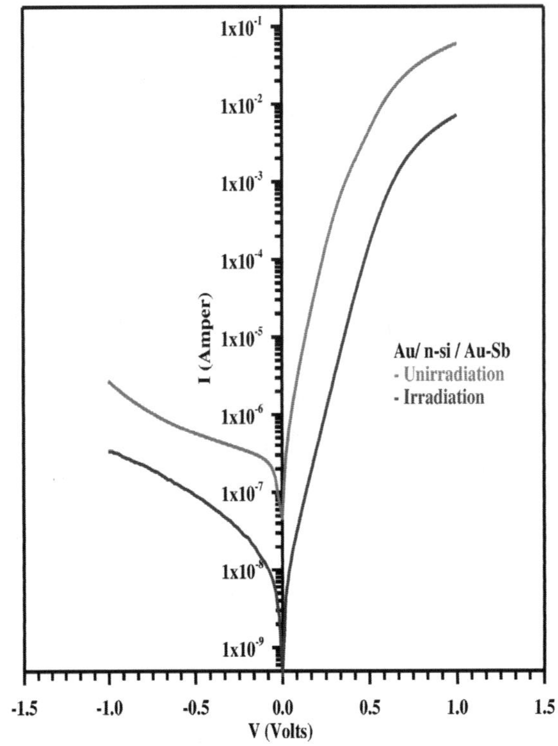


**Figure 1.** The I–V characteristics of the Au/n-Si/Au-Sb Schottky diode before and after irradiation.

The values of the barrier heights for before and after irradiation are calculated as 0.86 and 1.54 eV, respectively. Furthermore, the values of the ideality factors for before and after irradiation are calculated as 1.44 and 1.82, respectively. It is seen that the values of the ideality factor has increased after irradiation. Due to γ- radiation, the defects can be created in the crystal lattice and this causes an increase in the ideality factor. The increase in the barrier height is mainly responsible for the decrease in the reverse bias current of the diode. The radiation-induced degradation observed in the reverse I–V characteristics could be attributed to an increase in the interfacial defect density [15]. An increase in the ideality factor also indicates that for the diode, the current transport mechanism cannot be a thermionic emission type and it may be a tunneling type transport [16].

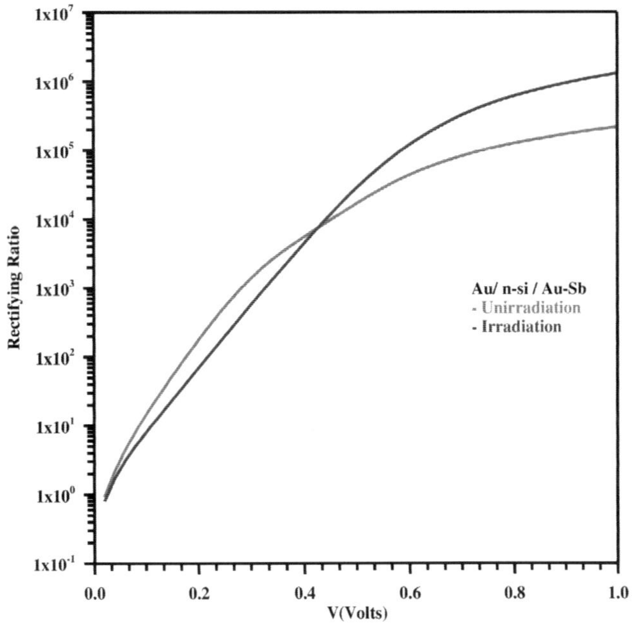


**Figure 2.** The rectifying ratio versus applied voltage characteristics of the Au/n-Si/ Au-Sb Schottky diode before and after γ-radiation.

Figs. 3 and 4 show the typical C–V characteristics for unirradiated and irradiated situations of the diode at various frequencies. It is seen from these figures that the values of the capacitance decrease with the irradiation and increasing frequency [9]. This irradiation effect can be attributed to the change in dielectric constant at the metal semiconductor interface or to the decrease in the net ionized dopant concentration with irradiation [17–19]. Furthermore, the acceptor-like defects can be used to explain the decrease in the C–V properties of the Au/n-Si/Au-Sb diode after γ-radiation.

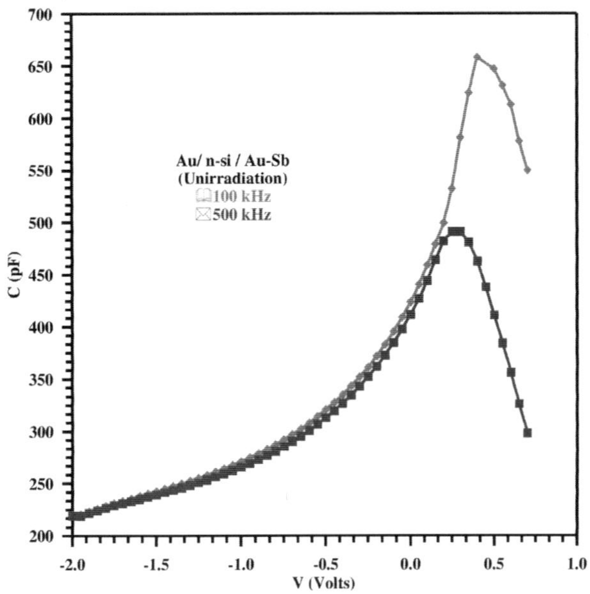


**Figure 3.** The forward and reverse bias C–V characteristics of the Au/n-Si/ Au-Sb Schottky diode before γ-radiation at various frequencies.

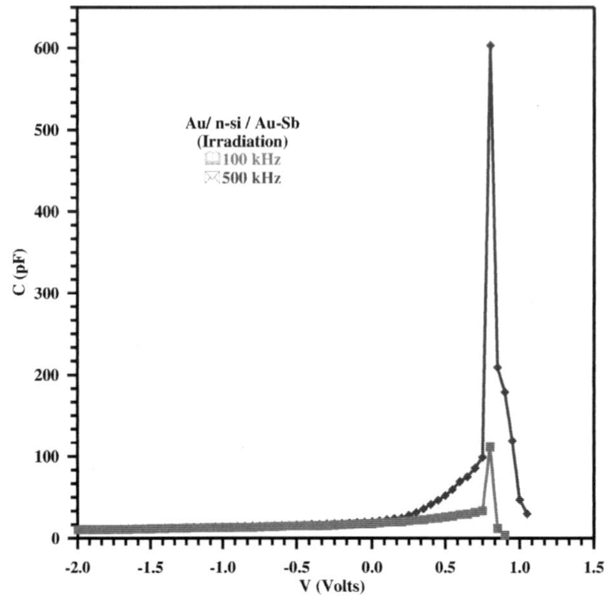


**Figure 4.** The forward and reverse bias C–V characteristics of the Au/n-Si/Au-Sb Schottky diode after γ-radiation at various frequencies.

Fig. 5 depicts the reverse bias $C^2$–V characteristics of the unirradiated diode, at various frequencies. It is seen that the $C^2$–V are linear. This can be explained by the absence of the excess capacitance [9].

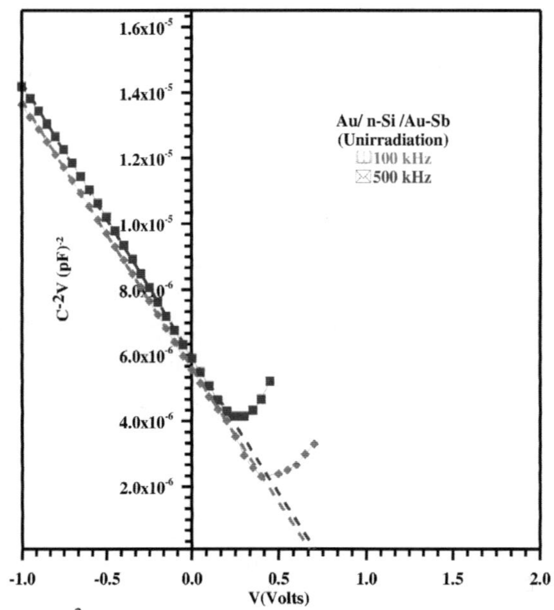


**Figure 5.** The reverse bias $C^{-2}$–V characteristics of the Au/n-Si/Au-Sb diode before γ-radiation at various frequencies.

Fig. 6 shows the reverse bias $C^2$–V characteristics of the irradiated diode at various frequencies. The lateral shift in the $C^2$–V characteristics is due to an increase in diffusion potential, which results in an increase in the barrier height [9].

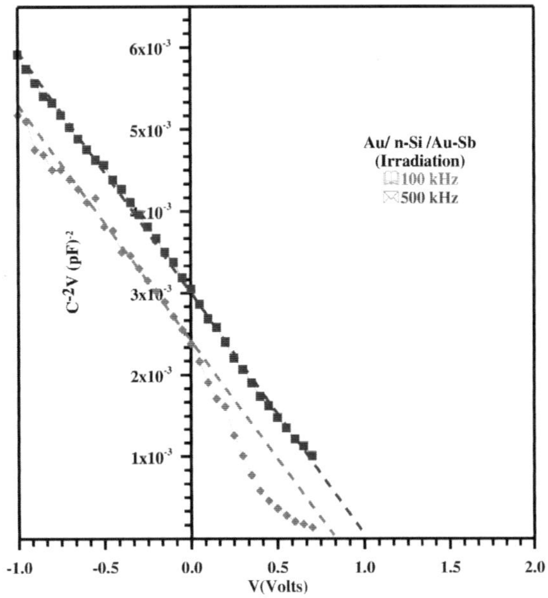


**Figure 6.** The reverse bias $C^{-2}$–V characteristics of the Au/n-Si/Au-Sb diode after γ-radiation at various frequencies.

**Table. 1** The characteristic parameters of the Au/n-Si/Au-Sb diode obtained from I-V and C-V characteristics before and after γ-radiation.

| Radiation situation | *I–V* | | | *C⁻²–V* | | | | |
|---|---|---|---|---|---|---|---|---|
| | *n* | $I_0$ *(A)* | $\Phi_b$ *(eV)* | $V_d$ *(eV)* | | $N_d$ *(cm⁻³)* | $E_f$*(eV)* | $\Phi_b$ *(eV)* |
| Unirradiation | 1.44 | 2.78x10⁻⁷ | 0.68 | 100 kHz | 0.68 | 2.38x10¹⁶ | 0.18 | 0.86 |
| | | | | 500 kHz | 0.71 | 2.33x10¹⁶ | 0.18 | 0.89 |
| Irradiation | 1.82 | 5.54x10⁻⁹ | 0.78 | 100 kHz | 0.85 | 69.37x10⁶ | 0.69 | 1.54 |
| | | | | 500 kHz | 1.03 | 66.98x10⁶ | 0.69 | 1.72 |

## 4. Conclusion

It was found that the electrical characteristics of the Au/n-Si/Au-Sb Schottky diode are very sensitive to γ-radiation. It was found that the values of Schottky barrier height and the ideality factor showed an increase after γ-radiation. The degradation in the Au/n-Si/Au-Sb diode properties may be due to the introduction of radiation-induced interfacial defects (between Au and n-Si), and lattice defects via displacement damage.

## References

[1] S.M. Sze, Physics of Semiconductor Devices, second ed., Wiley, New York, 1981.
[2] E.H. Rhoderick, R.H. Williams, Metal–Semiconductor Contacts, second. ed., Oxford University Press, Oxford, 1988.
[3] J.H. Warner, U. Rau, in: J.F. Luy, et al. (Eds.), Springer Series in Electronics and Photonics, vol. 32, 1994.
[4] A.Umapathi,V.Rajagopal Reddy.,Microelectron.Eng.114(2014) 31–37.
[5] M. Siada, A. Keffousb, S. Mammaa, Y. Belkacemb, H. Menari. 9 May 2004; accepted 9 May 2004.
[6] H C. Card, E.H. Rhoderick, J. Phys. D 4 (1971) 1589.
[7] H.A c¸etinkara, et al. Eur. Phys. J. Appl. Phys. 6 (1999) 89–94.
[8] M. Morita, et al., J. Appl. Phys. 68 (1990) 1272.
[9] E. Uğurel, Ş. Aydoğan, K. Şerifoğlu , A. Turut. Microelectronic Engineering 85 (2008) 2299-2303
[10] A. Mogro-Campero, M.F. Chang, J.L. Benjamin, J. Electrochem. Soc. 135 (1988) 172–176.
[11] E. Ntsoenzok, J.F. Barbot, P. Desgardin, J. Vernois, C. Blanchard, D.B. Isabelle, IEEE Trans. Nucl. Sci. 41 (1994) 1932–1936.
[12] E. Ntsoenzok, P. Desgardin, M. Saillard, J. Vernois, J.F. Barbot, J. Appl. Phys. 79 (1996) 8274–8277.
[13] M. Yamaguchi, S.J. Taylor, M.J. Yang, S. Matsuda, O. Kawasaki, T. Hisamatsu, J. Appl. Phys. 80 (1996) 4916–4920.
[14] C. Claeys, E. Simoen, Radiation Effects in Advanced Semiconductor Materials and Devices, Springer Verlag, Berlin, Heidelberg, Germany, 2002.
[15] G.A. Umana-Membreno, J.M. Dell, G. Parish, B.D. Nener, L. Faraone, U.K. Mishra, IEEE Trans. Electron Dev. 50 (2003) 12.
[16] F. Roccaforte, S. Libertino, V. Raineri, A. Ruggiero, V. Massimino, L. Calcagno, J. Appl. Phys. 99 (2006) 013515.
[17] O. Gullu, F. Demir, F.E. Cimilli, M. Biber, Vacuum 82 (2008) 789.
[18] S. Karatas, A. Turut, Nucl. Instrum. Meth. A 566 (2006) 584.
[19] P. Zukowski, J. Partyka, P. Wegierek, Phys. Stat. Sol. A 159 (1997) 509.

# InGaN thin film deposition on Si(100) and glass substrates by termionic vacuum arc

**E. Erdoğan[1,*], M. Kundakçı[2] and A. Mantarcı[1]**

[1]Department of Physics, Faculty of Art and Science, Muş Alparslan University, Muş, Turkey
[2]Department of Physics, Faculty of Science, Atatürk University, Erzurum, Turkey

E-mail: e.erdogan@alparslan.edu.tr

**Abstract.** Group-III nitride semiconductors covering infrared, visible and ultraviolet spectral range has direct band gaps changing from 0,7 eV (InN) to 3,4eV (GaN). LEDs emit red, blue, green light, ultraviolet (UV) laser diodes (LD), UV light detectors and high power electronic devices are obtained and commercialized based on group-III nitride materials. InGaN semiconductor can be deposited by different techniques such as molecular beam epitaxy (MBE), metal organic chemical vapor deposition (MOCVD). In this study, InGaN thin films were prepared on Si and glass substrates as well as on GaN layer by termionic vacuum arc (TVA) which is a plasma asisted thin film deposition technique. The film was deposited at $10^{-6}$ torr working pressure, 18A filament current. Plasma was produced at 200 V with 0,6A plasma current. The purpose of this research is to investigate the properties of InGaN thin films. X-ray diffraction (XRD) spectrophotometer was used to analyze microstructure of the deposited films. Scanning electon microscopy (SEM) were used for surface morphology characterizations. Compositional analysis was done by energy dispersive X-ray spectroscopy (EDAX).

## 1. Introduction

Semiconductors underlies the today's technology due to their tunable energy band gaps [1]. Elemental semiconductors are formed by same type af atoms, such as; silicon and germanium. Silicon technology has used in the construction of electronic circuits as the substrate for 20 years [2]. These conventional semiconductors are not suitable for fabrication of devices working on short wavelength optoelectronics. Devices based on these classical semiconductors are not tolerant on high temperatures.

III-V group semiconductors, including GaN(Gallium nitride), InN(Indium nitrite), and AlN (Aluminum nitride) are quite attracted attention and were examined separately. Group-III nitride semiconductors cover spectral range of infrared, visible and ultraviolet and they have direct band gap, ranging from 0.7eV (InN) to 6.2eV (AlN) [3]. Figure 1. shows the band gap versus lattice parameter of the group-III nitrides [4]. III-Nitrides have been used for a wide range of applications. Ligth emitting devices (LED), laser diodes, photovoltaics and high power electronic devices have been achieved based on III-Nitrides material systems [5,6]. There are two types of crystal structures of group III-Nitrides: cubic zincblende and hexagonal wurtzite. Both can coexist under various parameters for crsytal growth. Wurtzite structure is thermodinamically stable. InGaN material formed by GaN and InN has also wurtzite structure [7]. Lattice parameters of InGaN compund are given by the linear relationship between the lattice constant of the two compounds (GaN and InN).

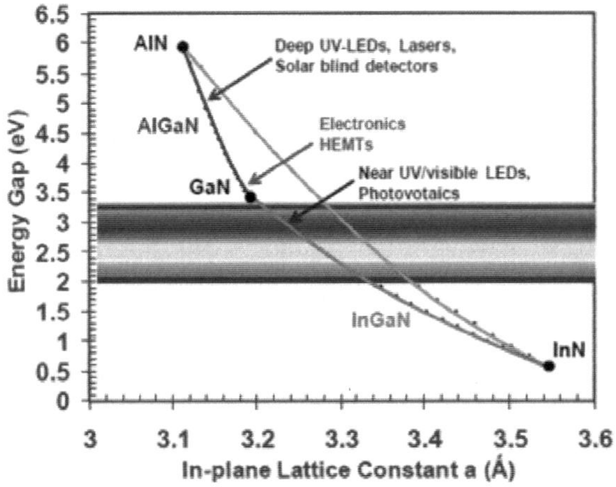


**Figure 1.** Band gap versus lattice constant of group III- Nitride material systems

In this study, InGaN thin films were prepared and deposited on Si and glass substrates as well as on GaN layer. The purpose of this research is to investigate the properties of InGaN thin films. X-ray diffraction (XRD) spectrophotometer was used to analyze microstructure of the deposited films. Scanning electon microscopy (SEM) were used for surface morphology characterizations. Compositional analysis was done by energy dispersive X-ray spectroscopy (EDAX).

## 2. Experimental

InGaN semiconductor can be deposited by different techniques such as molecular beam epitaxy (MBE) [8.9], metal organic chemical vapor deposition (MOCVD) [10-12], sputtering deposition [13-15]. Thermionic Vacuum Arc (TVA) is a thin film deposition technique which generates anodic plasma with fast deposition and within short times [16].

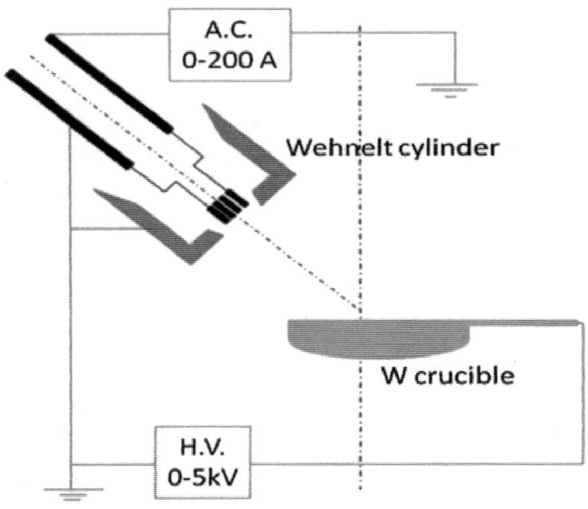


**Figure 2.** Schematic view of the TVA system [16]

Figure 2. represents a schematic view of the deposition technique. The film was deposited at $10^{-6}$ torr working pressure; hence there is no need to buffer gas for deposition at this high vacuum level. Plasma is established between the anode and cathode. The filament current is applied as 18 A to heat the cathode and the voltage is applied to move the electrons from the cathode the anode in order to melt the anode material. In this voltage value, the plasma current is established and is about 0.6 A. Then, melting material will be evaporated in the vacuum chamber. The ions in the chamber are combined with each other. Therefore, detected layers on substrate will be occurred [17].

## 3. Results and Discussion

### *3.1 Structural analysis*

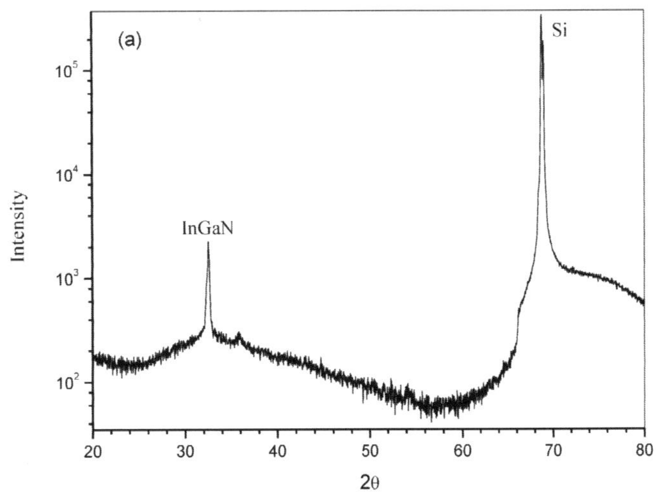


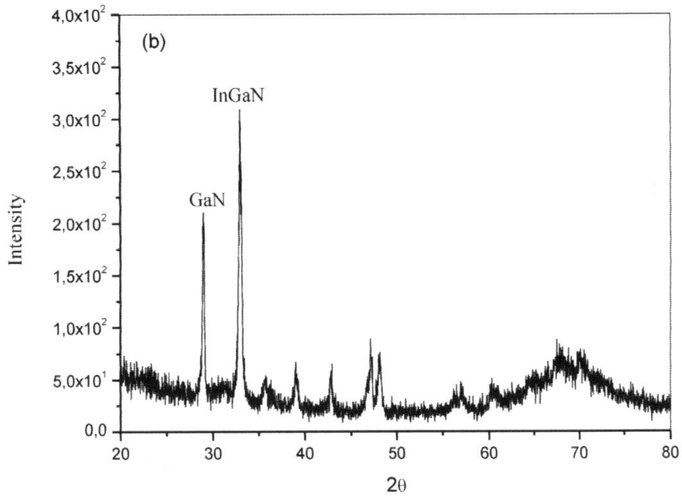

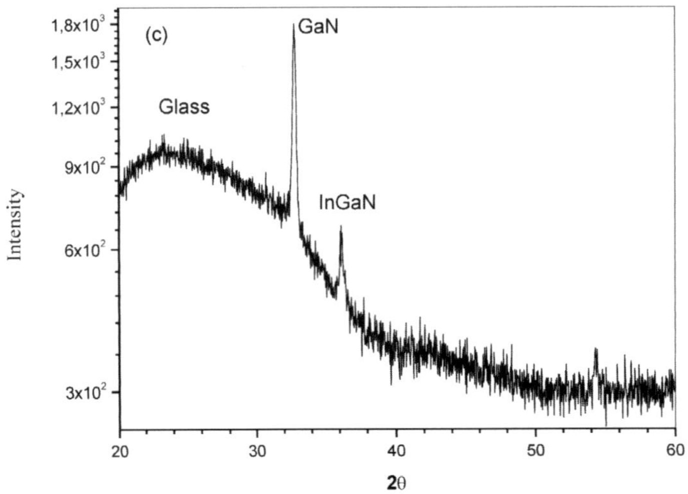


**Figure 3.** XRD spectras of deposited thin films (a) deposited InGaN on Si, (b) deposited GaN and INGaN on Si, (c) deposited GaN and InGaN on glass substrates

Figure 3. exhibits X-ray diffraction patterns of the produced films. It can be seen from Figure 3. (a), the diffraction peaks at 68.75 and 32.5 degrees for silicon and InGaN thin film, respectively. It is also seen from Figure 3. (b), diffraction peaks of deposited GaN and InGaN on silicon substrate are at 28.98 and 32.99 degrees, respectively. Deposited films of GaN and InGaN on glass substrate are also observed from the XRD spectra. It is apparent from Figure. 3 (c), GaN peak was observed at 32.75 degree, while InGaN peak was observed at 36.12 degree. In addition, the glass has a peak, eich are between the 20-25 degress. The GaN and InGaN peaks deposited on glass substrate are shifted to right on the XRD spectra, compared to that of the GaN and InGaN peaks deposited on silicon substrate. The observed diffraction peaks are sharp, indicating good crystalline quality.

### *3.2 Surface Morphological Analysis*

It is very impartant to investigate surface morphology of the films. For studying surface morphology, scanning electron microscopy (SEM, JEOL-JSM5600) was used. SEM images are shown in Figure 4. It is observed that the films are crystalline with uniform dimension of crystals. Figure 4. (a) and (b) indicates the SEM images of the deposited InGaN on silicon substrate for different magnitudes. Also, Figure 4. (c) shows the SEM image of the GaN and InGaN films on silicon substrate. As seen in image, InGaN thin film has a granular structure. Deposited films are dense, smooth, well adhered to the substrate and continue without any holes and cracks.

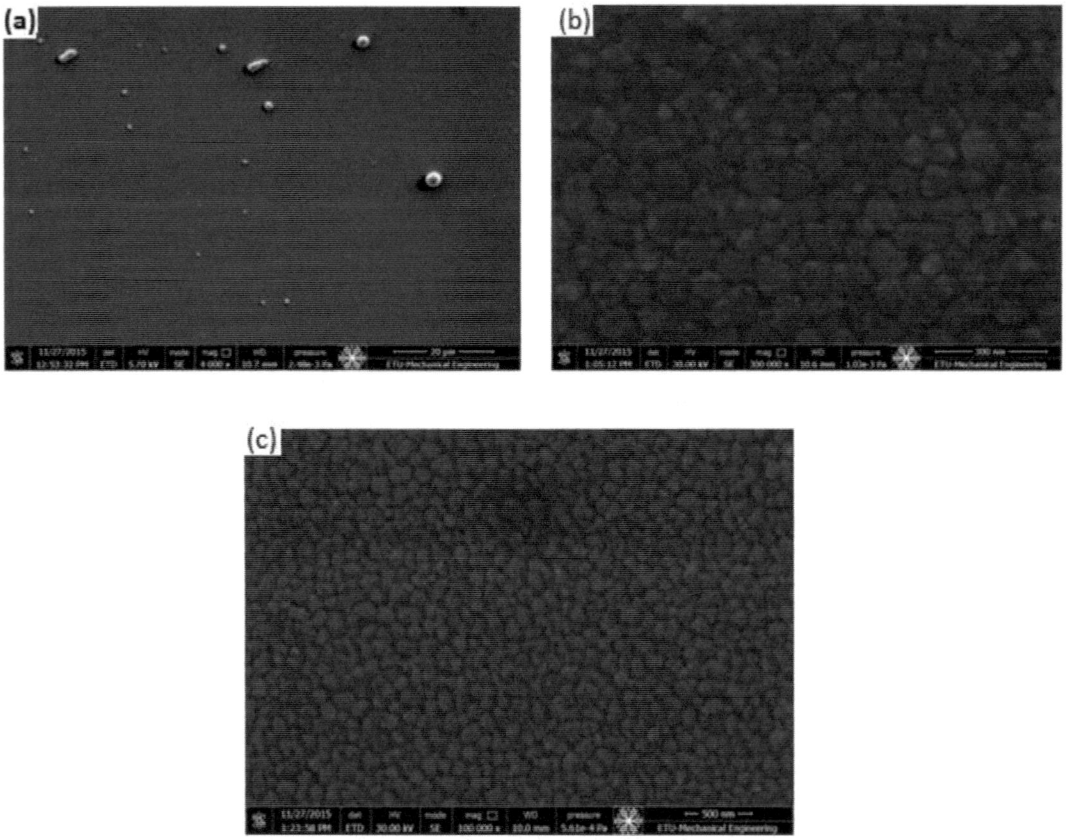

**Figure 4.** SEM images of deposited films spectras of deposited thin films (a) and (b) deposited InGaN on Si for different magnitudes, (c) deposited GaN and InGaN on Si substrates

### *3.3 Chemical Microanalysis*

**Table 1.** EDX analysis results of the GaN and InGaN deposited films on Glass

| Element | Weight ratio % | Atomic raito % |
|---|---|---|
| **Nitrogen** | 1.22 | 1.91 |
| **Oxygen** | 42.58 | 58.29 |
| **Silicon** | 48.90 | 38.13 |
| **Indium** | 5.07 | 0.97 |
| **Gallium** | 2.23 | 0.70 |

Energy dispersive X-ray spectroscopy (EDX) is a chemical microanalysis technique. Table-1 shows the weight and atomic ratio of elements which found in the films and substrates for deposited GaN and InGaN films on glass.

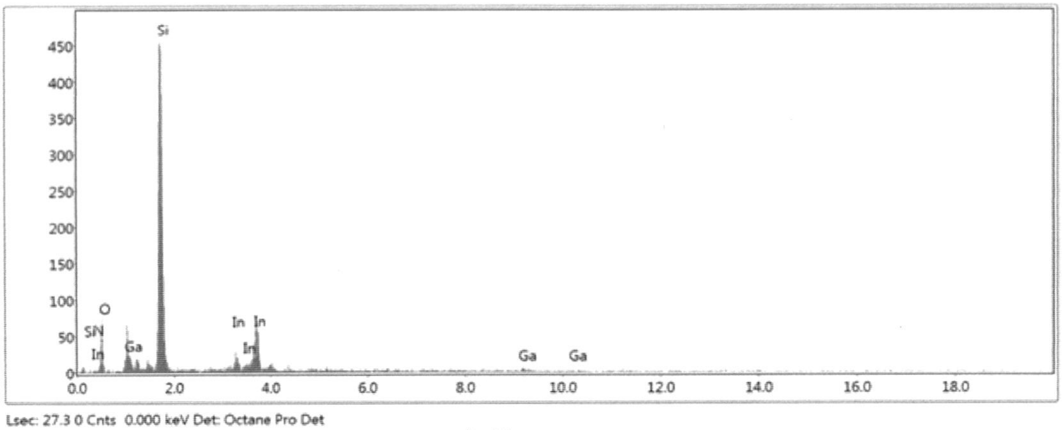


**Figure 5.** EDX spectrum of the deposited GaN and InGaN films on glass substrate

Figure 5. shows the EDX spectrum of GaN and InGan thin films grown on glass substrate. The elemental analysis results confirmed the presence of the Si, Ga, In, N and O in the film.

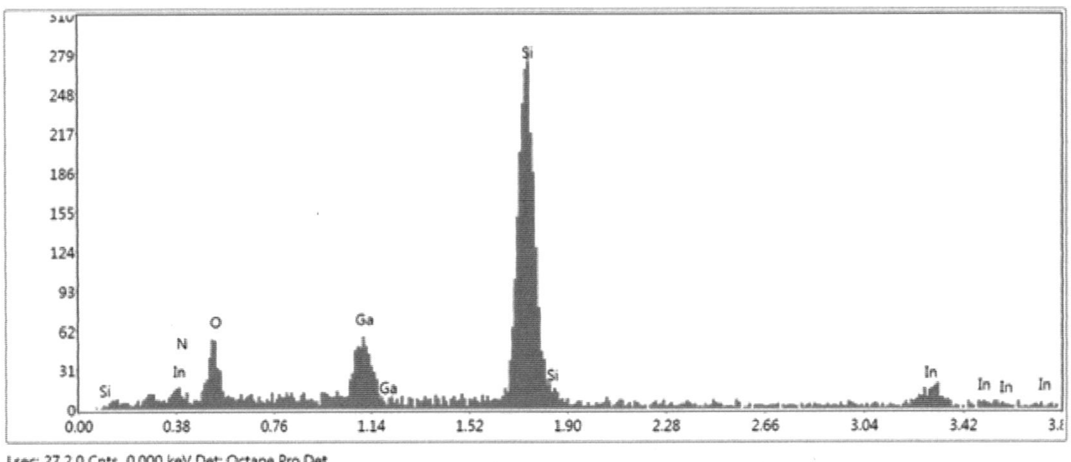


**Figure 6.** EDX spectrum of the deposited InGaN film on silicon substrate

Figure 6. indicates the EDX spectrum ogf the InGaN thin film on silicon substrate. As can be seen from Figure 6, presence of the Si, Ga. N and In atoms were detected by the EDX analysis.

## 4. Conclusion

InGaN thin films were succesfully deposited on silicon substrate and also deposited on glass substrate with GaN film using TVA technique. The main advantage of this deposition technique is its high deposition rate without any loss in the quality of the thin film. The material properties of the produced thin film were investigated. XRD peaks have been observed that the films have sharp diffraction peaks and good crystalline quality. SEM analysis of the produced films show that the thin films have grunalar structure and they are dense, homogenous and continue without any cracks and holes. EDX analysis are also investigated. From the EDX spectrum, presence of the atoms in the film is confirmed. The results show that TVA is suitable for deposition of thin films on silicon and glass substrates.

## References

[1] De Walle V and Gilbert C 2012 ed. Wide-band-gap Semiconductors. Elsevier
[2] William O, Herring R B, and. Hunt L P 2007 Handbook of semiconductor silicon technology. Crest Publishing House
[3] Morkoç H 2009 Handbook of Nitride Semiconductors and Devices, Materials Properties, Physics and Growth, John Wiley & Sons
[4] Yu E T 2003 III-Nitride semiconductors: Applications and devices, ser. Optoelectronic properties of semiconductors and superlattices **16** pp 161–192.
[5] Patrick M 2010 LED for lighting applications. **134** (John Wiley & Sons)
[6] Ezgi D and Medjdoub F 2015 Gallium Nitride (GaN): Physics, Devices, and Technology **47** p 227 (7 In GaN-Based Solar Cells)
[7] Shur M S and Gaska 2010 R IEEE Trans. Electron Devices **57** 12
[8] Godlewski M et al 2000 Atomic layer epitaxy of ZnO for applications in molecular beam epitaxy growth of GaN and InGaN Elektronika: konstrukcje, technologie, zastosowania **41** 11 pp 49-52.
[9] Akio Y et al 2010 Recent advances in InN-based solar cells: status and challenges in InGaN and InAlN solar cells Pphysica status solidi (c) **7** 5 pp 1309-1316.
[10] Seung K C et al 2008 Structural and optical properties of InGaN/GaN single quantum well grown via MOCVD Electronic Materials Letters 4 2 pp 67-70
[11] Sheng-Rui J et al 2008 Mechanical properties of InGaN thin films deposited by metal-organic chemical vapor deposition Materials Chemistry and Physics 109 2 pp 360-364.
[12] Chunlan M et al 2005 Growth and characterization of InGaN blue LED structure on Si (111) by MOCVD Journal of crystal growth **285** 3 pp 312-317.
[13] Cheng-Che L et al 2015 The effect of temperature on the growth and properties of green light-emitting $In_{0.5}Ga_{0.5}N$ films prepared by reactive sputtering with single cermet target Materials Science in Semiconductor Processing **29** pp 170-175.
[14] Qixin G et al 2010 Properties of InGaN Films Grown by Reactive Sputtering Japanese Journal of Applied Physics **49** 8R 081203
[15] Wang J et al 2013 Low-temperature growth of $In_xGa_{1-x}N$ films by radio-frequency magnetron sputtering Applied Surface Science **265** pp 399-404.
[16] Suat P et al 2015 GaN thin film deposition on glass and PET substrates by thermionic vacuum arc (TVA) Materials Chemistry and Physics **159** pp 1-5.
[17] Soner Ö et al 2015 The influence of voltage applied between the electrodes on optical and morphological properties of the InGaN thin films grown by thermionic vacuum arc Scanning

# Structural and Morphological Properties of Al doped ZnO Nanoparticles

**A Akdağ[1], H F Budak[1], M Yılmaz[2], A Efe[1], M Büyükaydın[1],M Can[1],G Turgut[3] and E Sönmez[1,4]**

[1]Department of Nanoscience&Nanoengineering, Graduate School of Natural and Applied Sciences, Ataturk University, Erzurum,25240, Turkey
[2]Department of Metallurgical and Materials Engineering, A.C. Faculty of Engineering, Necmettin Erbakan University, Seydişehir/Konya, 42370, Turkey
[3]Department of Physics, Faculty of Science, Erzurum Technical University, 25240, Erzurum, Turkey
[4]Department of Physics Education, K.K. Education Faculty, Atatürk University, Erzurum, 25240, Turkey

E-mail: aakdag09@gmail.com

**Abstract.** Zinc oxide nanoparticles have a wide area of use because of their unique properties such as catalytic, electrical, and optical properties and low cost. Since the suitable additive materials can be changed the electrical and optical properties of zinc oxide, the demand of the industrial commercial area to the zinc oxide increased. In this study, Al doped ZnO nanoparticles produced by using the methods of reduction thought having materials of the $Zn(NO)_3$, $AlCl_3$ and NaOH. XRD, SEM and EDS used for making analyzing of structural and dimensional of particles. The analyses show that the large amount of the $Al3^+$ ions did nut substitute with $Zn2^+$ successfully with the reduction method. XRD and EDS results confirm this situation.

## 1. Introduction

Zinc oxide (ZnO) is a material that widely studied for years on account of the fact that it exhibits semiconducting and piezoelectric dual properties. At the same time ZnO growth process is already working several decades [1, 2]. As well as a variety of unique properties such as including good transparency, high electron mobility, wide bandgap, and strong room-temperature luminescence it is the one of the promising materials for optoelectric applications such as transparent transistors and LEDs [3–6 ]. Otherwise, ZnO is a key technological material which is the control of the electrical conductivity, for example, p-type contribution is focused on the development of technologies for energy efficiency [7]. ZnO nanoparticles which different from those of their bulk materials because of the fact that quantum confinements make their properties unique so, ZnO is one of the most important semiconductor for nanostructure applications. ZnO can be doped different metal atoms for enhancement of their properties. There are some difficulties during the growth of the nanocrystals doped ZnO material.The reason is the limited amount of crystalline material growth techniques. The so-called hydrothermal method with high quality crystal doped ZnO bulk crystal is produced. However, this method is not suitable for producing single crystal doped. Because it is difficult to achieve simultaneous control of pH and impurity resolution [8].

Undoped ZnO shows n-type conductivity because of the existence of native defects [9–11]. ZnO has high electrical resistivity, but this disadvantage can be reduced by the doping with group III ions ($B^{3+}$, $Al^{3+}$, $Ga^{3+}$, and $In^{3+}$) [9–16]. These ions ensure extra electrons and improve ZnO optical, electrical, thermal, and magnetic properties. $Al^{3+}$ is most preferred dopant element due to its small ionic radius and low material cost. The substitution of $Zn^{2+}$ ions with $Al^{3+}$ in ZnO lattice causes that the increase of electrical conductivity due to the increase of charge carriers [17, 18]. This situation is suitable for fabrication of transparent electrodes in solar cells, gas sensors and also for ultrasonic oscillators [19]. Doping, however, creates disorder locations in the crystal structure. It causes that the mobility of charge carriers is affected by scattering. Therefore doped ZnO particles have to be pure phase for good electrical conducting.[20].

Al-doped ZnO (AZO) nanopowders use transparent electrode due to it is transparent in the visible region [21, 22]. Nowadays various methods have been reported for preparing AZO nanopowders such as sol–gel [23], spray pyrolysis [21], precipitation [24], and hydrothermal processes [25].

The aim of the present work is to achieve that the Al3+ ions doped to ZnO nanoparticles using reduction method so to obtain AZO nanoparticles.

## 2. Experimental procedure

AZO nanoparticles were synthesized using the reduction method. The materials used in the study include: $Zn(NO_3)_2.6H_2O$ and NaOH (Sigma Aldrich %99).Firstly, 0.1 M $Zn(NO_3)_2.6H_2O$ and concentration to be 2% by weight of deionized water using 100 ml $AlCl_3$ solution were prepared. The magnetic stirrer is thoroughly homogenizing the solution was stirred for half an hour. 150 ml 1 M NaOH solution per minute to mix 1.5 ml, continuously stirred on a magnetic stirrer at $60^0$C $Zn(NO_3)_2.6H_2O$ and $AlCl_3$ were added to the mixture dropwise. The resulting solution whether centrifuge at 4000 rpm for 5 times at 5 minute intervals, the residue obtained was washed with deionized water after each application. Finally synthesized Al:ZnO nanoparticles are dried for 4 hours at $80^0$C. The process is shown schematically in Scheme 1.

The crystal structure of the samples was analyzed using RIGAKU Part D/Max 2200 x-ray diffraction (XRD) system (Rigaku/Smartlab) with Cu Kα irradiation (λ=1.5418 Å) operated at 40 kV and 30 mA at the room temperature. The surface morphology of Al doped ZnO nanoparticles was investigated with a scanning electron microscope (SEM) (FEI Inspect 550). Furthermore, the ratio of Al as weighting in the Al-ZnO nanoparticles was investigated by Energy Dispersive X- ray spectrometer (EDS) analyzer.

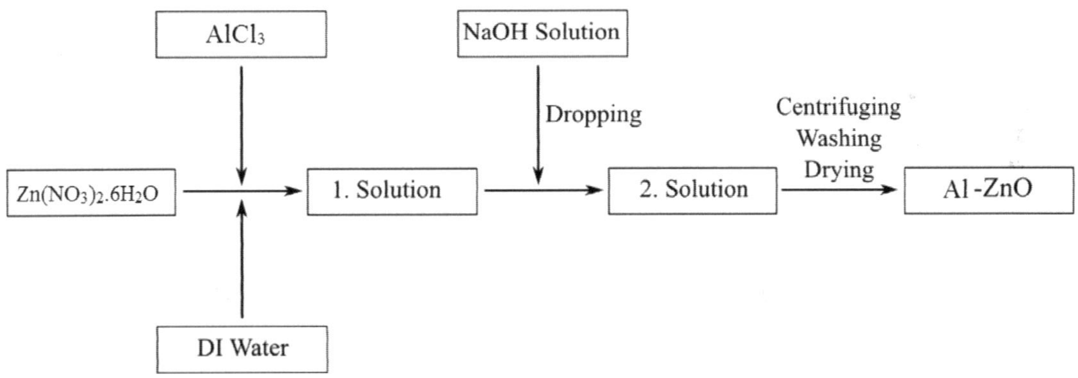


**Scheme 1.**

## 3. Results and Discussion

### 3.1 X-ray diffraction results

As mentioned in the experimental section AZO nanoparticles are investigated by XRD measurements as shown in Figure 1.

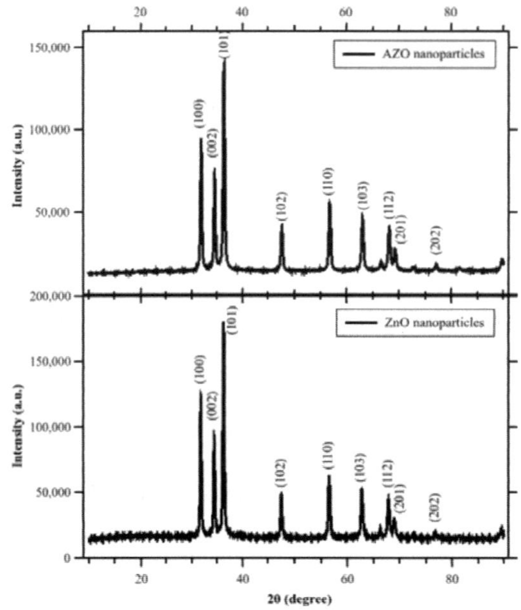


**Figure 1.** XRD analysis of Al-doped ZnO powder

XRD pattern of Al doped ZnO sample belongs to that crystal structure of pure and AZO nanoparticles are polycrystalline hegzagonal wurtzite structure with (101) orientation at 36.35 degree. The a, b, and c parameter of the pure and AZO nanoparticles are refined by Cohen's Method of Least Squares using XRD data. The a, b, and c parameters of pure ZnO are found 3.253 Å, 3.253 Å, and 5.213 Å respectively. These values consistent with the literature [26]. Similarly a, b, and c parameters of AZO nanoparticles are found 3.245 Å, 3.245 Å, and 5.202 Å respectively. Changing of a, b, and c paremeter of AZO shows that the doping process is successful. Otherwise, c/a parameters of pure and AZO nanoparticles are 1.602 and 1.603 respectively. The c/a parameters of samples very little changed. This result can interpret as a unchanged of wurtzite structure of nanoparticles. Preferred orientation of AZO is (101). The size of particles of crystal in the form of powder (D) and microstrain (ε) values over all peaks are determined by the Scherrer equations [24] as shown below.

$$D = \frac{0.9\lambda}{\beta cos\theta}$$

$$\varepsilon = \frac{\beta}{4tan\theta}$$

In these equations, β is full width at half of the peak maximum (FWHM) and calculated by Analysis of Variance (ANOVA), λ is the wavelength of the using X-ray and θ is the peak position which known as Bragg's angle. Calculated size of particles of nanocrystals and microstrain values of nanoparticles are given in Table 1.

**Table 1.** Structural parameters which calculated XRD patterns of pure and AZO nanoparticles.

| Pure ZnO nanoparticle | | | | | AZO nanoparticles | | | | |
|---|---|---|---|---|---|---|---|---|---|
| a,b (Å) | c (Å) | c/a | D (nm) | ε (x10$^{-3}$) | a,b (Å) | c (Å) | c/a | D (nm) | ε (x10$^{-3}$) |
| 3.253 | 5.213 | 1.602 | 30.21 | 1.95 | 3.245 | 5.202 | 1.603 | 25.48 | 2.31 |

Although XRD patterns of ZnO and AZO nanoparticles are same, their lattice parameters varried a little. At the same time nanoparticle size was decreased. Because atomic radius of Al atoms is slightly smaller than the atomic radius of Zn atoms and the covalent bond length of Al–O is estimated to be shorter than Zn–O, the length of the c-axis is shorten in the AZO nanoparticles. These factors result in a decrease in the unit cell dimensions of AZO. In addition, the position of the (002) zincite peak did not shift Al-doping concentration at in comparison with an undoped zincite powder. As a result small change in the crystal size of doped nanostructures shows that the substitution is not desired amount.

### 3.1 SEM and EDS results

**Figure 2.** SEM images of AZO nanoparticles.

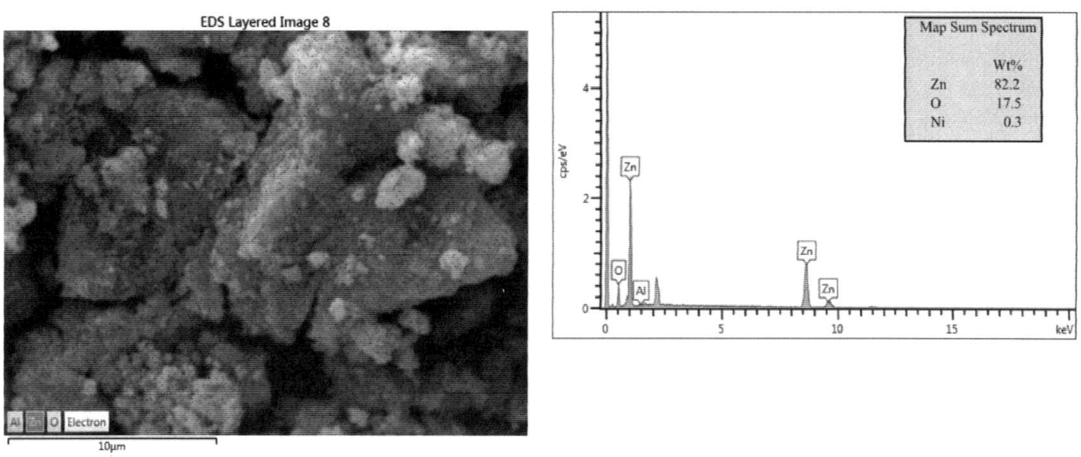


**Figure 3.** SEM images and EDS results on the nanoparticles with 2 wt.% AZO

The scanning electron microscopy (SEM) images of AZO nanoparticles given in Figure 2. The composition of the Al doped ZnO nanoparticles was determined by energy dispersive X-ray spectroscopy (EDS). EDS spectra of Al doped ZnO nanoparticles and the composition of elements in ZnO structure which viewing area are given in Figure 3.

SEM images show that the AZO nanoparticles have spherical shape. EDS spectra clearly confirm the existence of Zn, Al and O elements in the Al doped ZnO nanoparticles. But Al concentration is very low according to the other elements. This is evidence of that the large amount of the Al atoms did not enter the ZnO nanostructure. This situation compatible with XRD results.

## 4. Conclusion

In this study, AZO nanoparticles were synthesized using the reduction method and the success of doping is determined by XRD and SEM/EDS analyses. XRD measurements showed that Doping ZnO with aluminum reduced the particle size. The powder structure, shape and composition of weight of atoms were investigated by SEM/EDS. It was found that AZO nanocrystals have spherical shape. EDS analysis shows that the large amount of the $Al3^{+}$ ions did nut substitute with $Zn2^{+}$ succesfully.

## References

[1] R.A. Laudise, A.A. Ballman "Hydrothermal synthesis of zinc oxide and zinc sulfide" J. Phys. Chem. 64 (1960) 688.
[2] L.N. Demianets, D.V. Kostomarov "Mechanism of zinc oxide single crystal growth under hydrothermal conditions" ANN CHIM-SC. 26(1) (2001) 193.
[3] P.F. Carcia, R.S. McLean, M.H. Reilly, G. Nunes "Transparent ZnO thin-film transistor fabricated by rf magnetron sputtering" Appl. Phys. Lett. 82 (2003) 1117.
[4] A. Ohtomo, S. Takagi, K. Tamura, T. Makino, Y. Segawa, H. Koinuma, M. Kawasaki "Photo-Irresponsive Thin-Film Transistor with $Mg_xZn_{1-x}O$ Channel" Jpn. J. Appl. Phys. 45 (2006) Part 2, N.24-28, L694.
[5] J.-C. Sun, H.-W. Liang, J.-Z. Zhao, J.-M. Bian, Q.-J. Feng, L.-Z. Hu, H.-Q. Zhang, X.-P. Liang, Y.-M. Luo, G.-T. Du "Ultraviolet electroluminescence from n-ZnO:Ga/p-ZnO:N homojunction device on sapphire substrate with p-type ZnO:N layer formed by annealing in $N_2O$ plasma ambient" Chem. Phys. Lett. 460 (2008) 548.
[6] S.-J. Jiao, Z.-Z. Zhang, Y.-M. Lu, D.-Z. Shen, B. Yao, J.-Z. Zhang, B.-H. Li, D.-X. Zhao, X.-W. Fan, Z.-K. Tang "ZnO p-n junction light-emitting diodes fabricated on sapphire substrates" Appl. Phys. Lett. 88 (2006) 031911.
[7] A. Tsukazaki, A. Ohtomo, T. Onuma, M. Ohtani, T. Makino, M. Sumiya, K. Ohtani, S.F. Chichibu, S. Fuke, Y. Segawa, H. Ohno, H. Koinuma, M. Kawasaki "Repeated temperature modulation epitaxy for p-type doping and light-emitting diode based on ZnO" Nat. Mater. 4 (2005) 42.
[8] W. Lin, D. Chen, J. Zhang, Z. Lin, J. Huang, W. Li, Y. Wang, F. Huang "Hydrothermal Growth of ZnO Single Crystals with High Carrier Mobility" Cryst. Growth Des., 9 (10) (2009) pp 4378.
[9] B. Claflin, D. C. Look, S. J. Park, and G. Cantwell, "Persistent n-type photoconductivity in p-type ZnO," Journal of Crystal Growth, vol. 287, no. 1 (2006) pp. 16–22.
[10] A. Janotti and C. G. van de Walle, "Native point defects in ZnO," Physical Review B, vol. 76, no. 16 (2007) pp. 1–22.
[11] A. Janotti and C. G. van de Walle, "New insights into the role of native point defects in ZnO," Journal of Crystal Growth, vol. 287, no. 1 (2006) pp. 58–65.
[12] P. Jood, R. J. Mehta, Y. Zhang et al., "Al-doped zinc oxide nanocomposites with enhanced thermoelectric properties," Nano Letters, vol. 11, no. 10 (2011) pp. 4337–4342
[13] A. Favier, D. Muñoz, S. Martín de Nicolás, and P.-J. Ribeyron, "Boron-doped zinc oxide layers grown by metal-organic CVD for silicon heterojunction solar cells applications," Solar Energy Materials and Solar Cells, vol. 95, no. 4 (2011) pp. 1057–1061.

[14] V. Bhosle, J. T. Prater, F. Yang, D. Burk, S. R. Forrest, and J. Narayan, "Gallium-doped zinc oxide films as transparent electrodes for organic solar cell applications," Journal of Applied Physics, vol. 102, no. 2 (2007) pp. 1–5.

[15] P. S. Venkatesh, V. Ramakrishnan, and K. Jeganathan, "Vertically aligned indium doped zinc oxide nanorods for the application of nanostructured anodes by radio frequency magnetron sputtering," CrystEngComm, vol. 14, no. 11 (2012) pp. 3907–3914.

[16] J. M. Lin, Y. Z. Zhang, Z. Z. Ye et al., "Nb-doped ZnO transparent conducting films fabricated by pulsed laser deposition," Applied Surface Science, vol. 255, no. 13-14 (2009) pp. 6460–6463.

[17] Ü. Özgür, Y. I. Alivov, C. Liu et al., "A comprehensive review of ZnO materials and devices," Journal of Applied Physics, vol. 98, no. 4 (2005) pp. 1–103.

[18] T. Minami, "Transparent conducting oxide semiconductors for transparent electrodes," Semiconductor Science and Technology, vol. 20, no. 4 (2005) pp. S35–S44.

[19] Hong-ming Zhou, Dan-qing Yi, Zhi-ming Yu, Lai-rong Xiao, Jian Li "Preparation of aluminum doped zinc oxide films and the study of their microstructure, electrical and optical properties" Thin Solid Films, 515 (2007) pp. 6909–6914.

[20] Ç. Kiliç and A. Zunger, "Origins of coexistence of conductivity and transparency in SnO2," Physical Review Letters, vol. 88, no. 9 (2002) pp. 1–4.

[21] A. Verma, F. Khan, D. Kar, B. C. Chakravarty, S. N. Singh and M. Husain "Sol–gel derived aluminum doped zinc oxide for application as anti-reflection coating in terrestrial silicon solar cells" Thin Solid Films 518 (2010) pp.2649.

[22] E. Hamarberg, A. P. Schwab and C. Feldmann "Microwave-assisted polyol synthesis of aluminium- and indium-doped ZnO nanocrystals" J. Colloid. Interface Sci. 334 (2009) pp.29.

[23] H. Serier, M. Gaudon and M. Menetrier "Al-doped ZnO powdered materials, Al Solubility limit and IR absorption properties" Solid State Sci. 11 (2009) pp.1192-1197.

[24] S. F. Du, Y. J. Tian, H. D. Liu, J. Liu and Y. F. Chen "Calcination effects on the properties of gallium-doped zinc oxide powders" J. Am. Ceram. Soc. 89 (2006) 2440-2443.

[25] W. H. Zhang, W. D. Zhang and J. F. Zhou "Solvent thermal synthesis and gas-sensing properties of Fe-doped ZnO" J. Mater. Sci. 45(2009) pp.209-241.

[26] H. Schulz,, K.H., Thiemann "Structure parameters and polarity of the wurtzite type compounds Sic—2H and ZnO" Solid State Commun., 32 (1979) pp.783-785.

# Growth conditions effects on the $H_2$ and $CO_2$ gas sensing properties of Indium Tin Oxide

**S. Isik[1], O. Coban[1], C. Shafai[2], S. Tuzemen[1] and E. Gur[1]**

[1]Department of Physics, Faculty of Science, Ataturk University, Erzurum, Turkey.
[2]Department of Electrical and Computer Engineering, University of Manitoba, Winnipeg, Canada.

E-mail: emregur@atauni.edu.tr

**Abstract**. Indium Tin Oxide (ITO) thin films are transparent conducting wide bandgap oxide. In this study investigated optical, structural and morphological properties of sputtered ITO thin films using X-ray diffraction spectroscopy (XRD), Scanning electron microscopy (SEM), Energy Dispersive Spectroscopy (EDX) and optical absorption techniques. These measurements revealed that the oxygen gas percentage present in the ITO film deposited by RF magnetron sputter deposition showed systematic variation of its band gap, crystal orientation, growth rate, figure of merit (FOM) and dominant XRD peaks. All the thin films deposited at room temperature (RT). Once characterization of the films carried out, $H_2$ and $CO_2$ resistive gas sensors fabricated by depositing the ITO film on top of aluminium interdigitated contacts/electrode (IDE), that fabricated following lithography and etching processes. These devices showed reasonable sensitivity for pure $H_2$ and $CO_2$ at elevated temperature. A correlation found between the thin film properties of the ITO and its sensing capability for $H_2$ and $CO_2$, which these gases are important in many fields such as automotive, energy, biological and health-related applications.

## 1. Introduction

Gas sensor devices play mandatory role detecting, order less, toxic and explosive gases because of higher requirement and uses in day-to-day life from last few decades. Those gases extensively applied in industries such as transportation, industrial processes and monitoring environmental pollution and domestic safety [1]. For instance, Hydrogen ($H_2$) gas is widely used as fuel source in clean-energy transportations and power generation applications [3]. On the other hand, $CO_2$ concentrations are required in indoor spacing, agriculture field and bio-related processes [4]. Consequence to supply the demand of gas sensors powerful electrical sensing systems, high sensitivity, durable type of sensors are required to provide comprehensive detection of external physical parameters while maintaining easy of fabrication, simple operation, low production cost and miniaturization [1,2]. Especially gas sensors should design to achieve surface-to-volume ratios for better performance [5]. Recently, the indium tin oxide thin (ITO) film can be seen remarkable attention in gas sensors designing. Importantly ITO is a n-type semiconductor with a wider bandgap (3.5–4.06 eV). Furthermore ITO shows superior electrical conductivity, very high transmittance and high optical transparency. ITO also called as transparent conductive oxide (TCO) and its properties process dependent [6]. In this study resistive based simple ITO thin film sensitive layer prototype $H_2$, $CO_2$ sensor successfully fabricated and tested employing RF-magnetron sputtering the method, which characterized by high purity, low substrate temperature, good interfacial adhesion, high thickness uniformity, homogeneity qualities [7].

## 2. Experimental

### *2.1. Sample information and growth process*

First Indium tin oxide (ITO) thin films between 500nm-800nm thickness deposited on soda lime glass (SLG) substrate and interdigitated electrodes (IDE) patterned SLG substrate by reactive magnetron RF sputtering technique. To realize the deposition the used target was in 14 cm diameter ITO metal with high purity (ITO - 99.99%; 90% $In_2O_3$ and 10% $SnO_2$), in pure Argon and Ar/$O_2$ plasma. Furthermore, surface of the ITO thin film optimized by changing $O_2$ partial pressure that has impact on crystallographic structure and opto-electrical properties of the thin films and on sputtering chamber conditions. Prior to this process typical contaminants removed from the substrate. All of experimental conditions that grown ITO thin films tabulated in Table 1. Especially, throughout the study 2"x3" soda lime glass substrate used for thin film deposition and sensor device fabrication. In the middle of the experiment, Argon and $O_2$ percentage controlled to be form ITO with respect to the sputter total pressure.

Second, all of used chemicals in this study were commercially available products applied without further processing. Prior to the deposition of the thin films, the substrates cleaned to remove particles and contaminants on the surface. These cleaning procedures conducted two different ways for substrate utilized for thin film and substrate for device design. For the first case, the substrates washed in acetone for three minutes to remove organic impurities before rinsed in isopropanol (IPA) and methanol. Later de-ionized (DI) water used to soak the substrates from the solvents. After that substrates blown dry in nitrogen ($N_2$). All sample prepared for the thin film deposition cleaned by degassing one minute and 60-minute ultrasonic bath in the DI water then rinsed with acetone, followed by cleaning IPA, DI water and drying with nitrogen gas blow off. Soon after, all samples dried on the hot plate at 110°C for the 30 minutes duration. Nitrogen blow off dry and IPA solution used for clean sample holders tools and chucks whenever required. Then to second case that substrate used for device fabrication thoroughly cleansed by immersing in piranha solutions ($H_2SO_4$ - 98% and $H_2O_2$ -30% 4:1) in volume after following with ultrasonic cleansing for 10 min. Initially the substrates washed in acetone for three minutes to isopropanol alcohol and deionized (DI) water then dried using dry nitrogen flow. Cleaning process removed surface particles, any impurities organic contaminants, and native oxides and substrate converted to surfaces to be hydrophilic.

Third, in this study three types of patterns and three different sizes (¼ mm, ½ mm, and 1mm) selected to design for image of gas sensor device. Our focus was to give geometrically good surface volume ratio for the sensor device to enhance resistance. Later using computer aid L-edit standard CAD tool supported software from Taner EDA© designed the photo mask. A photo mask is a piece of glass that uses to transfer the design to a material layer that wish to pattern for processing. The physical properties of each mask fabricated using 5" x 5" soda lime glass 0.09" thick Chrome coated (approx. 70nm thick). Further the polarity of mask decided (Inverted polygons=chrome vs. Non-Inverted polygons=glass) for go along with 504 positive PR. As a next step, aluminum coated substrate is held on a spinner chuck by vacuum and resist is coated to uniform thickness by spin coating (3000-6000 rpm for 15-30 seconds). Typically 80-100, p=resist solids content in percent w=spinner rotational speed in rpm/1000. To proceed lithography 500nm Al sputtered thin film 2"x3" glass substrate moisturized by DI water. Soft baker tested for SLG surface temperature before bake state for SLG as it designed for Si substrate. Soft baker adjusted temperature 110°C for 2 minutes for the glass substrate and because of Al cooling very fast. Reasonable amount of 504 PR spread over Al deposited glass substrate by the spine coating. Resist 50/50 500 RAMO for 10 seconds, Spread 100 for 10 seconds and spine 300 for 30 seconds. The thickness measured of Al thin film after deposition that varied from 200nm to 900nm. More over photo resist thickness varied from 1.700μm to 3.1 μm. Later prebake (soft bake) used to evaporate the coating solvent and to densify the resist after spin coating. Consequent a narrow time-temperature window needed to achieve proper line width control. Further 504 PR coated sample glass kept under 210°C in soft bake hot plate for 2 minutes. The glass moisture by DI water and dried by $N_2$ gas blow. Then 504 PR coated glass kept on mask aligner holder and

Journal of Physics: Conference Series **707** (2016) 012021 doi:10.1088/1742-6596/707/1/012021

mounted the photo mask top of the glass holder then exposed 5 seconds for UV light. Once done that, developer arranged to the 10-second pre wet, 20 second develop, 10 seconds dwell, 20-second rinse and 20-second spine. Usually post bake used to stabilize and harden the developed photoresist prior to processing steps that will resist transferred pattern. Later the lithographed glass kept in hard baker (post bake) at least for 30 minutes at 110°C after we confirmed the patterns transfer in to the glass as expected. In this process, etch back method followed to etch unwanted Al film. So Al thin film cleaned out UV exposed areas of the glass substrate using the Al etch solution which heated 70°C, 138 Rpm rotating and on the 134 heated hot plate for two (2) minutes in the acid wet deck cleansed by DI water immediately and followed drying by $N_2$ gas. After etching, the Al of the well-dried glass substrate removed the PR by acetone in lithography wet-deck.

**Table 1.** Chamber Set up Conditions for ITO.

| | |
|---|---|
| $P_{Base}$ | $2.8x10^{-6}$Torr |
| $P_{Total}$ | $9.7x10^{-3}$ to $9.9x10^{-3}$Torr |
| Power | 300W |
| Target (d) | 14cm |
| Offset | Ar – 2.1 , $O_2$ - 1.7 |
| Time | 15 minutes for all Samples |

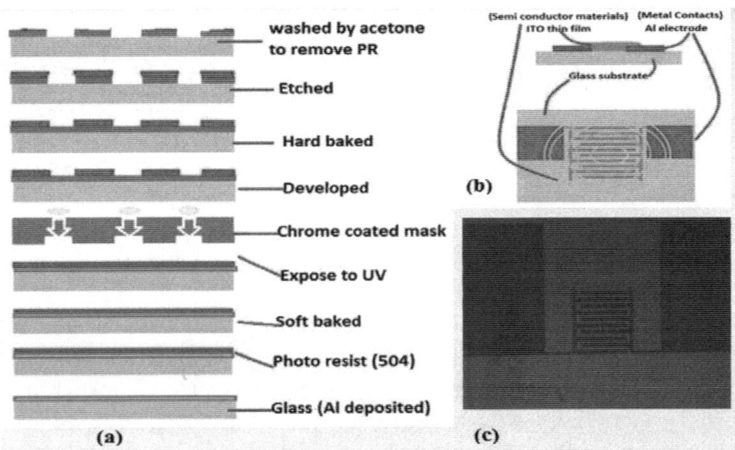


**Figure 1. (a)** Lithography process for ITO gas sensors; **(b)** Schematic drawing ITO deposited thin film device & cross sectional view; **(c)** Physical ITO deposited device under microscope.

**Table 2.** Bandgap variation with $P(O_2)$ %.

| $P(O_2)$ % | Bandgap |
|---|---|
| 2.7 | 3.68 |
| 6.9 | 3.79 |
| 25.0 | 4.13 |

### *2.2. ITO thin film sensor measurements*
All gas sensitivity measurements and performance evaluation of flexible Hydrogen and $CO_2$ Sensors carried out by a noncommercial available prototype gas measuring system (Figure 2). Before taking sensor measurement, the sensor Interdigitated Electrode/Contacts (IDE) tested for ohmic behavior at room temperature (RT), elevated temperatures, and then confirmed by I-V measurements. Once obtain expected ohmic behavior gas sensing evaluation performed by allowing inert and target gases into the chamber in alternative manner, and the target gas content adjusted by MFC inlet flow rate of 400 sccm. During the experiment, the substrate temperature was 200 °C, external bias voltage was 1.0 V supplied by a Keithley 487 picoammeter/voltage source. Each Hydrogen and $CO_2$ sensor fixed in a sealed chamber that flowed with hydrogen gas, $CO_2$ gas or Argon. The two probing electrodes of each sensor were mounted on a standard probe holder on an Al heater plate and connected to a computer controlled Keithley 487 which integrated by LabVIEW program. All measurements carried out at RT and elevated temperatures (30 $^0$C - 200 °C).

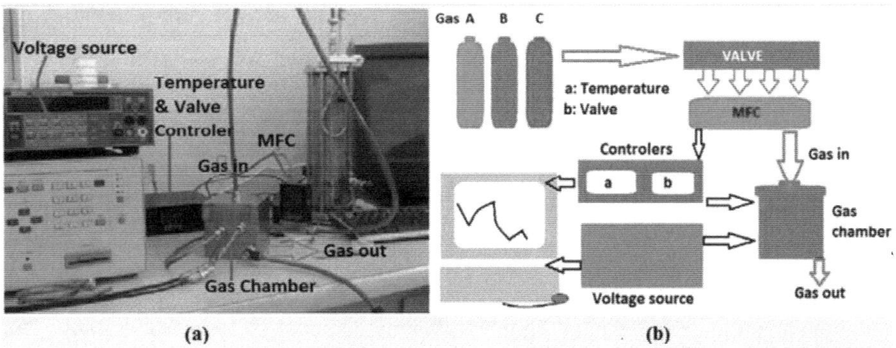


**Figure 2. (a)** Real time gas measurement system; **(b)** Schematic drawing of the gas delivery system.

Sensitivity, response time and recover time analysis and measurement calculation of gas sensors carried out according to method in the Figure 6. Response was determined using $R = R_a/R_g$ and response time and recovery time of the sensors defined as the time taken for the resistance to change by 90% on exposure to the target gas and Ar respectively. The response and recovery times defined as the time required for a change of the resistance to reach 90% of the equilibrium value after injecting and that for removing the detected gas, respectively by inlet and outlet. When air in standard cubic centimeter per minute (sccm) flow level target gas were flowed through the sensor element, the corresponding steady state resistances of the sensor in air ($R_{air}$) and in the air - gas target ($R_{gas}$) were recorded, respectively. Thus the sensor response (S) for oxidizing gas for instance CO is defined as the ratio of $R_{gas}/R_{air}$, while the response for reducing gas ( $H_2$) is defined as the ratio of $R_{air}/R_{gas}$. In all cases, flow of Ar (99.999%) and $H_2$ (99.9999%) was used to fill the chamber and the $H_2$ concentration was controlled using gas flow. Initial stabilization, prior to the measurements, was obtained by flowing 100% $H_2$ until the electrical current at a constant voltage became stable.

### *2.3. ITO thin film Analysis (XRD, SEM, EDAX and Absorption Techniques)*
The surface morphology of the ITO thin films grown by sputtering in this study was observed by SEM (Scanning Electronic Microscope - QUANTA 450 FEG:FEI Company branded) and data of XRD (X-Ray Diffraction) were taken in a computer controlled diffractometer and with X-ray generator Cu Kα radiation from an X-ray tube with normal focus X-ray diffraction (PANALYTICAL/Empyrean

branded XRD). Furthermore optical absorption/transmission measurements of the films measured in UV-visible-near infrared region by using a Perkin Elmer UV-VIS 2 Lambda spectrometer, which is available for wavelengths in the range of 200 nm to 1100.

## 3. Results and Discussion

### *3.1. XRD, SEM, EDAX and Absorption Analysis*

First, the Figure 3 (c) shows SEM image of RF sputtered ITO thin films deposited in $Ar/O_2$ plasma chamber that exhibit nano-sized grains. All those ITO thin film deposited in $O_2$ plasma chamber given rod like and more compact square shape grains. Examination and comparisons of these images clearly reveals that significant surface morphology induced by depositing with and without oxygen of ITO thin film resulted by sputtering technique.

Second in order to investigate behavior observed in the ITO sensor, we performed XRD analyses. The Figure 3 (a) shows that crystal formation of ITO thin film surface significantly affected by plasma chamber. The peak positions and full-width-at-half-maximum (FWHM) values of the diffraction peaks influenced ITO confirms increasing oxygen partial pressure and dominant peak found to be (400). Furthermore, the XRD results show the thin films generally poly crystals & multiple peaks. More over EDAX results confirmed the thin film composition for comparison the outcome.

Third Figure 4 (a) shows absorption measurements of ITO thin films grown at rich $O_2$ partial pressure that the causes to shift in the absorption spectrum of the ITO thin films. Absorption measurements plotted reference to the SLG substrate. Hence, the absorption spectrum of the SLG glass sample has shown to confirm the observed absorptions due to the grown thin films. Therefore plots that shift in the absorption spectrum in the range of 250 nm while the ITO films show the absorption spectra increase between 350-400 nm. More over the transmittance values are above 90% that gives lowest possible electrical resistivity with the optimized highest transparency helps sensitivity (FOM) of ITO thin film according to the values of Table 3. FOM also calculated and included in Table 3, which gave higher results that suitable for TCO gas sensors. Furthermore, we calculated bandgap with respect to $O_2$ partial pressure that increases $O_2$ given in plasma chamber. Figure 4(c) shows the absorption coefficient versus incoming light energy variation of the samples and the calculations of the bandgap from the square of the absorption coefficient. Figure 4(c) absorption coefficient and 4(d) Square of absorption coefficient vs energy graph for ITO (2.7% $O_2$) shows the fit to find the bandgap value of the thin film. Figure 4 (d) graph shows red line - fit made on the plot to calculate bandgap. The band gap for pure ITO 2.7% $O_2$ was Eg=3.68 eV obtained as in Table 2.

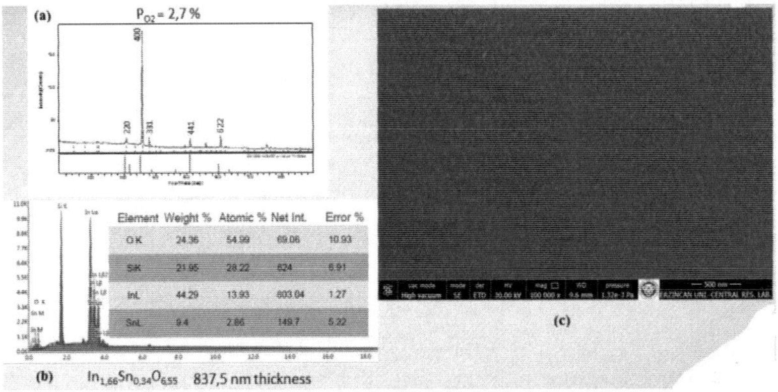


**Figure 3.(a)** XRD; **(b)** EDAX image & Composition analysis; **(c)** SEM image for P ($O_2$)=2.7%

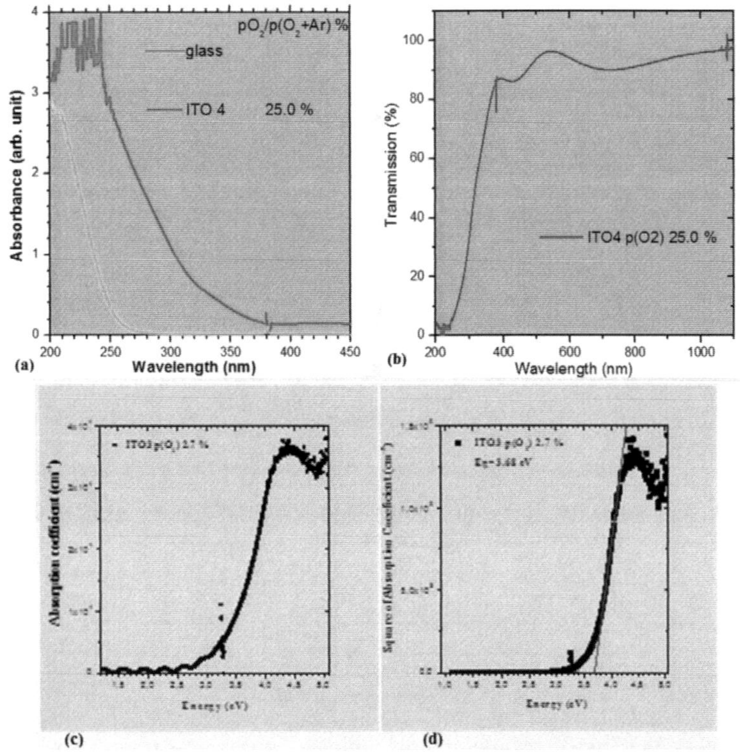


**Figure 4.(a)** Absorption measurements of the ITO thin films (200-450 nm); **(b)** high transmittance; **(c)** Absorption coefficient; **(d)** Square absorption coefficient vs energy graph for bandgap calculation ($P(O_2)$ = 2.7%.

**Table 3.** Effect of FOM with $P(O_2)$ %.

| $P(O_2)$ % | Transparency | FOM |
|---|---|---|
| 2.7 | 0.92 | $0.37x10^{-5}$ |

*3.2. Gas sensor measurements analysis*

The results obtained from gas sensor measurements system tabulated in Table 4 and Table 5 for ITO 3 device that contain 16.3% oxygen in the plasma chamber while depositing thin film. Reducing $H_2$ and $CO_2$ gases sensitivity observed by resistance versus time graph by integrated LabVIEW program in real time showed resistance changes. Nevertheless, adsorption mechanism of $CO_2$ different from reducing gas $H_2$, both type of sensors displayed better sensitivity and obeyed step response of first order sensor response. Hydrogen sensor showed highest sensitivity, response time and recovery time according to the calculations. Consequence $CO_2$ also showed similar sensitivity, response time and recovery time values for ITO 3 sensor device.

Journal of Physics: Conference Series **707** (2016) 012021 doi:10.1088/1742-6596/707/1/012021

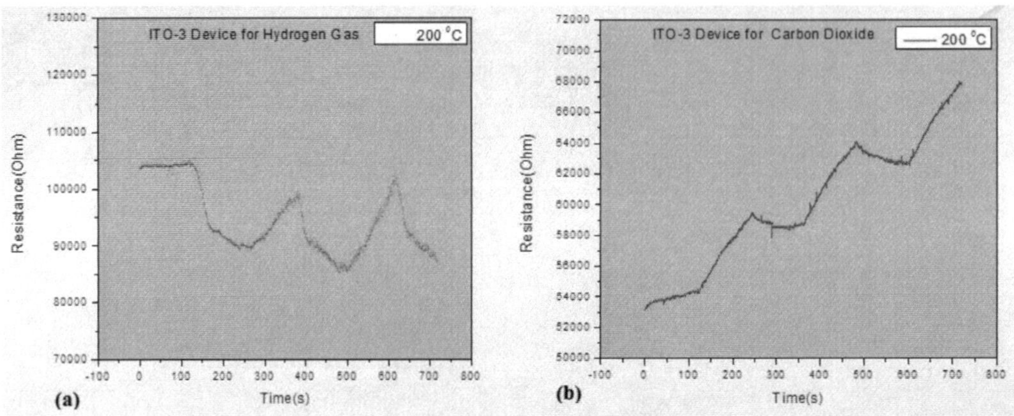


**Figure 5.** ITO 3 (16.3% $O_2$) Device at 200°C; **(a)** $H_2$; **(b)** $CO_2$.

**Table 4.** Responsivity, Response Time, Recovery time for $H_2$ at 200 $^0$C.

| Responsivity ((R1-R0)/R0)*100 | Response time(s) (1-1/e) | Recovery time(s) (1/e) |
|---|---|---|
| 10.9 | 32.8 | 36.4 |

**Table 5.** Responsivity, Response Time, Recovery time for $CO_2$ at 200 $^0$C.

| Responsivity ((R1-R0)/R0)*100 | Response time(s) (1-1/e) | Recovery time(s) (1/e) |
|---|---|---|
| 9.2 | 38.9 | 62.1 |

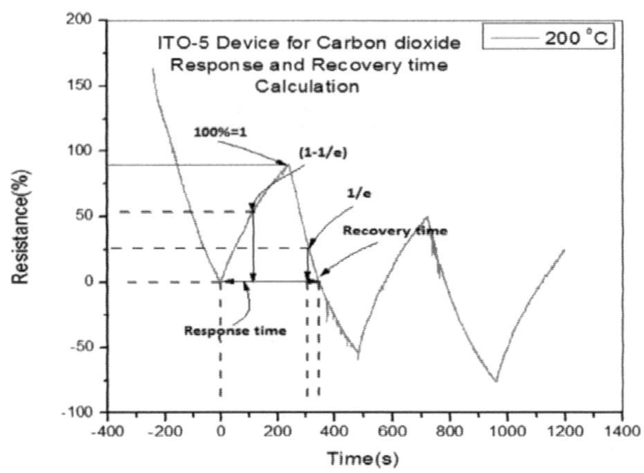


**Figure 6.** Calculating response time and recovery time.

## 4. Conclusions

The ITO (TCO) thin films prepared by RF magnetron sputtering technique are good gas sensing materials. It has showed reasonable sensitivity to reducing $H_2$ and $CO_2$ gases. The sensitivity of ITO thin films was remarkably affected by $O_2$ content introduced to growth plasma chamber. The optimal working temperature to detect $H_2$ and $CO_2$ was 200 ◦C. From the obtained results, the ITO thin film sensors shows potential for gas sensor application and further investigations are necessary to develop high performance gas sensors.

## 5. Acknowledgements

This work gas measurement system was financially supported by Atatürk university Bap project number 2015/93 which is gratefully acknowledged.

## References

[1] Nakamura Zhang J, Liu X, Neri G, and Pinna N 2015 Nanostructured Materials for Room-Temperature Gas Sensors *Adv. Mater.* 2015, DOI: 10.1002/adma.201503825 (WILEY-VCH Verlag GmbH & Co. KGaA, Weinheim)

[2] Huang L, Zhang Z, Li Z, Chen B, Ma X, Dong L, and Peng L Multifunctional Graphene Sensors for Magnetic and Hydrogen Detection *ACS Appl. Mater.* Interfaces 2015, 7, 9581−9588 DOI: 10.1021/acsami.5b01070

[3] Shin D H, Lee J S, Jun J, An J H, Kim S G, Cho K H and Jang J Flower-like Palladium Nanoclusters Decorated Graphene Electrodes for Ultrasensitive and Flexible Hydrogen Gas Sensing *Scientific Reports* 2015 5:12294 DOI: 10.1038/srep12294

[4] Kenji Obata a, Shizuko Kumazawa b, Shigenori Matsushima a, Kengo Shimanoe c, Noboru Yamazoe c NASICON-based potentiometric CO2 sensor combined with new materials operative at room temperature *Sensors and Actuators B 108* (2005) 352–358

[5] S. Xu, Y. Shi Low temperature high sensor response nano gas sensor using ITO nanofibers *Sensors and Actuators B 143* (2009) 71–75

[6] C.W. Lin et al. On an indium–tin-oxide thin film based ammonia gas sensor *Sensors and Actuators B* 160 (2011) 1481– 1484

[7] H.S. Al-Salman, M.J. Abdullah Hydrogen gas sensing based on ZnO nanostructure prepared by RF-sputtering on quartz and PET substrates *Sensors and Actuators B 181* (2013) 259– 266

# Electronic Structure of ABC-stacked Multilayer Graphene and Trigonal Warping:A First Principles Calculation

**Celal Yelgel**

Recep Tayyip Erdogan University, Faculty of Engineering, Department of Materials Science and Nanotechnology Engineering, 53100, Rize, Turkey

E-mail: celal.yelgel@erdogan.edu.tr

**Abstract.** We present an extensive density functional theory (DFT) based investigation of the electronic structures of ABC–stacked $N$–layer graphene. It is found that for such systems the dispersion relations of the highest valence and the lowest conduction bands near the **K** point in the Brillouin zone are characterised by a mixture of cubic, parabolic, and linear behaviours. When the number of graphene layers is increased to more than three, the separation between the valence and conduction bands decreases up until they touch each other. For five and six layer samples these bands show flat behaviour close to the **K** point. We note that all states in the vicinity of the Fermi energy are surface states originated from the top and/or bottom surface of all the systems considered. For the trilayer system, $N$=3, pronounced trigonal warping of the bands slightly above the Fermi level is directly obtained from DFT calculations.

## 1. Introduction

Graphene has become one of the most investigated systems due to its remarkable electronic, mechanical and thermal properties [1, 2]. Experiments reveal that the single layer graphene and stacks of graphene layers can be used in the design of new electronic devices [3, 4, 5]. In addition to the fascinating development in research on properties of graphene and bilayer graphene, few-layer graphene (FLG) has also gained recent attention [6, 7, 8, 9]. The ability of creating stacks of the individual graphene sheets in FLG provides an extra degree of freedom on electronic properties [10, 11, 12, 13]. It has been predicted and suggested that different stacking orders, with distinct lattice symmetries, have a forceful effect on the band structure and electronic properties of FLG [14, 15]. Experimentally, the strong influence of stacking order on the low–energy electronic structure of FLG was recently demonstrated by infrared (IR) spectroscopy [10]. The band structures of graphene from one to four layers were also measured using angle–resolved photoemission spectroscopy [16].

Although the FLG and graphite are mostly Bernal–stacked (AB), trilayer graphene has two different stacking orders: ABA (Bernal) and ABC (rhombohedral). These stackings for trilayer graphene are expected to be metal–like [17] and predicted as two stable configurations in Refs. [9, 11, 18]. Even though there are small cohesive energy differences between these two types of trilayer graphene, recent studies indicate that they exhibit very different electronic properties [11, 12, 19]. A recent study of the electronic structure of the ABA– and ABC–stacked multilayer graphene was reported by using tight–binding approach and density functional theory

[11, 20, 21, 22, 23]. The work in Ref. [22] shows that the type of interlayer coupling present in ABC–stacked multilayer with $N$ layers graphene gives rise to trigonal warping of the energy bands near the Fermi surface that is both qualitatively and quantitatively different from that in bilayer graphene.

In the present work, we investigate the band structure of ABC–stacked N–layer graphene by employing the plane wave pseudopotential method within the density functional scheme. The orbital natures of the highest occupied molecular orbital (HOMO) and lowest unoccupied molecular orbital (LUMO) states are investigated using partial charge density plots. We note that these states are characterised as surface states which form flat bands around the Fermi level, with an increasing the **k** space near the **K** point with the number of layers $N$. Our calculations indicate that the dispersion curve of the highest valence and lowest conduction bands for the ABC–stacked $N$-layer graphene are a mixture of cubic, quadratic, and linear behaviours. The nature of the trigonal warping of energy bands slightly above the Fermi level is examined for different layer thicknesses. This is achieved through an examination of the orbital origins of the highest occupied and lowest unoccupied bands at the Dirac (**K**) point. The trigonal warping effect of the energy bands near the Fermi level is also discussed using the constant electron energy countour plots.

## 2. Computational Method

Our total-energy and electronic-structure calculations were performed in the framework of the first-principles plane wave pseudopotential method within the local density approximation of the density functional theory (LDA-DFT). The ABC–stacked trilayer and multilayer graphene were modelled by using a repeated slab geometry. To avoid interaction between neighbouring slabs along the normal direction, a 15 Å vacuum region was used to separate them after several converge tests. The interlayer separation was found to be 3.33 Å with our theoretical lattice constant of 2.46 Å. The Perdew-Zunger exchange correlation scheme [24] was employed for electron-electron interaction. Norm-conserving [25] and fully separable [26] pseudopotentials for C were used to treat electron-ion interaction. Single-particle Kohn-Sham orbitals were expanded by using a plane-wave basis set with a kinetic energy cutoff of 40 Ryd. Self-consistent solutions of the Kohn-Sham equations were obtained by employing a $36 \times 36 \times 2$ **k**-points Monkhorst-Pack set [27] within the supercell Brillouin zone. Relaxed atomic positions were obtained by using the total energy and force minimisation methods. Details of the theoretical method employed in our work can be found in Refs. [28, 29].

## 3. Results

Before presenting results, we briefly discuss the appropriateness of the LDA-DFT method employed in our investigation. It has widely been mentioned that electron-electron interactions in graphene and graphite should be described by a combination of van der Waals (vdW) and density functional schemes. Indeed, there are some recent proposals to calculate vdW interactions within the DFT approach [30, 31, 32, 33, 34]. However, systematic improvement of results from the vdW-DFT theory has not yet been evidenced for all systems. For example, agreement between theory and experiment for interlayer spacing in graphite is less favourable from the vdW-DFT theory than from the LDA-DFT theory [35]. It has been concluded that for systems that LDA finds to be weakly bonded, the binding energies obtained with the vdW-DFT and LDA-DFT methods are very similar [36]. We thus expect that the LDA-DFT scheme will produce reasonably acceptable results for our system of the ABC–stacked $N$–layer graphene.

We first start with the ABC–stacked trilayer graphene (the $N = 3$ trilayer case). There are six basis carbon atoms in its unit cell, with two basis carbon atoms in each layer. As illustrated in Fig. 1, while the bottom and middle layers of the ABC-stacked trilayer graphene are arranged in Bernal (AB), the basis $C_{31}$ in the top layer lies exactly above the basis $C_{22}$ in the middle layer and the second basis $C_{32}$ on the top layer is at the centre of the hexagon formed by the carbon atoms in the bottom layer. The interlayer separation is found to be 3.33 Å by using our theoretical lattice constant of 2.46 Å after several total energy calculations. This result is in good agreement with experimentally measured interlayer distance of graphite [37].

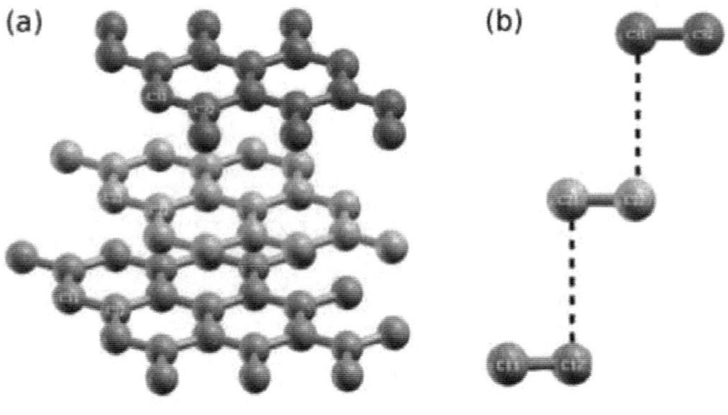


**Figure 1.** (a) Atomic structure of ABC–stacked trilayer graphene containing six basis carbon atoms in the primitive unit cell. (b) Schematic of the unit cell of ABC–stacked trilayer graphene.

The electronic band structure was examined close to the Dirac point (the **K** point in the Brillouin zone) for wavevectors lying in graphene plane. As represented in Fig. 2, along $k_x$ two bands lying close to the Fermi energy cross each other, thus indicating metallic behaviour of the system. These bands, however, do not cross each other for $-k_x$ values, indicating semiconducting

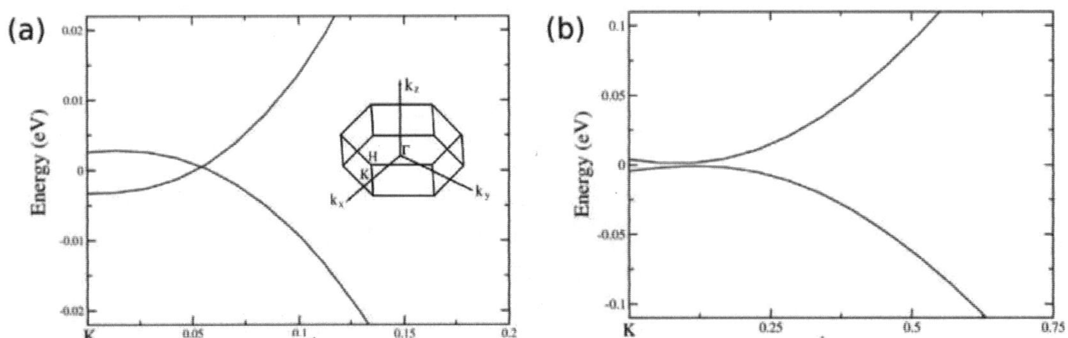


**Figure 2.** Band structure of the ABC–stacked trilayer graphene close to **K** point (a) along $k_x$ (the **KM** direction), (b) along $-k_x$ (the **K**Γ direction). The Fermi level is set to zero energy. The hexagonal Brillouin zone is also shown as an inset.

behaviour along the $\mathbf{K} - \mathbf{\Gamma}$ direction. A similar behaviour of these bands are also found in the work of Latil and Henrard [11]. A polynomial fit of the part of the curve in Fig. 3 (a) shows a mixture of linear, quadratic and cubic terms:

$$E = a_0 + a_1 k_x + a_2 k_x^2 + a_3 k_x^3, \tag{1}$$

where $k_x$ is expressed in nm$^{-1}$, $E$ in eV, and $a_0$=-0.004 eV, $a_1$=0.0027 eV nm, $a_2$=0.91 eV nm$^2$, $a_3$=4.524 eV nm$^2$.

We calculated the constant energy contours using a 50×50 mesh around the **K** point, which results in consideration of 2500 **k**-points. For the ABC–stacked trilayer graphene energy contour plots up to the energy values 0.85 eV above and 0.72 eV below the Fermi level and close to the **K** point are shown in Fig. 3 (b) and (c). In the Fig. 3 (b), the lowest energy contours appear to be located at 0.26 nm$^{-1}$ from the **K** point and 45 meV above the Fermi level. These energy contour plots clearly indicate prominent trigonal warping – stretching of equienergy lines along directions at 120° from each other.

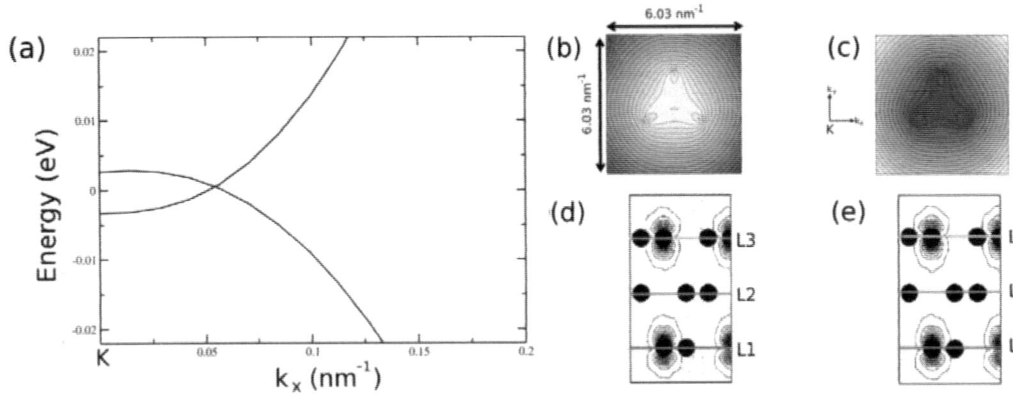


**Figure 3.** (a) Band structure of ABC–stacked $N = 3$ layer graphene close to **K** point with the energy zero set at the Fermi level. The electronic energy contour plots for (b) the lowest conduction band up to 0.85 eV above the Fermi level and (c) the highest valence up to 0.72 eV below the Fermi level. The minumum and maximum energies in the contour plots are shown as the lighest and darkest colours, respectively. Electronic charge density contour plots at the **K** point are also given for (d) the lowest conduction band and (e) the highest valance band.

We now model ABC–stacked multilayer graphene. In particular, we consider the $N$–layer stacking sequences ABCA, ABCAB, and ABCABC with $N = 4$, $N = 5$ and $N = 6$, respectively. For the $N = 4$ multilayer the top layer graphene sheet is exactly above the bottom layer of the ABC–stacked trilayer. In general, for $N > 3$ the top layer graphene sheet lies exactly above the $(N-3)^{\mathrm{rd}}$ layer(s).

The innermost pair of bands for $N = 4$ are shown in Fig. 4 (a). The two bands are degenerate at the **K** point at which the Fermi level lies. These bands split by a maximum of 2.5 meV slightly away from **K** along $k_x$ and then cross each other at a slightly larger value of the wavevector. This result is also in good agreement with the work in Ref. [11]. The pronounced trigonal warping seen for $N = 3$ system has been reduced for the $N = 4$ case. Figure 4 (b) and (c) show the energy contour plots up to 0.80 eV above and 0.65 eV below the Fermi level.

As shown in Fig. 5 (a), when the graphene thickness is increased to $N = 5$ layers, the innermost pair of bands become quite flat around the Fermi level very close to the **K** point. With the increase in the wavevector along $k_x$, these bands cross each other twice in the close vicinity of the **K** point. Between the two crossings there is a maximum splitting of approximately 10 meV between these bands. The amount of this splitting is larger than that for the $N = 3$ and $N = 4$ cases. The energy contour plots in 5 (b) and (c) reveal that the trigonal warping present for $N$=3 is almost lost for the $N$=5 system.

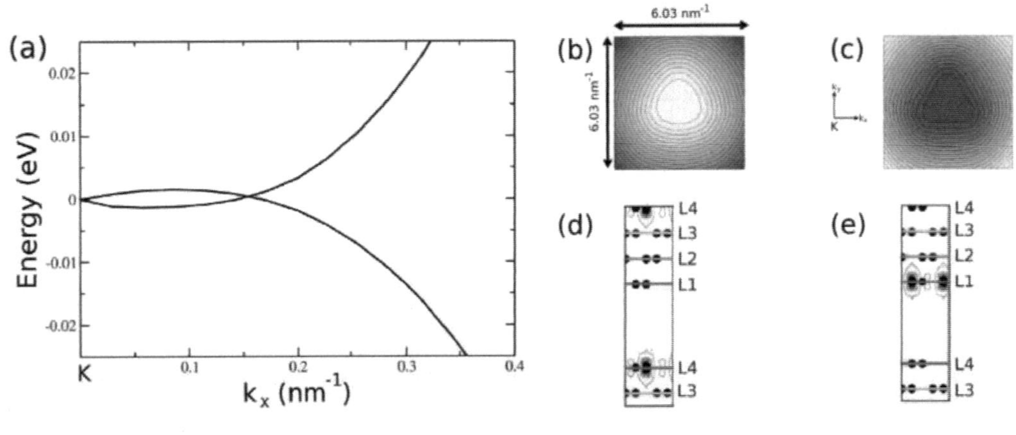


**Figure 4.** (a) Band structure of ABC–stacked $N = 4$ layer graphene close to **K** point with the energy zero set at the Fermi level. Energy contour plots for (b) low conduction bands up to 0.80 eV above the Fermi level, and (c) high valence bands up to 0.65 eV below the Fermi level. The minumum and maximum energies in the contour plots are shown as the lighest and darkest colours, respectively. Partial charge density plots at the **K** point also given for (d) the lowest conduction band and (e) the highest valence band.

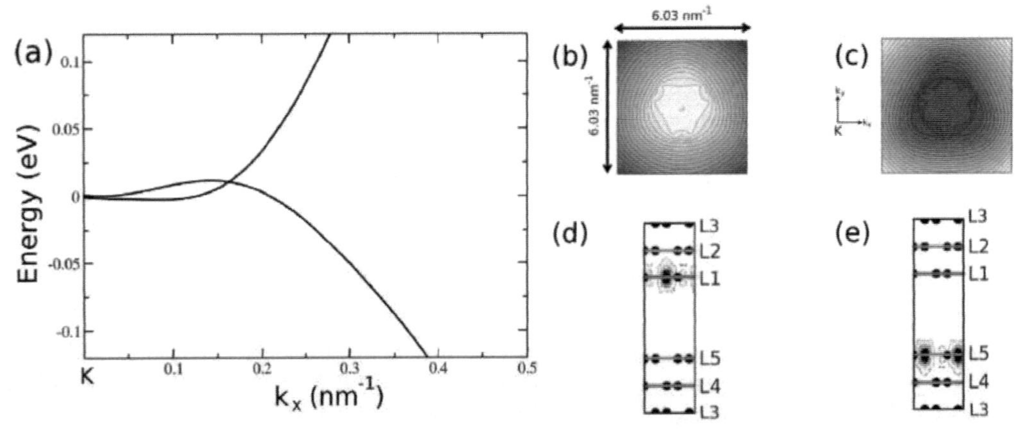


**Figure 5.** (a) Band structure of ABC–stacked $N = 5$ layer graphene close to **K** point with the energy zero set at the Fermi level. Energy contour plots for (b) low conduction bands up to 0.86 eV above the Fermi level and (c) high valence bands up to 0.61 eV below the Fermi level. The minumum and maximum energies in the contour plots are shown as the lighest and darkest colours, respectively. Partial charge density plots at the **K** point are also given for (d) the lowest conduction band and (e) the highest valence band.

The electronic band structure of the ABC–stacked 6–layer graphene is presented in Fig. 6 (a). There is a small amount of separation between the two bands around the Fermi level at the **K** point. These bands become degenerate slightly away from **K** point and remain so for some distance along $k_x$, in contrast with the $N = 3, 4, 5$ cases. The trigonal warping is found to be quite weak for this system, as seen in Fig. 6 (b) and (c).

In order to examine the origin of the HOMO and LUMO energy states at the **K** point, we made partial charge density plots in a vertical plane for ABC–stacked $N$–layer graphene. It is found that for $N$=3 and $N$=6 both HOMO and LUMO states are derived from the appropriate combinations of the $p_z$ orbitals on the first basis atom ($C_{11}$) on the bottommost layer and the second basis atom ($C_{32}$ or $C_{62}$) on the topmost layer. In contrast, for $N$=4 and $N$=5, both HOMO and LUMO have maximum contribution from the $p_z$ orbital of only one basis atom, either on the bottommost or the uppermost layer. The largest contribution for HOMO comes from $C_{11}$ for $N$=4 and from $C_{52}$ for $N$=5. The largest contribution for LUMO comes from $C_{42}$ for $N$=4 and from $C_{11}$ for $N$=5. These features can be seen in the contour plots presented in the Figs. 3-6 (d) and (e).

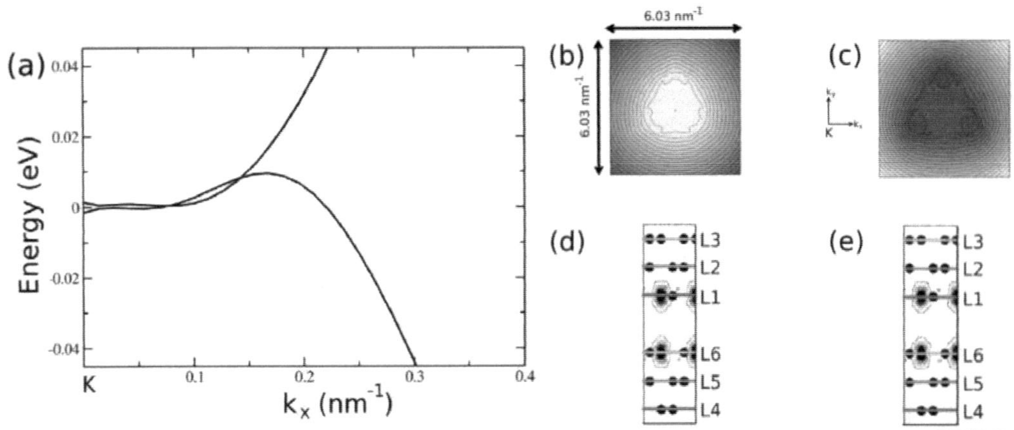


**Figure 6.** (a) Band structure of ABC–stacked $N = 6$ layer graphene close to **K** point with the energy zero set at the Fermi level. Energy contour plots for (b) low conduction bands up to 0.83 eV above the Fermi level and (c) high valence bands up to 0.61 eV below the Fermi level. The minumum and maximum energies in the contour plots are shown as the lighest and darkest colours, respectively. Partial charge density plots at the **K** point are also given for (d) the lowest conduction band and (e) the highest valence band.

We have noted that the HOMO and LUMO states for the ABC–stacked $N$–layer graphene systems studied here are surface states which form flat bands around the Fermi level. In fact any size of few–layer graphene system, thicker than the bi–layer graphene, can be viewed as a thin slab. Our analysis clearly suggests that the HOMO and LUMO states of the ABC–stacked $N$–layer graphene systems originate from the top and/or bottom surface of the slab. The surface feature extracted in our work has gone un-noticed in previous investigations of such systems.

## 4. Summary

In summary, we have studied the ABC–stacked $N$–layer graphene by using the density functional theory within the local approximation and the plane wave pseudopotential method. The dispersion of $\pi/\pi^*$ bands close to Dirac point is established to show a mixture of cubic, quadratic, and linear behaviours for the ABC–stacked trilayer graphene. Equienergy contour plots clearly indicate prominent trigonal warping in the ABC–stacked system– stretching of equienergy lines along directions at 120° from each other. ABC–stacked systems with layer numbers $N > 3$ are characterised with much reduced level of trigonal warping. Our first-principles results for the warping of equienergy contours have good agreement with the tight-binding results [22]. From an analysis of partial charge density plots, we have established that the highest occupied state and the lowest unoccupied state at the **K** point originate from the $p_z$ orbitals of carbon atoms in the bottom and uppermost layers of the $N$=3 and $N$=6 systems, but only from one of the layers for the $N$=4 and $N$=5 systems. Thus the HOMO and LUMO states of the $N$–layer graphene systems can be regarded as surface states.

## References

[1] Novoselov K S, Geim A K, Morozov S V, Jiang D, Zhang Y, Dubonos S V, Gregorieva I V and Firsov A A 2004 *Science* **306** 666
[2] Zhang Y, Small J P, Amori S and Kim P 2005 *Phys. Rev. Lett.* **94** 176803
[3] Berger C, Song Z M, Li T B, Li X, Ogbazghi A Y, Feng R, Dai Z, Marchenkov A N, Conrad E H, First P N and De Heer W A 2004 *J. Phys. Chem.* B **108** 19912
[4] Bunch J S, Yaish Y, Brink M, Bolotin K and McEuen P L 2005 *Nano Lett.* **5** 2887
[5] Novoselov K S, Jiang D, Schedin F, Booth T J, Khotkevich V V, Morozov S V and Geim A K 2005 *Proc. Natl. Acad. Sci.* **102** 10451
[6] Zhu W J, Perebeinos V, Freitag M and Avouris P 2009 *Phys. Rev.* B **80** 235402
[7] Bao W Z, Zhao Z, Zhang H, Liu G, Kratz P, Jing L, Velasco J, Smirnov D and Lau C N 2010 *Phys. Rev. Lett.* **105** 246601
[8] Ghosh S, Bao W Z, Nika D L, Subrina S, Pokatilov E P, Lau C N and Balandin A A 2010 *Nat. Mater.* **9** 555
[9] Guinea F, Castro Neto A H and Peres N M R 2006 *Phys. Rev.* B **73** 245426
[10] Mak K F, Shan J and Heinz T F 2010 *Phys. Rev. Lett.* **104** 176404
[11] Latil S and Henrard L 2006 *Phys. Rev. Lett.* **97** 036803
[12] Aoki M and Amawashi H 2007 *Solid State Comm.* **142** 123
[13] Norimatsu W and Kusunoki M 2010 *Phys. Rev.* B **81** 161410
[14] Avetisyan A A, Partoens B and Peeters F M 2008 *Phys. Rev. B* **80** 195401
[15] McCann E and Koshino M 2010 *Phys. Rev.* B **81** 241409
[16] Ohta T, Bostwick A, Seyller T, Horn K and Rotenberg E 2006 *Science* **313** 951
[17] Yuan S, De Raedt H and Katsnelson M I 2010 *Phys. Rev.* B **82** 235409
[18] Guinea F, Castro A H and Peres N M R 2007 *Solid State Comm.* **143** 116
[19] Nakamura M and Hirasawa L 2008 *Phys. Rev.* B **77** 045429
[20] Koshino M and McCann E 2009 *Phys. Rev.* B **79** 125443
[21] Koshino M and Ando T 2009 *Solid State Comm.* **27** 1123
[22] Koshino M and McCann E 2009 *Phys. Rev.* B **80** 165409
[23] Koshino M 2010 *Phys. Rev.* B **81** 125304
[24] Perdew J P and Zunger A 1981 *Phys. Rev.* B **23** 5048
[25] Gonze X, Stumpf R and Scheffler M 1991 *Phys. Rev.* B **44** 8503
[26] Kleinman L and Bylander D M 1982 *Phys. Rev. Lett.* **48** 1425
[27] Monkhorst H J and Pack J D 1976 *Phys. Rev.* B **13** 5189
[28] Srivastava G P 1999 *Theoretical Modelling of Semiconductor Surfaces* (Singapore:World Scientific)
[29] Srivastava G P 1993 *J. Phys. Condens. Matter* **5** 4695
[30] Grimme S 2006 *J. Comput. Chem.* **27** 1787
[31] Tkatchenko A and Scheffler M 2009 *Phys. Rev. Lett.* **102** 073005
[32] Ruiz V G, Liu W, Zojer E, Scheffler M and Tkatchenko A 2012 *Phys. Rev. Lett.* **108** 146103
[33] Dion M, Rydberg H, Schröder E, Langreth D C and Lundqvist B I 2004 *Phys. Rev. Lett.* **92** 246401
[34] Lee K, Murray E D, Kong L, Lundqvist B I and Langreth D C 2010 *Phys. Rev.* B **82** 081101
[35] Brako R, Sokcevic D, Lazic P and Atodiresei N 2010 *New J. Phys.* **12** 113016
[36] Vanin M, Mortensen J J, Kelkkanen A K, Garcia-Lastra J M, Thygesen K S and Jacobsen K W 2010 *Phys. Rev.* B. **81** 081408
[37] Zacharia R, Ulbricht H and Hertel T 2004 *Phys. Rev.* B **69** 155406

# Laterally inhomogeneous barrier analysis of cu/n-gap/al schottky devices

**K Çınar Demir[1], C Coşkun[2], S V Kurudirek, S Öz, Ş Aydoğan and M Biber[3]**

[1]Atatürk University, Oltu Earth Sciences Faculty, Department of Mining Engineering, 25400 Erzurum, Turkey
[2]Giresun University, Faculty of Arts and Sciences, Department of Physics, 28100, Giresun, Turkey
[3]Atatürk University, Faculty of Sciences, Department of Physics, 25240 Erzurum, Turkey

E-mail: kubra.cinar@atauni.edu.tr

**Abstract.** In this study, we examined the electrical parameters of Cu/n-GaP/Al Schottky structures at room temperature and examined the electrical characterization of these devices depending on and Capacitance-Voltage (C-V) and Current-Voltage (I-V) measurements. A statistical study on the experimental ideality factor (n) and BHs(barrier heights) values of the devices was stated. The n and BHs of all contacts have been determined from the electrical characteristics. Even though all of the diodes were conformably prepared, there was a diode-to-diode variation: the effective BHs changed from 0.988-0.07 to 1.216-0.07 eV, and the n from 1.01-0.299 to 2.16-0.299. The yielded results show that the mean electrical parameters of Schottky devices are different from one diode to another, even if they are identically prepared.It can be axplained that the lower BHs usher with the higher n values owing to inhomogeneities.

## 1. Introduction

InP compound is the first of III-V group compounds. Then, binary compounds between (III-a) with (V-a) group elements were formed by Huggens. In 1929, GaP, InSb, GaAs, GaSb, AlAs, AlSb, AlP and AlN compounds were performed by Hilsum [1]. III-V group compounds gained importance as a new family of semiconductors because of their small energy gap, direct band structure and high electron velocity. Superior circuit elements non-existing in Si circuit elements were achieved by using these semiconductors. Advanced technological devices were obtained by using these circuit elements. Additionally, the effects of dislocations on the electronic and interface states of Schottky devices based III-V semiconductor have been examined. Arrivals provided in circuit elements as diode, transistors gained great contributions to semiconductor technology and integrated circuit technology have been developed with the studies performed in this area [2]. GaP (III-V semiconductor) has a great potential for electronic circuit applications owing to wide and indirect band gap (2.26eV at room temperature) [3] and high ionization energy level.

GaP Schottky diodes can be used as opto-electronic circuit elements such as high efficiency detectors and LED (light emitting diode) in GaP Schottky technology The electronic parameters can be determined by different methods. The proceeding developments in GaP Schottky technologies in the production of high efficiency detector and LED's (light emitting diodes) have played an important role more researches and in the determination of their electronic parameters about MS and MIS structures. Electronic parameters can be determined by different methods. The most frequently used of these are the measurement methods with (I-V) and (C-V).

GaP (Gallium phosphide) has an indirect band gap (2.26eV) with the cubic crystal structure which are diamond and zinc blende. Yet, it has direction band gap with hexagonal (wurtzite) crystal structure when the band gap of this material changed [4]. GaP's lattice constant is calculated as 5.45 Å [5]. GaP semiconductor consists of the combination of the III groups G and the V groups P elements. Crystal structure of GaP is zinc sulphide (zinc blend) [4] structure. The only difference from the diamond structure of the zinc sulphide structure is diatomic and is a structure observed in semiconductors with two atoms. Electrical properties and growth of GaP semiconductor has attracted a lot of attention by scientists in recent years. GaP semiconductors grow with many techniques such MBE (Molecular Beam Epitaxy) [6], Liquid Phase Epitaxy [7], RF-Magnetron Sputtering [8] and Chemical vapour deposition [9]. Blank *et al.* Reported that single crystal n-GaP (100) deposited with Czochralsky method and it has properties such as $n = (2–4) \times 10^{17}$ $cm^{-3}$ and $\mu n = 100–110$ $cm^2$ $V^{-1}$ $s^{-1}$ (300 K). The dislocation density was $P_d \sim 10^6$ $cm^{-2}$ in crystals [10].

In recent years, electronic based on semiconductor materials has been made great progress. Semiconductor technology has entered into every phase of our daily live. For this reason, determination of the electrical and optical properties of semiconductor materials of all kinds will contribute to the expansion of the usage fields of these materials.

The preparation and production methods of the circuit elements produced from these materials and semiconductor materials have a significant influence on the electronic properties and parameters of these materials. The electronic industry is the basis of the electronic or opto-electronic devices obtained by using semiconductor crystals. The structure of these devices which are very small sizes, electrical characteristics and determination of operating conditions is very important in terms of the performance and of reliability these devices. Therefore, the examination of the electrical properties of metal-insulating-semiconductor (MIS), metal-oxide –semiconductor structures (MOS) and metal-semiconductor (MS) has gained importance. The parameters that affect the performance of such structures are the thickness and homogeneity of the insulating oxide layer, the series resistance of the structure, interfacial states, the homogeneity of the potential barrier formed between metal and semiconductor.

Semiconductor materials are being constantly investigated until today. In recent years, MS and MIS contacts have been heavily used in the semiconductor and optoelectronic technologies. High quality electronic and opto-electronic devices have been developed as a result of these researches. These devices have the wide application facilities in opto-electronic, integrated circuits, light and ultraviolet (UV) detectors, solar cells, defense and space technology.

In general, theoretical understanding of metal-semiconductor contacts has been after technological development. The most of the theoretical developments have been reviewed by scientists concerned with metal-vacuum systems. Initially, image force barrier lowering has been found owing to the electric field applied in metal-vacuum system. However, this was experimentally confirmed much later in MS system. Afterward, thermionic emission describes for electrons emitted into vacuum from the hot metal. This event demonstrates that can be applied to MS rectifiers, also.

Schottky has revealed the first time that a potential barrier has occurred in the interfacial of MS. Therefore, it is called Schottky diodes or Schottky contacts to MS contacts [11]. In the model developed by Schottky, electric field being in semiconductor increases linear with distance from depletion layer according to Gauss's law and potential decreases. Mott developed another a model associated with potential barrier. Mott adopted that a thin layer is in near of metal-semiconductor [11]. The electric field of this layer remains constant but potential varies linearly. According to Mott, potential barrier arises from the difference between the work functions of MS. Crowell and Size integrated the thermionic emission theory of Bethe and the diffusion theory of Schottky describing the current transmission mechanisms in MS contacts [12].

The first study on insulating layer between MS in Schottky contacts was studied by Cowley and Sze. They analyzed to Schottky BH with different metals [13]. Furthermore, many surface states in interfacial of semiconductor/oxide layer in the metal-semiconductor junction consist of depending on the oxidation process and thermal processes a disorder arising from the crystal structure during cleaning of crystal [14-15]. Card and Rhoderick discussed the intensity of the situation in interfacial and examined to effect of interfacial conditions on n calculated from forward bias I-V characteristics[16]. Tseng and Wu examined to effect on the behaviour of Schottky diode from forward bias I-V characteristics and the distribution of interfacial state densities in semiconductor energy-band gap in the existence of interfacial oxide layer [17]. There are numerous experimental and theoretical methods to obtained interfacial state densities and these methods have their own advantages and disadvantages [18]. However, the most practical and the fastest thing of them is the forward bias I-V method. This method contains a lnI-V plot that n and the potential BH changed to depend on the voltage [19-20].

In this work, our purpose is to experimentally examine the relationship between the effective BHs and n obtained from the forward bias I-V and reverse bias C-V characteristics of the Cu/n-GaP/Al Schottky devices at room temperature. The homogeneous BH value for the diode was yield from the linear relationship between the experimental effective ns and BHs. The characteristics of the Schottky diodes must be very well known for the correct and efficient use. The operation and reliability of these diodes depend on the characteristics and shaping of the insulating layer between MS, the distribution of surface states in semiconductor-insulator, homogeneity of contact. All of them are very important properties causing deviations from the the ideal situation of semiconductor device and should be considered in the calculations. The inter of metal-semiconductor may not be always ideal in Schottky contacts. Thus, a parameter called n takes place in characteristics at this stage. Ones can get knowledge about the properties of the structure to examination the capacity of contact region in metal-semiconductor contacts. The essential parameters such as the potential BH of the structure, built-in potential, carrier density can be determined with help I-V and C-V measurements of Schottky diodes. The statistical distribution of the characteristics of these structures was made with the Gaussian function.

## 2. Experimental

In this work, we used one side polished n-GaP samples. We have prepared Cu/n-GaP/Al Schottky diodes on n-GaP semiconductor with the thickness of 400 μm and carrier density of $2.4x10^{16}$ $cm^{-3}$. Initially, semiconductor samples were cleaned (with processes such as i. each sample was waited for 5 min using trichloroethylene, acetone and methanol and they were immersed into DI water, ii. samples were waited in $3H_2SO_4 + H_2O_2 + H_2O$ for 60 second and rinsed with DI water, iii. they were waited in $HF:H_2O$ (1:10) for 30 second and rinsed with DI water, then dried with unreacted nitrogen) [21]. Al metal was prefered to create ohmic contact on the back surface of the semiconductor sample in Univex-300 Pump system with a pressure of $4x10^{-5}$ Torr and then annealed under $N_2$ gas flow at 500°C for 3 minutes [22]. Schottky diode was fabricated by evaporation using Cu metal on the front surface of GaP wafers. A turbo molecular vacuum coating unit at about $10^{-6}$ mbar pressure were carried out the processes of evaporation. A Keithley 487 picoammeter device was used to measure the I-V characteristics of contacts and HP-4192 an impedance analyser used to obtain the C-V measurements at 100kHz, also.

## 3. Results obtained

Figures 1 shows the XRD patterns of the GaP The XRD measurements revealed the peaks corresponding to the GaP crystal planes of (200) and (400), by JCPDS files (The Joint Committee on Powder Diffraction Standard), indicating the polycrystalline nature of the sample. FWHM and grain size values of GaP(200) and GaP (400) directions were calculated as (0.0558-0.0883), (1490-1090), respectively. Figure 2 indicates both forward and reverse bias log (I-V) plot of Cu/n-GaP/Al Schottky. A statistical distribution on the experimentally BHs and n values of the devices was formed. The n and BHs of all contacts have been yielded from the electrical characteristics. Although the diodes were all identically prepared, there was a diode-to-diode variation: the effective BHs changed from 0.988-0.07 to 1.216-0.07 eV, and the n from 1.01-0.299 to 2.16-0.299.

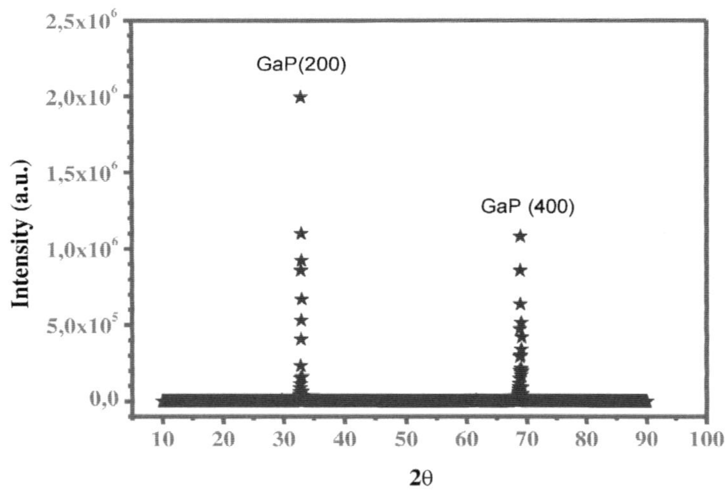


**Figure 1.** The XRD measurements of the n-GaP

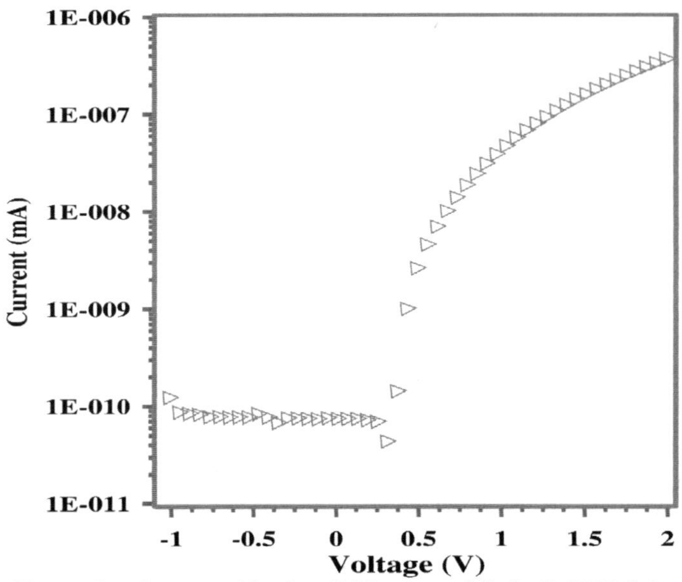


**Figure 2.** Forward and reverse bias log (I-V) plots of Cu/n-GaP/Al Schottky diode

It is estimated in MS devices that the current flow through a diode can be determined by thermionic emission (TE) theory as follows;

$$I = I_0 \exp\left(\frac{qV}{nkT}\right)\left[1 - \exp\left(-\frac{qV}{kT}\right)\right] \quad (1)$$

where $q$ is the electronic charge, $k$ Boltzmann constant, $T$ temperature, $V$ the applied voltage, which is given by:

$$n = \frac{q}{kT}\left(\frac{dV}{d(\ln I)}\right) \quad (2)$$

$I_0$ is the reverse saturation current which is obtained from the straight line intercept of $\ln I$ at $V = 0$ in Eq. (1) and is given:

$$I_0 = A^* A T^2 \exp\left(-\frac{q\Phi_{b0}}{kT}\right) \quad (3)$$

where $A^*$ is the effective Richardson constant, $A$ the diode area and $\Phi_{b0}$ the zero-bias BH. A* is the effective Richardson constant and is equal to 98.2 $Acm^{-2}K^{-2}$ for n-GaP. $\varepsilon_s$ is the dielectric constant of semiconductor and is equal to 10.75 for n-GaP [21], [23] and A, the diode area, is $1.963\times10^{-3}$ $cm^2$. The density of states in the conduction band is $Nc=1.27\times10^{19}$ $cm^{-3}$ for the calculation of contact parameters.
The basic *C-V* evaluation with respect to the semiconductor in-depth doping profile is based on the depletion approximation. Rhoderick stated to the capacitance of the space charge region within the depletion approximation [24]:

$$C = \frac{|\partial Q_{SC}|}{\partial V} = \sqrt{\frac{q\varepsilon_s N_D}{2(V_{bi} - V - kT/q)}} = \frac{\varepsilon_S}{W} \quad (4)$$

This equation can also be written as:

$$\frac{1}{C^2}=\frac{2\left(V_{bi}-V-kT/q\right)}{q\varepsilon_S N_d} \tag{5}$$

or

$$-\frac{d\left(1/C^2\right)}{dV}=\frac{2}{q\varepsilon_S N_d} \tag{6}$$

The capacitance of the space charge region provides important information about the metal/semiconductor interfacial in Schottky contacts. Elements such as the BH of rectifier contact, carrier concentration of the semiconductor, built-in potential ($V_{bi}$) which is obtained from intercept of $1/C^2$ -$V$ plot on bias axis and the Fermi level are assigned from capacitance depending on the change in voltage at reverse bias:

$$N_d=\frac{2}{q\varepsilon_S}\left[-\frac{1}{d\left(1/C^2\right)/dV}\right] \tag{8}$$

Where $Nd$ is the ionized donor concentrations, and if $Nd$ is constant throughout the depletion region, it can be made to fit the $1/C^2$ -$V$ plot. The SBH for semiconductor is given by:

$$\Phi_B(C-V)=V_{bi}+E_f \tag{9}$$

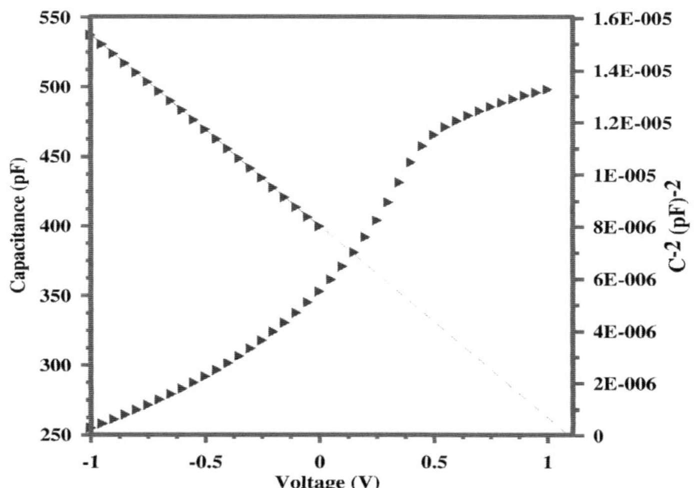

**Figure 3.** C-V and $C^{-2}$-V plots of Cu/n-GaP/Al Schottky diode at 100 kHz

Figure 3 demonstrates C-V and $C^{-2}$-V plots of Cu/n-GaP/Al Schottky diode. Built-in potential, BH, Fermi level and donor concentration values of Cu/n-GaP/Al Schottky contacts were obtained with the help of these curves. The values of these parameters were obtained as 1.094 eV, 1.180 V, 0.085eV and $4.639\times10^{17}$cm$^{-3}$ at 100 kHz, respectively.

Figure 4 indicates to the distribution of the ns obtained from the forward bias I-V characteristics of the Cu/n-GaP/Al Schottky devices. As shown, the Gaussian fit provides a mean n value of 1.30 with a standard deviation of 0.29. Figure 5 demonstrates the statistical distribution of BHs from forward bias I-V plots of the Cu/n-GaP/Al Schottky devices. And Figure 6 indicates the statistical distribution of BHs from the $(1/C)^2$-V plots of the same diodes. The experimental distribution of the effective BHs were fitted by the Gaussian function. At first, let us consider the I-V BHs. The statistical analysis obtained a mean BH value of 1.11 eV with a standart deviation of 0.069 eV. The distribution of the BHs obtained from the $(1/C)^2$-V curves of the Cu/n-GaP/Al Schottky diodes at 100 kHz. The statistical analysis obtained a mean BH value of 1.19 eV with a standart deviation of 0.029 eV. The statistical distribution of donor concentration values from the $(1/C)^2$-V plots of the same diodes is showed in Figure 7. The statistical calculations provided a mean donor concentration value of $4.278x10^{17}cm^{-3}$ with a standart deviation of $4.59x10^{16}$ $cm^{-3}$.

n, low series resistance and low reverse bias current values are important for the realization of the Schottky diodes. It is cleared as ideal diode that n is 1 at Schottky diode. The I-V characteristics of Schottky contacts do not only provide any knowledge regarding to the charge transport, but gives information relevant to the barrier formed in the MS interfacial. The electronic properties of Schottky barrier diode are determined by n and BH. The difference of BH on contacts changes with the thickness of the interfacial layer, the charge density in the interfacial layer, the oxidation of the semiconductor surface, the formation of the pollution atoms collected metal, the temperature and the radiation [25].

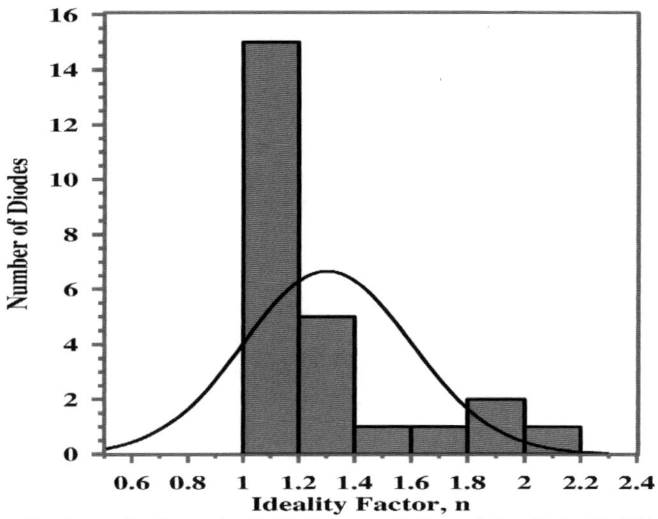


**Figure 4.** Gaussian distribution of n from the I–V characteristics of the Cu/n-GaP/Al Schottky structures

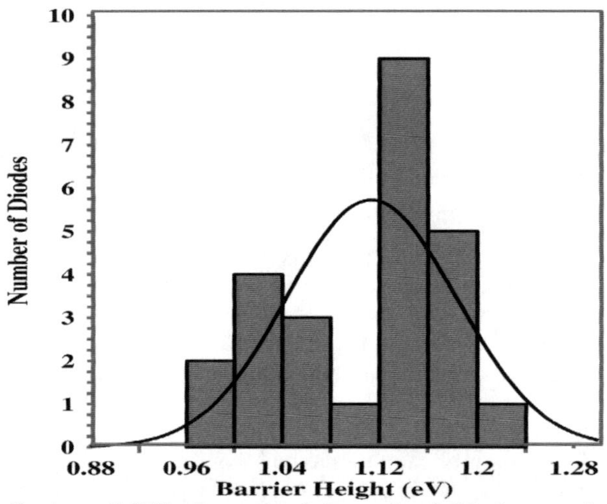


**Figure 5.** Gaussian distribution of BHs determined from the I–V characteristics of the Cu/n-GaP/Al Schottky structures

The BH obtained from I-V measurements is typically less than these from *C-V* measurements. The reason for this is arised from the difference in measurement methods. Furthermore, the many elements such as contaminants in interfacial, quantum mechanical tunnelling, leakage currents, image force decreasing and deep impurity levels focused on this difference [14], [26]. It can be stated by the presence a thin layer compensated in interfacial [27].

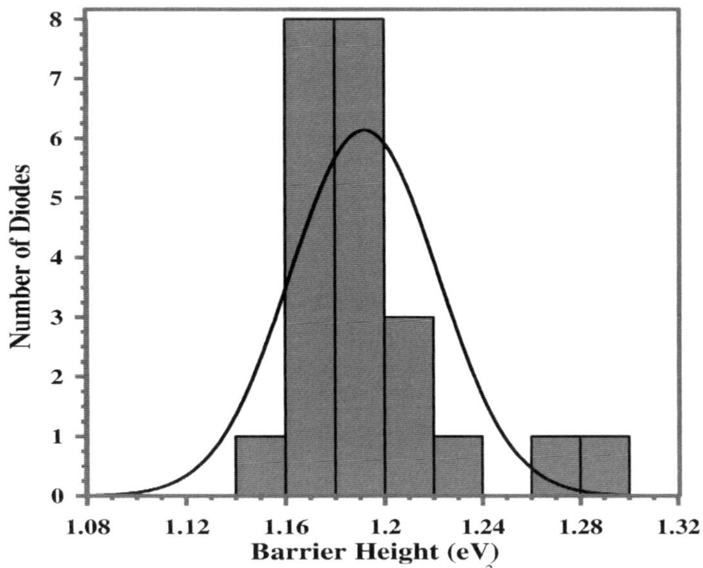


**Figure 6.** Gaussian distribution of BHs obtained from the $C^{-2}$-V characteristics of the Cu/n-GaP/Al Schottky structures

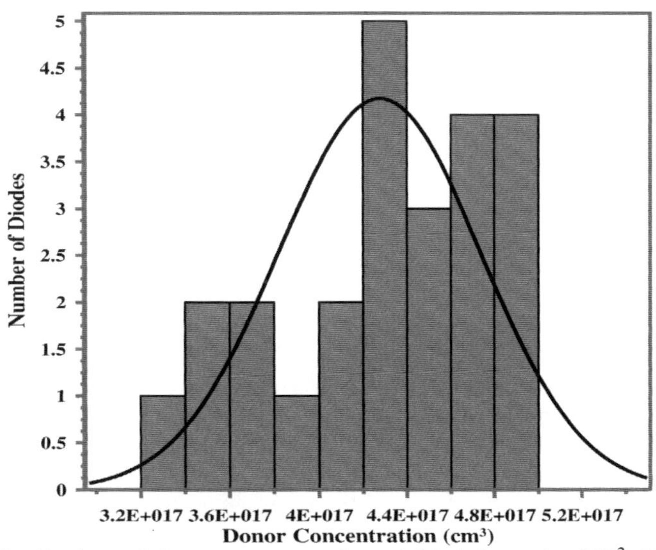


**Figure 7.** Gaussian distribution of donor concentration yielded from the $1/C^2$–$V$ characteristics of the Cu/n-GaP/Al Schottky structures

Cheung *et al.* reported that the series resistance values of Schottky devices can be obtained depending on TE theory [28];

$$I = AJ = \left[ AA^{*} T^{2} \exp\left( \frac{-e(\Phi_B - IR_s)}{kT} \right) \right] \cdot \left[ \exp\left( \frac{e(V - IR_s)}{nkT} \right) \right] \tag{10}$$

where the $IR_s$ term is the voltage drop across series resistance of diode. The values of the series resistance can be calculated from the following functions using eq. (10);

$$\frac{dV}{d\ln(I)} = \frac{nkT}{e} + IR_s \quad (11)$$

and H(I) is given as follows;

$$H(I) = V - \frac{nkT}{e}\ln\left(\frac{I}{AA^*T^2}\right) \quad (12)$$

$$H(I) = n\Phi_B + IR_s \quad (13)$$

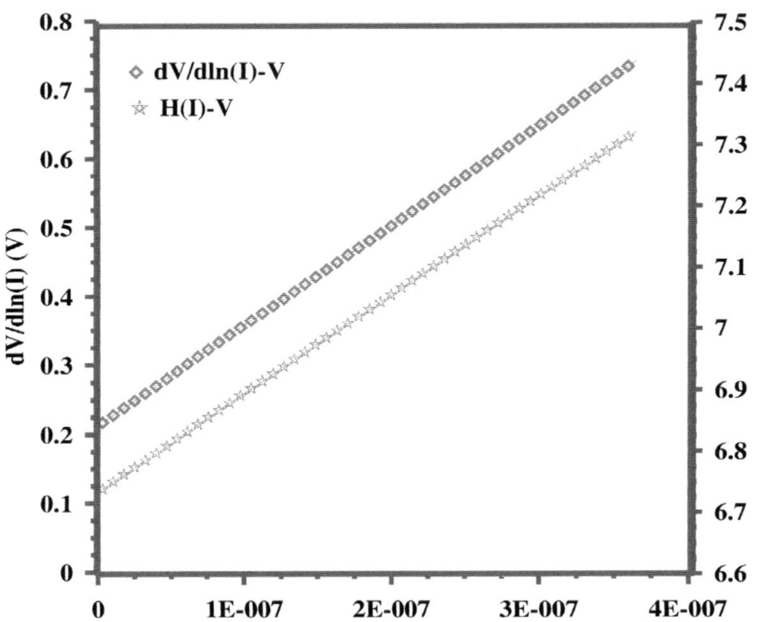


**Figure 8.** dV/dln(I) and H(I) vs. Voltage curves of Cu/n-GaP/Al Schottky diode

As shown Figure 8, series resistance, BH and n parameters of Cu/n-GaP/Al Schottky device were calculated with the help the slopes of the dV/dln(I) and H(I) versus voltage curves. n and series resistance values obtained from the dV/dln(I) versus voltage plots are 8.22 and 1148 kOhm. Also, the series resistance and BH values with the help of H(I) versus voltage plots were calculated as 1611 kOhm and0.819 eV, respectively. Semiconductor crystal surfaces used in semiconductor devices have usually organic soils and the natural oxide layer in a laboratory environment. The natural oxide layer in the cleaned semiconductor surfaces occurs chemically with exposure to the clean room air of semiconductor [29]. Thickness of the insulating layer between metal-semiconductor depends on chemically residual gases on the cleaned material and duration of exposure to the environment of semiconductor surface. The presence of this insulating layer affects the basic diode parameters such as surface states, Schottky BH, n and series resistance [30], [16].

Figure 9 shows the Cu/n-GaP/Al plot obtained with Norde method. Bohlin proposed using the modified Norde method as another way to find the Schottky contact parameters, in which its method is given as[31];

$$F(V) = \frac{V}{\gamma} - \frac{kT}{e} In\left(\frac{I(V)}{AA^*T^2}\right) \quad (14)$$

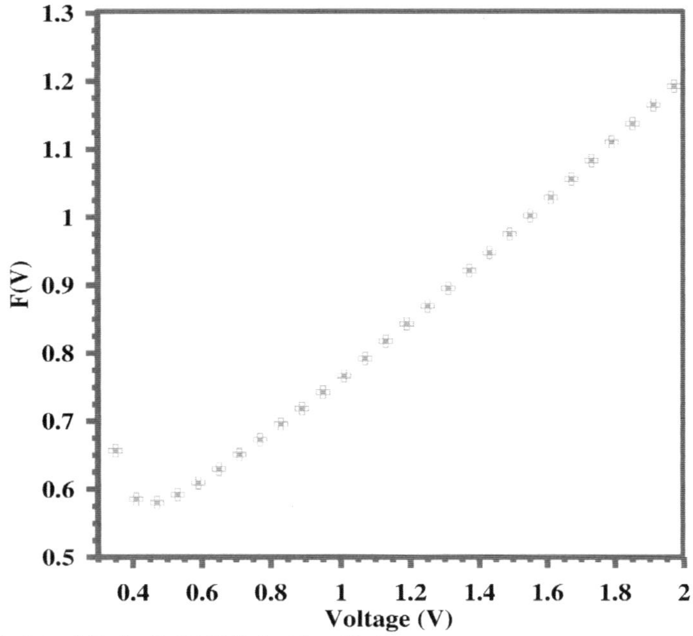


**Figure 9.** *F(V)* vs. *V* plot of Cu/n-GaP/Al Schottky diode

where n is the n calculated from the lnI–V plot and $\gamma$ is an integer greater than the n value. F(V) is the minimum point in the F(V)–V plot, V and I are the corresponding min. voltage and current, respectively. Also, the series resistance and BH values are determined follow as;

$$\Phi_B = Fm + \left[\frac{(\gamma - n)}{n}\right]\left[\frac{V_m}{\gamma} - \frac{kT}{e}\right] \tag{15}$$

$$\mathrm{R} = (\gamma - n)\left(\frac{kT}{eI_m}\right) \tag{16}$$

In 1967, Nicollian and Goetzberger reported that the frequency dependency of the series resistance can be determined with help the measurements of C–V–f and G/ο–V–f curves [32]:

$$s = \frac{G_{max}}{(G_{max}^2 + (wC_{max})^2)} \tag{17}$$

where $G_{max}$ and $C_{max}$ are values of conductance and capacitance performed at 100kHz. Figure 10 (a) and (b) show conductance-voltage and series resistance-voltage plots of the Cu/n-GaP/Al Schottky diode. Curves is observed conductivity and capacitance at an increase about 2V owing to the interface states at 100kHz. Series resistance is going through a peak due to this increase, and this peak also shows a parallel behaviour to the peak in the C-V curves.

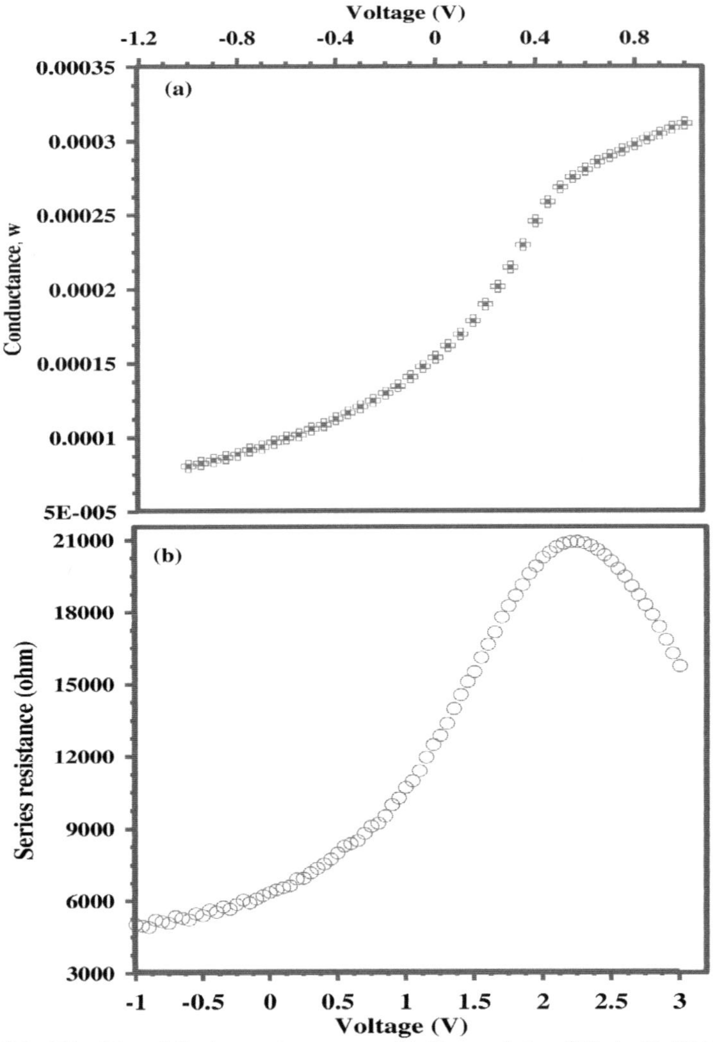


**Figure 10 (a) and (b).** G/w-V and Series resistance vs. Voltage plots of Cu/n-GaP/Al Schottky diode at 100kHz

## 4. Conclusions

We investigated the electrical properties on Cu/n-GaP/Al Schottky diodes. Its BH decreased when n of Cu/n-GaP/Al Schottky diode increased. Also, it is observed that the series resistance values which are calculated with three different methods increased for Cu/n-GaP/Al Schottky diodes. In addition, the parameters of Cu/n-GaP/Al Schottky diodes such as capacitance, built-in potential, donor concentration, Fermi level and BH obtained from $1/C^2$-V curves changed at room temperature. For this reason, there may be a mixture of different metallic phases with different BHs at diodes interfacial due to incomplete interfacial reaction. Thermionic field emission, Image force-lower, interfacial states, generation-recombination and are the mechanisms gave rise to the large values of the n. With the help of these calculated mechanisms can be obtained the absulate value of n.

Furthermore, the contamination at a interfacial is frequently exhibit at the interfacials of junction prepared by the routine processing methods used in the semiconductor electronics industries. These contaminants may act directly to introduce inhomogeneity or they may simply promote inhomogeneity, through the generation of defects, additional interfacial chemical phases and etc. BH inhomogeneity may be present even if the absence of chemical contaminants. Thus, interfacial roughness may contribute to the presence of BH inhomogeneity due to effectively increasing or decreasing the low-BH patches. Eventually, there are numerous, grain boundaries, stacking faults, dislocations, structural defects, at interfacials, and these may be the mean reason of BH inhomogeneity [33-34], [14].

**References**

[1] C Hilsum, A C Rose-Innes 1961 Pergamon Pres. Oxford, London.
[2] E H Nicollian, , J R Brews 1986 Nuclear Technology, 74, 235.
[3] A Owens, S Andersson, R den Hartog, F Quarati, A Webb, E Welter 2007 Nucl Instrum Meth A, 581, 709.
[4] S Assali, I Zardo, S Plissard, D Kriegner, M A Verheijen, G Bauer, A Meijerink, A Belabbes, F Bechstedt, J E M Haverkort, E P A M Bakkers 2013 Nano Lett, 13, 1559.
[5] L Bellaiche, S H Wei, A Zunger1996 Phys Rev B, 54, 17568.
[6] C H Lin, R J Hwu, L P Sadwick, D Heo, 2001 Ieee T Electron Dev, 48, 2205.
[7] A Ashery, A A M Farag, M Zeama, 2014 Superlattice Microst, 66, 136.
[8] J Gao, Q W Zhan, A M Sarangan 2011 Thin Solid Films, 519, 5424.
[9] F Cheng, K George, A L Hector, M Jura, A Kroner, W Levason, J Nesbitt, G Reid, D C Smith, J W Wilson 2011 Chem Mater, 23, 5217.
[10] T V Blank, Y A Goldberg, E A Posse 2009 Semiconductors, 43, 1164.
[11] A M Cowley, S M Sze 1965 J Appl Phys, 36, 3212.
[12] C R Crowell, S M Sze 1966 Solid State Electron, 9, 1035.
[13] J H Werner, H H Guttler 1991 Phys Scripta, T39, 258.
[14] R T Tung 1992 Phys Rev B, 45, 13509.
[15] E H Rhoderic 1982 Iee Proc-I, 129, 1.
[16] H C Card, E H Rhoderic 1971 J Phys D Appl Phys, 4, 1589.
[17] H H Tseng, and C Y Wu 1987 Solid State Electron, 30, 383.
[18] S M Sze 1981 Willey, New York, 245.
[19] Y P Song, R L Vanmeirhaeghe, W H Laflere, F Cardon 1986 Solid State Electron, 29, 633.
[20] D Mui, S Striteon, H Morkoç 1991 Solid State Electron, 34, 1077.
[21] I Orak, K Ejderha, E Sonmez, M Alanyalioglu, A Turut 2015 Mater Res Bull, 61, 463.
[22] S Duman, K Ejderha, O Yigit, A Turut 2012 Microelectron Reliab, 52, 1005.
[23] P Pipinys, A Rimeika, A Pipiniene 2002 Solid State Electron, 46, 1283.
[24] E H Rhoderick, R H Williams 1998 2nd edn. Clarendon, Oxford.
[25] M Saglam, F E Cimilli, A Turut 2004 Physica B, 348, 397.
[26] C Coskun, N Gedik, E Balci 2006 Semicond Sci Tech, 21, 1656.
[27] S Karatas, A Turut 2006 Nucl Instrum Meth A, 566, 584.
[28] S K Cheung, N W Cheung 1986 Appl Phys Lett, 49, 85.
[29] S J Fonash 1975 J Appl Phys, 46, 1286.
[30] A Sing, K C Reinhardt, and W A Anderson 1990 J Appl Phys, 68, 3475.
[31] K E Bohlin 1986 J Appl Phys, 60, 1223.
[32] E H Nicollian and A Goetzberger 1967 Bell System Technical Journal, 46, 1055.
[33] J P Sullivan, R T Tung, M R Pinto, W R Graham 1991 J Appl Phys, 70, 7403.
[34] M Soylu, F Yakuphanoglu, 2010 J Alloy Compd, 506, 418.

# Effect of annealing on photovoltaic performance of fabricated planar organic-inorganic perovskite solar cells

**Ali Baltakesmez[1,2], Mehmet Biber[1] and Sebahattin Tüzemen[1]**

[1]Department of Physics, Faculty of Science, Atatürk University, 25240 Erzurum, Turkey
[2]Department of Physics, Faculty of Science, Çankırı Karatekin University, 18100 Çankırı, Turkey

E-mail: baltakesmez@atauni.edu.tr

**Abstract.** We fabricated planar perovskite solar cells used $CH_3NH_3PbI_{3-x}Cl_x$ for light harvesting to investigate effect of annealing on photovoltaic performance of fabricated device. The devices have an architecture of Glass/ITO/Pedot:PSS/Perovskite/$PC_{61}BM$/Al. Layers of hole transport (Pedot:PSS), active and electron transport ($PC_{61}BM$) were prepared from solution based one step deposition method by a spin coater and standard annealing procedure. The current−voltage curves of devices were measured inside the glovebox using a Keithley 2400 sourcemeter. The cells were illuminated by a solar simulator have optical intensity value of 300 mW/cm$^2$. For the best cells, while PCE value of 5.78% before the annealing, photovoltaic efficiency was improved average 13% delivered a short-circuit current density of 3.20 mA/cm$^2$, open-circuit voltage of 0.82 V and fill factor of 0.74, leading to an efficiency of 6.54% with respect to prior to annealing.

## 1. Introduction

Organic-inorganic lead three or mixed-halide perovskite ($CH_3NH_3PbI_3$ or $CH_3NH_3PbI_{3-x}Cl_x$) based solar cells (PSCs) have been considered as a promising devices for next-generation due to their high power conversion efficiency (PCE) along with low cost and low temperature fabrication. As a light harvester, $CH_3NH_3PbI_3$ and $CH_3NH_3PbI_{3-x}Cl_x$ have several advantages such as a direct band gap, a large absorption coefficient, a low exciton bonding energy, high carrier mobility and long electron–hole diffusion lengths of 100 nm and 1 μm, respectively [1-2].

The first reported organometal perovskite based device with the PCE of 3.8% is a liquid electrolyte based cell [3]. Since the first report on a long-term durable perovskite active layer based solar cells were reported with ever-growing efficiency and the PCE climbed the certified non-stabilized efficiency of 20.1%, in 2014 [4]. According to the reported studies to obtain the high efficiency device, the all layers and the metal contact should be not only free from pinhole but also good contact between the adjacent layers to lower the series resistance and higher the shunt resistance [4-8].

In this study, we present the perovskite solar cells were fabricated by the one-step spin-coating process comprising changed performance parameters of the fabricated cell with the annealing. Optical and structural characterizations show that the coated perovskite thin films have suitable properties as a light absorber. We observed significant changing efficiency with the annealing after the fabrication.

## 2. Experimental

Indium tin oxide (ITO) coated glass substrates with a sheet resistance of ~15 Ωsq$^{-1}$ were cut in the dimensions of 1.25x1.25 cm. Then they were patterned by etching with 10wt% HCl solution for about 2 h. The patterned ITO substrates were cleaned by the following procedure: sonication in detergent (15 min), sonication in acetone (15min) and sonication in propanol (15 min), respectively and rinsed with distilled water before each the solvent. Afterward, for using as a hole transport layer (HTL), 5% DMSO added Poly(3,4-ethylenedioxythiophene) poly(styrene -sulfonate) (PEDOT:PSS (PH500)) was doi:10.1088/1742-6596/707/1/012024

spin coated onto the substrates and then the films were thermally annealing at 120 °C for 25 min in atmosphere.

The solutions and deposition of the other layers including the perovskite, the electron transport (ETL) and the cathode electrode were carried out in the Ar glovebox ($O_2$: <1 ppm and $H_2O$: <1 ppm). Methylammonim iodide ($CH_3NH_3I$ (Dyesol)) and lead (II) chloride ($PbCl_2$ (Aldrich)) were dissolved in a anhydrous N,N-dimethylformamide (DMF) to be used as the perovskite precursor which is 40wt% solution prepared from $PbCl_2$ (0.8 M) and $CH_3NH_3I$ (2.6 M) in a molar ratio of 1:3 and the solution stirred at 60°C for a day.

The perovskite layer was deposited by spin-casting 100 μl of the precursor solution onto the substrate of PEDOT:PSS/ITO/glass at 3000 rpm for 30 s. The films were thermally annealing at 90 °C for 90 min and then color of the films converted from yellowish to dark brownish. After the films cooled to the room temperature, PCBM layer was deposited on top of the perovskite films from 20 mg/ml chlorobenzene solution at 2000 rpm for 45 s. Finally, a 100 nm thick Al film was evaporated on top of the PCBM layer with rate of 1 A/s to form the back contact of the devices. The fabricated cells were annealed 65°C after the electrical measurements to observe effect of the annealing. The active area of the devices was 0.3 $cm^2$.

The devices were characterized under AM 1.5G solar simulator with light intensity of 300 $mV/cm^2$. The current density-voltage (J-V) curves were using a Keithley 2400 source-measurement system. The optical properties of the perovskite film were studied by Perkin-Elmer UV-visible spectrometry (LAMBDA 2S). The X-ray diffraction studies for structural perovskite film has been done by the diffractometer have a wavelength of 1.54059 A.

## 3. Results and discussion

Fig. 1(a) shows the optical absorbance of the perovskite film deposited on the PEDOT:PSS/ITO/glass substrate by spin-coating. It can be seen from this figure that the active layer absorbs a wide range of light from visible to the near infrared with an absorption onset at 780 nm. Additionally, the absorbance spectra is converted to absorption coefficients as a function of the photon energy hν and is plotted as shown in the Fig. 1(b). The optical band gap ($E_g$) estimated from Fig. 1(b) and is found to be 1.55 eV that is agreement with published results for $CH_3NH_3PbI_3$ or $CH_3NH_3PbI_{3-x}Cl_x$ in literature [9].

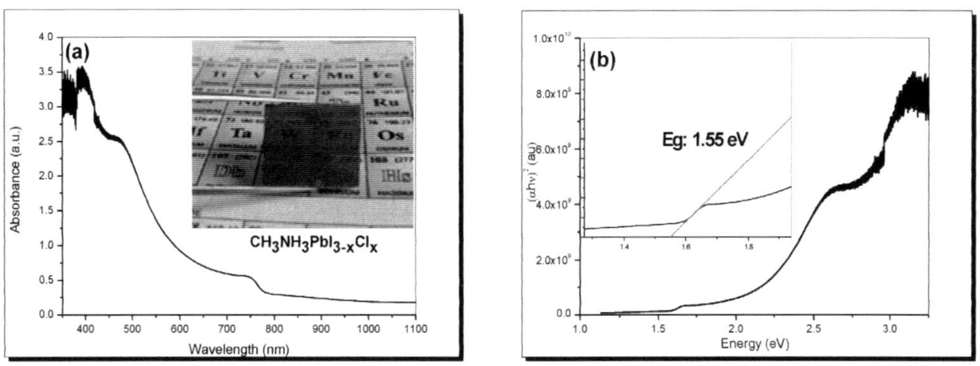


**Fig.1**. (a) Image of perovskite film on the PEDOT:PSS/ITO/glass and UV−visible absorbance spectra of the thin films. (b) The spectra of absorption coefficients as a function of the photon energy hν.

The structure of the resulting $CH_3NH_3PbI_{3-x}Cl_x$ mixed halide perovskite films was investigated by X-ray diffraction (XRD) as shown in Fig. 2. The XRD pattern of the spin coated film revealed the presence of two dominant peaks at the following 2θ values: 13.44° and 27.72°. These peaks correspond to the reflections from the (110) and (220) lattice planes that indicates the formation of the orthorhombic crystal structure [10].

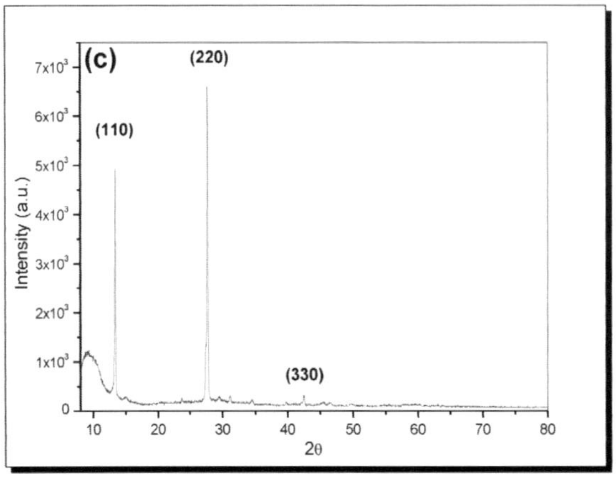


**Fig.2**. XRD pattern of a mixed halide perovskite $CH_3NH_3PbI_{3-x}Cl_x$ film.

Fig. 3 (a) shows a schematic diagram of the fabricated ITO/PEDOT:PSS/perovskite/PCBM/Al planar structured cells. As show in Fig 3 (b), the devices have the characteristic of a rectifying behavior in the dark. Moreover, under an illumination of 300 mW/cm$^2$, the current-voltage (J-V) characteristic of the devices is given in Fig 3 (c) and (d), that shows the change of the J-V characteristic before and after the annealing. The performance parameters of the devices are also summarized in Table 1.
The device I has a short-circuit current density ($J_{sc}$) of 2.36 mA/cm$^2$, an open-circuit voltage ($V_{oc}$) of 0.90 V and a fill factor (FF) of 71.5%, yielding an efficiency (η) of 5.07%. However, the annealed device I has a $J_{sc}$ of 3.04 mA/cm$^2$, a $V_{oc}$ of 0.84 V and a FF of 74.0%, leading to an efficiency of 6.31%. The effect of the annealing on the devices was significantly observed as improvement of current transport and the both resistance of the device as seen in the Table 1. The enhanced current transport is generally connected with decreased recombination of the photo-generated free carriers at the interfaces of the transport layers, passivated short-circuit current or the lower series resistance between the layer and the metal contact. In the our measurements for the device I, the value of FF increased from 71.5% to 74.0% dependent on the both resistances of the device while the $V_{oc}$ decreased 60 mV. Furthermore, the all performance parameters of the device II are better than before annealing, particularly at the FF and the efficiency. Therefore, the improvement of the performance can mainly attribute to the improved resistance of both series and shunt. Additionally, the carrier mobility might be enhanced between grains and also the interface states, which can be generated by a large density of crystal defects close to interface of the layers.

**Table 1.** The performance parameters of the cells after and before annealing.

| **Device** | **Annealing** | **$J_{sc}$ (mA/cm$^2$)** | **$V_{oc}$ (V)** | **FF (%)** | **η (%)** | **$R_{sh}$(kΩcm$^2$)** | **$R_s$(Ωcm$^2$)** |
|---|---|---|---|---|---|---|---|
| **I** | Before | 2.36 | 0.90 | 71.5 | 5.07 | 19.20 | 60 |
| | After | 3.04 | 0.84 | 74.0 | 6.31 | 39.24 | 43 |
| **II** | Before | 3.02 | 0.80 | 71.6 | 5.78 | 248.67 | 37 |
| | After | 3.20 | 0.82 | 74.9 | 6.54 | 313.22 | 32 |

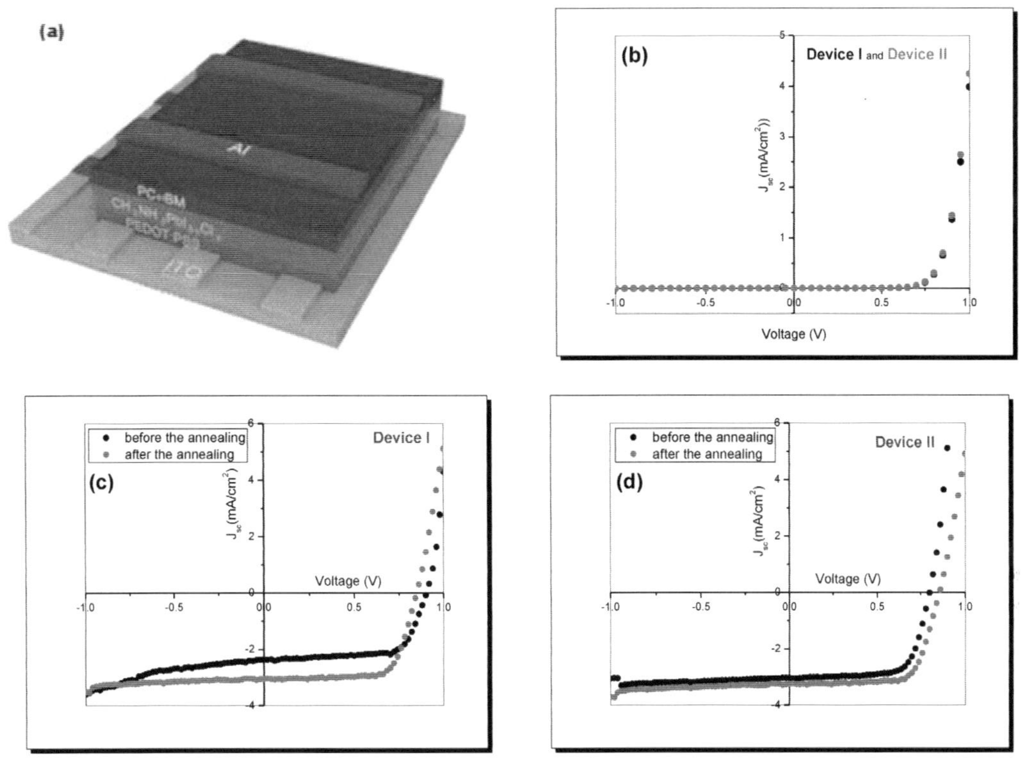


**Fig.3**. (a) Configuration of the solar cell in this study. (b) J – V curves of the cells in the dark after the annealing. (c) and (d) J – V curves of the cells under AM 1.5G simulated solar irradiation before and after the annealing.

## 4. Conclusion

In summary, we have developed the energy conversion efficiency of the organic-inorganic perovskite solar cells with the annealing after the fabrication. In spite of the photovoltaic efficiency was improved average 13%, which can be improved further optimization of the annealing processing. Therefore, we will systematically study effect of the annealing on the fabricated cells such as temperature and time.

## References

[1] Jeng J Y, Chen K C, Chiang T Y, Lin P Y, Tsai T D, Chang Y C, Guo T F, Chen P, Wen T C and Hsu J 2014, Advanced materials, 26, 4107–4113.
[2] Jung H S and Park N G 2015, small, 11(1), 10–25.
[3] Kojima A, Teshima K, Shirai Y and Miyasaka T 2009, Journal of the American Chemical Society, 131 (17), 6050–6051.
[4] http://www.nrel.gov/ncpv/images/efficiency chart.jpg.
[5] Wu Z, Bai S, Xiang J, Yuan Z, Yang Y, Cui W, Gao X, Liu Z, Jin Y and Sun B 2014, Nanoscale 6, 10505–10510.
[6] Seo J, Park S, Kim Y C, Jeon N J, Noh J H, Yoon S C and Seok S 2014, Energy Environ. Sci., 7, 2642.
[7] Barrows A T, Pearson A J, Kwak C K, Dunbar A D F, Buckleya A R and Lidzey D G 2014, Energy Environ. Sci., 7, 2944.
[8] Chen Q, Zhou H, Hong Z, Luo S, Duan H S, Wang H-H, Liu Y, Li G and Yang Y 2014, J. Am. Chem. Soc. 136, 622–625.
[9] Chae J, Dong Q, Huang J and Centrone A 2015, Nano Lett., 15(12): 8114–8121.

[10] Liu M, Johnston M B and Snaith H J 2013, Nature 501,395–398.

Journal of Physics: Conference Series **707** (2016) 012025 doi:10.1088/1742-6596/707/1/012025

# Temperature dependent electrical characteristics of Zn/ZnSe/n-GaAs/In structure

**M. Sağlam and B. Güzeldir**

Department of Physics, Faculty of Science, Atatürk University, 25240, Erzurum, Turkey

E-mail: msaglam@atauni.edu.tr

**Abstract**. We have reported a study of the I-V characteristics of Zn/ZnSe/n-GaAs/In sandwich structure in a wide temperature range of 80-300 K by a step of 20 K, which are prepared by Successive Ionic Layer Adsorption and Reaction (SILAR) method. The main electrical parameters, such as ideality factor and zero-bias barrier height determined from the forward bias I-V characteristics were found strongly depend on temperature and when the increased, the n decreased with increasing temperature. The ideality factor and barrier height values as a function of the sample temperature have been attributed to the presence of the lateral inhomogeneities of the barrier height. Furthermore, the series resistance have been calculated from the I-V measurements as a function of temperature dependent.

## 1. Introduction

Schottky barrier diodes (SBDs) are important research tools in the characterization of new semiconductor materials and, at the same time, the fabrication of these structures plays a crucial role in constructing some useful devices in technology [1-3]. SBDs with low barrier height have found applications in devices operating at cryogenic temperatures as infrared detectors and sensors in thermal imaging [4, 5]. The performance and reliability of an SBD are drastically influenced by the interface quality between the deposited metal and the semiconductor surface. The inhomogeneity of SBH may be due to the polycrystalline structure of metallic layer, the different crystallographic orientation of the grains together with possible different phase composition after some thermal treatments, and very often dielectric interfacial layers with non-homogeneous thickness [6-8]. The analysis of the I-V characteristics of the SBDs at room temperature does not give detailed information about their conduction process or the nature of barrier formation at the SBD interface. The temperature dependence of the I-V characteristics allows us to understand different aspects of conduction mechanisms [9,10].

In the SILAR technique the substrate is immerse into separate cation and anion precursor solutions and rinsed with purified water after each immersion [11]. The method involves an alternative immersion of the substrate in a solution containing a soluble salt of the cation and anion of the compounds to be grown. The substrate supporting the growing film is rinsed in high purity deionized water after each immersion in order to avoid homogeneous precipitation. The immersion and rinsing time periods can be experimentally determined.

In this study, some electrical properties of the Zn/ZnSe/n-GaAs/In sandwich structure have been investigated in the temperature range of 80-300 K by steps of 20 K.

## 2. Experimental process and data analysis

For fabricated Zn/ZnSe/n-GaAs/In structure, n-GaAs semiconductor wafer with relatively carrier density $2.5\times10^{17}$ $cm^{-3}$ and (100) orientation was used. The substrate was sequentially cleaned with trichloroethylene, acetone and methanol and then rinsed in deionized water. The native oxide on the surface was etched in sequence with acid solutions ($H_2SO_4$:$H_2O_2$:$H_2O$=3:1:1) for 60s, and HF:$H_2O$=1:1 for another 30s. After a rinse in deionized water of 18 MΩ and a blow-dry with $N_2$, ohmic contact on the back side of the sample was formed by evaporating indium metal at a pressure of about $10^{-5}$ Torr, followed by annealing at 425 °C for 3 min in $N_2$ atmosphere. After ohmic contact was made, the ohmic contact side and the edges of the n-GaAs semiconductor substrate was covered by wax so that the polished and cleaned front side of the semiconductor sample was exposed to the cationic precursor solution employed for SILAR method. Zinc selenide thin films were deposited by using $ZnCl_2$ and freshly prepared sodium selenosulphate ($Na_2SeSO_3$) solutions. The $Na_2SeSO_3$ solution was prepared by mixing 10 g selenium powder (99 % purity) with 100 gr anhydrous sodium sulfite in 500 ml of distilled water with constant stirring for 8-10 h at 80 $^0$C. It was sealed and kept overnight, since on cooling, a little selenium separated out from the solution. It was then filtered to obtain a clear solution. A single SILAR deposition cycle consisted of 20 s adsorption of $Zn^{+2}$ ions, 50 s rinsing with double distilled water, 30 s adsorption and reaction of $Se^{2-}$ ions with preabsorbed $Zn^{2+}$ ions on the substrate and 50 s rinsing with double distilled water. By repeating such deposition cycles 45 times, we obtained ZnSe thin films on GaAs substrate. Zn dots with diameter of about 1.0 mm were evaporated on the ZnSe thin film in vacuum coating unit at about $10^{-5}$ torr. In this way, Zn/ZnSe/n-GaAs/In sandwich structure was obtained.

The I-V characteristics of the devices were measured in the temperature range of 80-300 K using Leybold Heraeus closed-cycle hellium cryostat, HP4140B picoammeter and a HP model 4192A LF impedance analyser under dark conditions.

**Results and discussion**

Figure 1 shows the semi-logarithmic forward bias I-V characteristics of Zn/ZnSe/n-GaAs/In structure, measured at different temperatures over the range of 80-300 K. The parameters of the hetero-contact were obtained using a simple Schottky model which assumes a well-defined fixed potential barrier at the interface over which the electrons are thermionically emitted [13]. The I-V equation unrespect to the thermionic emission theory in the presence of interfacial layer is given by [1].

$$I = AJ = \left\{AA^*T^2\exp\left(\frac{-q\Phi_{B0}}{kT}\right)\right\}\left\{\exp\left(\frac{q(V-IR_s)}{nkT}\right)-1\right\} \quad (1)$$

where A is the effective area of diode, $A^*$ is the effective Richardson constant, $\Phi_{Bo}$ is the barrier height, T is the temperature, q is the elementary charge, k is the Boltzmann constant, $R_s$ is the series resistance and n is the ideality factor. Also, $I_0$ is the saturation current derived from the straight line intercept of lnI at V = 0 and is given by:

$$I_0 = AA^*T^2\exp\left\{\frac{-q\Phi_{B0}}{kT}\right\} \quad (2)$$

It can be seen that the I-V plots shift towards the slightly higher bias side with decreasing temperature. From Eq. (1), the ideality factor n can be written as:

$$n = \frac{kT}{q}\frac{dV}{d\ln I} \quad (3)$$

The ideality factor n refers to the linear part of characteristics where the effect of series resistance ($R_s$) can be negligible in the linear bias region. The $\Phi_{Bo}$ was calculated using theoretical value $A^*$ (8.16 $A/cm^2 K^2$) and extrapolated $I_0$ at each temperature according to [12]

$$\Phi_{B0} = (kT/q)\ln(AA^*T^2/I_o) \quad (4)$$

According to the Eqs. (3) and (4) the experimental values of n and $\Phi_{Bo}$ were calculated, respectively, and shown Figure 2 at various temperature. It is observed that both the values of n and $\Phi_{Bo}$ are strong functions of temperature. n=1 for an ideal diode. However, n has usually a value greater than unity.

The high value of ideality factor has been attributed to insulator layer (ZnSe) at metal/semiconductor interface and particular distribution of interface states localized at semiconductor/insulator (ZnSe/GaAs) interface. Also high values of *n* can be attributed to the presence of the interfacial layer and a wide distribution of low Schottky barrier height patches (or barrier in-homogeneities), and therefore, to the bias voltage dependence of the Schottky barrier height [13].

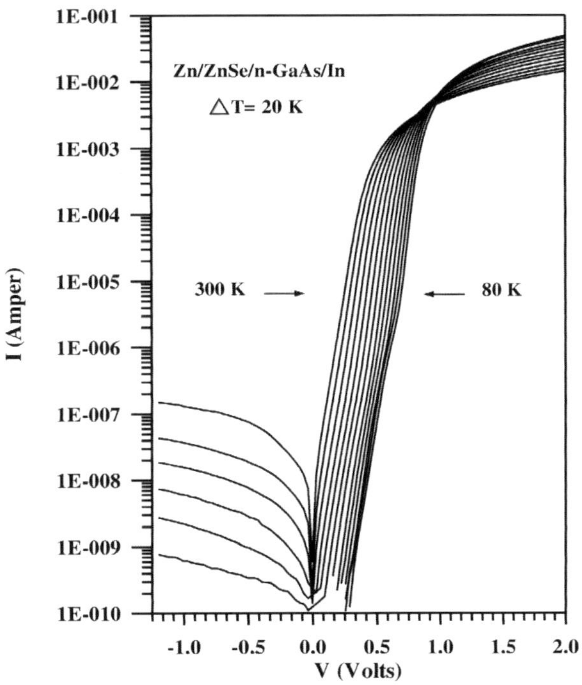


**Figure 1.** The semi-log reverse and forward bias current-voltage characteristics of Zn/ZnSe/n-GaAs/In sandwich structure at various temperature.

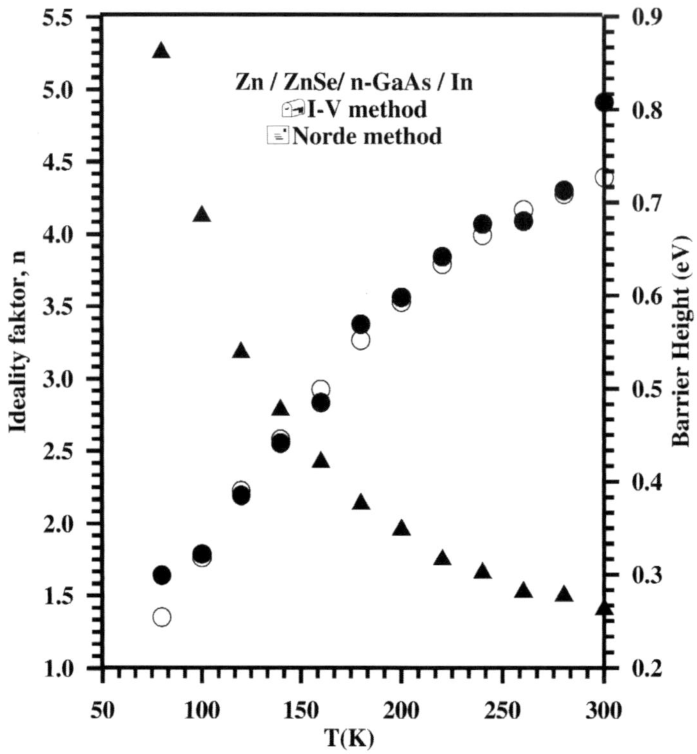


**Figure 2.** Temperature dependence of the barrier height and ideality factor.

The series resistance is an important parameter on the electrical characteristics of the rectifying contacts or junctions. Several methods to extract the series resistance $R_s$ of a device have been suggested [20]. For In our case, in order to calculate series resistance in the device with high ideality factor and series resistance, we used Norde method modified by Bohlin [13]. The Norde function, F (V), is defined as;

$$F(V) = \frac{V}{\gamma} - \frac{1}{\beta}\ln\left(\frac{I(V)}{AA^{*}T^{2}}\right) \tag{5}$$

where γ is an arbitrary constant greater than the ideality factor, I(V) is current obtained from the I–V curve and the other parameters are described above. Thus, the effective barrier height and series resistance can be determined by

$$\Phi_B = F_m + \left[\frac{(\gamma - n)}{n}\right]\left[\frac{V_m}{n} - \frac{kT}{q}\right] \tag{6}$$

$$R_S = (\gamma - n)\frac{kT}{qI_m} \tag{7}$$

Once the minimum of the F(V)-V plot is determined, the barrier height can be obtained from here, where $F_m$ is the minimum point of F(V) curve, and $V_m$ is the corresponding voltage, $I_m$ is the current corresponds to the minimum $V_m$. The F (V)-V plots of Zn/ZnSe/n-GaAs/In structure have been shown in Figure 3 as a function of temperature. Series resistance is influenced by the presence of the interface layer between the metal and the semiconductor and the series resistance ($R_s$) of device is responsible for the shape of non-linear I-V characteristics in forward bias current-voltage. When a forward bias V is applied across the device, the applied voltage V will be shared by the interfacial layer, the depletion layer and the series resistance combination of the structure. The high series resistance behavior can be attributed to the interface states and interface layer. Series resistance has been strongly changed with sample temperature which is shown in Figure 4 [13].

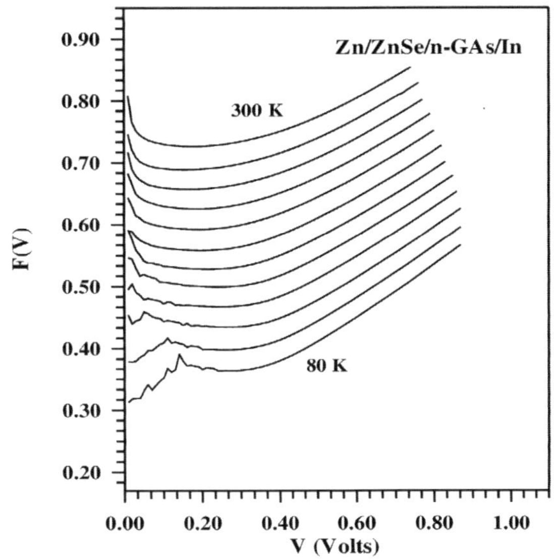


**Figure 3.** Experimental F(V) vs. V curves of Zn/ZnSe/n-GaAs/In structure at various temperature,

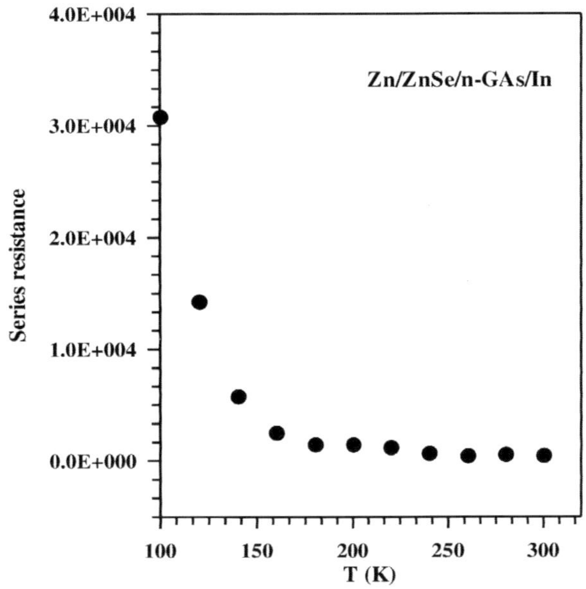


**Figure 4.** Temperature dependence of series resistance obtained from the experimental Norde plots of the Zn/ZnSe/n-GaAs/In structure.

In this study, the I-V characteristics of Zn/ZnSe/n-GaAs/In structure which is obtained by growing of ZnSe thin film on n-GaAs substrate by SILAR, have been investigated in the wide temperature range of 80-300 K by the steps of 20 K. Obtained from the forward bias I-V measurement values of n and $\Phi_{B0}$ have been shown strong temperature dependence. The values of n increases while $\Phi_{B0}$ decreases with decreasing temperature from 300 K to 80 K, respectively. The origin and nature of the increase in the ideality factor and decrease in the barrier height with decrease in temperature in the Zn/ZnSe/n-GaAs/In structure has been successfully explained on the basis of the thermionic emission with Gaussian distribution of the barrier heights. This behavior is attributed to Schottky barrier in-homogeneities by assuming a Gaussian distribution of barrier heights due to barrier heights in-homogeneities that prevail at metal-semiconductor interface.

**References**

[1] Rhoderick E H and Williams R H 1988 Metal-Semiconductor Contacts 2nd edn (Oxford: Clarendon)
[2] Henisch H K 1984 Semiconductor Contacts (London: Oxford University) p123
[3] Tung R T 2001 Mater. Sci. Eng. R. **35** 1
[4] McCafferty P G, Sellai A, Dawson P and Elabd H 1996Solid-State Electron. **39** 583
[5] Chand S and Kumar J 1996 Appl Phys A **63** 171
[6] Pakma O, Serin N, Serin T, Altindal S 2008 J Appl Phys **104** 014501.
[7] Osvald J, Dobrocka E 1996 Semicond Sci Technol **11** 1198.
[8] Tung R T 1992 Phys Rev B **45** 13509
[9] Osvalda J, Horvath Zs J 2004 Appl Surf Sci **234** 349.
[10] Huang S, Lu F 2006 Appl Surf Sci **252** 4027
[11] Nicolau Y F 1985 Appl Surf Sci **22** 1061.
[12] Altuntaş H, Altındal Ş, Shtrikman H, Özçelik S 2009 Microelectronics Reliability **49** 904.
[13] Sze S M 1981 Physics of Semiconductor Device second ed Wiley New York p. **245**.

# Substrate temperature effects on reactively sputtered $Cr_2O_3$/n-Si heterojunctions

**Yusuf Selim Ocak[1], Ali Ahmed Issa[2], Mustafa Fatih Genisel[1], Ahmet Tombak[3], Tahsin Kilicoglu[3]**

[1]Department of Science, Faculty of Education, Dicle University, Diyarbakir, Turkey
[2]Department of Science, Institute of Natural and Applied Science, Dicle University, Diyarbakir, Turkey
[3]Department of Physics, Faculty of Art & Science, Batman University, Batman, Turkey

E-mail: yusufselim@gmail.com

**Abstract.** To see the effects of substrate temperature on $Cr_2O_3$/n-Si heterojunctions, $Cr_2O_3$ thin films were formed on n-Si and glass substrates at 40, 150 and 250 °C by radio frequency (RF) reactive sputtering technique. High purity Cr was used as target and oxygen was used as reactive gas. Optical properties of $Cr_2O_3$/n-Si thin films were analyzed using UV-vis data. The band gaps of the films were compared. The electrical properties of $Cr_2O_3$/n-Si heterojunction were tested by their current voltage (*I-V*) measurements in dark. It was observed that the heterojunction which was fabricated by forming $Cr_2O_3$ thin film at 250 °C gave better rectification. The characteristic electrical parameters such as barrier height, ideality factor and series resistance were calculated by using its *I-V* data. The influence of light intensity on photovoltaic effect behavior of the device was also calculated, finally the barrier height value of the structure obtained from capacitance-voltage (*C-V*) data were compared with the one calculated from *I-V* measurements.

## 1. Introduction

Chromium oxide with $Cr_2O_3$ formula is a wide band p-type semiconductor. It has been used as protective coating because of its outstanding wear resistance. Kao et al (1989) formed $Cr_2O_3$ thin films by sputtering of $Cr_2O_3$ thin films for various substrate temperatures. It was showed that the $Cr_2O_3$ thin films formed at 150 °C in pure argon annealed at 300 °C had the hardness about 25 GPa [1].

Erdogan and Gullu [2] obtained $Cr_2O_3$ nanoparticles by hydrothermal technique. They formed $Cr_2O_3$ thin film on p-Si substrate by spinning using nanoparticle solution and annealed at 500 °C for 1 h. They studied optical, morphological and structural properties thin film. They used x-ray diffraction (XRD), non-contact atomic force microscopy (AFM), ultraviolet-visible spectroscopy (uv-vis) and photoluminescence (PL) spectroscopy. They reported that the thickness of the film is about 70 nm and the band gap of the film was 3.08 eV. They also formed Au/$Cr_2O_3$/p-Si metal-insulator-semiconductor (MIS) diodes to determine electronic parameters by means of *I-V* and *C-V* data. They calculated ideality factor and barrier height of the device as 2.14 and 0.74 eV, respectively. They also calculated interface state density distribution of the diode.

Formation of thin films to fabricate heterojunctions can be performed using many kinds of techniques such as spin coating, ultrasonic spray, thermal evaporation, pulsed laser deposition, electron beam and magnetron sputtering. It is very difficult to form thin films of refractory metals or ceramics. Magnetron sputtering technique is a unique method to obtain uniform thin films of ceramics refractory metals or ceramics [3].

In this study, $Cr_2O_3$ thin films were formed on n-Si semiconductor and substrate by means of reactive sputtering technique at 40, 150 and 250 °C. A chromium metal target used as source and $O_2$ used as reactive gas. Optical properties of the thin films were determined using UV-vis data. Au metal was evaporated on $Cr_2O_3$/n-Si structures. Electrical properties of the $Cr_2O_3$/n-Si heterojunctions were examined using their *I-V* measurements in dark at room temperature. It was seen that the device obtained using the $Cr_2O_3$ formed at 250 °C gave an excellent rectification property. The characteristic electrical parameters including barrier height, ideality factor and series resistance of the rectifying diode were determined. The effect of the light on photoelectrical parameters such as sensitivity to light, open circuit voltage and short circuit current were determined for various illumination intensities. Furthermore, capacitance-voltage properties of the structure were also analyzed.

## 2. Experimetal Procedures

In the study, an n-Si wafer with (100) orientation and 1-10 Ωcm resistivity is used. The ohmic contact was formed on unpolished side of n-Si wafer as in literature. $Cr_2O_3$ thin films were growth on n-Si semiconductor and glasses by reactive sputtering technique using Nanovak NVTS 400 vacuum system. Cr metal was used as target and $O_2$ was used as reactive gas. For all processes, firstly, pressure of the vacuum system was decreased at $10^{-6}$ Torr. Argon gas was send to system and the pressure was increased to 20 mTorr. The flow of the Argon gas was about 4 ccm. 80 W power was applied to high purity Cr target and 0.4 ccm oxygen was send to vacuum system to make reactive sputtering process. This process continued 10 minutes. The substrate temperatures were 40, 150 and 250 °C during sputtering process. To increase crystallianity, $Cr_2O_3$ thin films were annealed at 500 °C for 30 minutes in air ambient. Au metal was evaporated to obtain front contact. The circular diameters of the Au contacts were about 1.5 mm. To make circular front contacts a shadow mask is used. Electrical and photoelectrical characteristics of Au/$Cr_2O_3$/n-Si structures were analyzed using its current-voltage (*I-V*) and capacitance-voltage (*C-V*) measurements by the help of Keithley 2400 sourcemeter and Agilent 4294 A impedance analyzer at room temperature in dark, respectively. To see the effect of the light on the photoelectrical properties of the device an Oriel 9600 solar simulator was used. Current-voltage (*I-V*) measurements were repeated under solar simulator between 40-100 mW/cm$^2$ with 20 mW/cm$^2$ intervals. Optical properties of the thin film on quartz were analyzed using Shimadzu 3600 Uv-vis-nir spectrophotometer.

## 3. Results and Discussion

### *1.1. Optical properties of $Cr_2O_3$ thin films on glasses*

Optical properties of the $Cr_2O_3$ thin films were analyzed by means of UV-Vis spectra. Figure 1 shows the absorbance vs. wavelength plots of $Cr_2O_3$. As seen from the figure, the absorbance of $Cr_2O_3$ thin films decreases sharply till 500-600 nm band then decreases

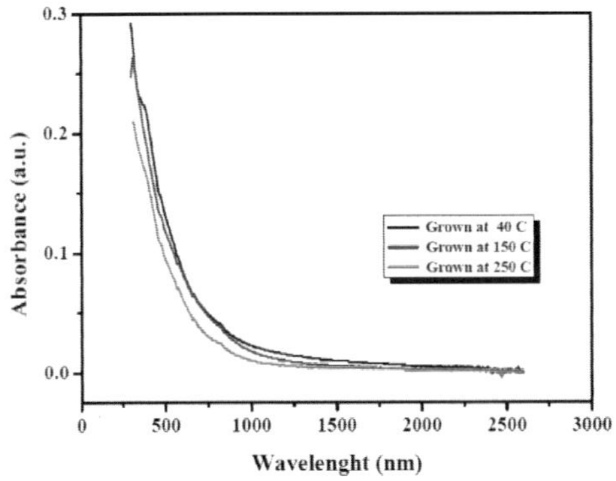


**Figure 1** Absorbance vs wavelength plots of $Cr_2O_3$ thin films on glasses

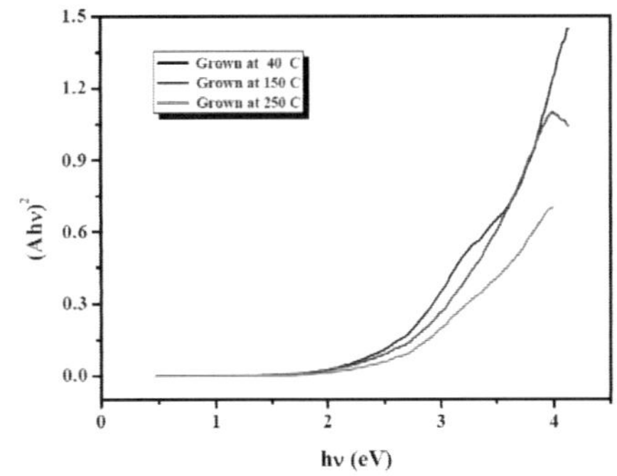


**Figure 2.** $(Ahv)^2$–$hv$ plots of $Cr_2O_3$ thin films on glasses

slightly till 1000 nm. After this value, the absorbance is nearly constant. It means that the $Cr_2O_3$ thin film has small absorbance in infrared region. Furthermore, the absorbance value of thin film decreases with increasing the growth temperature. The film deposited at 250 °C had the lowest absorbance values and the thin deposited at 40 °C had the highest absorbance values. The optical band gap ($E_g$) of the $Cr_2O_3$ can be calculated through the equation

$$\alpha h\nu = B(h\nu - E_g)^m \tag{1}$$

Where $\alpha$ is the absorption coefficient, $B$ is a constant, $h$ is the Plank constant, $m$ is ½ for direct band gap and 2 for indirect band gap. $Cr_2O_3$ has absorbance constant higher than $10^4$. Therefore, it can be said that $Cr_2O_3$ a direct band gap. In this study, the band gap of sputtered $Cr_2O_3$ thin films were calculated by the help of $(Ahv)^2$–$hv$ plot using equation 1 using. Figure 2 shows the $(Ahv)^2$–$hv$ sputtered $Cr_2O_3$ thin films on glasses and The $E_g$ of the $Cr_2O_3$ thin films were extracted as 2.99, 3.05 and 3.10 eV for the film deposited at 40, 150 and 250 °C, respectively. The results show the strong effect of substrate temperature on optical properties of $Cr_2O_3$ thin films.

### *1.2. Current-voltage properties of $Cr_2O_3$/n-Si heterojunctions*

The current-voltage measurements of $Cr_2O_3$/n-Si heterojunctions are presented in figure 3. As it is seen from the figure, the diodes fabricated using the film deposited at 40 and 150 °C showed nearly ohmic properties and the device fabricated using the film deposited onto n-Si at 250 °C had an excellent rectification. These behaviors presented the substrate temperature effects of electrical properties of $Cr_2O_3$/n-Si. The perfect rectification for the heterojunction fabricated by deposition of $Cr_2O_3$ onto n-Si substrate at 250 °C may because of the high quality of $Cr_2O_3$ thin film. *I-V* relationship of a heterojunction for qV>>3kT (where $q$ is the electron charge, $V$ is the applied voltage, $k$ is the Boltzmann constant and $T$ is the absolute temperature) is owing to thermionic current. According to thermionic emission theory, the current-voltage relationship of a diode with a series resistance can be written as [4]

$$I = I_0 \exp\left[\frac{q(V - IR_s)}{nkT}\right] \tag{2}$$

where, $R_S$ is the series resistance, $n$ is dimensionless ideality factor and $I_0$ is the saturation current which can be expressed

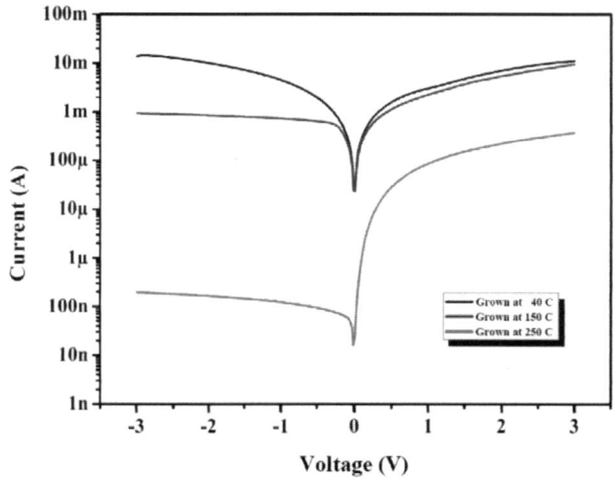


**Figure 3.** ln*I*-*V* plots of $Cr_2O_3$/n-Si heterojunctions at various substrate temperatures

$$I_0 = SA^*T^2 \exp\left(-\frac{q\Phi_b}{kT}\right) \quad (3)$$

where $S$ is the diode area, $A^*$ Richardson constant, $\Phi_b$ is the barrier height. The ideality factor of a diode can be calculated from the slope of the linear region of the In$I$-$V$ plot of the diode using the equation

$$n = \frac{q}{kT}\frac{dV}{d\ln(I)} \quad (4)$$

The ideality factor of Au/$Cr_2O_3$/n-Si heterojunction formed by deposition of $Cr_2O_3$ thin film was calculated as 1.29. Table 1 presents some electrical parameters of the Au/$Cr_2O_3$/n-Si diode. The ideality factor of an ideal diode 1.01 when image force lowering of the diode is taken into account. The ideality greater than this value shows a non-ideal diode behavior. The non-ideal diode behavior of the Au/$Cr_2O_3$/n-Si may be because of the effect on native oxide layer on n-Si semiconductor. It is known that except for special fabrications, a very thin native oxide layer occurs during fabrication process.

Another important electrical parameter of a diode is barrier height. The barrier height of the junction can be calculated using $I_0$ value obtained from interception of In$I$-$V$ plot to $I$ axis by the help of the equation

$$\phi_b = \frac{kT}{q}\ln\left(\frac{SA^*T^2}{I_0}\right) \quad (5)$$

The barrier heights of the device were calculated using saturation current value of Au/$Cr_2O_3$/n-Si by the help of the equation 5. The room temperature barrier height value of Au/$Cr_2O_3$/n-Si device determined as 0.761 eV. Saglam et al fabricated 24 Au/n-Si Schottky barrier dots and reported the barrier height values between 0.789 and 0.819 eV [5]. Therefore, Au/$Cr_2O_3$/n-Si diode has smaller barrier height value than conventional Au/n-Si MS diode. This result implies the strong effect of $Cr_2O_3$ thin film on the performance of the device.

The curvature in the plot of Au/$Cr_2O_3$/n-Si heterojuncton fabricated using the film deposited at 250 °C for high voltages is because strong effect of the series resistance on the current-voltage properties. The series resistance results from the contact wires or bulk resistance of the $Cr_2O_3$ and n-Si semiconductors. The series resistances of the Au/$Cr_2O_3$/n-Si structure were calculated by means of Norde functions given below as [6]

**Table 1.** Some electrical parameters of $Cr_2O_3$/n-Si heterojunction at room temperature

| ln*I-V* | | *Norde* | | *C-V* | |
|---|---|---|---|---|---|
| n | $\phi_b$ (eV) | $\phi_b$ (eV) | $R_S$ (Ω) | $\phi_b$ (eV) | $N_d$ ($cm^{-3}$) |
| 1.29 | 0.761 | 0.758 | 9169 | 0.718 | 7.434 x$10^{14}$ |

$$F(V) = \frac{V}{\gamma} - \frac{kT}{q}\left(\frac{I(V)}{AA^*T^2}\right) \tag{6}$$

where γ is the first integer greater than ideality factor and $I(V)$ is the current obtained from *I–V* measurements. The barrier height of device can be calculated using the minimum value of *F* vs. *V* plot using the equation

$$\phi_b = F(V_0) + \frac{V_0}{\gamma} - \frac{kT}{q} \tag{7}$$

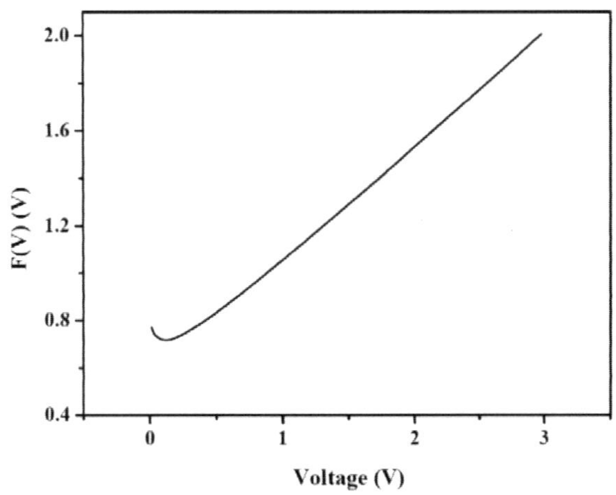

**Figure 4.** Norde plot of $Cr_2O_3$/n-Si heterojunction

where $F(V_0)$ is the minimum $F(V)$ value and $V_0$ is the corresponding voltage value. F(*V*)–*V* curve of Au/$Cr_2O_3$/n-Si heterojunction structure is depicted in Figure 4. The series resistance of the device is determined through the relation

$$R_S = \frac{kT(\gamma - n)}{qI_{\min}} \tag{8}$$

Where $I_{min}$ the corresponding current value at $V_0$. The barrier height and the series resistance values of the heterojunction are also depicted in table 1. As it is seen from the table, the series resistance and barrier height of the diode calculated as 9.169 kΩ and 0.758 eV. The barrier height values of the diode obtained from both ln*I-V* and F(*V*)-*V* plots are nearly same.

Furthermore, figure 5 shows the current-voltage plots of Au/$Cr_2O_3$/n-Si junction for various light intensities. As seen from the figure, light intensity has a strong

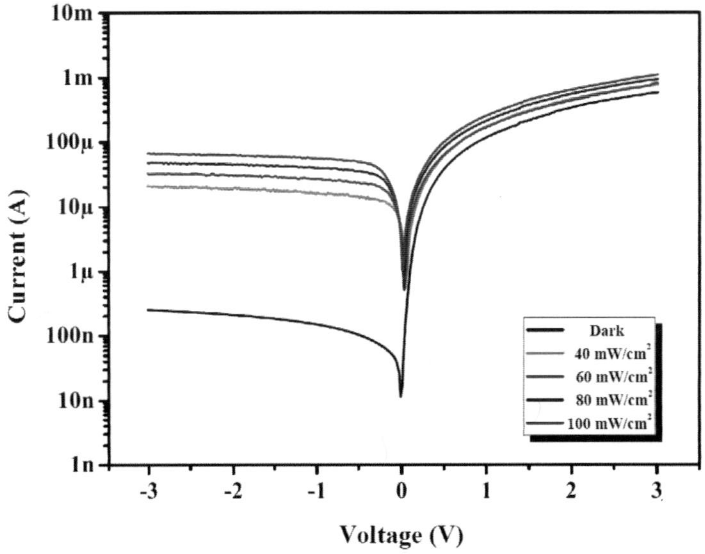


**Figure 5.** ln*I*-*V* plots of $Cr_2O_3$/n-Si heterojunction in dark and various illumination conditions

effect on photoelectrical performance of the device. Some photoelectrical parameters including sensitivity to light at -3V (illuminated current/dark current), open circuit voltage and short circuit current values are given in table 2. As seen from the table, all photoelectrical parameters increase with the increase in light intensity. This property showed that the device had photodiode characteristics which are attributable to the formation of electron–hole pairs following optical absorption. Similar results have been reported for various metal oxide/silicon heterojunctions. For instance, Lee et al [7] optimized n-ZnO/p-Si heterojunctions for photodiode applications. They deposited n-ZnO thin films onto p-Si semiconductor at various temperatures in $Ar/O_2$ ambient. They reported that all structures had typical rectifying properties and maximum amount of photocurrent observed for n-ZnO/p-Si heterojunction when the ZnO film was deposited at 480 °C. The results showed that film growing conditions had strong effects on photoelectrical characteristics. Similarly, Elsayed et al [8] prepared $CuNiO_2$ thin film by sol gel method onto p-Si substrate. They determined the size of $CuNiO_2$ nanoparticles as 115 nm and the band gap of the film as 2.4 eV. They showed that the Al/p-Si/$CuNiO_2$/Al heterojunction good rectification property in dark and high photosensitivity value of $1.02 \times 10^3$ under 100 mW/cm$^2$.

**Table 2.** Photoelectrical parameters of $Cr_2O_3$/n-Si heterojunction

| Light intensity mW/cm$^2$ | $I_{SC}$ (μA) | $V_{OC}$ (mV | Sensitivity to light (Times) |
|---|---|---|---|
| 40 | 3.79 | 10 | 83 |
| 60 | 4.08 | 30 | 130 |
| 80 | 4.34 | 30 | 190 |
| 100 | 5.14 | 50 | 263 |

## 4. Capacitance-voltage properties of Au/$Cr_2O_3$/n-Si heterojunction

Capacitance-voltage measurements Au/$Cr_2O_3$/n-Si heterojunction for various frequencies are depicted in figure 6. As it is seen from the figure, capacitance value of the structure decreased for forward bias values with the increase in frequencies. It means that at sufficiently high frequencies, charge carriers at the interface can no longer follow the AC signal. This situation showed the strong effect of interface states to capacitance of the structure. The depletion region capacitance of a diode is given [4]

$$\frac{1}{C^2} = \frac{2(V_{bi} + V)}{q\varepsilon_S S^2 N_d}, \tag{9}$$

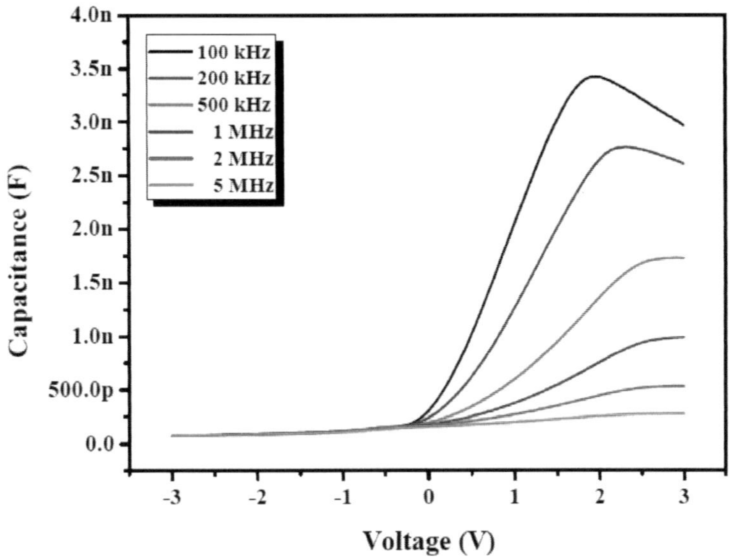


**Figure 6.** *C-V* plots of Au/$Cr_2O_3$/n-Si heterojunction for various frequencies

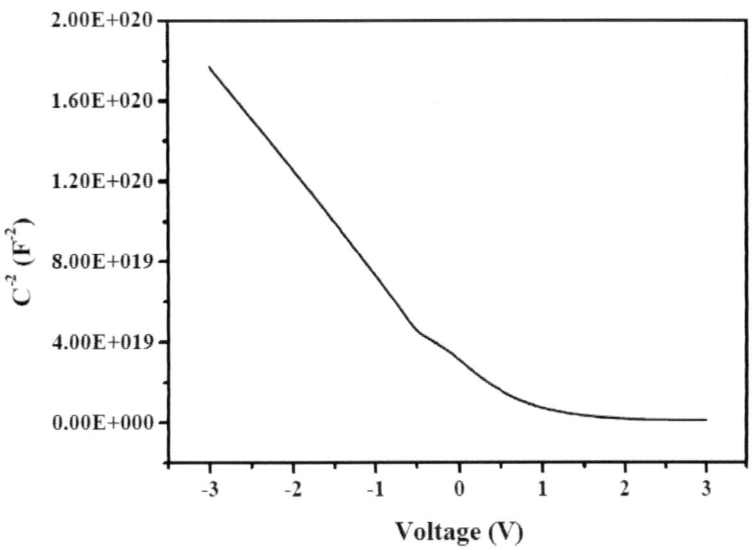


**Figure 7.** $C^{-2}$-*V* plot of Au/$Cr_2O_3$/n-Si heterojunction at 1 MHz

Where $V_{bi}$ is the built-in potential, $\varepsilon_S$ is the dielectric constant of the semiconductor and $N_d$ is the donor concentration. $N_d$ was determined from the intercept of a plot of $C^{-2}$ as a function of $V$, and $V_{bi}$ was determined from the gradient of this plot, as shown in Fig. 6.7. The built-in potential, the barrier height and the donor concentration of the Au/$Cr_2O_3$/n-Si diode were executed as 0.46 V, 0.718 eV and 7.434 x$10^{14}$ $cm^{-1}$, respectively. The barrier height value of the diode is inconsistent with the value obtained from ln*I*-*V* and Norde plots. Erdogan and Gullu [2] reported the carrier density, the built-in potential and the barrier height of the Au/$Cr_2O_3$/p-Si MIS diode as $5.28 \times 10^{14}$ $cm^{-3}$, 0.73 V and 1.10 eV, respectively. The obtained result for carrier density is very near to the one obtained in this study.

## 5. Conclusion

In this study, three $Cr_2O_3$ thin films deposited on n-Si and glass substrates 40, 150 and 250 °C by reactive sputtering technique. Optical properties of the $Cr_2O_3$ thin films on were analyzed by means of UV-Vis spectra. It was seen that the absorbance value of thin decreased with increase in deposition temperature. The band gaps of sputtered $Cr_2O_3$ thin films were calculated by the help of $(Ahv)^2$–$hv$ plot using UV-Vis data. The results showed the strong effect of substrate temperature on optical properties of $Cr_2O_3$ thin films. The electrical properties of $Cr_2O_3$/n-Si heterojunctions were examined using current-voltage measurements. The results showed that the diodes fabricated using the film deposited at 40 and 150 °C had nearly ohmic properties and the device fabricated using the film deposited onto n-Si at 250 °C had an excellent rectification. These behaviors presented the substrate temperature effects of electrical properties of $Cr_2O_3$/n-Si devices. The perfect rectification for the heterojunction fabricated by deposition of $Cr_2O_3$ onto n-Si substrate at 250 °C may be because of the high quality of $Cr_2O_3$ thin film. The electrical properties of $Cr_2O_3$/n-Si device fabricated by deposition of $Cr_2O_3$ onto n-Si substrate at 250 °C analyzed by means of both current voltage and capacitance-voltage measurements. Furthermore, current-voltage measurements of Au/$Cr_2O_3$/n-Si junction were taken under a solar simulator with AM1.5 Global filter for various light intensities. It was seen that light intensity had a strong effect on photoelectrical performance of the device. Some photoelectrical parameters including sensitivity to light at -3V (illuminated current/dark current), open circuit voltage and short circuit current values were calculated. It was seen that all photoelectrical parameters increase with the increase in light intensity. This property showed that the device had photodiode characteristics which are attributable to the formation of electron–hole pairs following optical absorption.

## 6. References

[1] Kao A S, Doerner M F, Novotny V J 1989 *J Appl Phys* ***66*** **(11)** 5315-5321.
[2] Erdoğan I Y, Gullu O 2010 *Appl Surf Sci* **256 (13)** 4185-4191.
[3] Ocak Y S, Genisel M F, Kilicoglu T 2010 *Microelectron Eng* **87** 2338–2342
[4] Rhoderick E H, Willams R H 1988, Metal–Semiconductor Contacts, Clarendon, Oxford
[5] Saglam M, Cimilli F, Turut A 2004 *Physica B* **348(1)** 397-403.
[6] Norde H 1979 *J Appl Phys* **50** 5052–5053.
[7] Lee J Y, Choi Y S, Kim J H, Park M O, Im S 2002 *Thin Solid Films* **403** 553-557.
[8] Elsayed I A, Cavas M, Gupta R, Fahmy T, Al-Ghamdi A A, Yakuphanoglu F 2015 *J Alloy Comp* **638** 166-171.

# The effect of Sn doping Urbach Tail and optical absorbtion measurements of InSe crystal

**Songul Duman[1], Bekir Gurbulak[2], Seydi Dogan[3] and Fikriye Seyma Ozcelik[2]**

[1]Department of Basic Sciences, Faculty of Sciences, Erzurum, Technical University, 25240, Erzurum, Turkey
[2]Department of Physics, Faculty of Sciences, Atatürk University, 25240, Erzurum, Turkey
[3]Faculty of Engineering and Architecture, Balıkesir University, Balıkesir, Turkey
E-mail: songul.duman@erzurum.edu.tr

**Absract.** InSe and n-InSe:Sn crystals grown by Bridgman-Stockbarger method. Absorption measurements have been carried out in InSe and InSe:Sn samples in the temperature range 10–320 K with a step of 10 K. Sn doping to InSe have increased the absorption intensity and caused both increasing in the Urbach energy, decreasing in the steepness parameters and shifting of the absorption edge towards the shorter wavelengths. The steepness parameters and Urbach energy values for InSe and InSe:Sn samples have also increased with increasing sample temperature in the range 10-320 K.

## 1. Introduction

In recent years, owing to very interesting physical properties, the binary semniconductor crystals such as InSe, GaSe have raised considerable deal of research interest followed by technological application in the fields solar cells [1-3], batteries [4], optoelectronics [5] and photoconductive [6, 7] as well as in commercial and industrial applications in micro-devices where the compounds are used as solid-state ionic conductors [8].

Measurements of absorbance and transmittance in $In_XSe_{1-x}$ thin films in the spectral range of 1.1 to 3.9 eV carried out at room temperature [9]. Gürbulak et al. have grown InSe:Ho single crystals by a method which is similar to the direct freezing. The first exciton energies for n=1 were calculated as 1.315 eV in n-InSe:Ho, 1.331 eV in InSe and 1.326 eV in InSe:Dy at 10 K [10, 11].

Ateş et al. reported that the absorption edge shifted towards longer wavelengths and a decrease of intensity in absorption spectra occurred under an electric field of 5 kV/cm. At 10 and 320 K, in InSe the first exciton energies were calculated as 1.350 and 1.311 eV for zero voltage and 1.334 eV 1.301 eV for electric field, respectively [12].

The cleaved surfaces do not need any additional treatment at p–n-junction formation and are chemically inert at ambient conditions [3, 13]. Bonding within the layers is ionic-covalent while the layers are bonded together by Van der Waals-like interactions. Owing to these relatively weak interlayer forces, there are different stacking arrangements of the layers leading to the four polytypes $\gamma, \varepsilon, \beta$ and $\delta$ [14] and also cleaving InSe and InSe:Sn ingots into the perpendicular planes can easily be done. Each layer is formed in packets of two In and two Se sublayers in the sequences of Se-In-In-Se. [15]. InSe single crystals can be grown as both p and n type depending on dopant elements added during the growth process. Detailed information on impurity levels in InSe is important in view of its possible application in high-quality devices. There are few studies available on the impurity levels of n-InSe doped with group IV atoms of Sn, Si and Ge [16, 17]. The behaviour of semiconductor optical properties in the presence of electric field is an important issue, both from the standpoint of understanding the basic physics involved and in terms of potential applications in opto-electronic devices. The modification of the band edge absorption under the application of an external electric field was studied independently by Franz [18] and Keldysh [19] known as Franz–Keldysh effect (FKE). The FKE occurs in bulk semiconductors when a strong dc electric field is applied, resulting in a red-shift of the fundamental absorption edge and oscillations in the optical absorption above the band gap.

In the present paper, the effects of both Sn doping and electrical field were investigated on the absorption edge, the steepness parameter and Urbach energy for InSe and InSe:Sn crystals in the temperature range 10-320 K. To our knowledge, this is the first results for the effect of an electric field on the absorption edge of Sn-doped InSe single crystals.

## 2. Experimental procedures

InSe and InSe:Sn (0.03 wt% Sn) monocrystals were grown in our growth laboratory by the modified Bridgman-Stockbarger at the same time. For the constituent of monocrystal $A^{III}B^{VI}$, the basic elements having high purity (5N) elements were weighed in a stoichiometric ratio accurate to 0.1 mg. The total mass of the elements was about 20 g. The growth programme for InSe and InSe:Sn single crystals are given in Fig. 1. The melting point of InSe which is 660±5 °C [20], was determined from the phase diagram. The detailed knowledge about the quartz ampoules used this study has been given Ref. [21].

The grown InSe and InSe:Sn single crystal ingots were cleaved into perpendicular planes of naturally cleaved planes using a razor blade and cut into 48±5 μm thicknesses had no cracks or voids on the surface. Since the cleaved faces of the grown ingots had the natural mirror-like cleavage faces, no polishing or cleaning treatments were carried out on the cleaved faces of these crystals. It was found that our samples exhibited n-type conductivity by the means of hot probe technique.

The optical absorption measurements as a function of temperature were made in closed-cycle He cryostat. A Perkin Elmer UV/Vis Lambda 2S Spectrometer with spectral range of 190-1100 nm and better than ±0.3 nm wavelength accuracy was used. Therefore $E_g$ can be calculated with an accuracy of approximately better than 0.6 meV considering the wavelength accuracy of our spectrometer. The sample temperature was varied from 10 K to 320 K with a step of 10 K.

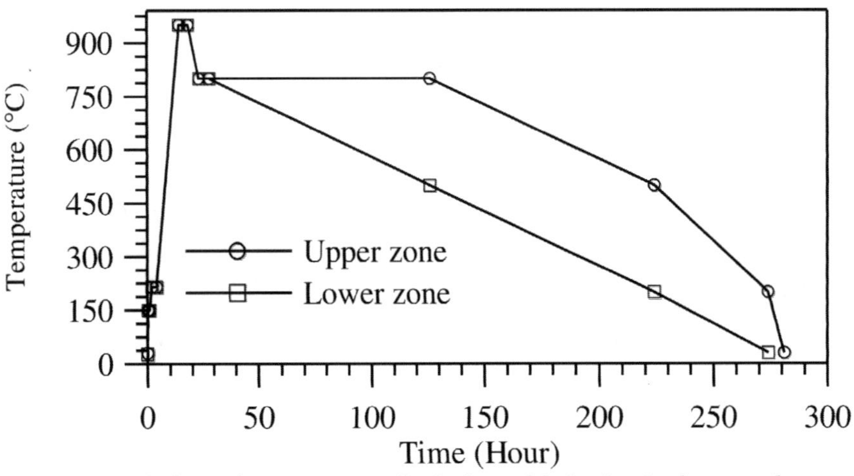


**Figure 1.** Growth programme for InSe and InSe:Sn single crystals

## 3. Basic equations

The optical absorption coefficients ($\alpha$) of the samples were determined from the following transmission data relationship [22]

$$T = (1 - R)^2 \exp(-A) = (1 - R)^2 \exp(-\alpha d)$$

(1)

where R is the reflectivity of the sample, $A$ is absorbance and $d$ is thickness.
The change of the band gap with temperature is given as following:

$$\frac{dE_g}{dT} = \left(\frac{dE_g}{dT}\right)_{l-ex} + \left(\frac{dE_g}{dT}\right)_{e-p} \tag{2}$$

where the first and second terms depend on crystal lattice thermal expansion and electron-phonon interaction, respectively.

In most semiconductors, the following empirical expression suggested by Varshni is used to describe the temperature dependence of the excitonic transition energy [23]:

$$E_{ex}(T) = E_{ex}(0) - \delta \frac{T^2}{T+\beta} \tag{3}$$

where $E_{ex}(T)$ is the excitonic energy at sample temperature $T$, $E_{ex}(0)$ is the excitonic transition energy at 0 K, and δ and β are parameters being characteristics of the material. The parameter $\beta$ is usually used to estimate of Debye temperature ($\theta_D$). The temperature variation of excitonic peak or band gap energy of semiconductors have been fitted by Cody as below [24]:

$$E_{ex}(T) = E_{ex}(0) - B\left[((\theta_E / T) - 1)^{-1}\right] \tag{4}$$

where B is a constant and $\theta_E$ is the Einstein characteristic temperature and corresponds to the average phonon temperature. It is known that $\theta_E$ is closely related to the $\theta_D$ by the relation $\theta_D \approx \frac{4}{3}\theta_E$ [25].

That absorption edge increases exponentially in a number of insulators, including ionic crystals, semiconductors, and organic crystals follows the below empirical expression [25]:

$$\alpha(E,T) = \alpha_0 \exp\left[\frac{\sigma(T)}{k_B T}(E - E_0)\right] \tag{5}$$

where $E = \hbar w$ is the photon energy, $\alpha_0$ and $E_0$ are characteristic parameters of the material, $\sigma(T)$ is the steepness parameter, and $k_B$ is the Boltzmann constant . Eq. (5) implies that logarithm of $\alpha$ plotted as a function of $E$ can be approximated by a straight line in energies just below the fundamental absorption edge (AE). The extrapolations of those lines for various temperatures usually converge at the same point $(E_0, \alpha_0)$, which is called the 'converging point'. This empirical rule was found by Urbach [26] for silver halides and established by Martienssen [27] for alkali halides.

The steepness parameter $(\sigma(T))$, which is characterizing the steepness of the straight line near the absorption edge, is expressed empirically as a function of temperature by [26-28] as

$$\sigma(T) = \sigma_o \left( \frac{2k_B T}{\hbar w_p} \right) \tanh \left( \frac{\hbar w_p}{2kT} \right) \tag{6}$$

where $\sigma_0$ is a parameter, non depending on the temperature and depending on the material, being inversely proportional to the strength of the coupling between excitons and phonons [29]. Some researchers have stated that $\hbar w_p$ corresponds to the energy of phonons associated with the Urbach tail [29, 30]. $kT/\sigma$ which can be derived from above equation is called Urbach's energy ($E_u$). $E_u$ represents the thermal disorder or the occupancy level of phonon states in crystals [28]. $E_u$ can be modelled as an Einstein oscillator taking into account contributions of structural and thermal disorder in the sample by considering Skettrup's theory [31]. According to this model, Urbach's energy can be expressed by:

$$E_u(X,T) = [E_u(X) + E_u(T)] = A + B\left( \frac{1}{e^{\theta_E/T} - 1} \right) \tag{7}$$

where A and B are the parameters related to structural and thermal disorders.

## 4. Result and discussion

Figs. 2 shows the absorption spectra of InSe and InSe:Sn at 10, 80, 160, 240 and 320 K under zero electric field (ΔU= 0 V). Sn doping to InSe caused an increasing in absorption peak intensities. As can be seen in Fig. 2 the absorption edge shifts towards to shorter wavelengths and band gap energy increases by Sn doping.

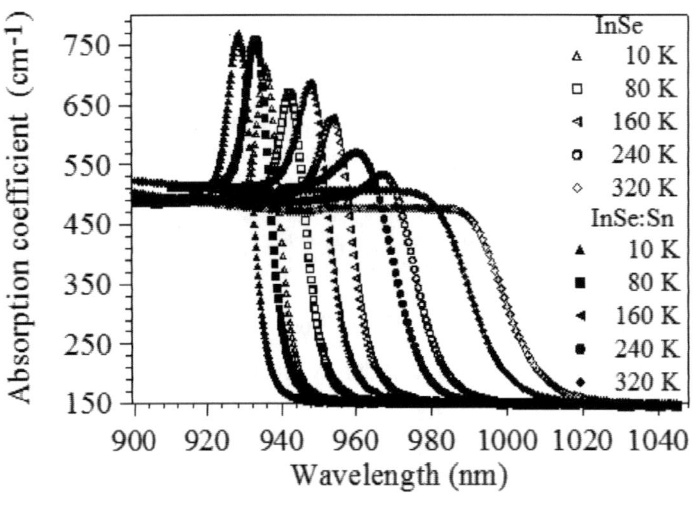


**Figure 2.** Absorption spectra of InSe and InSe:Sn crystals obtained at 10, 80, 160, 240 and 320 K.

The Urbach tail was observed for both samples between 10-320 K. The absorption spectra for both samples obey the Urbach rule and typical Urbach tails for various temperatures are shown in Fig. 3. Sn doping into InSe has changed both $E_0$ and logarithm of absorption coefficient values.

Figs. 4 and 5 show the $E_u$ which corresponds to the inverse of a slope of the Urbach tails of the InSe and InSe:Sn samples and steepness parameters (σ) of these samples as a function of temperature, respectively. Fig. 4 represents the experimental data together with the solid curves representing the curves fitted to Eq. (7) for Urbach's energy. A, $B$ and $\theta_E$ parameters of InSe and InSe:Sn crystals were obtained from a fit to Eq. (7). These obtained parameters are given in the Table 1. By taking into the account these parameters, it has been seen that the structural disorder contribution has increased with Sn doping and thermal disorder contribution decreases.

**Table 1.** Values of the parameters A, B and $\theta_E$ obtained from a fit to Eq. (7).

| Fit to Eq. (17) | A (meV) | B (meV) | $\theta_E$ (K) |
|---|---|---|---|
| n-InSe | 7.980 | 19.280 | 260±10 |
| n-InSe:Sn | 9.370 | 18.250 | 260±10 |

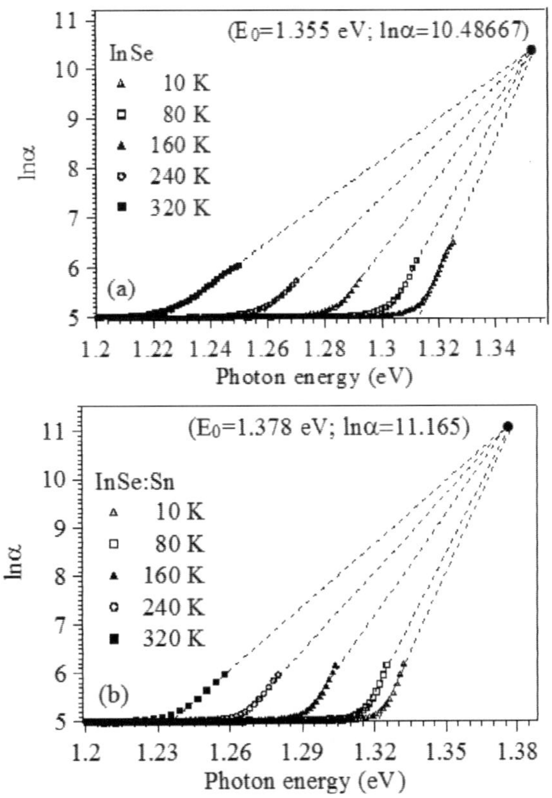


**Figure 3.** $\ln\alpha$ vs. photon energy ($\hbar w$) plots for (a) InSe, (b) InSe:Sn

The Sn doping to InSe decreased steepness parameter and increased Urbach energy. To compare the experimental σ(T) data with theoretical values, the theoretical steepness parameter σ(T) was estimated using Eq. (6). $\sigma_0$ and $\hbar w_p$ parameters of InSe and InSe:Sn crystals were obtained from a fit to Eq. (6). The obtained parameters are given in the Table 2.

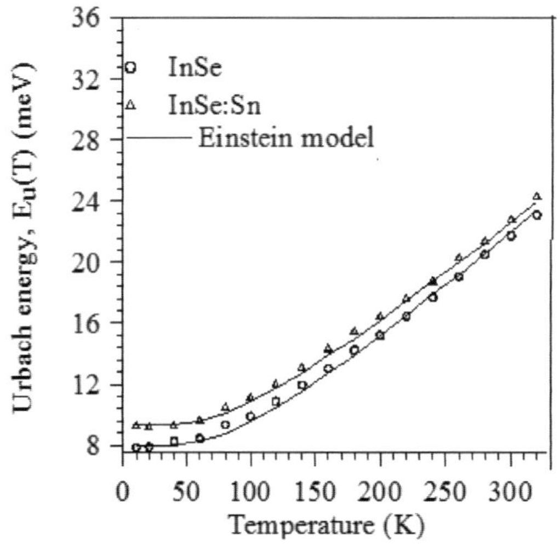


**Figure 4.** The Urbach energy of InSe and InSe:Sn as a function of temperature.

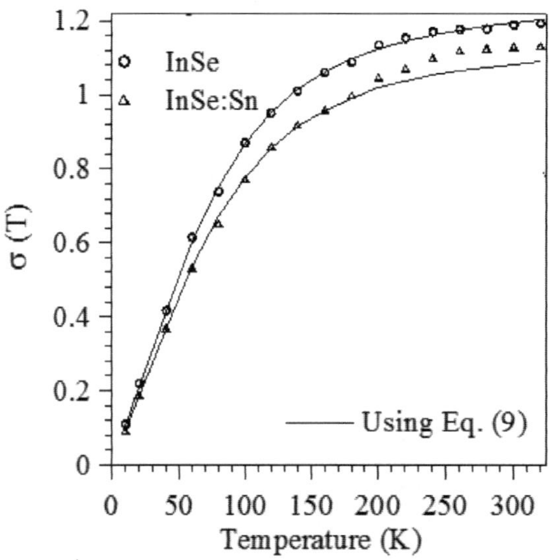


**Figure 5.** The stepness parameter of InSe and InSe:Sn as a function of temperature.

**Table 2**. Values of the parameters $\sigma_0$ and $\hbar w_p$ obtained from a fit to Eq. (6).

| Fit to Eq. (6) | $\sigma_0$ | $\hbar w_p$ (meV) |
|---|---|---|
| InSe | 1.260 | 21.0 |
| InSe:Sn | 1.145 | 21.5 |

As can be seen in Figs. 4 and 5 the experimental data and theoretical fittings are correlated with each others. The steepness parameters and Urbach energy of InSe and InSe:Sn increased with increasing sample temperature in the range 10-320 K. The steepness parameter of InSe is greater than that of InSe:Sn sample at the same temperature. It is the first time that the effect of Sn on the absorption measurements, Urbach rule and Steepness parameter of InSe sample has been investigated. As a result, Sn doping (0.03%) have caused a shift to shorter wavelength and an increase in intensity of absorption spectra. The Urbach energy has increased with Sn doping and temperature.

## 4. Conclusion

InSe and InSe:Sn single crystals used in this research were grown by using the modified Bridgman/Stockbarger method. The optical absorption measurements of semiconductor crystals n-InSe and n-InSe:Sn were performed as a function of temperature with 10 K increments in the range 10-320 K. The effect of Sn doping on the optical absorption measurements, Urbach energy and stepness parameters for InSe:Sn has been investigated. Sn doping to InSe crystal increased the absorption intensity and caused the shifting of the absorption edge towards the shorter wavelength of the spectrum.

## References

[1] M A Kenawy, H A Zayed, A M A El-Soud, *J. Mater. Sci., Mater. Electron.* **1,** 1990, 115
[2] S Shigetomi and T Ikari: *J. Appl. Phys.* **88**, 2000, 1520
[3] Z D Kovalyuk, V M Katerynchuk, A I Savchuk, O M Sydor, *Materials Science and Engineering B* **109**, 2004, 252
[4] C Julien, E Hatzikraniotis, A Chevy, K Kambas, *Mater. Res. Bull.* **20**, 1985, 287
[5] G Micocci, A Tepore and R Rella, *Phys. Stat. Sol.* **148**, 1995, 431
[6] K Ando, A Katsui, *Thin Solid Films,* **141,** 1981, 76
[7] N Benramdane, JP Guesdon, C Julien, *Phys. Stat Solidi (a),* **146,** 1994, 675
[8] G Micocci, A Tepore, R Rella, P Siciliano, *Phys Stat Solidi,* **148** ,1995, 431
[9] A B Abd El- Moiz: *Physica B* **191**, 1993, 293
[10] B Gürbulak, M Yıldırım, B Abay, S Tüzemen, M Alieva, and Y K Yoğurtçu, *Phys. Stat. Sol. A* **168,** 1998, 495
[11] B Gürbulak, *Solid State Commun.* **109**, 1999, 665
[12] A Ateş, B Gürbulak, M Yıldırım, S Doğan, *Physica E,* **16**, 2003, 274
[13] S Duman, B Gürbulak, and A Türüt, *Appl. Surface Science* **253**, 2007, 3899
[14] A Kuhn, A Chevy and R Chevalier, *Phys. Status Solidi A,* **31**, 1975, 469
[15] A J Mcevoy, A Parker, K Solt and R Bicchsel, *Thin Solid Films,* **15**, 1980, 69
[16] S Shigetomi and T Ikari, *Jpn. J. Appl. Phys.,* **42**, 2003, 6951
[17] S Shigetomi, T Ikari, *Journal of Applied Physics,* **93**, 2003, 2301
[18] W Franz, Z *Naturforsch* **13A**, 1958, 484
[19] LV Keldysh: *Zh. Eksp. Teor. Fiz.* **34** (1958) **1138.**
[20] K Imai, K Suzuki, T Haga, Y Hasegawa and Y Abe, *J. Cryst Growth* **54**, 1981, 501
[21] B Gürbulak, S Duman and A Ateş, *Czechoslovak Journal of Physics,* **54**, 2004, 857
[22] T S Moss, *Optical prosess in Semiconductors* (Butterworts. London), **247,** 1959
[23] Y P Varshni, *Physica* **34**, 1967, 149
[24] G D Cody, in: J I Pankkov (Ed.), *Semiconductor and Semimetals*, vol. **21(b),** Academic Press, New York, p. 11, 1984
[25] F Urbach, *Phys. Rev.* **92**, 1953, 1324
[26] P Lautenschlager, M Garriga, S Logothetidis, and M Cardona, *Phys. Rev. B,* **35**, 1987, 9174
[27] H W Martienssen, *J. Phys. Chem. Solids,* **2** ,1957, 257
[28] G D Cody, T Tiedje, B Abales, and Y Goldstein, *Phys. Rev. Lett.,* **47**, 1981, 1480
[29] M V Kurik, *Phys. Status Solidi A,* **8**, 1971, 9
[30] H Sumi, Y Toyozowa, *J. Phys. Soc. Jpn.* **31**, 1971, 342
[31] D J Dunstan, *Solid State Commun.* **43**, 1982, 341
[32] A Ates, B Gürbulak, M Yildirim, S Tüzemen, *Czechoslovak Journal of Physic,* **54**, 2004, 377
[33] Y W Tsang and M L Cohen, *Phys. Rev. B,* **3**, 1970, 1254
[34] V A Aliev, E F Bagirzade, N Z Gasanov and G D Guseinov, *Sov. Phys. Solid State,* **29**, 1987, 1

# Magnetron sputtered $WS_2$; optical and structural analysis

**Y Koçak[1], Y Akaltun[2] and Emre Gür[1]**

[1]Department of Physics, Faculty of Science, Atatürk University, 25240 Erzurum, Turkey
[2]Department of Electric Electronic Engineering, Erzincan, Turkey

E- mail: emregur@atauni.edu.tr

**Abstract.** Remarkable properties of graphene have renewed interest in inorganic, Transition Metal Dichalgogenits (TMDC) due to unique electronic and optical properties. TMDCs such as $MoS_2$, $MoSe_2$, $WS_2$ and $WSe_2$ have sizable bandgaps that change from indirect to direct in single layers, allowing applications such as solar cells, transistors, photodetectors and electroluminescent devices in which the graphene is not actively used. So, fabrication and analysis of these films are important for new generation devices. In this work, polycrystalline $WS_2$ films were grown by radio frequency magnetron sputtering (RFMS) on different substrates like n-Si(100), n-Si(111), p-Si(100), glass and fused silica. Structural, morphological, optical and electrical properties were investigated as a function of film thickness and RF power. From XRD analysis, signals from planes of (002), (100), (101), (110), (008) belong to the hegzagonal $WS_2$ were obtained. Raman spectra of the $WS_2$ show that there are two dominant peaks at ~351 $cm^{-1}$ (in-plane phonon mode) and ~417$cm^{-1}$ (out-of-plane phonon mode). XPS analysis of the films has shown that binding energy and the intensity of tungsten 4f shells shifts by depending on the depth of the films which might be due to the well-known preferential sputtering.

## 1. Introduction

Unique optical and electrical properties of 2D materials permit many important device applications. One of the most well-known 2D material graphene attracts attention for semiconductor device applications due to its strong interaction with photons in a wide energy range and its high carrier mobility, in a broad wavelength range [1-3]. However, field effect transistors produced from graphene cannot be effectively switched off and have low on/off switch ratios due to lack of bandgap [4]. This brings up the importance of other 2D materials. Many other 2D materials are known such as TMDCs (transition metal dicalcogenides), $MX_2$, M;transition metal (Mo, W, i.e) and X; chalcogenide (S, Se, i.e) [4, 5]. TMDC materials such as MoS2 and WS2 are covalently bonded to six chalcogenide neighbors within the same plane, but weakly linked to sheets above and below by van der Waals force. The layers are stacked along the c-axis. If the thin films grown with the c-axis parallel to substrate, i.e. the X-M-X layers are perpendicular to the substrate, they will referred to as " I I c ". The "⊥c" orientation has the advantage of photovoltaic materials, with c axis perpendicular and layers parallel to the substrate [6].

Important properties of 2D materials are [2]; 1) their surfaces are naturally self-passivated without any dangling bonds, making the integration of 2D materials with photonic structures easy, for example such as waveguides and cavities. 2) It is possible to raise vertical hetero-structures using different 2D materials without the conventional lattice mismatch, due to sheets with different lattice constants in hetero-structures are only weakly bonded by van der Waals force as in layered materials. 3) Even if the thin films exist atomically thin, many 2D materials interact with light strongly. For example a single layer tungsten disulfide ($WS_2$) absorbs roughly %10 at exitonic resonances (615nm) [7]. 4) 2D materials can cover a very wide electromagnetic spectrum, since their diverse electronic properties [2].

In this paper, $WS_2$ thin films grown by sputtering are investigated depending on the film thickness and substrate. The structural analysis has been performed by using X-ray diffraction (XRD). Raman spectroscopy used to the phonon modes that can provide a fingerprint for $WS_2$ thin films. X-ray photoelectron spectroscopy (XPS) was used in order to investigate the elemental analysis together with the depth profile analysis.

## 2. Experimental Produce

WS2 thin films were deposited by Radio Frequency Magnetron Sputtering (RFMS), using %99.99 $WS_2$ target. Substrates were chemically cleaned prior to the insertion in the vacuum chamber. Base pressure of $2x10^{-7}$ Torr was reached in the sputtering chamber before the growth process. Growth pressure was 8.6mT. Five different substrates were put in the chamber in order to investigate the substrate effect on the WS2 films. The substrates temperature was 350 $^0$C during the deposition. Details of the deposition parameters are summarized in table 1.

**Table 1**. Parameters of sputtering system

| | |
|---|---|
| Target (purity) | $WS_2$ (99.99%) |
| Target diameter | 2 inc |
| Target to substrate distance | 80 mm |
| Base pressure | $2\times10^{-7}$ torr |
| Sputtering pressure | 8.6 mtorr |
| Substrate temperature | 350 $^{O}$C |
| Substrates | Si(100), Si(111), fused silica, quartz, glass |

Structure of the thin films was studied by x-ray diffraction (XRD) with Empyrean, PANalytical with Cu $K_\alpha$ radiation. The measurements were performed at grazing incidence angle of 0.5$^o$ degree. Further evidence for composition of the grown thin films was obtained using XPS which was performed with a Thermo Scientific K-Alpha–Monochromated high-performance XPS spectrometer. Raman measurements were conducted with Ar+ laser wavelength of 514.5 nm with a CCD camera attached to the 1 m long monochromotor.

## 3. Results and discussion

A comparison of the XRD profiles of $WS_2$ thin films grown in different RF power is illustrated in Fig. 1. Polycristalline nature of thin films is observed in multiple peaks in the XRD spectra. Three distinct peak clearly observed in the figure. The first one appeared around 13.24$^0$ degree which is belong to the (002) plane. Second peak observed in the XRD spectra of the thin films attributed to the combined (10*l*) (*l is* 0,1,2..), third peak which is (103) observed only in the thin film grown at 6 WRF power38.08 degree and forth peak is observed at 60.9 degree which is (11*l*). Dominant crystal orientation in the grown thin films has changed by changing the RF power from (002) to (101) as shown in Fig. 1. The degree of preferred orientation, which can be inferred from the peak heigh of the (002) peak is dramatically decreased by increasing RF power in the growth process. FWHM (full widh half maximum) of the all peaks observed in the XRD spectra given aboveshows that FWHM inreases while increasing of RF power (i.e. FWHMs of (002) and (112) peaks are increasing from 3.30 to 6.31 and from 2.35 to 2.84 respectively but FWHMs of (100)+(101) peak are decreasing from 2.24 to 1.68, while RF power increased).

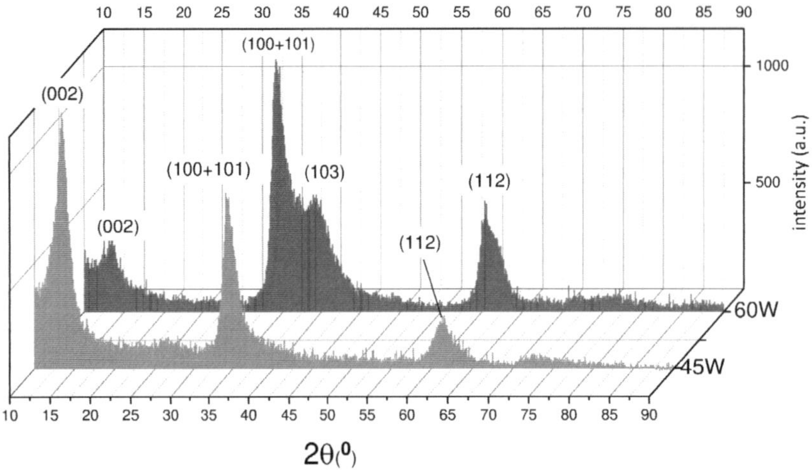


**Fig. 1.** XRD-patterns in different RF power of WS2 films on n-Si(111) substrate.

This might be due to the observation of some extra peaks, namely (103), plane as observed in the thin film grown at 60W. Also, (002), (112) and combined (100)+(101) peaks are slightly has shifted smaller angles as a function of RF power. If diffraction occurs off of the horizontal planes (tensile strain), and the Bragg peak shifts to higher angle as the interplanar spacing decreases [8, 9]. Then it can be confirm that the (002), combine (101) + (101) and (112) plane exposure to compresive strain.

Raman Spectroscopy was used to characterize the samples' structural quality and vibrational properties. Since it is non-devastating technique and it is widely used to characterize the structural and electronic properties of monolayer materials such as graphene, $WS_2$ and $MoS_2$ [10-12]. To reason out this observation we fitted the Raman peaks using a Voigh function and positions $E^1_{2g}$ and $A_{1g}$ modes depending on the type of the substrate. The clear observation of the raman modes indicates good quality of the material. The Raman spectra for the various substrate (p-Si(100), n-Si(111) and fused silica) in 310nm $WS_2$ thin films are shown in Fig. 2. It is confirmed that the $WS_2$ exhibit first order modes at wavenumber of 348 $cm^{-1}$ and 418 $cm^{-1}$ separated by about 70 $cm^{-1}$, corresponding to the $E^1_{2g}$ and $A_{1g}$ modes from in plane and out-of plane vibrations, respectively.

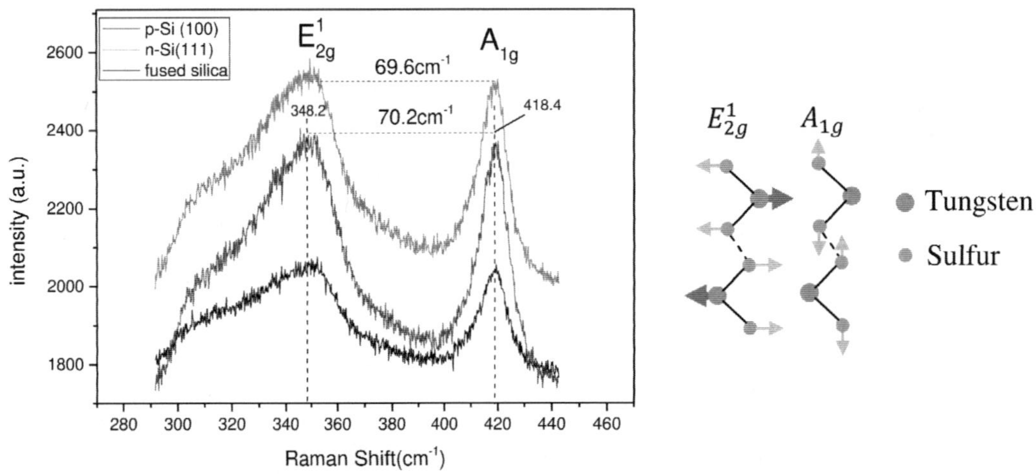


**Fig. 2.** Room-temperature Raman spectra from a monolayer WS2 region; 310nm thickness of $WS_2$ thin films on various substrate (45W RF Power).

The peak of $E_{2g}^{1}$ mode is broad due to consisting of another mode so called 2LA(M) (second-order longitudinal acoustic mode) [11, 13, 14]. In generally, 2LA(M) peak which matches in energy with the density of states absorption peak, is prevalent as a small left portion of the $E_{2g}^{1}$ peak [14]. small changes observed depending on the substrate type in Raman modes. This might be due to the different strain in which the grown $WS_2$ thin films are subjected to. In addition Raman peak shifts are showing a trend to decrease in frequency of the optical $E_{2g}^{1}$ phonon mode and an increase in frequency of $A_{1g}$ phonon mode with increasing thickness of the thin films (Fig. 2.). These variations in the Raman frequency of the $A_{1g}$ and $E^{1}_{2g}$ modes depending on the different substrate can be explained by weak interlayer substrate can be explained by weak interactions and the reduced long-range Coulomb interaction between the effective charges caused by an increase in the dielectric screening, respectively [13, 15, 16]. No raman signal corresponding to the $WO_3$ has been observed.
XPS measurements performed in order to find out the concentration of the elements. As shown in Fig. 3, XPS survey spectrum of $WS_2$ shows that only C, O, W and S is observed all of the films in. To identify the chemical composition of $WS_2$ thin films, XPS was used to measure the binding energy of W and S as well as carbon and oxygen.

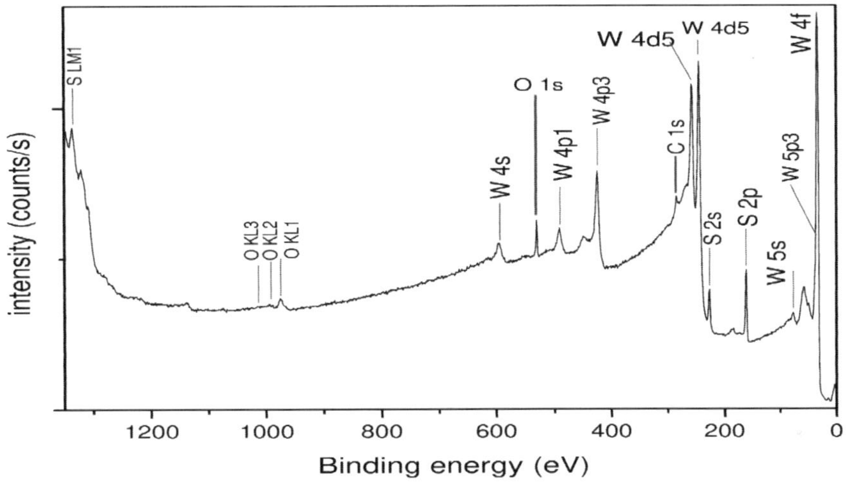


**Fig. 3.** XPS survey spectrum of $WS_2$ thin film after 100s etched.

Depth resolved XPS measuements were used to understand the elemental profile of W and S. Ar gun was used to achive etching procesXPS data was acquired 120s of sputter between every measurement. The characteristic W4f and S2p XPS spectrum of $WS_2$ thin film is shown in figure 4.The binding energy profile for $W4f_{7/2}$ and $W4f_{5/2}$ is represented by two peak at 31.5eV and 33.8eV, respectively, corresponding to W in $WS_2$ [17-19], after the film is etched for the first 120s. the W 4f spectrum has consisted of 4 peaks for the last three XPS measuments (see fig. 4.a). The appeared peaks shows a shift to higher binding energy values i.e. 35.7eV and 37.6 which might be corresponding to tungsten in $WO_3$ ($W^{6+}$) [17, 20] While $W4f_{7/2}$ peak intensity of $WS_2$ decreases with the depth profile, possible observed peaks belong to the $WO_3$ starts to increase, as shown in Fig. 4. a. As shown in figure 4.d, the S2p spectrum is deconvoluted to a dublet consisting of $S2p_{3/2}$ and $S2p_{1/2}$ peaks centered at binding energy values of 161.88eV and 163.18eV which correspond to $S^{2-}$ in $WS_2$.
In order to find the peak energy position of the W 4f belonging to $WS_2$, Lorentz function were used to fit observed data (see fig.4.c). Binding energy shifts has observed in 4f features as shown in the inset of figure 4c.

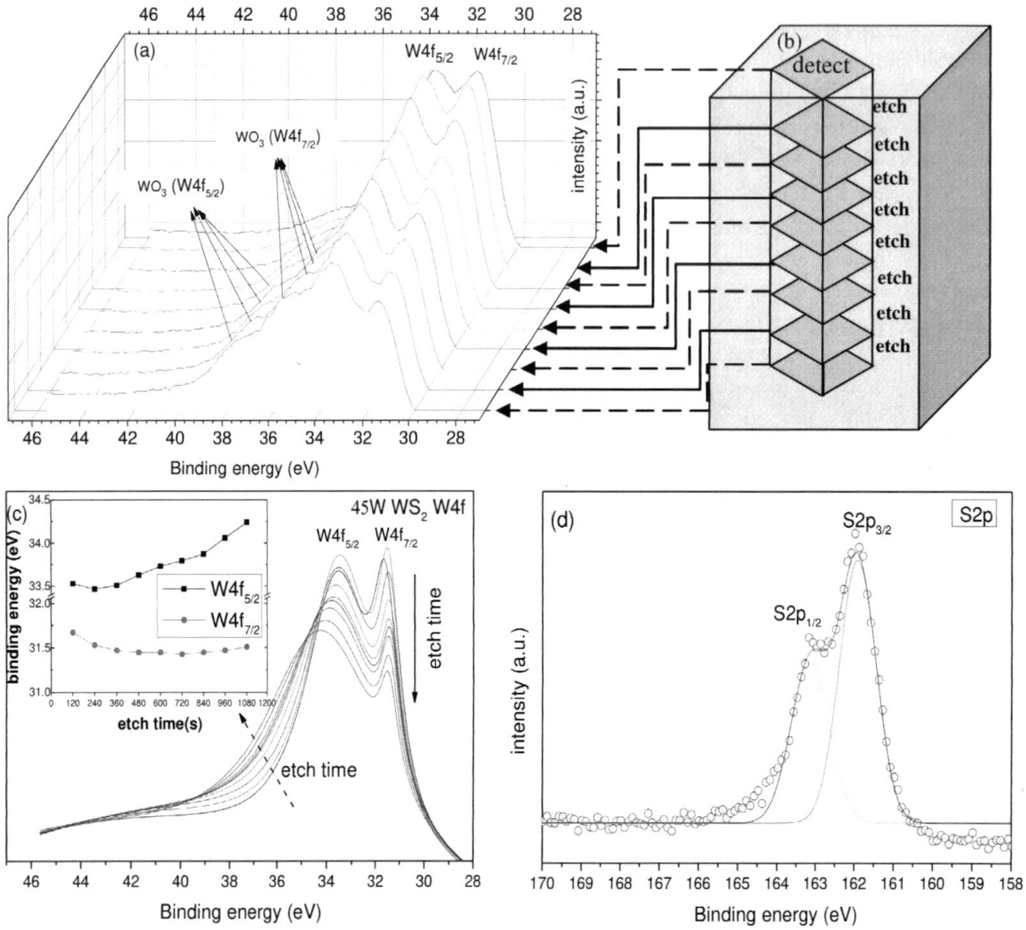


**Fig. 4.** XPS analysis of $WS_2$ thin film, a) $W4f_{7/2\text{-}5/2}$ b) Schematic illustration of grid zone in an XPS depth profile c) the Lorentz fit data graph of $WS_2$ thin film at fig 4.a and also shown is the binding shifts while the etch time is increasing (inset) d) S2p binding energy.

Figure 5 shows the UV visible absorption spectrum of $WS_2$ thin film.

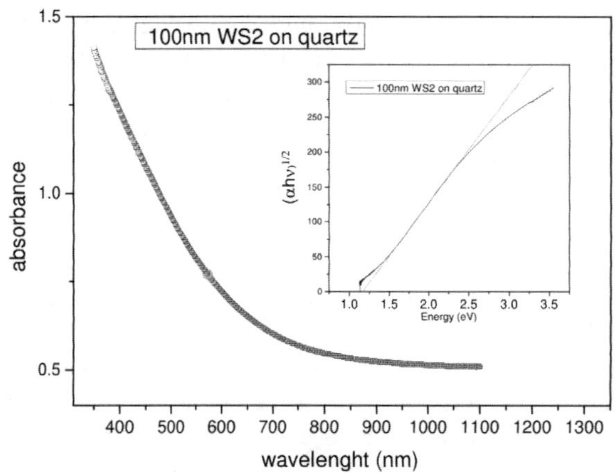


**Fig. 5.** Optical absorption of $WS_2$ thin film in a wavelength range of 350-1100 nm with an computed optical bandgap ~1.17 eV from the calculated at the lineer fit of optical absorbance plot data, as shown in the inset.

The figure inset shows the bandgap energy obtained from plotting the $(\alpha h\nu)^{1/2}$~$(h\nu\text{-}E_g)$ graph, where $\alpha$ is the absorption coefficient of the $WS_2$, hν is the incident photon energy, and Eg is the optical bandgap. Bandgap energy of the $WS_2$ was obtained from the linear fit as shown in the inset of the figure 5 as 1.17 eV which is very close to the bulk indirect bandgap of the $WS_2$.

## 4. References

[1] Mak KF, Ju L, Wang F, Heinz TF, 2012. *Solid State Commun.*;**152**:1341-9.
[2] Xia F, Wang H, Xiao D, Dubey M, Ramasubramaniam A, 2014 *Nature Photonics*.**8**:899-907.
[3] Xia F, Yan H, Avouris P, 2013 *Proceedings of the IEEE*.**101**:1717-31.
[4] Wang QH, Kalantar-Zadeh K, Kis A, Coleman JN, Strano MS, 2012 *Nature nanotechnology*.**7**:699-712.
[5] Fiori G, Bonaccorso F, Iannaccone G, Palacios T, Neumaier D, Seabaugh A, et al., 2014 *Nature nanotechnology*.**9**:768-79.
[6] Martin-Litas I, Vinatier P, Levasseur A, Dupin J, Gonbeau D, Weill F, 2002 *Thin Solid Films*.**416**:1-9.
[7] Eda G, Maier SA, 2013 *Acs Nano*.**7**:5660-5.
[8] Regula M, Ballif C, Moser J, Levy F, 1996 *Thin Solid Films*.**280**:67-75.
[9] Hashmi S. 2014 Comprehensive Materials Processing: Newnes.
[10] Calizo I, Balandin A, Bao W, Miao F, Lau C, 2007 *Nano Lett.*;**7**:2645-9.
[11] Su L, Yu Y, Cao L, Zhang Y, 2015 *Nano Research*.**8**:2686-97.
[12] Su L, Zhang Y, Yu Y, Cao L, 2014 *Nanoscale*.**6**:4920-7.
[13] Berkdemir A, Gutiérrez HR, Botello-Méndez AR, Perea-López N, Elías AL, Chia C-I, et al., 2013. *Scientific reports*.**3**:
[14] Zhao W, Ghorannevis Z, Amara KK, Pang JR, Toh M, Zhang X, et al., 2013 *Nanoscale*.**5**:9677-83.
[15] Molina-Sanchez A, Wirtz L, 2011 *Physical Review B*.**84**:155413.
[16] Song J-G, Park J, Lee W, Choi T, Jung H, Lee CW, et al., 2013 *ACS nano*.**7**:11333-40.
[17] Deepthi B, Barshilia HC, Rajam K, Konchady MS, Pai DM, Sankar J, et al., 2010 *Surf. Coat. Technol.*;**205**:565-74.
[18] Sundberg J, Lindblad R, Gorgoi M, Rensmo H, Jansson U, Lindblad A, 2014 *Appl. Surf. Sci.*;**305**:203-13.
[19] Yin G, Huang P, Yu Z, He D, Tu J, 2006 *Tribology Letters*.**22**:37-43.
[20] Wong K, Lu X, Cotter J, Eadie D, Wong P, Mitchell K, 2008 *Wear*.**264**:526-34.

## Acknowledgments

We appreciate Hasan Feyzi Budak for the XRD measurements This project was supported by Atatürk University Scientific Resarch Project Unit under contract No: 2015/343. We would like to thank to Enver Kahveci for the XPS measurements. We are thankful to Prof. Dr. Atilla Aydınlı for discussion on the Raman measurements.

# Indoor radon measurements in Erzurum province of Turkey

**R. Durak, D. Kiran, E. Kavaz and N. Ekinci,**

Ataturk University, Science Faculty, Department of Physics, 25240, Erzurum, Turkey

E-mail: rdurak@atauni.edu.tr

**Abstract.** Indoor radon measurements were carried out in dwellings in Erzurum province during the winter months of February 2012 to early April 2012 and the summer months of July 2012 to early September 2012. Nuclear track detector LR-115 was used for the measurements. According to the results of investigations, it was understood that the indoor radon concentration averages in dwellings are in the range of $11 \pm 6$ Bq $m^{-3}$ - $380 \pm 91$ Bq $m^{-3}$ in winter season and $8 \pm 3$ Bq $m^{-3}$ - $356 \pm 64$ Bq $m^{-3}$ in summer season. We found that the $^{222}$Rn effective dose values in the studied dwellings in winter season range from 0.278 to 9.59 mSv $y^{-1}$. Also, the $^{222}$Rn effective dose values in the studied dwellings in summer season range from 0.202 to 8.98 mSv $y^{-1}$. These values are within the ICPR recommended values. The radon activity has not been found to vary with seasonal changes, but also with the age, the construction mode of houses, the ventilation conditions and with specific sites and geological materials.

## 1. Introduction

The increasing awareness concerning radon ($^{222}$ Rn) health hazards on organic structures obligates researchers to investigate the sources and impacts within the environment. There is a direct correlation between radon exposure and some types of cancer. The risk of lung cancer from the inhalation of radon is the most significant health hazards. Radon is a radioactive gas produced by the radioactive decay of uranium and thorium. The detection of indoor radon levels is extremely difficult. It is a colorless, odorless, tasteless gas detectable by utilizing special equipment. The main natural sources of indoor radon are the soil gas, building materials, tap water and natural gas used for cooking. Also, a number of anthropogenic activities (mining operations, the use of phosphate fertilizers, burning of fossil fuels) has increased the radon exposure to general public.

It is necessary to monitor radon. Therefore, both active and passive techniques have been developed. Active methods are usually used for short term measurements of radon and for detailed investigations of ındividual sites under inspection. Passive methods are more suitable for the assessment of radon exposure over long time scales and can be used for large scale surveys at moderate cost. For that reason, many countries have performed large-scale radon surveys using passive monitoring devices, which helps to assess the public exposure, and adopted appropriate actions for protection against radon. In this work, LR- 115 nuclear track detector was used for large-scale surveys of environmental radon.

There has been an increasing interest in the measurement of indoor radon concentration levels throughout the world. The International Commission on Radiological Protection (ICRP), with over a hundred publications devoted to radon studies, gives both human and environmental recommendations for radiological protection. Further, they have researched and identified geographic areas, where houses are likely to have high radon risk and should therefore be considered as radon prone.

Recently several studies underscore the attention and severity of radon concerns. For example: Concentration of radioactive radon gas ($^{222}$ Rn) were measured using passive monitors based on LR115 solid state track detectors during June- July 2010 indoor air of dwelling in the Nui Beo coal mining area, mostly in Cam Pha and Ha Long coastal towns (Nhan et al., 2012). In a study, Indoor radon measurements were carried out in elementary schools in Tunis, the capital city of Tunisia, during the winter months of December 2008 to early March 2009. Two classrooms, one each from ground floor and first floor were chosen from each school making a total of 60 classrooms. Nuclear track detector type LR- 115 (Kodalpha) was used for the measurements (Labıdı et al., 2010). In another work, radon / thoron and their progeny concentrations were measured in different types of dwellings at different locations around industrial areas, cities and rural areas of Brahmaputra Valley of Assam by using LR-115(type- II) plastic detector. Radon levels of different dwellings were analysed with reference to the nature building materials, ventilation patterns and the types of underlying soil. The results were discussed under the light of exposure limits set by ICPR (Deka et al., 2003).

In addition, radon ($^{222}$Rn) and thoron ($^{220}$ Rn) $\alpha$ activities per unit volume were measured inside and outside different building materials by using two types of solid state nuclear track detectors (SSNTD) (CR-39 and LR-115 type II). The radon and thoron emanation coefficients of the studied materials were evaluated (Misdaq et al., 2000). Furthermore, the results of radon activity recorded in 70 dwellings of Nurpur area in India are reported. LR -115 type II films in bare mode were exposed for four seasons of three months each covering a period of one year for the measurement of indoor radon levels (Singh et al., 2004). Indoor radon measurement survey has been carried out in properly selected houses of the city of Islamabad (Rahman et al., 2008). The ındoor radon activity level and radon effective dose rate were carried out in Moroccan dwellings using SSNTDs (Oufni et al, 2005).

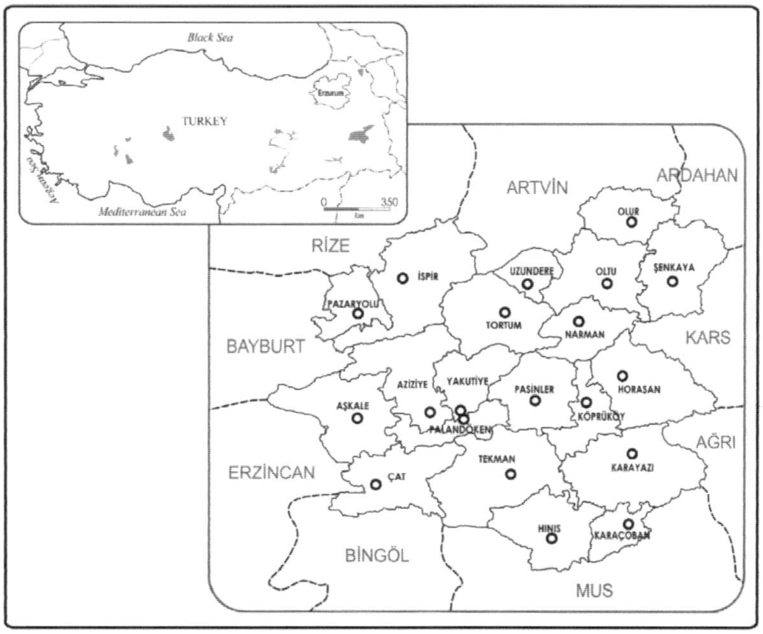


**Fig 1**. Map showing the studied area of Erzurum Province in Turkey.

Therefore, we want to recognize the importance of the accurate evaluation of radon concentration and exposure. Although Turkey has conducted Radon surveys, the city of Erzurum has not. Therefore this area is the target of our study to assess any health risk from radon. Erzurum is a city in eastern Turkey. The city is found with 39° 54' 35" N, 41° 16' 32"E coordinates (see Fig 1.) Erzurum has a continental climate. Summer is brief, but summer days are warm, though with cool nights. Winters are very cold and long, with an average minimum during January of around -16 °C (3 °F); temperatures fall below -30 °C (-22 °F) most years.

## 2. Material and method

Measurements over several months are better than short-term measurements for estimating annual average radon levels. The plastic track etch detectors are the most common method to measure radon. For this reason, the indoor radon activity concentrations were measured by LR-115 nuclear track detectors.

**Fig. 2.** Nuclear track detector LR-115.

A total of 110 of these units were placed in houses at center of Erzurum city and extending to the surrounding areas. Rooms that are rarely used were avoided to accurately reflect people's true exposure to radon. Therefore, detectors were placed in rooms where people spend most of the time such as living room and bedrooms. Average exposure time of LR-115 detectors was 60 days. Detectors were placed in the selected houses, at least 1m from the floor and away from doors or windows. LR-115 type II detectors are given in Fig.2. LR -115 is a commercial name of cellulose nitrate ($C_6H_2O_9N_2$) detector made by Kodak-Path, France. After an exposure time of 2 months, detector films were removed and etched in a NaOH solution (2.5 N at 60 °C during 100 min for LR-115 type II films) in a constant temperature bath. Then these SSNTDs were washed, dried and scanned under a binocular microscope for track density measurements. LR -115 has an active layer of 12 μm red-dyed cellulose nitrate film coated on inert polyester base of 100 μm thickness. The thickness of 12 μm is designed for spark counting of α- particle tracks recorded in the film. LR-115 is welcomed by scientist for the reason that α- particles emitted from radon progenies attached onto the surface of detector cannot developed by chemical etching for short etching times. The energies of the α-particles from the short-lived progenies are all higher than 6.0 MeV, which are beyond the energy window 0.06-6 MeV of LR-115. In contrast, all the α- particles from the progenies can form etchable tracks in CR-39 by a short etching time. This is an advantage of LR- 115 over CR-39 (L'Annunziata, 2003).

## 3. Results and discussions

In the present investigations, the indoor $^{222}$Rn concentration has been studied in 110 dwelling of Erzurum. The houses were chosen randomly in such a way that the dwellings constructed with different types of building materials such as brick, cement, and concrete. Indoor radon concentration levels have been determined for ground floor and first floor since floor level significantly affects the radon concentration. Radon levels vary significantly between seasons, so the measurements for indoor radon concentration levels were carried out during the winter months of February 2012 to early April 2012 and the summer months of July 2012 to early September 2012. The results of these measurements are shown in Table 1 and Table 2. The value of standard deviations is also given in Table1 and 2.

**Table1.** The results of the radon concentration measurements with the value of standard deviations in winter.

| Detector Number | Starting date | Ending date | No.days | Exposure kBq.h/m$^3$ | Avg. radon conc. Bq/m$^3$ | Uncertainty 2 σ | |
|---|---|---|---|---|---|---|---|
| | | | | | | ± % | ± Bq/m$^3$ |
| 384267 | 16.02.12 | 16.04.12 | 60.0 | 28 | 19 | 46% | 9 |
| 384248 | 16.02.12 | 16.04.12 | 60.0 | 35 | 24 | 44% | 11 |
| 384160 | 11.02.12 | 11.04.12 | 60.0 | 34 | 24 | 44% | 11 |
| 384156 | 11.02.12 | 11.04.12 | 60.0 | 25 | 17 | 48% | 8 |
| 384162 | 11.02.12 | 11.04.12 | 60.0 | 14 | 10 | 56% | 6 |
| 384161 | 11.02.12 | 11.04.12 | 60.0 | 29 | 20 | 46% | 9 |
| 384245 | 16.02.12 | 16.04.12 | 60,0 | 27 | 19 | 47% | 9 |
| 384155 | 11.02.12 | 11.04.12 | 60,0 | 56 | 39 | 38% | 15 |
| 384166 | 11.02.12 | 11.04.12 | 60.0 | 55 | 38 | 39% | 15 |
| 384167 | 11.02.12 | 11.04.12 | 60.0 | 23 | 16 | 49% | 8 |
| 384168 | 11.02.12 | 11.04.12 | 60.0 | 111 | 77 | 32% | 25 |
| 384165 | 11.02.12 | 11.04.12 | 60.0 | 30 | 21 | 45% | 9 |
| 384169 | 11.02.12 | 11.04.12 | 60.0 | 36 | 25 | 43% | 11 |
| 384150 | 11.02.12 | 11.04.12 | 60.0 | 18 | 12 | 53% | 6 |
| 384152 | 11.02.12 | 11.04.12 | 60.0 | 16 | 11 | 54% | 6 |
| 384151 | 11.02.12 | 11.04.12 | 60.0 | 31 | 22 | 45% | 10 |
| 384163 | 11.02.12 | 11.04.12 | 60.0 | 39 | 27 | 42% | 11 |
| 384220 | 15.02.12 | 15.04.12 | 60.0 | 50 | 35 | 39% | 14 |
| 384222 | 15.02.12 | 15.04.12 | 60.0 | 46 | 32 | 40% | 13 |
| 384229 | 15.02.12 | 15.04.12 | 60.0 | 44 | 31 | 41% | 13 |
| 384230 | 15.02.12 | 15.04.12 | 60.0 | 28 | 19 | 46% | 9 |
| 384231 | 15.02.12 | 15.04.12 | 60.0 | 62 | 43 | 37% | 16 |
| 384232 | 15.02.12 | 15.04.12 | 60.0 | 27 | 19 | 47% | 9 |
| 384233 | 15.02.12 | 15.04.12 | 60.0 | 37 | 26 | 43% | 11 |
| 384223 | 15.02.12 | 15.04.12 | 60.0 | 47 | 33 | 40% | 13 |
| 384224 | 15.02.12 | 15.04.12 | 60.0 | 114 | 79 | 32% | 25 |
| 384226 | 15.02.12 | 15.04.12 | 60.0 | 292 | 203 | 26% | 53 |
| 384227 | 15.02.12 | 15.04.12 | 60.0 | 157 | 109 | 30% | 33 |
| 384228 | 15.02.12 | 15.04.12 | 60.0 | 130 | 90 | 31% | 28 |
| 384234 | 15.02.12 | 15.04.12 | 60.0 | 60 | 41 | 38% | 16 |
| 384221 | 15.02.12 | 15.04.12 | 60.0 | 49 | 34 | 40% | 14 |
| 384157 | 11.02.12 | 11.04.12 | 60.0 | 38 | 27 | 42% | 11 |
| 384158 | 11.02.12 | 11.04.12 | 60.0 | 30 | 21 | 45% | 9 |
| 384154 | 11.02.12 | 11.04.12 | 60.0 | 22 | 15 | 50% | 8 |
| 384202 | 14.02.12 | 14.04.12 | 60.0 | 16 | 11 | 54% | 6 |
| 384201 | 14.02.12 | 14.04.12 | 60.0 | 50 | 34 | 40% | 14 |
| 384178 | 14.02.12 | 14.04.12 | 60.0 | 330 | 229 | 26% | 60 |
| 384179 | 14.02.12 | 14.04.12 | 60.0 | 103 | 72 | 33% | 24 |
| 384203 | 14.02.12 | 14.04.12 | 60.0 | 274 | 190 | 27% | 51 |
| 384204 | 14.02.12 | 14.04.12 | 60.0 | 391 | 271 | 25% | 68 |
| 384239 | 15.02.12 | 15.04.12 | 60.0 | 60 | 42 | 38% | 16 |
| 384238 | 15.02.12 | 15.04.12 | 60.0 | 24 | 17 | 48% | 8 |
| 384237 | 15.02.12 | 15.04.12 | 60.0 | 58 | 40 | 38% | 15 |
| 384236 | 15.02.12 | 15.04.12 | 60.0 | 68 | 47 | 36% | 17 |
| 384235 | 15.02.12 | 15.04.12 | 60.0 | 136 | 95 | 31% | 29 |
| 384205 | 14.02.12 | 14.04.12 | 60.0 | 88 | 61 | 34% | 21 |

**Table1.** Continued.

| | | | | | | | |
|---|---|---|---|---|---|---|---|
| 384206 | 14.02.12 | 14.04.12 | 60.0 | 30 | 21 | 46% | 10 |
| 384171 | 13.02.12 | 13.04.12 | 60.0 | 71 | 50 | 36% | 18 |
| 384172 | 13.02.12 | 13.04.12 | 60.0 | 22 | 16 | 49% | 8 |
| 384174 | 13.02.12 | 13.04.12 | 60.0 | 85 | 59 | 34% | 20 |
| 384176 | 13.02.12 | 13.04.12 | 60.0 | 42 | 29 | 42% | 12 |
| 384177 | 13.02.12 | 13.04.12 | 60.0 | 30 | 21 | 45% | 9 |
| 384209 | 14.02.12 | 14.04.12 | 60.0 | 44 | 31 | 41% | 13 |
| 384208 | 14.02.12 | 14.04.12 | 60.0 | 24 | 17 | 48% | 8 |
| 384207 | 14.02.12 | 14.04.12 | 60.0 | 22 | 16 | 49% | 8 |
| 384173 | 13.02.12 | 13.04.12 | 60.0 | 279 | 194 | 27% | 52 |
| 384175 | 13.02.12 | 13.04.12 | 60.0 | 47 | 33 | 40% | 13 |
| 384254 | 16.02.12 | 16.04.12 | 60.0 | 251 | 174 | 27% | 47 |
| 384255 | 16.02.12 | 16.04.12 | 60.0 | 25 | 17 | 48% | 8 |
| 384256 | 16.02.12 | 16.04.12 | 60.0 | 42 | 29 | 42% | 12 |
| 384257 | 16.02.12 | 16.04.12 | 60.0 | 86 | 60 | 34% | 20 |
| 384258 | 16.02.12 | 16.04.12 | 60.0 | 344 | 239 | 26% | 62 |
| 384259 | 16.02.12 | 16.04.12 | 60.0 | 233 | 162 | 28% | 45 |
| 384249 | 16.02.12 | 16.04.12 | 60.0 | 63 | 44 | 37% | 16 |
| 384250 | 16.02.12 | 16.04.12 | 60.0 | 58 | 40 | 38% | 15 |
| 384251 | 16.02.12 | 16.04.12 | 60.0 | 32 | 23 | 44% | 10 |
| 384252 | 16.02.12 | 16.04.12 | 60.0 | 87 | 61 | 34% | 21 |
| 384253 | 16.02.12 | 16.04.12 | 60.0 | 25 | 17 | 48% | 8 |
| 384260 | 17.02.12 | 17.04.12 | 60.0 | 167 | 116 | 30% | 35 |
| 384261 | 17.02.12 | 17.04.12 | 60.0 | 444 | 309 | 25% | 77 |
| 384262 | 17.02.12 | 17.04.12 | 60.0 | 124 | 86 | 32% | 28 |
| 384263 | 17.02.12 | 17.04.12 | 60.0 | 183 | 127 | 29% | 37 |
| 384264 | 17.02.12 | 17.04.12 | 60.0 | 216 | 150 | 28% | 42 |
| 384265 | 17.02.12 | 17.04.12 | 60.0 | 241 | 168 | 27% | 45 |
| 384266 | 17.02.12 | 17.04.12 | 60.0 | 38 | 26 | 43% | 11 |
| 384268 | 17.02.12 | 17.04.12 | 60.0 | 32 | 22 | 45% | 10 |
| 384159 | 12.02.12 | 12.04.12 | 60.0 | 172 | 119 | 29% | 35 |
| 384180 | 12.02.12 | 12.04.12 | 60.0 | 29 | 20 | 46% | 9 |
| 384191 | 12.02.12 | 12.04.12 | 60.0 | 51 | 35 | 39% | 14 |
| 384219 | 12.02.12 | 12.04.12 | 60.0 | 25 | 17 | 48% | 8 |
| 384183 | 12.02.12 | 12.04.12 | 60.0 | 34 | 24 | 44% | 11 |
| 384184 | 12.02.12 | 12.04.12 | 60.0 | 32 | 22 | 45% | 10 |
| 384185 | 12.02.12 | 12.04.12 | 60.0 | 25 | 17 | 48% | 8 |
| 384186 | 12.02.12 | 12.04.12 | 60.0 | 22 | 15 | 50% | 8 |
| 384218 | 12.02.12 | 12.04.12 | 60.0 | 30 | 21 | 46% | 10 |
| 384187 | 13.02.12 | 13.04.12 | 60.0 | 36 | 25 | 43% | 11 |
| 384198 | 13.02.12 | 13.04.12 | 60.0 | 110 | 77 | 32% | 25 |
| 384188 | 13.02.12 | 13.04.12 | 60.0 | 45 | 31 | 41% | 13 |
| 384189 | 13.02.12 | 13.04.12 | 60.0 | 69 | 48 | 36% | 17 |
| 384190 | 13.02.12 | 13.04.12 | 60.0 | 41 | 28 | 42% | 12 |
| 384192 | 13.02.12 | 13.04.12 | 60.0 | 28 | 19 | 46% | 9 |
| 384193 | 13.02.12 | 13.04.12 | 60.0 | 28 | 19 | 46% | 9 |
| 384243 | 16.02.12 | 16.04.12 | 60.0 | 32 | 22 | 45% | 10 |
| 384242 | 16.02.12 | 16.04.12 | 60.0 | 43 | 30 | 41% | 12 |
| 384194 | 13.02.12 | 13.04.12 | 60.0 | 49 | 34 | 40% | 14 |

**Table1.** Continued.

| | | | | | | | |
|---|---|---|---|---|---|---|---|
| 384195 | 13.02.12 | 13.04.12 | 60.0 | 49 | 34 | 40% | 14 |
| 384196 | 13.02.12 | 13.04.12 | 60.0 | 548 | 380 | 24% | 91 |
| 384197 | 13.02.12 | 13.04.12 | 60.0 | 84 | 58 | 35% | 20 |
| 384210 | 14.02.12 | 14.04.12 | 60.0 | 77 | 53 | 35% | 19 |
| 384211 | 14.02.12 | 14.04.12 | 60.0 | 37 | 26 | 43% | 11 |
| 384212 | 14.02.12 | 14.04.12 | 60.0 | 81 | 56 | 35% | 20 |
| 384213 | 14.02.12 | 14.04.12 | 60.0 | 62 | 43 | 37% | 16 |
| 384215 | 14.02.12 | 14.04.12 | 60.0 | 29 | 20 | 46% | 9 |
| 384214 | 14.02.12 | 14.04.12 | 60.0 | 39 | 27 | 42% | 11 |
| 384216 | 14.02.12 | 14.04.12 | 60.0 | 38 | 27 | 42% | 11 |

**Table 2.** The results of the radon concentration measurements with the value of standard deviations in summer.

| Detector Number | Starting date | Ending date | No.days | Exposure kBq.h/m$^3$ | Avg. radon conc. Bq/m$^3$ | Uncertainty 2 σ | |
|---|---|---|---|---|---|---|---|
| | | | | | | ± % | ± Bq/m$^3$ |
| 384270 | 12.07.12 | 12.09.12 | 62,0 | 27 | 18 | 31% | 6 |
| 384271 | 12.07.12 | 12.09.12 | 62,0 | 29 | 19 | 31% | 6 |
| 384282 | 12.07.12 | 12.09.12 | 62,0 | 36 | 24 | 29% | 7 |
| 384283 | 12.07.12 | 12.09.12 | 62,0 | 143 | 96 | 21% | 20 |
| 384286 | 12.07.12 | 12.09.12 | 62,0 | 19 | 13 | 34% | 4 |
| 384284 | 12.07.12 | 12.09.12 | 62,0 | 25 | 17 | 32% | 5 |
| 384285 | 12.07.12 | 12.09.12 | 62,0 | 25 | 17 | 32% | 5 |
| 384278 | 12.07.12 | 12.09.12 | 62,0 | 22 | 15 | 33% | 5 |
| 384277 | 12.07.12 | 12.09.12 | 62,0 | 48 | 32 | 27% | 9 |
| 384276 | 12.07.12 | 12.09.12 | 62,0 | 28 | 19 | 31% | 6 |
| 384279 | 12.07.12 | 12.09.12 | 62,0 | 196 | 132 | 20% | 26 |
| 384275 | 12.07.12 | 12.09.12 | 62,0 | 18 | 12 | 34% | 4 |
| 384273 | 12.07.12 | 12.09.12 | 62,0 | 17 | 11 | 35% | 4 |
| 384274 | 12.07.12 | 12.09.12 | 62,0 | 42 | 29 | 28% | 8 |
| 384281 | 12.07.12 | 12.09.12 | 62,0 | 53 | 36 | 26% | 9 |
| 384280 | 12.07.12 | 12.09.12 | 62,0 | 36 | 24 | 29% | 7 |
| 384272 | 12.07.12 | 12.09.12 | 62,0 | 65 | 43 | 25% | 11 |
| 384303 | 12.07.12 | 12.09.12 | 62,0 | 62 | 41 | 25% | 10 |
| 384304 | 12.07.12 | 12.09.12 | 62,0 | 36 | 24 | 29% | 7 |
| 384306 | 12.07.12 | 12.09.12 | 62,0 | 28 | 19 | 31% | 6 |
| 384307 | 12.07.12 | 12.09.12 | 62,0 | 91 | 61 | 23% | 14 |
| 384305 | 12.07.12 | 12.09.12 | 62,0 | 71 | 48 | 24% | 12 |
| 384309 | 12.07.12 | 12.09.12 | 62,0 | 24 | 16 | 32% | 5 |
| 384296 | 12.07.12 | 12.09.12 | 62,0 | 73 | 49 | 24% | 12 |
| 384297 | 12.07.12 | 12.09.12 | 62,0 | 150 | 100 | 21% | 21 |
| 384299 | 12.07.12 | 12.09.12 | 62,0 | 23 | 16 | 32% | 5 |
| 384298 | 12.07.12 | 12.09.12 | 62,0 | 64 | 43 | 25% | 11 |
| 384301 | 12.07.12 | 12.09.12 | 62,0 | 13 | 9 | 38% | 3 |
| 384293 | 12.07.12 | 12.09.12 | 62,0 | 44 | 30 | 27% | 8 |
| 384294 | 12.07.12 | 12.09.12 | 62,0 | 33 | 22 | 30% | 7 |
| 384302 | 12.07.12 | 12.09.12 | 62,0 | 14 | 10 | 37% | 4 |

**Table2.** Continued.

| | | | | | | | |
|---|---|---|---|---|---|---|---|
| 384287 | 12.07.12 | 12.09.12 | 62,0 | 13 | 9 | 38% | 3 |
| 384288 | 12.07.12 | 12.09.12 | 62,0 | 62 | 41 | 25% | 10 |
| 384291 | 12.07.12 | 12.09.12 | 62,0 | 241 | 162 | 19% | 31 |
| 384289 | 12.07.12 | 12.09.12 | 62,0 | 168 | 113 | 20% | 23 |
| 384290 | 12.07.12 | 12.09.12 | 62,0 | 530 | 356 | 18% | 64 |
| 384327 | 13.07.12 | 13.09.12 | 62,0 | 29 | 19 | 31% | 6 |
| 384325 | 13.07.12 | 13.09.12 | 62,0 | 17 | 11 | 35% | 4 |
| 384319 | 13.07.12 | 13.09.12 | 62,0 | 52 | 35 | 26% | 9 |
| 384330 | 13.07.12 | 13.09.12 | 62,0 | 36 | 24 | 29% | 7 |
| 384321 | 13.07.12 | 13.09.12 | 62,0 | 41 | 27 | 28% | 8 |
| 384317 | 13.07.12 | 13.09.12 | 62,0 | 295 | 199 | 19% | 38 |
| 384316 | 13.07.12 | 13.09.12 | 62,0 | 22 | 15 | 33% | 5 |
| 384324 | 13.07.12 | 13.09.12 | 62,0 | 18 | 12 | 34% | 4 |
| 384322 | 13.07.12 | 13.09.12 | 62,0 | 120 | 81 | 22% | 18 |
| 384315 | 13.07.12 | 13.09.12 | 62,0 | 25 | 17 | 32% | 5 |
| 384314 | 13.07.12 | 13.09.12 | 62,0 | 14 | 10 | 37% | 4 |
| 384329 | 13.07.12 | 13.09.12 | 62,0 | 17 | 11 | 35% | 4 |
| 384320 | 13.07.12 | 13.09.12 | 62,0 | 106 | 71 | 22% | 16 |
| 384348 | 13.07.12 | 13.09.12 | 62,0 | 41 | 27 | 28% | 8 |
| 384346 | 13.07.12 | 13.09.12 | 62,0 | 21 | 14 | 33% | 5 |
| 384347 | 13.07.12 | 13.09.12 | 62,0 | 57 | 38 | 26% | 10 |
| 384345 | 13.07.12 | 13.09.12 | 62,0 | 453 | 304 | 18% | 55 |
| 384341 | 13.07.12 | 13.09.12 | 62,0 | 28 | 19 | 31% | 6 |
| 384343 | 13.07.12 | 13.09.12 | 62,0 | 42 | 29 | 28% | 8 |
| 384344 | 13.07.12 | 13.09.12 | 62,0 | 23 | 15 | 33% | 5 |
| 384342 | 13.07.12 | 13.09.12 | 62,0 | 182 | 122 | 20% | 24 |
| 384359 | 13.07.12 | 13.09.12 | 62,0 | 43 | 29 | 27% | 8 |
| 384333 | 13.07.12 | 13.09.12 | 62,0 | 92 | 62 | 23% | 14 |
| 384334 | 13.07.12 | 13.09.12 | 62,0 | 103 | 69 | 22% | 15 |
| 384335 | 13.07.12 | 13.09.12 | 62,0 | 222 | 149 | 20% | 30 |
| 384332 | 13.07.12 | 13.09.12 | 62,0 | 62 | 41 | 25% | 10 |
| 384336 | 13.07.12 | 13.09.12 | 62,0 | 225 | 151 | 20% | 30 |
| 384337 | 13.07.12 | 13.09.12 | 62,0 | 63 | 43 | 25% | 11 |
| 384338 | 13.07.12 | 13.09.12 | 62,0 | 73 | 49 | 24% | 12 |
| 384339 | 13.07.12 | 13.09.12 | 62,0 | 73 | 49 | 24% | 12 |
| 384369 | 14.07.12 | 14.09.12 | 62,0 | 81 | 54 | 24% | 13 |
| 384368 | 14.07.12 | 14.09.12 | 62,0 | 54 | 36 | 26% | 9 |
| 384367 | 14.07.12 | 14.09.12 | 62,0 | 17 | 12 | 35% | 4 |
| 384376 | 14.07.12 | 14.09.12 | 62,0 | 28 | 19 | 31% | 6 |
| 384373 | 14.07.12 | 14.09.12 | 62,0 | 39 | 26 | 28% | 7 |
| 384372 | 14.07.12 | 14.09.12 | 62,0 | 15 | 10 | 36% | 4 |
| 384371 | 14.07.12 | 14.09.12 | 62,0 | 27 | 18 | 31% | 6 |
| 384370 | 14.07.12 | 14.09.12 | 62,0 | 40 | 27 | 28% | 8 |
| 384374 | 14.07.12 | 14.09.12 | 62,0 | 17 | 11 | 35% | 4 |
| 384375 | 14.07.12 | 14.09.12 | 62,0 | 135 | 91 | 21% | 19 |
| 384311 | 13.07.12 | 13.09.12 | 62,0 | 50 | 33 | 26% | 9 |
| 384312 | 13.07.12 | 13.09.12 | 62,0 | 25 | 17 | 32% | 5 |
| 384357 | 14.07.12 | 14.09.12 | 62,0 | 39 | 26 | 28% | 7 |
| 384354 | 14.07.12 | 14.09.12 | 62,0 | 20 | 13 | 34% | 4 |

**Table2.** Continued.

| | | | | | | | |
|---|---|---|---|---|---|---|---|
| 384356 | 14.07.12 | 14.09.12 | 62,0 | 34 | 23 | 29% | 7 |
| 384313 | 13.07.12 | 13.09.12 | 62,0 | 41 | 28 | 28% | 8 |
| 384310 | 13.07.12 | 13.09.12 | 62,0 | 12 | 8 | 39% | 3 |
| 384353 | 14.07.12 | 14.09.12 | 62,0 | 28 | 19 | 31% | 6 |
| 384351 | 14.07.12 | 14.09.12 | 62,0 | 41 | 27 | 28% | 8 |
| 384352 | 14.07.12 | 14.09.12 | 62,0 | 76 | 51 | 24% | 12 |
| 384350 | 14.07.12 | 14.09.12 | 62,0 | 31 | 21 | 30% | 6 |
| 384361 | 14.07.12 | 14.09.12 | 62,0 | 20 | 14 | 34% | 5 |
| 384362 | 14.07.12 | 14.09.12 | 62,0 | 25 | 17 | 32% | 5 |
| 384360 | 14.07.12 | 14.09.12 | 62,0 | 13 | 9 | 38% | 3 |
| 384366 | 14.07.12 | 14.09.12 | 62,0 | 54 | 36 | 26% | 9 |
| 384365 | 14.07.12 | 14.09.12 | 62,0 | 32 | 22 | 30% | 7 |
| 384364 | 14.07.12 | 14.09.12 | 62,0 | 14 | 10 | 37% | 4 |
| 384363 | 14.07.12 | 14.09.12 | 62,0 | 28 | 19 | 31% | 6 |

From these tables we find that the average indoor radon concentration value for houses in this area was found to vary from $11 \pm 6$ Bq m$^{-3}$ to $380 \pm 91$ Bq m$^{-3}$ in winter season and from $8 \pm 3$ Bq m$^{-3}$ to $356 \pm 64$ Bq m$^{-3}$ in summer season. The highest average indoor concentration was $380 \pm 91$ Bq m$^{-3}$ in winter season and $356 \pm 64$ Bq m$^{-3}$ in summer season. These values are upper the recommended threshold of 200-300 Bq/m$^3$ in summer and winter season. The lowest average indoor concentration was $11 \pm 6$ Bq m$^{-3}$ in winter season and $8 \pm 3$ Bq m$^{-3}$ in summer season. The radiation from radon and its daughter products, is considered the second leading cause of lung cancer after smoking, according to a 1999 report the National Academy of Science (NAS Reports, 1998).

The World Health Organization (WHO, 2009) has suggested that homeowners take actions when radon levels exceed 100 Bq/m$^3$. This is a much more conservative figure than the Environmental Protection Agency (EPA) action level of 148 Bq/m$^3$ (EPA, 1991), which has been the USA standard for many years (Mowlavi, 2012). The upper limit value of radon by TAEK is 400 Bq/m$^3$.

In this study, we calculated annual effective dose utilizing UNSCEAR's guidelines (UNSCEAR, 2000). Their suggestions are as follows:

- An indoor radon decay product equilibrium factor of EF = 0.4
- A radon effective dose coefficient factor of EDCF = 9nSv/(Bq h m$^{-3}$)
- An indoor occupancy factor of OF = 0.8, which is the fraction time that people spend indoors, but not essentially in their homes. Therefore, during one year (T= 365x24 h), people spend about 7,008 hours in home and office environments.

Annual effective dose value is given in Eq.3.

$$D= (C_{Rn}) \times (EF) \times (EDCF) \times (OF) \times (T) \tag{3}$$

$$D= [11\ \mathrm{Bq/m^3}] \times [0.4] \times [9\times10^{-9}\ (\mathrm{Sv})/(\mathrm{Bq\ h\ m^{-3}})] \times [0.8] \times [8760\ (\mathrm{h})]$$

D=0.278 mSv

The calculated values of average annual effective dose for study area in winter season vary from 0.278 mSv to 9.59 mSv. The calculated values of average annual effective dose for study area in summer season vary from 0.202 mSv to 8.98 mSv. These annual effective dose values are less than even the lower limit of suggested action level (3-10 mSv). Therefore, the calculated average annual effective dose values do not exceed the Turkish average. No difference was found when the results of the study were compared with the data acquired from other provinces of Turkey (Canbazoğlu, 2012). Further, the average annual effective dose values is more than the accepted value of 1.3 mSv , as set by UNSCEAR in 1993, but on the lower side of recommendation level of (3-10 mSv). For this reason, these average values will pose none serious health risk.

These winter season radon measurements are expected to be higher than other times of the year, especially in poorly ventilated houses. The distribution of indoor radon levels among number of dwellings in Erzurum is shown in Fig. 3 in winter season and in Fig. 4 in summer season.

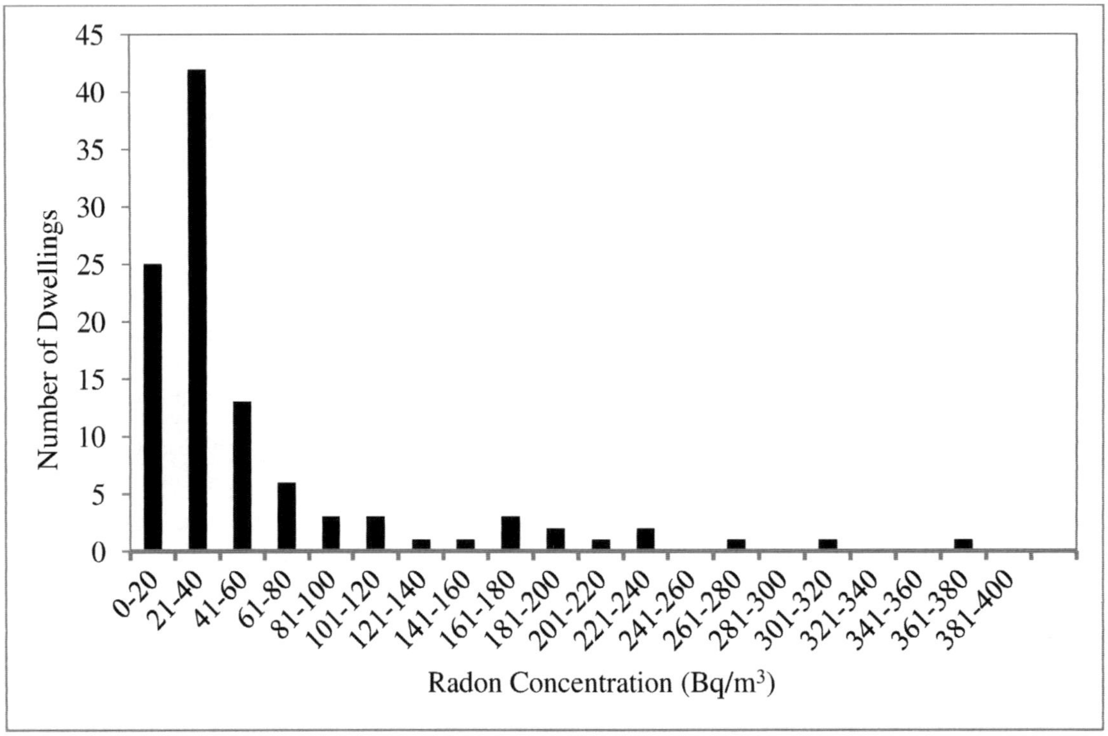


**Fig. 3.** Distribution of radon concentration among various dwellings in winter season

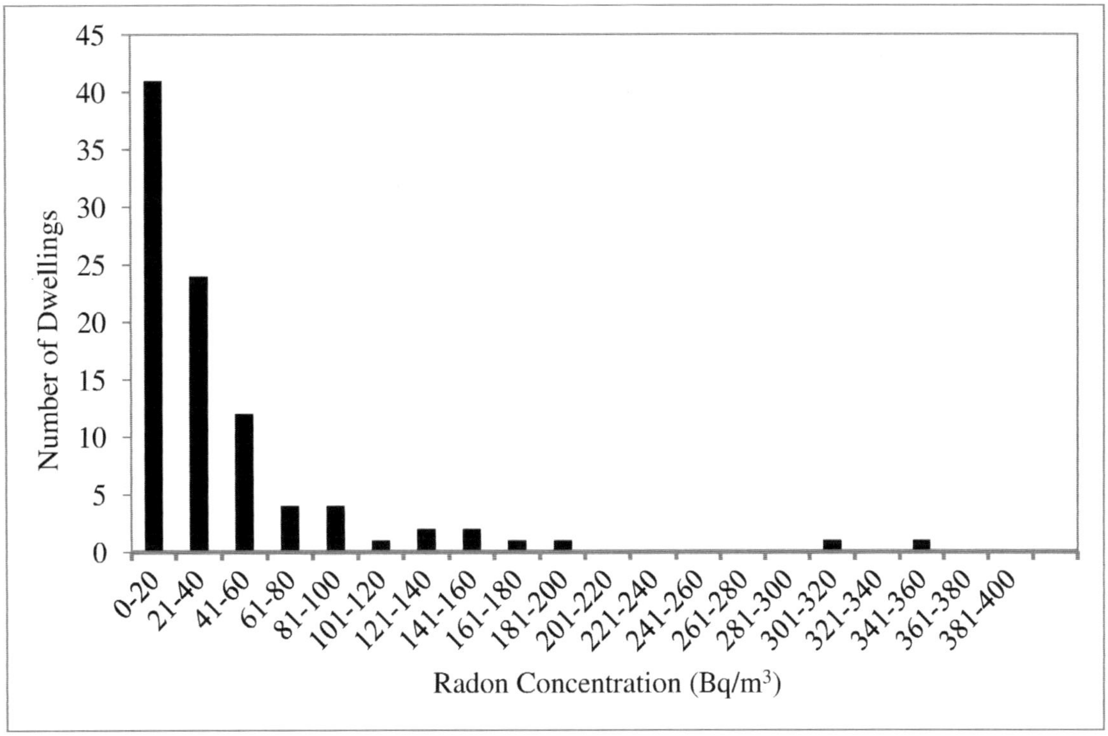


**Fig 4.** Distribution of radon concentration among various dwellings in summer season.

In winter season, radon concentrations in 43.6 % of houses range between 0 and 20 Bq/m$^3$.
25.5 % of them range between 21 and 40 Bq/m$^3$.
12.8 % of them range between 41 and 60 Bq/m$^3$.
4.3 % of them range between 61 and 80 Bq/m$^3$.
4.3 % of them range between 81 and 100 Bq/m$^3$.
1.06 % of them range between 101 and 120 Bq/m$^3$.
2.13 % of them range between 121 and 140 Bq/m$^3$.
2.13 % of them range between 141 and 160 Bq/m$^3$.
1.06 % of them range between 161 and 180 Bq/m$^3$.
1.06 % of them range between 181 and 200 Bq/m$^3$.
1.06 % of them range between 301 and 320 Bq/m$^3$.
1.06 % of them range between 341 and 360 Bq/m$^3$.

In summer season, radon concentrations 23.8 % of houses range between 0 and 20 Bq/m$^3$ .

40 % of them range between 21 and 40 Bq/m$^3$.
12.4 % of them range between 41 and 60 Bq/m$^3$.
5.7 % of them range between 61 and 80 Bq/m$^3$.
2.85 % of them range between 81 and 100 Bq/m$^3$.
2.85 % of them range between 101 and 120 Bq/m$^3$.
0.95 % of them range between 121 and 140 Bq/m$^3$.
0.95 % of them range between 141 and 160 Bq/m$^3$.
2.85 % of them range between 161 and 180 Bq/m$^3$.
1.9 % of them range between 181 and 200 Bq/m$^3$.
0.95 % of them range between 201and 220 Bq/m$^3$.

1.9 % of them range between 221 and 240 Bq/m$^3$.
0.95 % of them range between 261 and 280 Bq/m$^3$.
0.95 % of them range between 301 and 320 Bq/m$^3$.
0.95 % of them range between 361 and 380 Bq/m$^3$.

It is found that the radon levels depend upon many factors inside in dwellings. The radon levels in dwellings vary with seasons. Ventilation plays an important role. During the winter months door and windows tend to be closed, concentrating the radon counts. Radon levels are usually higher in winter compared to summer. Also, radon concentrations vary with the floor levels. Therefore, people in basement would have a greater risk for exposure.

In conclusion, the result of present study provides a database on indoor radon level in Erzurum, Turkey.

**Acknowledgment**
This project was supported by Atatürk University Scientific Resarch Project Unit under contract No: 2011/381.

**References**

[1] D. D. Nhan, C. P. Fernando, N. Thi Thu. Ha, N. Quang Long, D. D. Thuan, H. Fonseca, Jour. Environ. Radioac. 110 (2012) 98-103
[2] S. Labıdı, D. Al-azmi, H. Mahjoubı, R. Ben Salah, Radioprotection. 45 (2010) 209-217.
[3] P.C. Deka, S. Sarkar, B. Bhattacharjeee, T. D. Goswami, B. K. Sarma, T. V. Ramachandran, Radiat. Measu. 36 (2003) 431-434.
[4] M.A. Misdaq, A. Ktata, A. Bakhchi, Radiat. Measu. 32 (2000) 35-42.
[5] S. Singh, D.K. Sharma, A. Kumar, Jour. Of Environ. Radioac. 76 (2004) 369-376.
[6] S. U. Rahman, J. Anwar, Matiullah, Radiat. Measu. 43 (2008) 401-404.
[7] L. Oufni, M.A. Misdaq, M. Amrane, Radiat. Measu. 40 (2005) 118-123.
[8] M. L'Annunziata, Handbook of Radioactivity Analysis, Academic Press, 2003.
National Academy of Science (NAS) Reports, 1999. Health Effects of Exposures to Radon, Committee on Health Risks of Exposures to Radon (BEIR VI), National Academy Press.
[9] WHO, World Health Organization, 2009.Sets radon Action level of 2.7 – less lung cancer risk than EPA 4.0. Global Press Release Distribution. The website:www. PRLog.org.
[10] EPA: U.S. Environmental Protection Agency, 1991. National primary drinking water regulations: radio nuclides; proposed rules. Federal Register, 56(138), 33050. The website: www.epa.org.
[11] A. A. Mowlavi, M.R. Fornasier, A. Binesh, M. Denaro,. Environ Monit Assess. 184 ( 2012) 1085-1088.
[12] UNSCEAR, 2000.United Nations Scientific Committee on the Effects of Atomic Radiation Sources, effects and risks of ionizing radiation. New York: United Nations.
[13] B. Canbazoğlu, N. Albayrak, N. Çelebi, M. Doğru, J. Radioanal. Nucl. Chem. 292 ( 2012) 1059-1063.

# Removal and Recovery of Some From Simulated Nuclear Waste Radionuclides Using Amberlite XAD-16-Pyridylazo Resorcinol (AXAD-PR) Composite

**Sevil Akçağlar**

Dokuz Eylül University, Engineering Faculty, Mechanical Engineering Department, Kaynaklar Campus, Buca-İzmir, Turkey

E-mail: sevil.akcaglar@deu.edu.tr

**Abstract.** In this study, Amberlite XAD-16-Pyridylazo resorcinol composite (AXAD-PR) were developed under laboratory conditions to detect the removals and recoveries of U(IV) and Am(III) from simulated nuclear wastes by adsorption and desorption processes. The effects of contacting time (5, 15, 25 and 200 min), of AXAD-PR concentrations (5, 25, 50 and 100 mg/l), of temperatures (15, 30 and 45°C), increasing U(IV) and Am(III) concentrations (1, 5, 10, 15, 20, 30 and 40) and pH levels (acidic, basic and neutral conditions) on the removals and recoveries of U(IV) and Am(III) radionuclides were investigated. The data obtained showed that the maximum adsorption yields of U(IV) and Am(III) were found to be 96% and 95%, respectively, after 25 min using 25 mg/l AXAD-PR at 30°C at an acidic pH of 5. The adsorption kinetic of radionuclides to AXAD-PR isotherm was in accordance with Langmuir. The AXAD-PR can be reused and can be used again. The regeneration of the composite is possible and can be attained a desorption efficiency of 99% using 0.4 M $HNO_3$ acid solution from 25 mg/l prepared AXAD-PR.

## 1. Introduction

Uranium (U-VI) and Americium (Am-III) are the most important strategic elements that have got wide applications which mainly include the nuclear industry and where its significant commercial uses, apart from manufacturing the nuclear fuel for energy production, include the electron microscopy and the glass and pottery glaze industries. However, uranium has a serious long-term potential environmental hazard because of its long half-life as well as its high radiological and biological toxicity [1]. In nature, uranium assays an average of about 4 mg/l in the earth's crust; however, it forms several economic deposits in various rock types. Uranium is one of the most important radionuclide as nuclear fuel for energy production. Due to insufficient conventional energy sources in Turkey (coal, natural gas, thermal and sun and wind energies) the nuclear energy centrals has been developed. However, the radionuclides such as Uranium (U-IV) and Americium (Am-III) in the nuclear wastes are very toxic pollutants in the environment due to U(IV) and Am(III) have serious long-term potential environmental hazard because of theirs long half-life and high radiological and biological inhibitions. Extensive efforts have thus been recently focused on new solid phase extraction techniques using adsorbents, ion exchangers and other technologies for the separation and recovery of uranium (IV) from its aqueous solutions [2, 3]. In this regard, Amberlite XAD-16 resin series (polystyrene-divinely benzene) such as Amberlite XAD-1, XAD-2, XAD-4 and XAD-16 have a high hydrophobic character and have no ion-exchange capacity, however, these resins are quite important from both the economic and environmental points of view for solid phase extraction after their proper functionalization [4, 5]. Although several methods have been proposed in literature on techniques for removal of radionuclides; adsorption process appears to be the most applicable method for treatment and recovery of used composite adsorbent .The adsorption efficiency of AXAD-PR polymeric adsorbent on 2, 4, 6 trinitrofenol was investigated by Uslu and Demir (2012) [6]. 1.5 gr of AXAD-PR was found to be as the optimum those for maximum 2, 4, 6 trifenol adsorption. Borin and Pavko

(2009) investigated the adsorption of vancomycin antibiotic on AXAD-PR in a packed bed column with a maximum adsorption capacity of 53.76 Kg/m$^3$ at pH 7 [7]. Anaia et all.,(2014) studied the adsorption of 1-(2-thiazolylazo)-2-naphtol on Amberlite XAD-7 [8]. Hosseini-Bandegharaei (2016) investigated the adsorption of Th(IV) on XAD-7 [9]. Hosseini-Bandegharaei et all., (2013) investigated separation of U(VI) from polluted environmental wastewaters using Amberlite XAD-2010 with a maximum adsorption capacity of 0.632 mmol/g [10]. Ghaedi et all., (2014) researched the 50 μg/l heavy metal adsorption on XAD-7 modified by chlorophenol and imidazole moieties with a yield of 80% [11]. Prabhakaran and Subramanian (2004) used AXAD-PR modified with hydroxyarsinoyl methyl osphonic acid to extract the U(VI), Th(IV) and La(III), concentrations varying between 18-23 ng/ml with a sorption yield of 80% from acidic wastewaters under 15 min contacting time [12].

Uranium is one of the most important radionuclide as nuclear fuel for energy production. Due to insufficient conventional energy sources in Turkey (coal, natural gas, thermal, sun and wind energies) the nuclear energy centrals has been developed. However, the radionuclides such as Uranium (U-IV) and Americium (Am-III) in the nuclear wastes are very toxic pollutants in the environment due to U(IV) and Am(III) have serious long-term potential environmental hazard because of theirs long half-life and high radiological and biological inhibitions [1].

In this study AXAD-PRs were developed under laboratory conditions to detect the removals and recoveries of U(IV) and Am(III) from simulated nuclear wastes by adsorption and desorption processes. The effects of contacting time (5, 15, 25, 200 min), of AXAD-PR concentrations (5, 25, 50 and 100 mg/l), of U(IV) and Am (III) concentrations ( 1, 5, 10, 15, 20, 30 and 40 mg/l), of temperatures (15, 30 and 45 $^{o}$C), and pH levels (acidic, basic and neutral conditions) on the removals and recoveries of U(IV) and Am(III) radionuclides and the sorption isotherms were investigated.

## 2. Material and methods

### 2.1. Composition of synthetic nuclear wastewater composition

The synthetic composition of the nuclear wastewater used in this study is given in Table 1.

**Table 1. Synthetic composition of the nuclear wastewater**

| Name of radionuclide | Concentration(mg/l) |
|---|---|
| Pu-239 | 19 |
| Sr-90 | 7 |
| Co-60 | 2.3 |
| Y-90 | 1.2 |
| Cs-134 | 12 |
| Cs-137 | 13 |
| Eu-154 | 9 |
| Eu-155 | 10 |
| Pm-147 | 17 |
| Pr-144 | 3.5 |
| C-14 | 25 |
| H-3 | 1.5 |
| S-35 | 2.5 |
| Np | 4.5 |
| U(IV) | 30 |
| Am(III) | 25 |

### 2.2. Batch sorption studies

Simulated synthetic nuclear power industry wastewater containing different U(IV) and Am(III) concentrations were mixed with increasing concentrations of AXAD-PR using an shaker at a shaking rate of 200 rpm. The adsorption studies were performed at batch wise using 400 ml glass quartz reactors. 100 ml of synthetic nuclear wastewater containing increasing U(IV) and Am (III) concentrations and increasing AXAD-PR were added to the glass quartz reactors. They were mixed using magnetic stirrers at a rpm of 200. The reactors were operated at different retention times, different pH levels and at different temperatures. The temperature was adjusted using water baths. The Ph of the samples were adjusted by using weak acid and alkaline. The solid material was separated from the liquid after 15 min sedimentation. The radionuclide analysis was performed in the supernatant samples. The sorption efficiency, *R* (%), can be calculated by using the Eqs. 1 and 2.

$$q_e = (Ci - Cf)\,V/m \qquad (1)$$

Sorption capacity, $q_e$ (mg/g) was obtained using Equation 2

$$A\% = (Ci - Cf)\,100/Ci \qquad (2)$$

Where $C_0$ and $C_e$ are the initial and equilibrium concentrations of the U(IV) and Am(III) (mg /l), respectively. *V* is the volume of the testing solution (L), and *m* is the amount of the AXAD-PR (g) and A is the sorption efficiency (%).

### 2.3. Reusability of AXAD-PR

The regeneration–reusability property of AXAD-PR is important for an appropriate adsorbent at an optimum pH range, which implied that the loaded U(IV) and Am(III) radionuclides might be desorbed from spent sorbent. Therefore, the reusability experiments were performed using 1.0 mol/l HCl as desorbing agent.

### 2.4. Teflon coated quartz glass reactors for sorption processes

Sorption experiments were carried out in self-designed quartz glass reactors. The dimensions of the reactors were 38 and 3.5 cm. The experiments were performed at room temperature and the pH of the reaction mixture was adjusted from 2 to 8 using 1 mol/l of $H_2SO_4$ and NaOH solution.

### 2.5. Operational conditions in batch reactors

The effects of contacting time (5, 15, 25, 200 min), of AXAD-PR concentrations (5, 25, 50 and 100 mg/l), of U(IV) and Am(III) concentrations ( 1, 5, 10, 15, 20, 30 and 40 mg/l), of temperatures (15, 30 and 45 $^{o}$C), and pH levels (acidic, basic and neutral conditions) on the removals and recoveries of U(IV) and Am(III) radionuclides and the sorption isotherms were investigated.

### 2.6. Analytical procedures

#### 2.6.1. Measurements of U(IV) and Am (III)

The measurements of U(IV) and Am(III) were performed in Lawrence Berkeley Laboratory in Canada. Radionuclide analyzes were performed using an ion chromatography device comprising a scintillation detector with a cell flow. This device is an CS5 separation column providing the separation of cation and anion in the waste water and the detector with a cell flow is scintillated from the glass bed activated with cerium. The studies were performed by taking 0.5 ml sample. The Chromatographic unit consisting from the ion exchanging column, from the gradient liquid pump and from an scintillation detector providing the wastes nonhazardous and from an computer. The chromatographic unit is an pre-concentration column and is consists from and cation exchange resin (ionex CC2) containing sulfonic acid and an analytical column (from an ionex CS5 containing both anion exchanging resin and both the sulfonic acid groups and alkyl quaternary's amine groups). The radiation detector, consist from a chilled tube (solid scintillation apparatus) containing a glass bead activated with cerium. This detector is placed between two pairs of tubes increasing the rays. The system is run with the stages given in Table 2 and the sample volumes are taken from 0.5 ml up to 5 ml.

**Table 2. Examples of the radionuclide analysis stages**

| | Time (min) |
|---|---|
| The sample is rinsed with 4 ml of water | 0.5 |
| 0.09 M oxalic acid | 5.17 |
| Washing with water | 17.18 |
| 0.1M diglycolic acid | 18.29 |
| Washing with water | 29.30 |
| 1M nitric acid | 30.50 |
| Washing with water | 50.55 |
| Flow rate | 1 ml/min |

After separation process, the actinides are sequentially going to the scintillation detector with a flow cell. The flow cell detector is calibrated with Am(III) and U(IV) at very low concentrations in National Institute of Standards and Technology. In order to determine the retention time of radionuclides the media was acidized by using 0.1 M $HNO_3$.

**Performed studies:** Fluent → sampling → passing from the pre-concentration column passing from → isolated column passing from the → separation column ------ → passing from the detector cell flow.

**Calculation of Area:** It is shown in the Chromatograph given below (Figure 1). From here it is passing from the radionuclide concentration

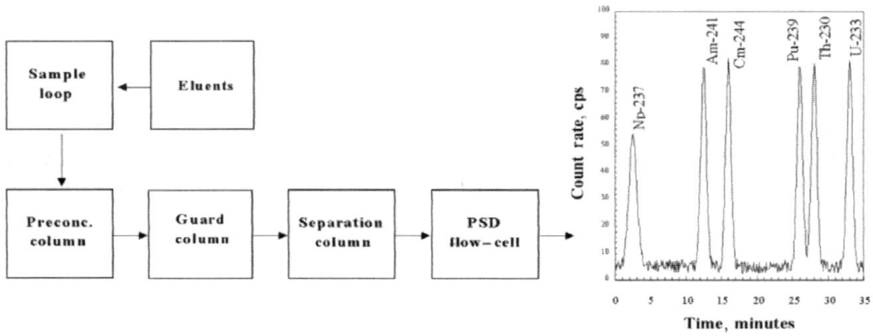


**Figure 1.** Studies for radionuclide measurements and standby time of radionuclide

### 2.6.2. Sorption Isotherm

**Langmuir isotherm**

The Langmuir isotherm is one of the most important monolayer sorption models, and it is based on the assumption that the sorption occurs on a perfect surface consisting of equivalent and independent active sites. According to this model, the sorption cannot proceed beyond monolayer coverage. The linear form of Langmuir model can be represented as follows:

$$C_e/q_e = C_e / q_{max} + 1 / b \times q_{max} \quad (3)$$

Where $b$ is the Langmuir constant relating to the sorption (l/mg) and $q_{max}$ is the maximum sorption capacity (mg /mg). Where $C_e$ (mg/l) is the equilibrium concentration of U(IV) or Am (III), $q_e$ (mg/mg) is the sorption capacity at equilibrium, $q_{max}$ (mg /mg) is the maximum sorption capacity and b (l/mg) is the Langmuir constant which is related to the sorption energy. A linear plot of $C_e/q_e$ vs $C_e$ was obtained.

### 2.6.3 . Radionuclide Sorption Kinetics

First-Order Kinetics

$$log(q_e - q_t) = log\, q_e = k_1 \, x \, (t / 2{,}303) \quad (4)$$

Where $q_e$ and $q_t$ (mg/mg) are the amounts of U(IV) or Am(III) sorbet at equilibrium and at time t (min), respectively, $k_1$ (1/min) is the rate constant of the pseudo-first order sorption and t (min) is the contact time. The adsorption rate constants ($k_1$) and the sorption capacity ($q_e$) for the U(IV) or Am(III) sorption by AXAD-PR were calculated from the slope and intercept of the plots of log ($q_e$-$q_t$) against t.

### 2.6.4. Synthesis of AXAD-16-PR resin

The AXAD-PR-pyridylazo resorcinol sorbent (AXAD-PR) is prepared by stirring 10 g of AXAD-PR powder 200 mesh size, with a nitration mixture(20 mL conc. HNO3 and 50 mL conc. H2SO4) at 60 $^{o}$C in a water bath for 1 h. The reaction mixture is then poured into ice-water followed by filtration and washing for several times with distilled water until acid-free. The nitrated resin is afterwards reduced with a mixture of 40 g SnCl2, 45 mL conc. HCl and 50 mL ethanol and refluxed at 60 $^{o}$C for 12 h. The product is filtered off, washed with water followed by washing with 2 M NaOH to obtain the free amino polymer. The latter is then treated with 2 M HCl and the excess of which is removed by water washing before it is suspended in 350 mL of ice-water. Subsequently, the reaction mixture is stirred with 1 mL aliquots of 1 M NaNO2 and 1 M HCl until a durable blue color is formed on the starch iodide paper as indicator. The polymer has thus been converted to the diazonium salt and which is washed with cold water followed by it streatment with 250 mL of 10% NaOH containing 5g pyridylazo resorcinol at 0–5 $^{o}$C and keep at this temperature for 24 h until completion of the reaction. The obtained dark brown powder is filtered off and is finally washed with distilled water. The reactions sequence for the various preparation steps of the AXAD-PR resin functionalized with pyridylazo resorcinol.

## 3. Results

### 3.1 Effect of increasing AXAD-PR concentrations on the sorption efficiencies of 25 mg/l U(IV) and Am(III)

As the AXAD-PR concentrations were increased from 5 mg/l up to 30 mg/l the sorption efficiency of 25 mg/l U(IV) increased from 57% to 99% in the U(IV) sorption process at pH =5,3 after 25 min contacting time at 29 $^{o}$C temperature (Figure 2). Further increase of AXAD-PR concentration to 40 and 50 mg/l decreased the U(IV) sorption yield to 90% throughout 25 mg/l U(IV) sorption. As the AXAD-PR concentrations were increased from 5 mg/l up to 40 mg/l, the sorption efficiency of 25 mg/l Am (III) increased from 57% to 99% in the Am(III) sorption process at pH =5,6 after 25 min contacting time at 29 $^{0}$C temperature. After 40 mg/l AXAD-PR concentration, the increasing of AXAD-PR to 50 mg/l decreased the 25 mg/l Am(III) sorption efficiency to 85% from 99%. U(IV) sorption reached to the maximum at low AXAD-PR concentration in the comparison to the sorption of Am(III) to AXAD–PR for the same operational conditions. Hosseini-Bandegharaei (2016) also found that by increasing the XAD-7 concentrations the removal percentage of Th(IV) increased while further increase of adsorbent dose from 80 to 140mg/l decrease the Th(IV) adsorption yield [9].

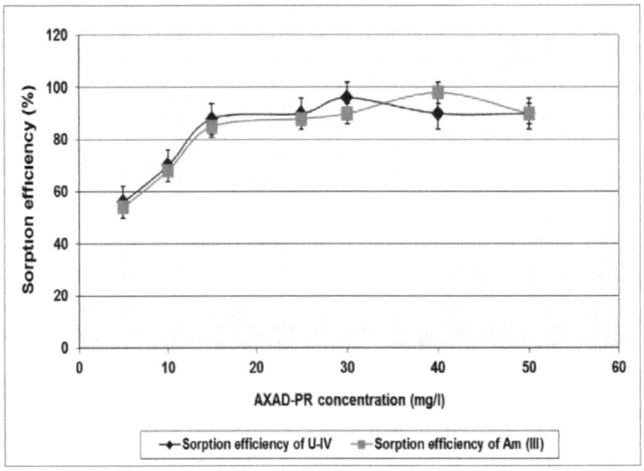


**Figure 2.** Effect of AXAD-PR concentrations on the sorption of U(IV) and Am(III) sorption efficiencies at pH =5,3 after 25 min contacting time at 29 °C temperature.

### 3.2. Effects of U(IV) and Am(III) concentrations on the sorption efficiencies on AXAD-PR

The maximum sorption efficiency of U(IV) (97%) was obtained at a U(IV) concentration of 20 mg/l while the maximum sorption efficiency of Am(III) (99%) was obtained at an Am(III) concentration of 30 mg/l after 25 min contacting time at 29 $^0$C temperature (Figure 3). The maximum sorption efficiency of U(IV) radionuclide was obtained at low U(IV) concentration compared to the Am(III) sorption. The maximum sorption efficiency of Am(III) was obtained (99%) at higher Am(III) concentration compared to U(IV) sorption. The reached sorption efficiency of Am(III) is higher compared to U(IV) sorption efficiency. As mentioned above at high U(IV) and Am(III) concentrations the adsorption yields of these radionuclides degreased. Similar results were found at Th(VI) adsorption on XAD-7 [9]. It should be mentioned that at higher radionuclide concentrations the lower adsorption yields can be attributed to the lower ratio of initial numbers of radionuclide concentrations to the available adsorbing sides existing on AXAD-PR surface and, also, the higher removal percentages at the low initial Am(III) and U(IV) concentrations.

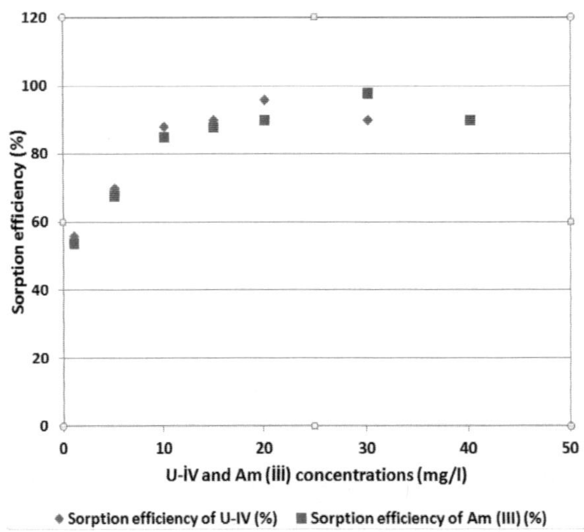


**Figure 3.** Effects of U(IV) and Am(III) concentrations on the sorption efficiencies on 30 mg/l AXAD-PR at pH =5,3 after 25 min contacting time at 29 °C temperature.

**3.3. Effect of sorption time on the sorption efficiency of U(IV)**

The maximum sorption efficiency of 20 mg/l U(IV) was found to be 97% at a contacting time of 30 min at a pH of 5,2 at 29 $^{0}$C temperature for 25 mg/l AXAD-PR concentration (Figure 4). Low sorption efficiencies was found for the lower AXAD–PR concentration: for example around 79% sorption efficiencies for 30 mg/l U(IV) was found for both 5 and 15 mg/l AXAD-PR concentrations. For higher AXAD-PR concentrations (>25 mg/l) the sorption removal of 30 mg/l U(IV) decreased to around to 45% and 25%. Demir and Uslu 2012 found that 120 min sorption time was the optimum duration for maximum adsorption of 14 gr/l 2, 4, 6 trinitrofenol using 1.5 g/l AXAD–PR [6]. In the comparison of the results it can be concluded that small amount of AXAD-PR is enough to remove the U(IV) from the nuclear wastes although the studied adsorbent and pollutant parameters are different. Hosseini-Bandegharaei et all., (2016) found that the contact time between the Th(VI) and XAD-7 can play important role in the adsorption process [9]. They found that the uptake of Th(VI) increased up to 60 min contacting time and then further increase of time did not affect the adsorption yield .

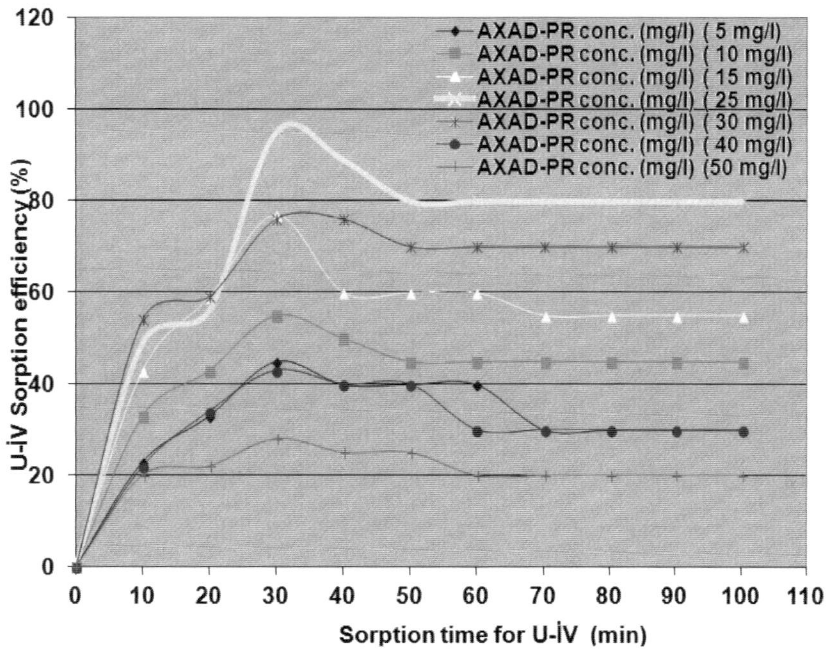


**Figure 4.** Effect of sorption time on the sorption efficiency of U(IV) at a contacting time of 30 min at a pH of 5,2 at 29 $^{0}$C temperature for 25 mg/l AXAD-PR concentration.

**3.4. Effect of increasing sorption time on the sorption efficiency of Am(III) versus increasing AXAD-PR concentrations.**

The maximum sorption efficiency of 40 mg/l Am(III) was found to be 98% after 40 min contacting for sorption to 25 mg/l AXAD-PR concentration at pH=5.3 at 29 $^{0}$C temperature (Figure 5). Low sorption efficiencies was found for the lower AXAD–PR concentration: for example around 40-55% sorption efficiencies of 40 mg/l Am(III) was found for both 5, 10 and 15 mg/l AXAD-PR concentrations. For higher AXAD-PR concentrations (>25 mg/l) for 40-50 mg/l AXAD-PR; the sorption removal of 40 mg/l Am(III) decreased to around to 50% and 40%, respectively. The maximum 10 µg/l U(VI) removal was obtained after 25 min and the loading half time was less than 6 min [10]. In the study also lower U(IV) removals was obtained at higher sorption time. This behavior is due to ionic strength of U(VI) is enhanced and the sorption process is inhibited.

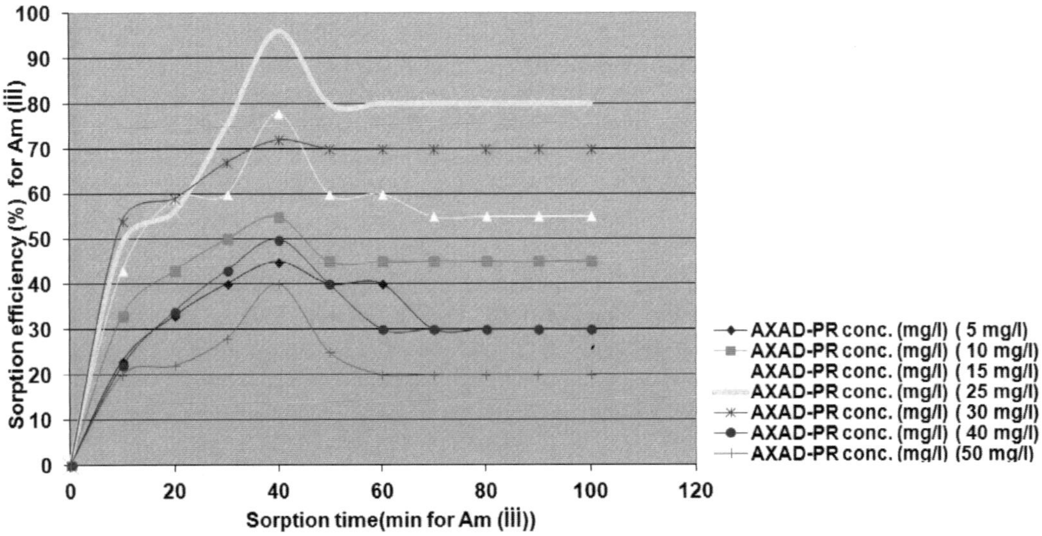


**Figure 5.** Effect of increasing sorption time on the sorption efficiency of Am(III) versus increasing AXAD-PR concentrations at pH=5.3 at 29 $^0$C temperature.

**3.5.Effect of increasing temperature on the sorption efficiency of U(IV)**

As the temperature was increased from, 20 $^o$C to 30 $^o$C, the sorption yield of 20 mg/l U(IV) increased from 49% up to 90% at 25 mg/l AXAD-PR concentration after 30 min sorption time (Figure 6) at pH=5.3. At temperatures > 30 $^0$C the sorption yields of 30 mg/l U(IV) radionuclide to 25 mg/l AXAD-PR decreased from 75% to 60% for 25 mg/l AXAD-PR concentration. The maximum U(IV) sorption efficiency was obtained at 30 $^0$C temperature. The sorption mechanism is dependent on the nature of adsorbent. On the type of pollutant as mentioned above, on the pH on the temperature. At high temperatures (60-80 $^0$C) degrease the adsorption yields can be attributed to the increase of wastewater viscosity containing the U(IV) and Am(III) radionuclides. Under these situations the radionuclide ion mobility decreases. Hosseini-Bandegharaei et all., (2016) found different results between Th(VI) sorption and XAD-7. Thus, at Th(VI) concentration >50mg/l higher sorption capacities was observed due to higher ratio of the initial Th(VI) concentrations [9].

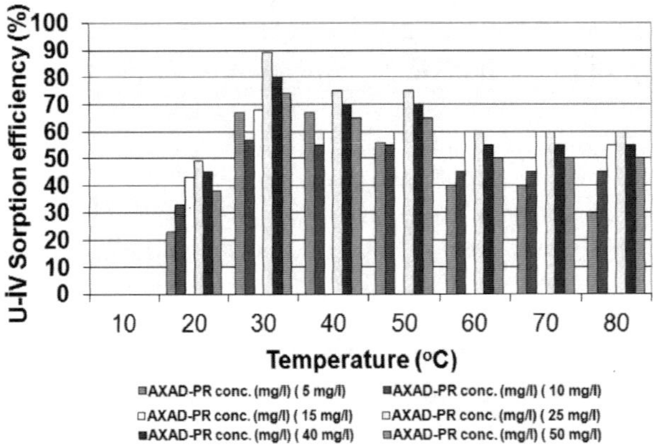


**Figure 6.** Effect of increasing temperature on the sorption efficiency of U(IV) at 25 mg/l AXAD-PR concentration after 30 min sorption time.

### 3.6. Effect of increasing temperature on the sorption efficiency of Am(III)

As the temperature was increased from 20 °C to 30 °C and 40 $^0$C the sorption yield of 30 mg/l Am(III) increased from 49% up to 94% at 25 mg/l AXAD-PR concentration after 40 min sorption time at a pH of 5.3 (Figure 7). At temperatures > 30 °C the sorption yields of 40 mg/l Am(III) radionuclide to 25 mg/l AXAD-PR decreased from 75% to 60% for 25 mg/l AXAD-PR concentration. The maximum Am(III) sorption efficiency was obtained at 40 $^0$C temperature. As the temperature increased the raised diffisubility may contribute available Am(III) at the subsurface for more adsorption of radionuclide the decrease in adsorption efficiency at 50, 60, 70 and 80 $^0$C temperature may be due to less favorable adsorption resulting in low adsorption yields of Am(III). Uslu and Demir (2012) found that the adsorption capacity of the XAD-16 is decreasing with increasing temperature from 24.85 $^0$C and 44.85 $^0$C [6].

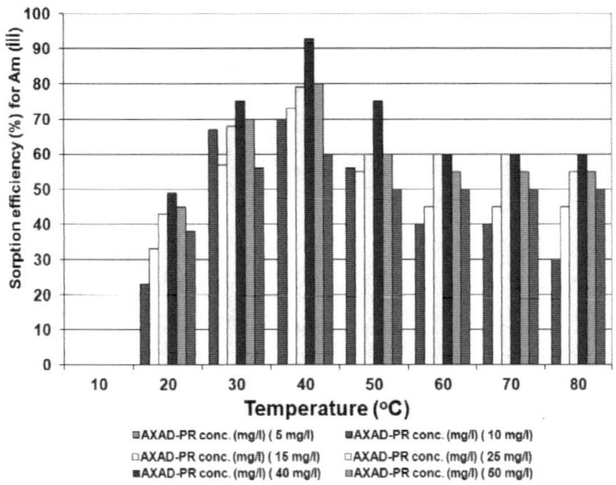


**Figure 7.** Effect of increasing temperature on the sorption efficiency of Am(III) at 25 mg/l AXAD-PR concentration after 40 min sorption time at a pH of 5.3

### 3.7. Effect of pH on the sorption efficiency of Am(III) and U(IV)

As the pH of the process increased from 2 up to 5; The sorption efficiencies Am(III) and U(IV) increased from 39% and 35% up to 93% and 95%, respectively (Figure 8). The increasing of pH from acidic (pH=6) up to alkaline (pH=8) level reduced the sorption efficiencies of both radionuclides to around 30% and 12%. The optimum pH for maximum Am(III) and U(IV) sorption yields was found to be 5. Hosseini-Bandegharaei (2016) observed that the degree adsorption of Th(IV) on XAD-7 was close to %90 at the pH range of 4-5 [9]. Therefore in this study and in our study adsorption was directly dependent to the pH . Hosseini-Bandegharaei et all., (2013) measured the pH effect on the sorption of U(VI) on XAD-2010.the sorption increases with increasing pH of sorptive simulated wastewater solution and attains a maximum value at pH 4.2-4.7 [10]. This indicated that the sorption process involves the release of $H^+$ ions to allow the firm complexation of U(IV)and Am(III) ions to AXAD-PR at the optimum pH. At the low pH, the release of $H^+$ ions was not efficient because of adsorbent could not be used efficiently for preconcentrations of U(IV) AND Am(III) in the low concentration range. Gaedi et all., (2014) studied the heavy metal adsorption at lower pH (pH > $pH_{zcp}$) [11]. In these pH values the XAD-7 has positive charge and at higher value it takes negative charge. Therefore, it seems that the pollutants are significantly and thermodynamically more fissile to accumulate on XAD at pH higher than $pH_{zcp.}$ The approximate $pH_{zcp}$ of modified AXAD-PR was found to be around 5. Therefore, it has neutral charge at this pH and confirms that the main mechanism for binding radionuclides and their enrichment is complexation with proposed method at pH >5 the competition of h+ ions with radionuclides for binding to reactive centers of XAD leads to significant reduction of adsorption of radionuclides. Therefor pH 5 was recommended to improve the sorption of radionuclides on XAD.

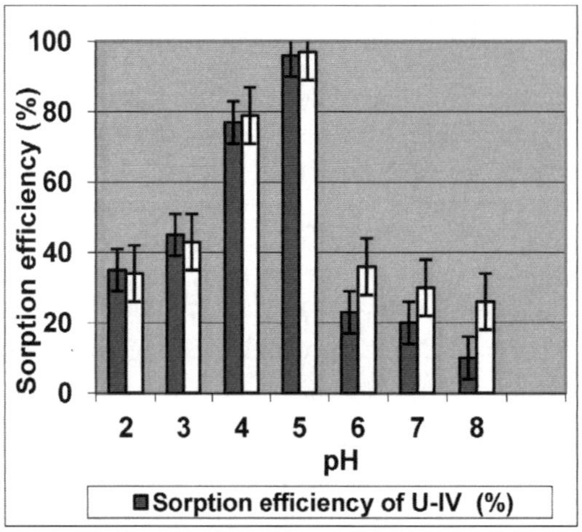


**Figure 8.** Effect of pH on the sorption efficiency of Am(III) and U(IV) at 25 mg/l AXAD-PR concentration after 40 min sorption time.

### 3.8. First order reaction kinetics of U(IV)

The sorption studies of 20 mg/l U(IV) to AXAD-PR showed a first order sorption kinetic with $R^2$ values of 0.99 and first order reaction kinetic constants ($K_1$) of $1x10^{-4}$, $2x10^{-4}$ and $3x10^{-2}$ 1/min for 15, 30 and 50 mg/l AXAD-PR concentrations after 30 min sorption time at a temperature of 30 $^0$C and a pH of 5.3 (Figure 9). Hosseini-Bandegharaei et all., (2016) also found that the Th(VI) sorption on XAD-7 follows first order reaction with a $q_{max}$ level of 53.4 mg/g [9].

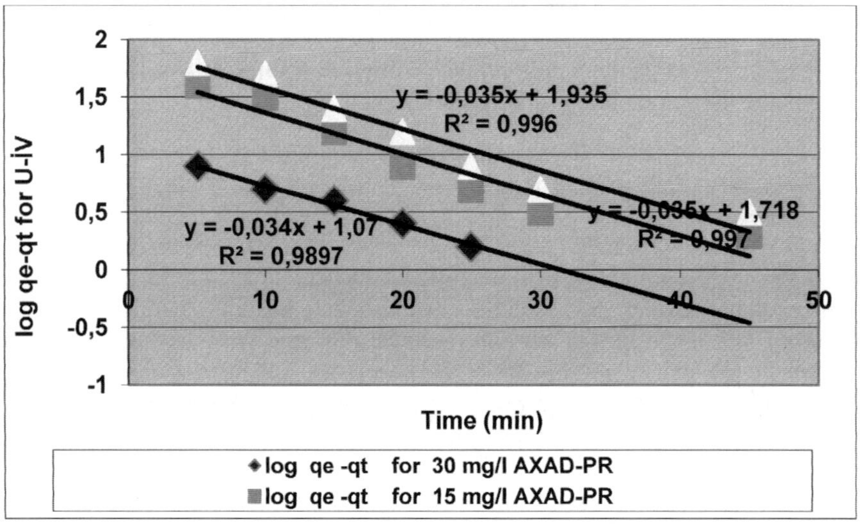


**Figure 9.** First order reaction kinetics of U(IV) at 20 mg/l U(IV); 15, 30 and 50 mg/l AXAD-PR concentrations after 30 min sorption time at a temperature of 30 °C and a pH of 5.3.

### 3.9. First order reaction kinetics of Am(III)

The sorption studies of 30 mg/l Am(III) to AXAD-PR showed a first order sorption kinetic with $R^2$ values of 0.99 and first order reaction kinetic constants ($K_1$) of $1,3x10^{-4}$ , $2,98x10^{-4}$ and $3.67x10^{-2}$ 1/min for 15, 30 and 50 mg/l AXAD-PR concentrations after 30 min sorption time at a temperature of 30 $^0$C and a pH of 5.3 after 40 min contacting time (Figure 10). Uslu and Demir (2012) obtained that the adsorption of 2, 4, 6, trinitrofenol on XAD-16 fallows pseudo second order reaction kinetics

[6]. The differences of adsorption kinetic can be attributed to the different wastewater type pollutant kind, to the pH and to the different environmental studied.

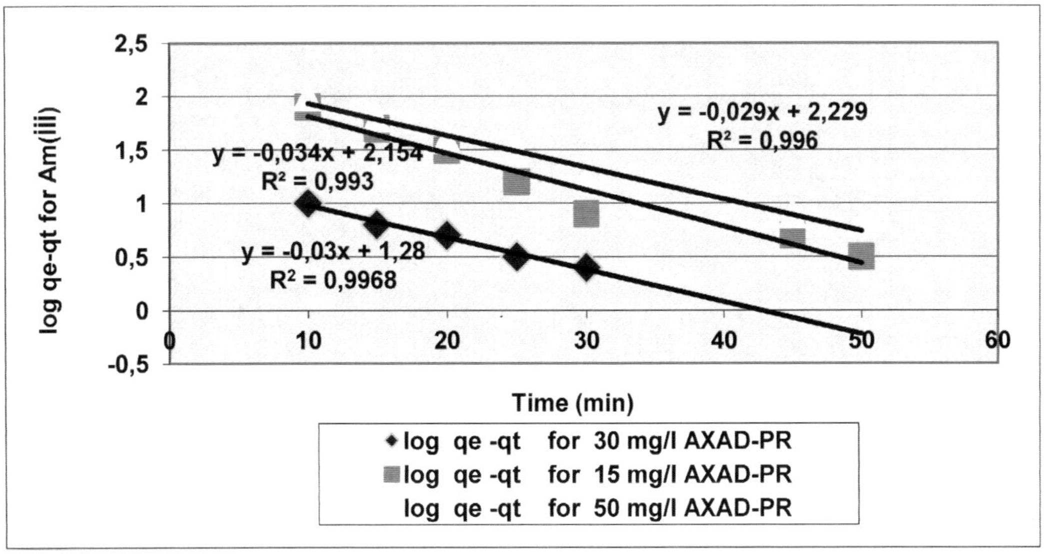


**Figure 10.** First order reaction kinetics of Am(III) for 30 mg/l Am(III); 50, 15 and 30 mg/l AXAD-PR concentrations at a temperature of 30 °C and a pH of 5.3 after 40 min contacting time

### 3.10. Langmuir sorption isotherms for U(IV) and Am(III)

The sorption data were used to evaluate the sorption kinetic of U(IV) and Am(III). Both radionuclides were sorbet according to Langmuir isotherm with high $R^2$ values and acceptable Langmuir Kinetic Constants ($K_L$) (Figures 11a and 11b). values for U(IV) and Am(III) were calculated as 1,09 l/mg and 1.01 l/mg, respectively, while $q_{max}$ values were calculated as 19.8 mg/mg and 23.6 mg/mg, respectively. The Langmuir adsorption isotherm is based on the assumption that all adsorption sides are equivalent and that adsorption is an active side, is independent of whether the adjacent sides are occupied or not. The fact that, the Langmuir isotherm fits the experimental data very well may be due to homogenous distribution of active chelating sides on the AXAD-PR since, the Langmuir equations assumes that the adsorbent surface is homogenous. Uslu and Demir (2012) also observed that the AXAD-PR exhibited Langmuir isotherm with a $K_L$ level of 0.0410 l/mg [6]. These results indicate that AXAD-PR shows similar responses to the adsorption process although the pollutants are quite different. Borin and Pavko (2009) also found that adsorption kinetic of vancomycin on AXAD-PR can be definite with Langmuir isotherm [7]. Anaia et all., (2014) found that 1-(2-thiazolylazo)-2-naphtol sorbet to Amberlite xad-7 by fallow the kinetic proposed by Langmuir [8].

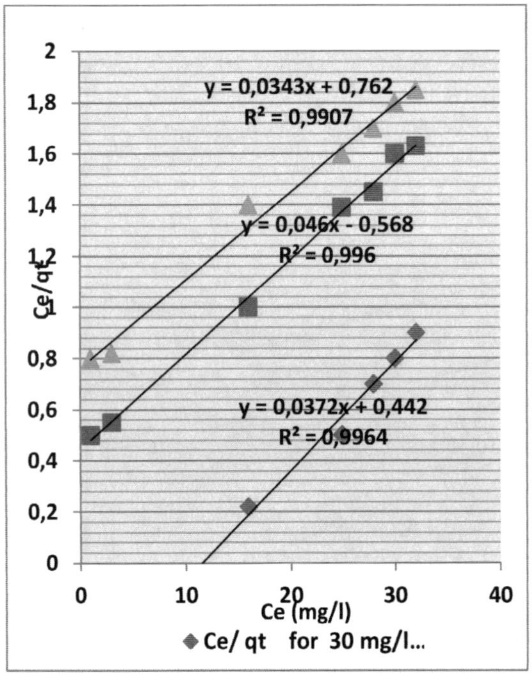


**Figure 11a.** Langmuir isotherm for 30 mg/l Am(III) at 15, 30 and 50 mg/l AXAD-PR concentrations at a temperature of 30 °C at a pH=5.3 at a temperature of 30 °C after 40 min sorption time

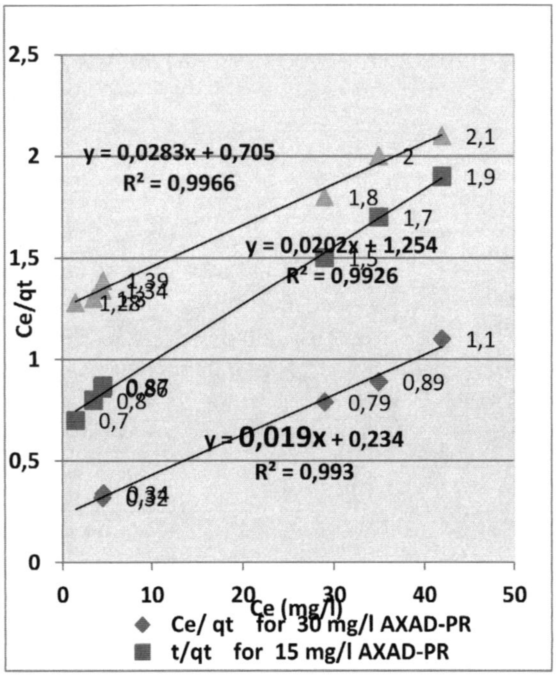


**Figure 11b.** Langmuir isotherm for 20 mg/l U(IV) at 15, 30 and 50 mg/l AXAD-PR concentrations at a temperature of 30 $^0$C at a pH=5.3 at a temperature of 30 °C after 30 min sorption time for Am(III) after 30 min contact time

### 3.11. Recovery of U(IV) and Am(III)

The sorption efficiencies in both radionuclides were not reduced significantly, after four times sequential utilization of AXAD-PR (Table 3). Hosseini-Bandegharaei et all., (2015) also found higher Amberlite XAD-7 recoveries in the U(IV) and Th(VI) sorption [13] while Hosseini-Bandegharaei et all., (2016) obtained % 70 AXD-7 desorption-recovery yield in the sorption studies of Th(VI) [9].

**Table 3.** Sequential treatment of 20 mg/l U(IV) and 30 mg/l Am(III) in the presence of 30 mg/l AXAD-PR concentration for U(IV) and 40 mg/l AXAD-PR concentration for Am(III)

| **Sorption efficiency (%)** | **Cycles** | | | |
|---|---|---|---|---|
| | **First** | **second** | **third** | **Fourth** |
| **20 mg/l U(IV)** | **96%** | **92%** | **88%** | **85%** |
| **30 mg/l Am(III)** | **98%** | **96%** | **90%** | **88%** |
| **50 mg/l U(IV)** | **75%** | **70%** | **65%** | **60%** |
| **50 mg/l Am(III)** | **78%** | **73%** | **70%** | **68%** |

## 4. Conclusions

Among the kinetic studies it was found that the adsorption kinetic was first order ($K_1 = 2x10^{-3}$ $min^{-1}$) while the adsorption of radionuclides to AXAD-PR isotherm was in accordance with Langmuir. The AXAD-PR can be reused again with high yields after four sequential utilization times. The regeneration of the composite is possible and can be attained a desorption efficiencies of 85% and 88% for U(IV) and Am(III) removals using 0.4 M $HNO_3$ acid solution using 20 and 30 mg/l prepared AXAD-PR. The maximum adsorption yields of U(IV) and Am(III) were found to be 96% and 95%, respectively, after 25 min contacting time using 30 and 40 mg/l AXAD-PR, respectively, at 30 $^{o}$C at a pH of 5.

## References

[1] V.Neck, J.İ.KİM, Radiochim Acta 89, ( 2001), 1-16, doi: 10.1524/ract.2001.89.1.001
[2] P.K.Tewari, A.K. Singh, Talanta 53 823 (200) doi:http://dx.doi.org/10.1016/S0039-9140(00)00572-5. 18968172
[3] R. Qadeer, J. Hanif, M. Khan, M. Saleem, Radiochim. Acta 68, 197 (1995) http://www.degruyter.com/view/j/ract.1995.68.issue-3/issue-files/ract.1995.68.issue-3.xml doi: 10.1515/ract-2016-frontmatter1
[4] P.Metilda, K. Sanghamitra, J.M. Gladis, G.R.K. Naidu, T.P. Rao, Talanta 65 (1) (2005) 192-200. Doi:http://dx.doi.org/10.1016/j.talanta.2004.06.005.18969783.
[5] J.M. Gladis, T. Prasada Rao, Anal. Bional. Chem. 373 (8) (2002) 867-887, doi: http://dx.doi.org/10. 1007/s00216-002-1387-7. 12194052.
[6] H. Uslu, G. Demir, 39:1-3, 88-94, (2012) http://dx.doi.org/10.1080/19443994.2012.669163
[7] B. Borin and A. Pavko, *Chem. Biochem. Eng. Q.* 23 (4) 479–483 (2009),doi :
[8] *G. C. Anaia, P. A. M. Freitas, M.E. V. Suárez-Iha and F. R. P. Rocha, Chem. Soc.*, Vol. 25, No. 4, 648-657, (2014), http://dx.doi.org/10.5935/0103-5053.20140012
[9] A. Hosseini-Bandegharaei, A. Allahabadi, A. Rahmani-Sani, A. Rastegar, R. Khamirchi, M. Mehrpouyan, R. Hekmat-Shoar, Z. Pajohankia, Radioanal Nucl Chem, (2016), doi : 10.1007/s10967-015-4689-x
[10] A. Hosseini-Bandegharaei, M. Sarwghadi, A. Heydarbeigi, S. H. Hosseini, and M. Nedaie, Journal of Chemistry Volume 2013, Article ID 671564, 10 pages http://dx.doi.org/10.1155/2013/671564
[11] M. Ghaedi, B. Karami, Sh. Shamsaldini, M. Soylak, 1319-6103, (2014) http://dx.doi.org/10.1016/j.jscs.2014.01.012
[12] D. Prabhakaran, M.S. Subramanian, Anal and Bioanalytical Chem., 379, 3, 519-525(2004), doi:10.1007/s002-004-2600-7
[13] A. Hosseini-Bandegharaei, Intern J Environ Anal Chem 95:277–290 (2015) doi : 10.1080/03067319.2015.101.6009

# Statistical thermal model analysis of particle production at LHC

**A. Karasu Uysal[1] and N. Vardar[1*]**

[1]KTO Karatay University, Konya, Turkey

E-mail: necati.vardar@karatay.edu.tr

**Abstract.** A successful description of the particle ratios measured in heavy-ion collisions has been achieved in the framework of thermal models. In such a way, a large number of observables can be reproduced with a small number of parameters, namely the temperature, baryo-chemical potential and a factor measuring the degree of strangeness saturation. The comparison of experimental data at and the model estimations has made possible to define the thermodynamic parameters of strongly interacting matter at chemical freeze-out temperature. The detailed study of hadron and meson production including resonances using the statistical-thermal model is discussed. Their ratios are compared with the existing experimental data and predictions are made for pp and heavy-ion collisions at RHIC and LHC energies.

## 1. Introduction

The statistical-thermal model has been proved successful in applications to relativistic collisions of both heavy ions and elementary particles. The achievement of this approach has pioneered to the design of various software codes [1-2]. THERMUS [3], a thermal model analysis package, has been developed for incorporation into the object-oriented ROOT framework.

THERMUS is able to make calculations within three statistical ensembles; grand-canonical ensemble which allows to fluctuate the conserved quantum numbers -Baryon (B), Strangeness (S) and Charge (Q)- on average; a canonical ensemble imposing all the quantum numbers to be conserved exactly and finally a mixed or strangeness canonical ensemble which is a combination of the canonical treatment of strangeness and grand canonical treatment of quantum number B and Q.

## 2. Analysis

In order to calculate the thermal properties of a system, one starts with an evaluation of its partition function. The form of the partition function obviously depends on the choice of ensemble. THERMUS is able to make calculations within three statistical ensembles; the grand canonical ensemble (GCE), the canonical ensemble (CE) and the strangeness canonical ensemble (SCE), respectively.

In all ensemble THERMUS use the different parameters:

**Table 1.** Parameters used in various canonical ensembles.

| GCE | CE | SCE |
|---|---|---|
| $T, \mu_B, \mu_S, \mu_Q, \mu_C, \gamma_S, \gamma_C, R$ | $T, B, S, Q, \gamma_S, R$ | $T, \mu_B, \mu_Q, \gamma_S, R_C, R$ |

$T$ presents the temperature of the systems, $\mu_B$, $\mu_S$ and $\mu_Q$ are the baryon, strangeness and charge chemical potentials. $\gamma_S$ is strangeness saturation factor, $\gamma_C$ is charm saturation factor and $R$ is the radius of the fireball, $B$, $S$ and $Q$ are baryon, strangeness and charge quantum numbers, $R_C$ is the correlation or the canonical radius, respectively.

The STAR detector [5], one of the experiments at Relativistic Heavy Ion Collider (RHIC), primarily designed to study the formation of quark–gluon plasma (QGP). In this proceeding, STAR pp, dAu and AuAu collisions at the energy of 200 GeV are discussed.

Figure 1(a) shows the comparison of STAR data in pp collisions at $\sqrt{s} = 200$ GeV and THERMUS results. CE is preferred since there is poor particle production in the given collision system. The analysis of STAR data from dAu collisions at $\sqrt{s_{NN}} = 200$ GeV and the particle ratios from the STAR AuAu data at $\sqrt{s_{NN}} = 200$ GeV are given in figure 1(b) and figure 1(c), respectively. In all figures, the solid lines show the THERMUS predictions while closed circles refer to the experimental data.

In figure 1(a), CE is preferred since there is a poor particle production in the given collision system. The fit parameter we obtain for the radius of the fireball ($R$), is $4.14 \pm 0.43$ fm, and also the quantum numbers B and Q are equal to 2 while S is fixed to 0. In figure 1(b), the radius of the fireball fixed to 7 fm in both GCE and the SCE analysis while some of the thermal model parameters are listed as: $\mu_S = 0.1 \pm 0.6$ MeV, $\mu_Q = 5 \pm 1.6$ MeV for GCE analysis, $\mu_Q = 3 \pm 2.8$ MeV, $R_C = 5.28 \pm 1.35$ fm for SCE analysis. Also in figure 1(c), two different ensembles are compared while R is fixed to 10 fm. Some fit parameters taken from GCE analysis are $\mu_S = 5.2 \pm 3.5$ MeV, $\mu_Q = -1.9 \pm 3.1$ MeV and ones from SCE analysis are as follows: $\mu_Q = -1.9 \pm 3.8$ MeV, $R_C = 2 \pm 2.3$ fm.

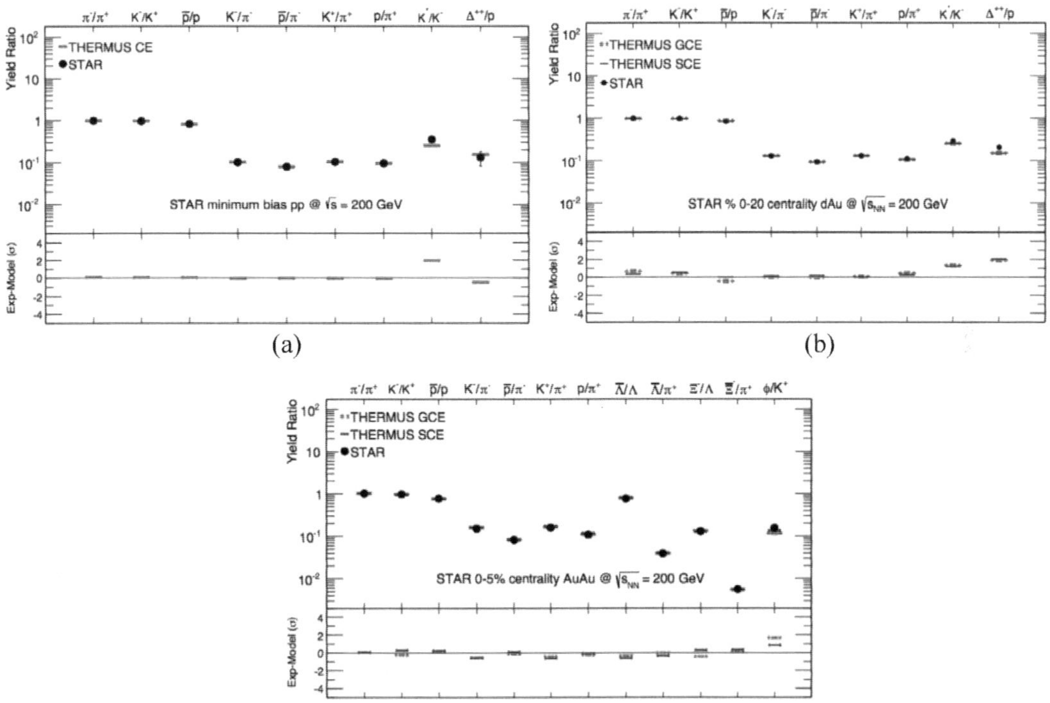


(a) (b) (c)

**Figure 1.** Comparison of STAR data and THERMUS results [4]

The ALICE detector [6] in The Large Hadron Collider (LHC) is also designed to engage the QGP formation and its physical processes. In this analysis, the experimental data from both elementary particles and heavy-ion collisions are included.

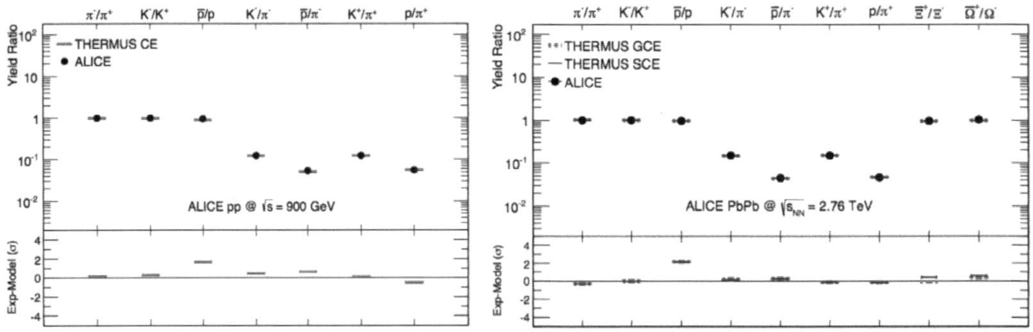


(a) (b)

**Figure 2.** Comparison of ALICE data and THERMUS results [4]

In figure 2(a) shows the comparison of ALICE data in pp collisions at $\sqrt{s} = 900$ GeV and THERMUS results and comparison of ALICE data in PbPb collisions at $\sqrt{s_{NN}} = 2.76$ TeV and THERMUS results are given in figure 2(b).

The radius R is extracted as $8 \pm 0.2$ fm and the quantum numbers B and Q are fixed to 2 and S equals to 0 as well in figure 2(a). In figure 2(b), the analysis is achieved by using two ensembles while R fixed to 10 fm. The model parameters extracted by the GCE are $\mu_S = 2.3 \pm 0.8$ MeV and $\mu_Q = -2.7 \pm 0.8$ MeV whereas the ones coming from SCE analysis are: $\mu_Q = -2.7 \pm 0.8$ MeV and $R_C = 3.38 \pm 3.21$ fm.

## 3. Results and Conclusions

Both the STAR and the ALICE data together with their THERMUS predictions and the temperature dependencies of statistical thermal parameters are studied. The canonical ensemble is used for pp system, and SCE is implemented for not only pp system at higher energies but also for dAu and PbPb systems. Finally, GCE is chosen for AuAu and PbPb collisions and some cases for dAu system. In this study is the first one give predictions for the STAR data in pp, dAu and AuAu collisions at the energy of 200 GeV and also ALICE data in pp system at $\sqrt{s} = 900$ GeV and PbPb system at $\sqrt{s_{NN}} = 2.76$ TeV are successfully reproduced by a thermal model.

In the analysis presented in this proceeding, the statistical-thermal parameters are determined at the chemical freeze-out point which gives information about the QGP medium. The parameters, $T_{ch} = 166 \pm 3.8$ MeV and $T_{ch} = 173 \pm 5.8$ MeV are extracted from the analysis of STAR data in dAu collisions and in AuAu collisions at the energy of 200 GeV, respectively. Also, $T_{ch} = 146 \pm 4$ MeV is from the analysis of ALICE data in pp collisions at $\sqrt{s} = 900$ GeV and $T_{ch} = 163 \pm 0.5$ MeV is from the analysis of ALICE data in PbPb collisions at $\sqrt{s_{NN}} = 2.76$ TeV have been extracted.

**Acknowledgment**

This work was supported by The Scientific and Technological Research Council of Turkey (TUBITAK, Project No 113F221).

**References**

[1] Torrieri, G., Steinke, S., Broniowski, W., Florkowski, W., Letessier, J. and Rafelski, J., SHARE: Statistical Hadronization with Resonances, *Comput. Phys. Commun.*, 2005 **167** 229

[2] Torrieri, G., Jeon, S., Letessier, J. and Rafelski, J., SHAREv2: Fluctuations and a Comprehensive Treatment of Decay Feed-down, *Comput. Phys. Commun.*, 2006 **175**, 635

[3] S. Wheaton, J. Cleymans, and M. Hauer, *Comput. Phys. Commun.*, 2009 **180** 84-106

[4] C. Yasar, Statistical Thermal Model Description of Hadron Yields in High Energy

Collisions, M.Sc. Thesis, ITU 2014.
[5] STAR detector overview, K.H.Ackermann et.al., *Nucl. Instrum. Meth. A* 2003 **499** 624-632
[6] The ALICE experiment at the CERN LHC, K. Aamodt et al., *JINST* 2008 **3** S08002

# Three-dimensional imaging in degraded visual field

**A. Oran[1], S. Ozharar[2] and I. Ozdur[3],**

[1]*Abdullah Gül University, Kayseri, Turkey,*
[2]*Bahçeşehir University, Istanbul, Turkey,*
[3]*Abdullah Gül University, Kayseri, Turkey,*

E-mail: *abdullah.oran@agu.edu.tr*

**Abstract.** Imaging at degraded visual environments is one of the biggest challenges in today's imaging technologies. Especially military and commercial rotary wing aviation is suffering from impaired visual field in sandy, dusty, marine and snowy environments. For example during landing the rotor churns up the particles and creates dense clouds of highly scattering medium, which limits the vision of the pilot and may result in an uncontrolled landing. The vision in such environments is limited because of the high ratio of scattered photons over the ballistic photons which have the image information. We propose to use optical spatial filtering (OSF) method in order to eliminate the scattered photons and only collect the ballistic photons at the receiver. OSF is widely used in microscopy, to the best of our knowledge this will be the first application of OSF for macroscopic imaging. Our experimental results show that most of the scattered photons are eliminated using the spatial filtering in a highly scattering impaired visual field. The results are compared with a standard broad area photo detector which shows the effectiveness of spatial filtering.

## 1. Introduction, state-of-the-art, objectives and overview of the action

One of the biggest challenges in today's imaging technology is three-dimensional (3-D) imaging in degraded visual fields (DVF). Especially military and commercial rotary wing aviation is suffering from degraded visual environments in sandy, dusty, marine and snowy environments. In critical mission phases such as take-off and landing accidents are caused by reducing visibilities to near zero due to brownout, whiteout or water spray conditions from obscurant clouds created by the downwash of the rotor blades. Not only in landing but also in flying scenarios heavy fog and snow can easily impair the vision of the pilot. Some examples of degraded visual environments are shown in Figure 1.

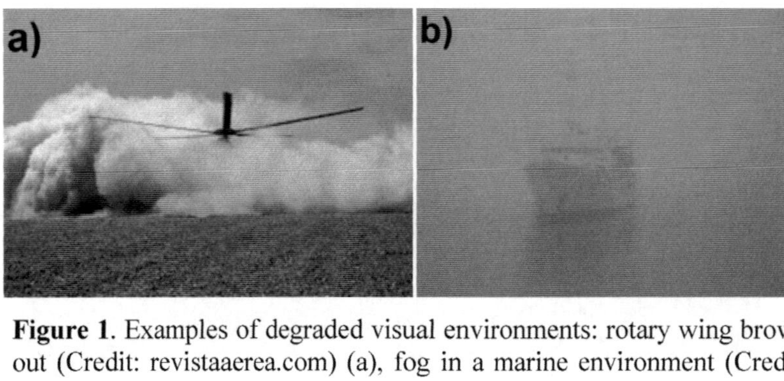


**Figure 1**. Examples of degraded visual environments: rotary wing brown out (Credit: revistaaerea.com) (a), fog in a marine environment (Credit: NOAA) (b).

This vision impairment in rotary wing aviation is resulted in many accidents and life losses in different countries. For example, the UK experienced 24 accidents in the 5 year period of 2005-2009, France has experienced 8 brownout mishaps in the past 15 years, German defense forces has experienced more than 30 mishaps in association with dust or snow. US department of defense report 60 crew members have lost their lives since 1990 due to the visual impairments in degraded visual environments. Many other countries (Canada, Sweden, Norway, and Netherlands) are also report losses and mishap due the degraded visual environments [1].

Degraded visual environments do not only effects rotary wind aviation; people are also affected by fog and snow in daily life especially while driving. According to a report prepared in US, in a 10 year average (2002-2012), 11,812 people are injured, 511 people are dead in 31,385 crashes *annually* in US [2].

One of the notable accidents due to fog was a maritime disaster occurred in 1956, a collision of the ocean liners the SS Andrea Doria and MS Stockholm. As a consequence of the collision 46 people died and SS Andrea Doria capsized and sank [3]. Heavy fog related some other notable accidents are the 1945 crash of a B-25 Mitchell into the Empire State Building [4] and the 1977 collision of Pam Am and KLM planes at the Tenerife Airport [5].

Degraded visual environments are also a problem for subsea investigations. Even for very short distances (a few meters) the visibility can be very poor [6]. These aforementioned cases are main motivations for developing systems to improve the eyesight of pilots, drivers, search and rescue teams, and firefighter

There are mainly 3 different sensor technologies that offer potential solutions to imaging at degraded visual environments; these technologies are based on RADAR, LIDAR and passive sensors.

RADAR related technologies are based on dust penetration capabilities of mm wave signals (35-95 GHz). Their current limitation is low resolution due to the relatively long wavelength. Because of the strong demand from avionics industry there are commercial products based on this technology, however the research is still going on for better resolution and longer ranges [7]–[9]. The second sensor technology is LIDAR based 3D sensors. These sensors transmit an optical pulse to the area that will be imaged and then detects the reflected optical pulse with fast electronics. When compared with RADAR based sensors, LIDAR based 3D sensor have potential to offer higher resolution and there are promising products from different vendors [10], [11]. The third and the last sensor technology is the passive sensors. Passive sensors rely on thermal imaging but operating at longer wavelengths. Current technologies operate at 94 GHz and exploit naturally reflected and emitted radiation [12].

The LIDAR systems are especially common in 3-D applications such as forest management and planning, flood modeling, determining structural characteristic of land [13] and the Google self-driving car [14]. It can receive an image with high spatial resolution and accuracy remotely; it can detect both ballistic and scattering photons as well. Since; there is no problem of reaching saturation similar with a detector. However; in DVF, the scattering photons may become dominant over the ballistic photons and prevent original image because the LIDAR cannot distinguish the difference between the ballistic and scattering photons. Herein, this paper presents a method of improving the imaging capabilities in degraded visual environments using a *LIDAR based 3D sensor*. In order to achieve this, we investigated the *optical spatial filtering (OSF)* properties for *coherent* 3D LIDAR architectures.

## 2. Research methodology and approach

Imaging through degraded visual environments is a very challenging problem and research and development is still going on. Our approach on the problem is 3D coherent LIDAR technology with optical spatial filtering. Optical spatial filtering is a technique in which a lens focuses an image/beam to a pinhole. In this technique only the beams with the correct wavefront can pass through the pinhole.

Any disturbances on the wavefront will be filtered. One of the main applications of the pinholes is cleaning up laser beams from any unwanted intensity variations [15], Figure 2a. Another application of optical spatial filtering is to filter out the out-of-focus light. This property is used in confocal microscopy where a laser beam is focused to a focal plane and then a lens and pinhole pair is used to collect light *only from the focal plane* [16], Figure 2*b*. The wavefront coming from out-of-focus does not have the correct wavefront for the light to pass through the pin hole.

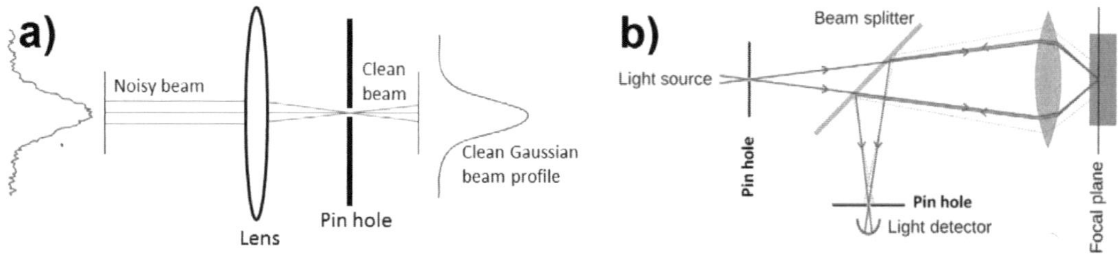


**Figure 2.** Application of pin holes: cleaning up Gaussian beams (a) and collecting light only from the focal plane (b).

In this paper, the intensity noise and out-of-focal light filtering effect was used in order to filter out the scattering light from the receiver.

In a typical 3D imaging LIDAR system, the signal-to-noise ratio is limited by the amount of collected light. In these systems the link loss can be as high as 100 dB, so the amount of receiver optical power is very small and the noise floor is limited by the shot noise or noise of sub-components [17]. In order to increase the amount of collected beam, broad area detectors and multimode fibers are used at the receiver. At degraded visual environments the signal-to-noise ratio (SNR) is not limited by the amount of collected light or sub-components but it is limited by the scattered light, that's why increasing the collection efficiency in such systems does not help to solve the problem. In order to eliminate the scattered light we used optical spatial filtering at degraded visual environments.

In the experimental setup, shown in Figure 3, an ultra-narrow linewidth ($<$ 1 kHz ) laser, working at 1550 nm is used as the light source. An isolator is placed after the laser in order to eliminate any unwanted back reflections. A power splitter is used to measure the power of the laser. The laser beam is sent to the DVF medium using a free space coupler and survived light beams from DVF are measured by different detectors at the back of the DVF medium.

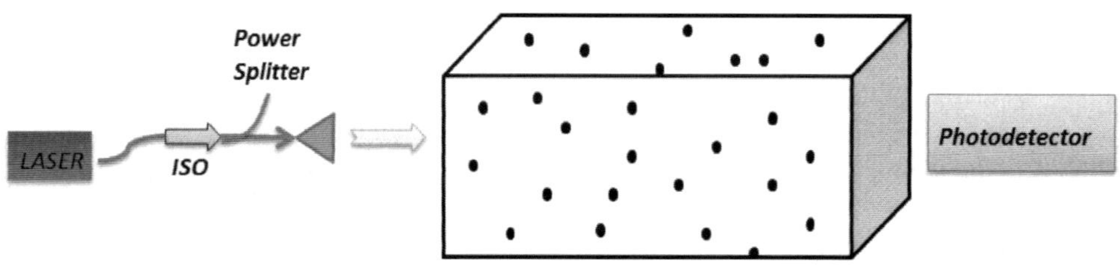


**Figure 3.** Experimental setup (iso: isolator).

The experiments were done in a homemade degraded visual environment. Cement powder is used to create a DVF effect, and particle size distribution of the powder is given in Figure 4. The experiments were repeated on two separate occasions when the tank is empty and filled with the cement material in order to measure the effectiveness of the optical spatial filtering.

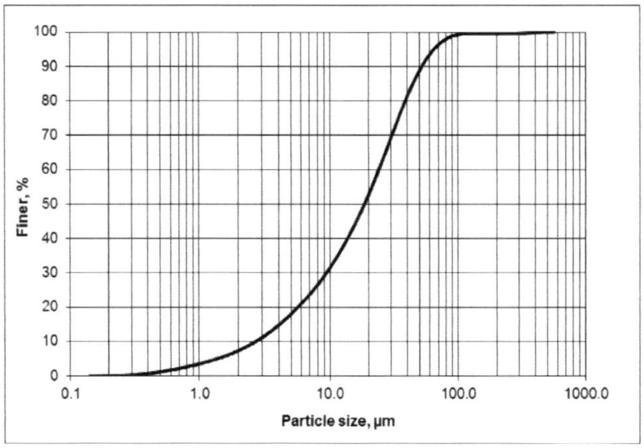


**Figure 4.** Particle size of cement used in the experiment.

In a degraded visual environment there are two photon types to deal with, the scattering photons and the ballistic photons. The scattering photons reach to the detector after bouncing many times from the particles in the environment. Whereas the ballistic photons arrive directly to the detector without experiencing any scattering. A broad area (or multimode) receiver does not distinguish the scattering and ballistic photons and collect all of them. In Figure 5, the receiver architecture with a broad area detector is shown. In this setup, all the photons coming from the degraded visual environments, including scattering photon 1 and 2 are detected by the detector which is not desirable for high SNR.

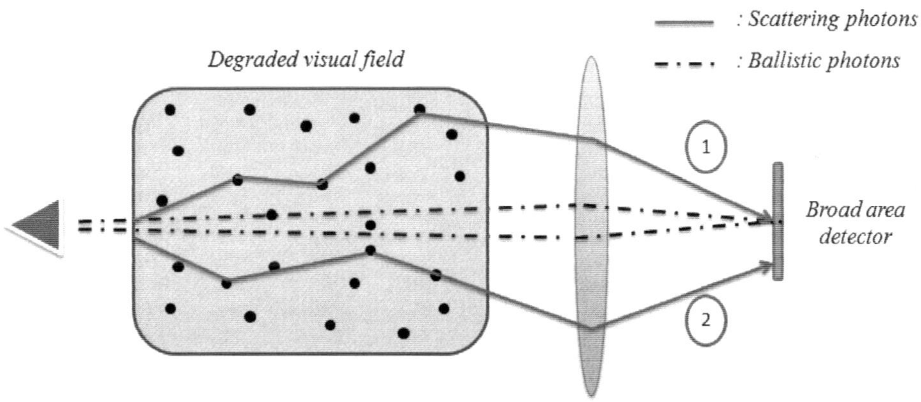


**Figure 5.** Receiver architecture with broad area detector (the size is exaggerated) for visual aid.

When the tank is empty and using broad area detector, all the detected photons are ballistic photons and there are no scattering photons. The distribution of the ballistic photons can be seen in Figure 6 black curve, in this figure the x-axis is the distance from the on axis. If the medium is filled with particles then both the ballistic and scattered photons are present at the detection plane. The distribution of the ballistic and scattered photons can be seen in Figure 6 red curve. As expected the collected photons have a broader distribution from the former case which is a result of broad area detection.

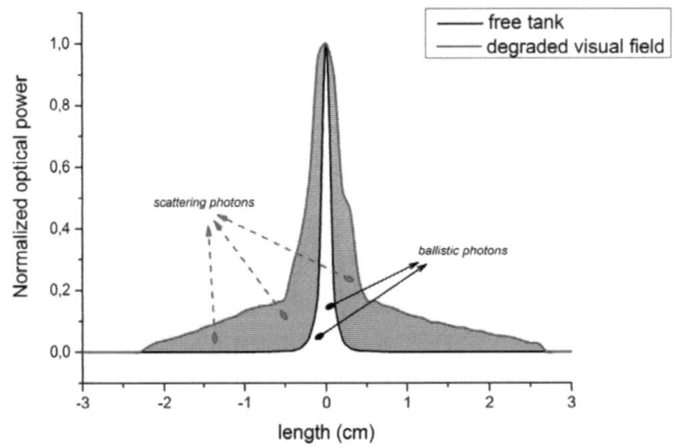


**Figure 6.** Optical power of broad area detector for two cases: free tank and DVF.

To eliminate effects of scattering photons, another architecture is prepared, shown in Figure 7. The transmitted beam is shown with blue lines. If the transceiver system is focused on the target (as shown in the Figure 7) only the light that is coming from the path of the beam will be coupled back to the fiber in other words out-of-focus light filtering will be achieved. In coherent LIDAR transceiver architecture less photons are coupled back to the fiber that is why it is not preferable in 3D LIDAR systems. However for 3D imaging at degraded visual environment systems the signal-to-noise ratio of the system is not limited by the amount of light coupled back to the fiber but it is limited by the ratio of scattering photons.

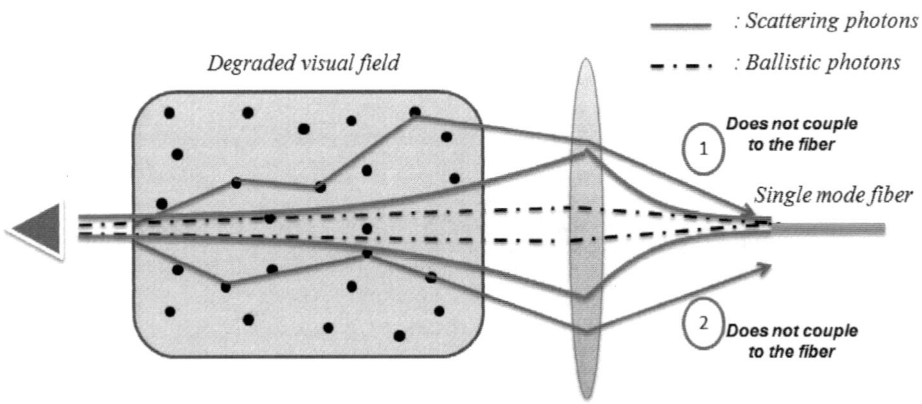


**Figure 7.** Receiver architecture with single mode fiber.

In this architecture the scattered light is less likely to couple to the fiber. For example the scattered photon 1 does not couple to the fiber even though it falls on the fiber as it does not have the right $\vec{k}$ vector (or the incoming angle is bigger than the acceptance angle) and photon 2 does not couple to the fiber simply because it does not fall on the fiber. The ballistic beams on the other hand come directly to the fiber and can easily couple to the fiber. In this setup the detected number of ballistic photons are not changing however detected scattered photon number decreases significantly, which will result in a better signal-to-noise ratio. Measured optical power is plotted for different off axis distances in Figure 8. From the figure we observe that most of the scattering photons are eliminated by optical spatial filtering method.

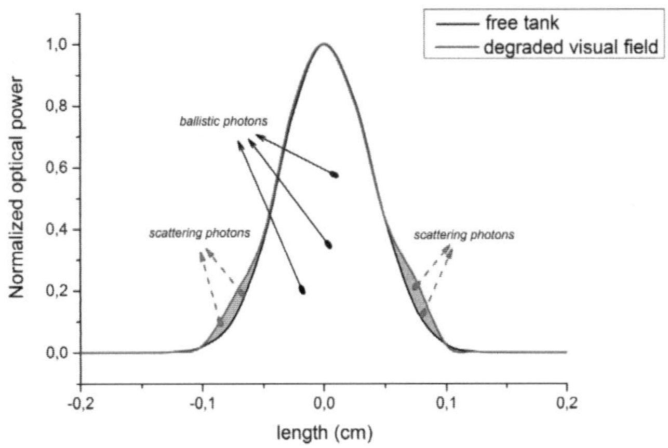


**Figure 8.** Optical power of SMF for two cases: free tank and DVF.

## 3. System Comparison

### *3.1. Error Results:*
The relative error is defined as the ratio of the absolute error, which is defined as the magnitude of the difference between the DVF and the free tank over the magnitude of the ballistic photons. According to values of two Figures (3 and 6); error rate are respectively 0.830 and 0.056.

### *3.2. Correlation Results:*
All two architecture systems and two cases were simulated and compared based on a projected reach. However, for highest confidence we analyzed two performance comparisons with MATLAB.

**Tablo 1.** Significant coefficients of two system performances.

| | BAD: free tank | BAD: DVF | SMF: free tank | SMF: DVF |
|---|---|---|---|---|
| **BAD: free tank** | 1 | 0.676<br>0.000 | x | x |
| **BAD: DVF** | x | 1 | x | x |
| **SMF: free tank** | x | x | 1 | 0.997<br>0.000 |
| **SMF: DVF** | x | x | x | 1 |

According to the Table 1, there are two hypotheses that can be made comparison for **BAD: free tank** *(data 01)* ~ **BAD: DVF** *(data 02);*

- $H_0$: There is no significant correlation between the *data 01* and *data 02.*
- $H_1$: There is a significant correlation between the *data 01* and *data 02.*

The returned value of significant = 0,000<0,05 indicates that $H_0$ hypothesis is rejected . As a result, there is a significance correlation statistically between *data 01* and *data 02* at the % 95 confidence interval as represented in the equation below:

$$\square_{data01,data02}=0.676 \quad (1)$$

For second performance of our system is compared same idea; **SMF: free tank** *(data 03)* ~ **SMF: DVF** *(data 04).*

- $H_0$: There is no significance correlation between the *data 03* and *data 04.*
- $H_1$: There is a significance correlation between the *data 03* and *data 04.*

The returned value of significant= 0,000<0,05 indicates that $H_0$ hypothesis is rejected . As a result, there is a significance correlation statistically between *data 03* and *data 04* at the 95% confidence interval as represented in the equation below:

$$\square_{data03,data04}=0.997 \quad (2)$$

*3.3. T-Test Results:*

**Tablo 2.** Distributions of two comparisons.

| | Mean | Variance |
|---|---|---|
| **BAD: free tank** | 39700.5701 | 7.292e-10 |
| **BAD: DVF** | 5.3533 | 124.335 |
| **SMF: free tank** | 273938.7574 | 2.694e-11 |
| **SMF: DVF** | 292.5565 | 280416.89 |

Hypothesis test result is returned as a logical value. H=ttest2(x,y) performs a t-test of the hypothesis that two independent samples, in the vectors *BAD: DVF* and *SMF: DVF* ( from Table 2 ) represented as x and y , come from distributions with equal means, and returns the result of the test in h.

- If h = 1, this indicates the rejection of the null hypothesis (means $\mu_1= \mu_2$) at the 5% significance level.
- If h = 0, this indicates a failure to reject the null hypothesis at the 5% significance level.

Test the null hypothesis that the pairwise difference between data vectors of **BAD: in case of DVF** and **SMF: in case of DVF** has a mean equal to zero;

```
[h,p] = ttest2(x,y)
>>h =1
>>p =  1.9197e-07
```
(3)

The returned value h=1 indicates that *ttest* rejects the null hypothesis at the default 5% significance level. That means, there is significance different statistically between the BAD: in case of DVF and SMF: in case of DVF at the 95% confidence interval.

## 4. Conclusions

In this work we used optical spatial filtering method in order to eliminate scattering photons for degraded visual fields. We compared two different detection schemes, one of them is a broad area detector and the other one is a single mode fiber which will have optical spatial filtering effect. Our results show that using the optical spatial filtering effect the ratio of scattering photons are decreased to 5 % from 83 %. The filtering of scattering photons in a scattering medium will improve the imaging performance significantly.

This work is supported by TUBİTAK (The Scientific and Technological Research Council of Turkey) BİDEB Project number 113C029.

**References**

[1] R. T. O. T. Report, *Rotary-Wing Brownout Mitigation : Technologies and Training ( Remèdes contre le phénomène de brownout*, vol. 323, no. January. 2012.
[2] Booz-Allen-Hamilton, "Ten-year averages from 2002 to 2012 based on nhtsa data," *US Dep. Transp. - Fed. Highw. Adm.*, 2012.
[3] Wikipedia, "SS Andrea Doria," 2016. [Online]. Available: https://en.wikipedia.org/wiki/SS_Andrea_Doria. [Accessed: 10-Feb-2016].
[4] Wikipedia, "B-25 Empire State Building crash," 2016. [Online]. Available: https://en.wikipedia.org/wiki/B-25_Empire_State_Building_crash. [Accessed: 10-Feb-2016].
[5] Wikipedia, "Tenerife airport disaster," 2016. [Online]. Available: https://en.wikipedia.org/wiki/Tenerife_airport_disaster. [Accessed: 10-Feb-2016].
[6] D. McLeod and J. Jacobson, "Autonomous Inspection using an Underwater 3D LiDAR," ... , *San Diego Mar. ...*, 2013.
[7] J. Fritz, A. J. Gasiewski, and K. Zhang, "3D surface imaging through visual obscurants using a sub-THz radar," vol. 9087, p. 908702, 2014.
[8] B. Sykora, "BAE Systems Brownout landing aid system technology ( BLAST ) system overview and flight test results," pp. 1–15, 2009.
[9] R. D. Visual and E. Solution, "Enhanced Vision Solution for Degraded Visual Environment Enhanced Vision Solution for Degraded Visual Environment."
[10] E. Trickey, P. Church, and X. Cao, "Characterization of the OPAL obscurant penetrating LiDAR in various degraded visual environments," *Proc. SPIE*, vol. 8737, no. 613, p. 87370E–87370E–9, 2013.
[11] J. T. Murray, J. Seely, J. Plath, E. Gotfreson, J. Engel, B. Ryder, N. Van Lieu, R. Goodwin, T. Wagner, G. Fetzer, N. Kridler, C. Melancon, K. Panici, and A. Mitchell, "Dust-Penetrating (DUSPEN) 'see-through' lidar for helicopter situational awareness in DVE," vol. 8737, p. 87370H–87370H–8, 2013.
[12] T. E. Dillon, C. a. Schuetz, R. D. Martin, D. G. Mackrides, S. Shi, P. Yao, K. Shreve, C. Harrity, and D. W. Prather, "Passive, real-time millimeter wave imaging for degraded visual environment mitigation," vol. 9471, p. 947103, 2015.
[13] Wikipedia, "Lidar." [Online]. Available: https://en.wikipedia.org/wiki/Lidar. [Accessed: 10-Feb-2016].
[14] Wikipedia, "Google self-driving car." [Online]. Available: https://en.wikipedia.org/wiki/Google_self-driving_car. [Accessed: 10-Feb-2016].
[15] "Thorlabs.com - Precision Pinholes." [Online]. Available: http://www.thorlabs.de/NewGroupPage9_PF.cfm?Guide=10&Category_ID=22&ObjectGroup_ID=1400. [Accessed: 08-Feb-2016].
[16] K. Shi, P. Li, S. Yin, and Z. Liu, "Chromatic confocal microscopy using supercontinuum light," *Opt. Express*, vol. 12, no. 10, p. 2096, 2004.
[17] P. Toliver, I. Ozdur, A. Agarwal, and T. K. Woodward, "Comparison of LIDAR system performance for alternative single-mode receiver architectures: modeling and experimental validation," in *Proceedings of SPIE - The International Society for Optical Engineering*, 2013, vol. 8731, p. 87310W.

Journal of Physics: Conference Series **707** (2016) 012033 doi:10.1088/1742-6596/707/1/012033

# On The Exact and JWKB Solution of 1D Quantum Harmonic Oscillator by Mathematica

**Coşkun Deniz**

Adnan Menderes University, Faculty of Engineering, Department of Electrical and Electronics Engineering, Aytepe Central Campus-09100 Aydın-TURKEY

E-mail: cdeniz@adu.edu.tr

**Abstract.** Although being the fundamental semiclassical approximation method mainly used in quantum mechanics and optical waveguides, JWKB method along with the application of the associated JWKB asymptotic matching rules is known to give exact solutions for the Quantum Harmonic Oscillator (QHO). Asymptotically matched JWKB solutions are typically accurate or exact in the entire domain except for a narrow domain around the classical turning points where potential energy equals the total energy of the related quantum mechanical system. So, one has to cope with this diverging behavior at the classical turning points since it prohibits us from using continuity relations at the related boundaries to determine the required JWKB coefficients. Here, a computational diagram and related mathematica codes to surmount the problem by applying parity matching for even and odd JWKB solutions rather than boundary continuities are being presented. In effect, JWKB coefficients as well as the conversion factor for the dimensionless form of the Schrodingers equation, which is common to both exact and JWKB solutions, is being successfully obtained.

## 1. Introduction

$(JWKB)_1$ method[1] is conventionally known to be a strong and effective semiclassical approximation method enabling accurate analytical solutions in quantum mechanical systems, described by the Time Independent Schroedinger's Equation (TISE), given by:

$$\varphi''(x) + f(x)\varphi(x) = 0; f(x) = k^2(x) = \frac{2m}{\hbar^2}\left[E - U(x)\right] \quad (1)$$

where these terms are in usual meanings: $m$ represents mass, $\hbar$ represents Planck's constant divided by $2\pi$, $E$ represents total energy, and $U(x)$ represents function of potential energy, namely, i.e., [1–5, 7]. Although being an approximation method, JWKB method is known to give exact solutions for the Shape Invariant Potentials (SIP) involving the QHO where the potential in (1) is given by [1–3]:

$$U(x) = \frac{1}{2}mw^2x^2 \quad (2)$$

Although given exact solutions for the QHO, the $(JWKB)_1$ method has a typical property that both complementary $(JWKB)_1$ solutions (and hence the general $(JWKB)_1$ solution) diverge at a small region around the classical turning points where $f(x) = 0 \Rightarrow E = V(x)$ in eqn. (1),

[1] We simply refer to "$n$th order JWKB (or WKB)" by a simple abbreviation: $(JWKB)_n$ here.

i.e., [1, 2, 4, 5]. So, it does not enable us to apply the continuity relations for the wavefunctions at these turning points to obtain the necessary constant JWKB coefficients. Moreover, although the general $(JWKB)_1$ wavefunction is accurate for a given potential (and exact for the QHO) in the Classically Accessible Region (CAR), it needs asymptotic matching in the Classically Inaccessible Region (CIR) for accurate (or exact for the QHO) $(JWKB)_1$ solutions [1, 2, 2, 4]. CAR is the region where the particle can classically exist since its potential energy is smaller than its total energy: $f(x) > 0 \Rightarrow E > V(x)$, and CIR is the region where it can not classically exist since its potential energy is greater than its total energy: $f(x) < 0 \Rightarrow E < V(x)$.

In this work, we study the JWKB solutions of the QHO in 1D by mathematica and present the results of the applications of parity conditions of wavefunctions rather than their continuity at the turning points to surmount the problem similar to [2, 4]. By following this procedure, we also find the conversion factor for the dimensionless form TISE.

## 2. Exact solution

Exact solution of the QHO by different conventional approaches (such as analytical solution, or algebraic methods, etc.) is given in any fundamental textbooks, i.e., [4–6]. Results of the solutions can be summarized as follows:
TISE for the QHO in 1D (by using (2) in (1)):

$$\varphi''(x) + \frac{2m}{\hbar^2}(E_n - \frac{1}{2}mw^2x^2)\varphi(x) = 0 \tag{3}$$

Change of variable:

$$y(x) = \beta x = \sqrt{\frac{mw}{\hbar}}x \tag{4a}$$

TISE for the QHO in 1D in dimensionless form:

$$\varphi''(y) + k^2[\lambda(E_n), y]\varphi(y) = 0 \Leftrightarrow \left\{k^2[\lambda(E_n), y] = \lambda^2 - y^2, \lambda^2 = \frac{2E_n}{\hbar w}\right\} \tag{4b}$$

Exact eigenenergies:

$$\Lambda_{EX} =: \Lambda = \lambda^2 = 2n + 1 \Rightarrow E_n = (n + \frac{1}{2})\hbar w, n = 0, 1, 2, ... \tag{4c}$$

Exact eigenfunctions (wavefunctions) in $y$:

$$\varphi_n(y) = \sqrt{\frac{1}{\sqrt{\pi}2^n n!}}H_n(y)\mathrm{e}^{-\frac{y^2}{2}}, n = 0, 1, 2, ... \tag{4d}$$

And by using (4a), we have the wavefunctions in $x$:

$$\varphi_n(x) = \sqrt{\frac{\beta}{\sqrt{\pi}2^n n!}}H_n(\beta x)\mathrm{e}^{-\frac{(\beta x)^2}{2}}, n = 0, 1, 2, ... \tag{4e}$$

## 3. Conventional 1st order JWKB approximation ($(JWKB)_{n=1}$) method

*(i)* JWKB Eigenenergies of the dimensionles form in (4b):
They can be found by applying the Bohr-Sommerfeld quantization formula [1–6]:

$$\int_{y_1}^{y_2} k[\widetilde{\lambda}(E_n), y]dy = (n + \frac{1}{2})\pi \tag{5}$$

where $\widetilde{\lambda}$ represents JWKB eigenvalues and $y_1 \& y_2$ are the classical turning points of the TISE for the QHO in dimensionless form given in (4b), which are solutions of $k[\lambda(E), y] = 0$ for $y$.

*(ii)* JWKB Eigenfuntions (wavefunctions) of the dimensionless form in (4b):
Conventional $n$ = 1st order JWKB ($=(JWKB)_1$) eigenfunctions involve only the first two terms ($n = 0$ and up to $N = 1$ only) of the JWKB expansion terms, $S_{ij}(i = 0, 1, 2, ...; j = 1, 2)$, to give the famous JWKB wavefunction formula as follows:

$$\begin{array}{l} (\varphi_{JWKB})_n(y) =: \widetilde{\varphi}(\lambda, y) = \exp[\frac{1}{\delta}S_0(\lambda, y) + S_1(\lambda, y)] \ , \ \delta = \hbar/i \to 0 \\ \Rightarrow \left\{ \begin{array}{l} \widetilde{\varphi}_1(\lambda, y) = \exp[\frac{S_{01}(\lambda,y)}{\delta} + S_{11}(\lambda, y)] \\ \widetilde{\varphi}_2(\lambda, y) = \exp[\frac{S_{02}(\lambda,y)}{\delta} + S_{12}(\lambda, y)] \end{array} \right\}, \delta = \hbar/i \to 0 \\ \Rightarrow \widetilde{\varphi}(\lambda, y) = c_1\widetilde{\varphi}_1(\lambda, y) + c_2\widetilde{\varphi}_2(\lambda, y) \\ \Rightarrow \widetilde{\varphi}(\lambda, y) = \frac{c_1}{\sqrt{k(\lambda,y)}} \exp\left[-i \int^{y_t,y} k(\lambda, y)dy\right] + \frac{c_2}{\sqrt{k(\lambda,y)}} \exp\left[i \int^{y_t,y} k(\lambda, y)dx\right] \end{array} \tag{6}$$

where $y_t$ is either of the classical turning points (either $y_{t1}$ or $y_{t2}$) and $c_1\&c_2$ are arbitrary constants [1,2,7]. One of the advantages of the JWKB method is that once the solution of one region has been obtained, the other adjacent region, is not required to be calculated, they can directly be written via the following conventional $(JWKB)_1$ connection formulae as in [1,2,4,8]:

$$\begin{array}{l} \frac{2}{\sqrt{k(\lambda,y)}} \sin\left[\int^{y_t,y} k(\lambda, y)dy + \frac{\pi}{4}\right] << - > \frac{1}{\sqrt{\kappa(\lambda,y)}} \exp\left[-\int_{y_t,y} \kappa(\lambda, y)dy\right] \\ \frac{1}{\sqrt{k(\lambda,y)}} \cos\left[\int^{y_t,y} k(\lambda, y)dy + \frac{\pi}{4}\right] < - >> \frac{1}{\sqrt{\kappa(\lambda,y)}} \exp\left[\int_{y_t,y} \kappa(\lambda, y)dy\right] \end{array} \tag{7}$$

where $\kappa^2(\lambda, y) = -k^2(\lambda, y)$ as usual. The integrals here are the definite integrals whose upper and lower values include the turning point $y_t$ and the variable $y$ whose sequence is chosen in such a way that the connection spans a continuous integration through the $y$ axis. JWKB connections are made in the double arrow direction, but the connections in the reverse direction are normally not allowed. To make a connection in the reverse direction, one has to interfere a small phase term as in [2, 4]. For example, the reverse connection of the first one of (7) can be achieved as follows:

$$\begin{array}{l} \frac{2}{\sqrt{k(\lambda,y)}} \sin\left[\int_y^{y_t} k(\lambda, y)dy + \frac{\pi}{4} + \varepsilon\right] = \frac{2}{\sqrt{k(\lambda,y)}} \sin\left[\mu(\lambda, y) + \varepsilon\right] \\ \qquad \left(= \frac{2}{\sqrt{k}}\left[\sin\mu(\lambda, y)\cos\varepsilon + \cos\mu(\lambda, y)\sin\varepsilon\right]\right) - >> \\ \frac{\cos\varepsilon}{\sqrt{\kappa(\lambda,y)}} \exp\left[-\int_{y_t}^{y} \kappa(\lambda, y)dx\right] + \frac{2\sin\varepsilon}{\sqrt{\kappa(\lambda,y)}} \exp\left[\int_{y_t}^{y} \kappa(\lambda, y)dy\right] \end{array} \tag{8}$$

In this case, the solution of the classically allowed region should have taken to have a solution of a *sine* function with an extra small phase, $\varepsilon$, added to the present phase of $\pi/4$ to have a total phase of $\pi/4 + \varepsilon$ to give the right-hand-side of (8) in order to make a successful JWKB connection in the reverse direction of the double arrow in (7) [2,4].

*(iii)* Asymptotic Matching in the CIR: Conventional $(JWKB)_1$ asymptotic matching rules require either of the complementary solutions in (6) to be cancelled in the CIR as follows [1,2,4,7]:

$$\widetilde{\varphi}^{m.}(y) = \left\{ \begin{array}{l} \widetilde{\varphi}(y) \text{ for CAR:} f(y) > 0 \\ \text{either } \widetilde{k}_1\widetilde{\varphi}_1(y) \text{ or } \widetilde{k}_2\widetilde{\varphi}_2(y) \text{ for CIR:} f(y) < 0 \end{array} \right. \tag{9a}$$

so that the CIR should meet the following requirement regarding the asymptotical matching:

$$\widetilde{\varphi}^{m.}(y) \to \left\{ \begin{array}{l} \lim_{y\to-\infty} \widetilde{\varphi}(y) = 0 \text{ if CIR lies on the LHS} \\ \lim_{y\to\infty} \widetilde{\varphi}(y) = 0 \text{ if CIR lies on the RHS} \end{array} \right. \tag{9b}$$

where $\widetilde{\varphi}^{m.}(y)$ represents the asymptotically matched general $(JWKB)_1$ solution. In other words, asymptotically diverging term for $y \to \pm\infty$ in the CIR should be cancelled in the general solution so that (9b) can hold. Alternative JWKB asymptotic matching rule regarding the JWKB expansion terms is also available in [2, 7].

## 4. JWKB solution

### *4.1. Dealing with the discontinuities at the classical turning points*

Since JWKB method is useless at the classical turning points, they diverge and hence, we have discontinuities at these points binding two adjacent regions via the unidirectional JWKB connection formulae given in (7). Although we can find the structure of the wavefunctions in two adjacent regions via these connection rules, we can not use continuity requirements of quantum mechanical wavefunctions to determine the energy dependent coefficients for these wavefunctions $(A(\lambda)\&B(\lambda)$ in Sect. in 4.3). Let us follow the parity conditions for even potentials as in [2, 4] but by considering (4d) and (4e) in multivariables as follows:

$$\varphi_n[\beta,\lambda(n),x] = \sqrt{\frac{\beta}{\sqrt{\pi}2^n n!}}H_n(\beta x)\mathrm{e}^{-\frac{\beta^2x^2}{2}} = \sqrt{\beta}\psi_n[\lambda(n),y]\Big|_{y\to\beta x}, n = 0,1,2,... \tag{10a}$$

$$\psi_n[\lambda(n),y] = \sqrt{\frac{1}{\sqrt{\pi}2^n n!}}H_n(y)\mathrm{e}^{-\frac{y^2}{2}} = \left\{\varphi_n[\beta,\lambda(n),x]\Big|_{x\to y/\beta}\right\}\Big|_{\beta\to 1} = \varphi_n[1,\lambda(n),y], n = 0,1,2,... \tag{10b}$$

By using the parity conditions (E.P.: even parity wavefunctions, O.P.: odd parity wavefunctions) we get:

$$\text{E.P.} : \varphi_n(\beta,\lambda,-x) = \varphi_n(\beta,\lambda,x), n = 0,2,4,... \Rightarrow \begin{cases} i)\ \varphi_n(\beta,\lambda,0) = \pm p, p>0 \\ ii)\ \frac{\partial\varphi_n(\beta,\lambda,x)}{\partial x}\Big|_{x=0} = 0 \end{cases}$$

$$\Rightarrow \begin{cases} i)\ \varphi_n(\beta,\lambda,0) = \left[\sqrt{\beta}\varphi_n(1,\lambda,x)\right]\Big|_{x=0} = \sqrt{\beta}\psi_n(\lambda,y)|_{y=0} = \pm p \Rightarrow \boxed{\psi_n(\lambda,0) = \frac{\pm p}{\sqrt{\beta}}} \\ ii)\ \frac{\partial\varphi_n(\beta,\lambda,x)}{\partial x}\Big|_{x=0} = \frac{\partial}{\partial x}\left[\sqrt{\beta}\psi_n(\lambda,y)\right]\Big|_{x=0} = 0 \Rightarrow \boxed{\frac{\partial\psi_n(\lambda,y)}{\partial y}\Big|_{y=0} = 0} \end{cases} \tag{11a}$$

$$\text{O.P.} : \varphi_n(\beta,\lambda,-x) = -\varphi_n(\beta,\lambda,x), n = 1,3,5,... \Rightarrow \begin{cases} i)\ \varphi_n(\beta,\lambda,0) = 0 \\ ii)\ \frac{\partial\varphi_n(\beta,\lambda,x)}{\partial x}\Big|_{x=0} = \pm q, q>0 \end{cases}$$

$$\Rightarrow \begin{cases} i)\ \varphi_n(\beta,\lambda,0) = \left[\sqrt{\beta}\varphi_n(1,\lambda,x)\right]\Big|_{x=0} = \sqrt{\beta}\psi(\lambda,y)|_{y=0} = 0 \Rightarrow \boxed{\psi_n(\lambda,0) = 0} \\ ii)\ \frac{\partial\varphi_n(\beta,\lambda,x)}{\partial x}\Big|_{x=0} = \frac{\partial}{\partial x}\left[\sqrt{\beta}\psi_n(\lambda,y)\right]\Big|_{x=0} = \sqrt{\beta}\frac{\partial\psi_n}{\partial y}\frac{\partial y}{\partial x}\Big|_{x=0} = \pm q \Rightarrow \boxed{\frac{\partial\psi_n(\lambda,y)}{\partial y}\Big|_{y=0} = \frac{\pm q}{\sqrt{\beta^3}}} \end{cases} \tag{11b}$$

from which we can also find a relation with the conversion factor for O.P. and E.P. wavefunctions as follows:

$$\boxed{\beta_{E.P.} = \left[\frac{\pm p}{\varphi_n(1,\lambda,y)\Big|_{y=0}}\right]^2 \equiv \widetilde{\beta}_{E.P.} = \left[\frac{\pm p}{\widetilde{\varphi}_n(1,\lambda,y)\Big|_{y=0}}\right]^2, n = 0,2,4,...} \tag{12a}$$

$$\boxed{\beta_{O.P.} = \left[\frac{\pm q}{\frac{\partial\varphi_n(1,\lambda,y)}{\partial y}\Big|_{y=0}}\right]^{2/3} \equiv \widetilde{\beta}_{O.P.} = \left[\frac{\pm q}{\frac{\partial\widetilde{\varphi}_n(1,\lambda,y)}{\partial y}\Big|_{y=0}}\right]^{2/3}, n = 1,3,5,...} \tag{12b}$$

where $p\&q$ are positive real constants regarding the EP and OP initial values of the physical system. We already have from (10a)-(10b):

$$\widetilde{\psi}(\lambda, y) = \frac{1}{\sqrt{\beta}}\widetilde{\varphi}(1, \lambda, y/\beta) \equiv \psi(\lambda, y) = \frac{1}{\sqrt{\beta}}\varphi(1, \lambda, y/\beta), y \neq y_t \tag{12c}$$

where symbols with tilde represent JWKB terms. Although we derived these results by using exact wavefunctions ($\varphi$), it should also be valid for the JWKB wavefunctions ($\widetilde{\varphi}$) since the parity conditions require the same initial values for exact and JWKB wavefunctions at $y = 0$, namely:

$$\left\{ \begin{array}{l} \varphi_n(1,\lambda,x)\Big|_{x=0} = \widetilde{\varphi}_n(1,\lambda,x)\Big|_{x=0} = \left\{ \begin{array}{l} 0 \text{ for O.P., n=1,3,5,...} \\ \pm p \text{ for E.P., n=0,2,4,...} \end{array} \right. \\ \frac{\partial\varphi(1,\lambda,x)}{\partial x}\Big|_{x=0} = \frac{\partial\widetilde{\varphi}(1,\lambda,x)}{\partial x}\Big|_{x=0} = \left\{ \begin{array}{l} \pm q \text{ for O.P., n=1,3,5,...} \\ 0 \text{ for E.P., n=0,2,4,...} \end{array} \right. \end{array} \right\} \tag{13}$$

For the JWKB solutions we assume that the asymptotic matching in the CIR (and also the parity matching in the entire domain) has been achieved, however it is not necessary for $x = 0$ since it is not in the CIR. Mathematica codes for the determination of $A(\lambda)\&B(\lambda)$ are given in Fig. 1.

### *4.2. JWKB Eigenenergies*

JWKB quantization rule given in (5) for the QHO gives [1, 2, 4, 5]:

$$\int_{y_1}^{y_2} k[\widetilde{\lambda}(E_n), y]dy = \int_{y_1=-\lambda}^{y_2=\lambda} \sqrt{\widetilde{\lambda}^2 - y^2}dy = (n + \frac{1}{2})\pi \tag{14}$$

which gives exactly the exact eigenenergies given in (4c):

$$\boxed{\begin{array}{l} \lambda^2_{JWKB} =: \widetilde{\lambda}^2 = 2n + 1 \equiv \lambda^2_{EX} =: \lambda^2 \\ \Rightarrow E_{n(JWKB)} =: \widetilde{E}_n = (n + \frac{1}{2})\hbar w = E_n := E_{JWKB} \end{array}, (n = 0, 1, 2, ...)} \tag{15}$$

Simple mathematica codes to determine JWKB eigenenergies is given in Fig. 1

### *4.3. JWKB Eigenfunctions*

One can find the E.P. and O.P. JWKB wavefunctions by working only on the first quadrant of the $f - y$ graph using the connection formulae given in (7)-(8) and then by applying the symmetry relations to extend it to the second quadrant, i.e., in [1, 2, 4, 5, 7]. We also apply both asymptotic matching (modification) and parity matchings given in [1, 2, 4]. We study here the specific cases for $p = q = 1$:

$$p = q = 1 \Rightarrow \psi(\lambda, y) =: \psi^{(1)}(\lambda, y) \tag{16a}$$

from which the other cases with $p \neq 1 \wedge q \neq 1$ can be determined as follows:

$$p \neq 1 \wedge q \neq 1 \Rightarrow \psi(\lambda, y) =: \psi^{(2)}(\lambda, y) = \psi^{(1)}(\lambda, y) \times \left\{ \begin{array}{l} p \text{ for EP} \\ q \text{ for OP} \end{array} \right. \tag{16b}$$

```
λ =.; Clear; k[λ_, y_] = Sqrt[λ^2 - y^2];
Assuming[y∈Re & λ∈Re & 0 < y < λ, int[λ_, y_] = ∫_0^y k[λ, y] dy];
λ = λ /. Assuming[n∈Re & n > 0, Solve[2 * int[λ, λ] == (n + 1/2) * π, λ]];
Energyjwkb = 1/2 ħ w λ^2; Print["λn=" λ, ", En(JWKB)=", Energyjwkb]

{-λn= √(1 + 2 n), λn= √(1 + 2 n)}, En(JWKB)={1/2 (1 + 2 n) w ħ, 1/2 (1 + 2 n) w ħ}
```

**Figure 1.** Mathematica codes to determine JWKB eigenenergies.

*4.3.1. Even Parity (EP) Wavefunctions* For the EP wavefunctions, if we start with a cosine-like wavefunction in region II:

$$\widetilde{\varphi}_{II}(\widetilde{\lambda}_n = \lambda_n, y) = \frac{A(\lambda_n)}{\sqrt{k(\lambda_n, y)}} \cos\left[\int_0^y k(\lambda_n, y)dy\right], \text{ for } 0 < y < \lambda_n \tag{17a}$$

and by applying the JWKB connection formulae, we get:

$$\widetilde{\varphi}_{III}(\widetilde{\lambda}_n = \lambda_n, y) = \frac{A(\lambda_n)}{\sqrt{\kappa(\lambda_n, y)}} \left\{ \begin{array}{l} \cos[\alpha(\lambda_n)] \exp[\zeta(\lambda_n, y)] \\ + \frac{1}{2}\sin[\alpha(\lambda_n)] \exp[-\zeta(\lambda_n, y)] \end{array} \right\}, \text{ for } \lambda_n < y < \infty \tag{17b}$$

However, we have to use the asymptotically matched (modified) wavefunction in region III via (9a)-(9b) of [1, 2, 4]:

$$\widetilde{\varphi}^{(m.)}_{III}(\widetilde{\lambda}_n = \lambda_n, y) = \frac{A(\lambda_n)}{2\sqrt{\kappa(\lambda_n, y)}} \sin[\alpha(\lambda_n)] \exp[-\zeta(\lambda_n, y)], \text{for } \lambda_n < y < \infty \tag{17c}$$

So far, we have obtained the general structure of the JWKB wavefunctions in the first quadrant ($0 < x < \infty$) and once the coefficients are determined, they can be extended to the second quadrant ($-\infty < x < 0$) by using the parity conditions of the wavefunctions. However, we can not apply the continuity of wavefunctions between region II and III since the JWKB solutions diverge at the classical turning points coinciding this interface here. So, we apply the parity conditions in (11b)-(11a) to determine the coefficients residing in the wavefunction formulae obtained above. For the EP wavefunctions, we have:

$$\widetilde{\varphi}^{(m.)}_{(E.P.)}(\lambda_n, y) = P_{(E.P.)}(\lambda_n) \times \left\{ \begin{array}{l} \widetilde{\varphi}^{(m.)}_{III}(\lambda_n, -y) \text{ for } -\infty < x \leq -\lambda \\ \widetilde{\varphi}_{II}(\lambda_n, -y) \text{ for } -\lambda < x \leq 0 \\ \widetilde{\varphi}_{II}(\lambda_n, y) \text{ for } 0 < x \leq \lambda \\ \widetilde{\varphi}^{(m.)}_{III}(\lambda_n, y) \text{ for } \lambda \leq x < \infty \end{array} \right. \tag{17d}$$

where $P_{(E.P.)}(\lambda_n)$ is for parity matching and defined by:

$$P_{(E.P.)}(\lambda_n) = (-1)^{(\lambda_n^2 - 1)/4} \tag{17e}$$

Abbreviations we use for EP JWKB solutions here (and also for the OP solutions in the next subsection) are as follows:

$$\begin{cases} \alpha(\lambda_n) = \int\limits_0^{\lambda_n} k(\lambda_n, y)dy + \frac{\pi}{4} = \eta(\lambda_n, 0) + \frac{\pi}{4} \\ \eta(\lambda_n, y) = \int\limits_y^{\lambda_n} k(\lambda_n, y)dy \rightarrow= \frac{\lambda_n^2 \pi}{4} - \frac{\lambda_n^2}{2}\sin^{-1}(\frac{y}{\lambda_n}) - \frac{y}{2}\sqrt{\lambda_n^2 - y^2} \\ \zeta(\lambda_n, y) = \int\limits_{\lambda_n}^{y} \kappa(\lambda_n, y)dy \rightarrow= \frac{y}{2}\sqrt{y^2-\lambda_n^2} - \frac{1}{2}\lambda_n^2 \ln\left|\frac{y+\sqrt{y^2-\lambda_n^2}}{\lambda_n}\right| \end{cases} \tag{17f}$$

However, since the determination of constant $A(\lambda_n)$ requires the values of $\beta$ in (11a) and it is not convenient to apply (11a) for all $\beta$ values repeatedly (we might also not have found them yet), we have a little trick here in our algorithm as follows:

$$\begin{array}{l} \text{our IVP: } \psi(\lambda, y)\Big|_{y=0} = \frac{\pm p}{\sqrt{\beta}} \Rightarrow A(\lambda_n) =? \\ \text{modified IVP : } \psi(\lambda, y)\Big|_{y=0} = \pm p\Big|_{p=1} \Rightarrow A(\lambda_n) =? \end{array} \tag{18a}$$

Once we find $A(\lambda_n)$ from the modified IVP, then by using (11a) and (12a) we see that we should replace it by $p\beta^{1/2}A(\lambda_n)$ but the $\beta$ value does not change, namely:

$$A(\lambda_n) =: p\beta^{-1/2}A(\lambda_n) \ , \ \beta =: \beta \tag{18b}$$

Simple mathematica codes to determine the conversion factor $\beta$ for the EP wavefunctions according to the algorithm in (11a)-(12a) with $p = 1$ is given in Fig. 2, the related mathematica codes to find $A(\lambda_n)$ via this algorithm are given in Fig. 3, and the graphs of some exact and JWKB EP wavefunctions calculated according to the prescription discussed here are given for $p = 1$ in Fig. 4.

*4.3.2. Odd Parity (OP) Wavefunctions* For the OP wavefunctions, if we start with a sine-like wavefunction in region II:

$$\widetilde{\psi}_{II}(\widetilde{\lambda}_n = \lambda_n, y) = \frac{A(\lambda_n)}{\sqrt{k(\lambda_n, y)}} \sin\left[\int\limits_0^y k(\lambda_n, y)dy\right] \ , \text{ for } 0 < y < \lambda_n \tag{19a}$$

and by applying the JWKB connection formulae along with the asymptotic matching in region III, we similarly get:

$$\widetilde{\psi}_{III}(\widetilde{\lambda}_n = \lambda_n, y) = \frac{B(\lambda_n)}{\sqrt{\kappa(\lambda_n, y)}} \left\{ \begin{array}{l} \sin\left[\alpha(\lambda_n)\right] \exp[\zeta(\lambda_n, y)] \\ -\frac{1}{2}\cos\left[\alpha(\lambda_n)\right] \exp[-\zeta(\lambda_n, y)] \end{array} \right\} \ , \text{ for } \lambda_n < y < \infty \tag{19b}$$

$$\widetilde{\psi}_{III}^{(m.)}(\lambda_n, y) = -\frac{B(\lambda_n)}{2\sqrt{\kappa(\lambda_n, y)}} \cos\left[\alpha(\lambda_n)\right] \exp[-\zeta(\lambda_n, y)] \ , \text{ for } \lambda_n < y < \infty \tag{19c}$$

We have the following results in the entire domain for the OP wavefunctions:

$$\widetilde{\varphi}_{(O.P.)}^{(m.)}(\lambda_n, y) = P_{(O.P.)}(\lambda_n) \times \begin{cases} -\widetilde{\psi}_{III}^{(m.)}(\lambda_n, -y) \text{ for } -\infty < x \le -\lambda \\ -\widetilde{\psi}_{II}(\lambda_n, -y) \text{ for } -\lambda < x \le 0 \\ \widetilde{\psi}_{II}(\lambda_n, y) \text{ for } 0 < x \le \lambda \\ \widetilde{\psi}_{III}^{(m.)}(\lambda_n, y) \text{ for } \lambda \le x < \infty \end{cases} \tag{19d}$$

```
g =.; r = {}; λ =.; β =.; Clear [β, φexact];
φexact[β_, n_, x_] = (√(β/(√π * 2^n * n!))) * HermiteH[n, β*x] * Exp[-0.5 * (β*x)^2];
Print["n", "      λ=√(2 n + 1)", "       β"];
Table[g = N[(φexact[1, m, x])^(-2) /. {x → 0.}]; r = Join[r, {g, 0}];
 Print[m, "      ", N[√(2*m + 1), 7], "       ", N[r[[m + 1]]], 7]; g =., {m, 0, 30, 2}]
```

| n | $\lambda=\sqrt{2n+1}$ | $\beta$ |
|---|---|---|
| 0 | 1.000000 | 1.772457 |
| 2 | 2.236068 | 3.544917 |
| 4 | 3.000000 | 4.726547 |
| 6 | 3.605551 | 5.671857 |
| 8 | 4.123106 | 6.482127 |
| 10 | 4.582576 | 7.202357 |
| 12 | 5.000000 | 7.857117 |
| 14 | 5.385165 | 8.46157 |
| 16 | 5.744563 | 9.02567 |
| 18 | 6.082763 | 9.556527 |
| 20 | 6.403124 | 10.05957 |
| 22 | 6.708204 | 10.53857 |
| 24 | 7.000000 | 10.99677 |
| 26 | 7.280110 | 11.43667 |
| 28 | 7.549834 | 11.86027 |
| 30 | 7.810250 | 12.26917 |

**Figure 2.** Mathematica codes to determine conversion factor $\beta$ for EP wavefunctions.

```
Print[
 "Determination of A(λ) for E.P. wf. in Region II by even parity condition
   at x=0:"]; A =.; λ =.;
ψII[λ_, y_] = Assuming[{λ, y} ∈ Re & 0 < y < λ, A/√(k[λ, y]) * Cos[int[λ, y]]];
A = A /. A -> Quiet[A = A /. Solve[ψII[λ, 0] == 1, A]]
Determination of A(λ) for E.P. wf. in Region II by even parity condition at x=0:
{(λ^2)^(1/4)}
```

**Figure 3.** Simple Mathematica codes to determine JWKB coefficient $A$ for EP wavefunctions.

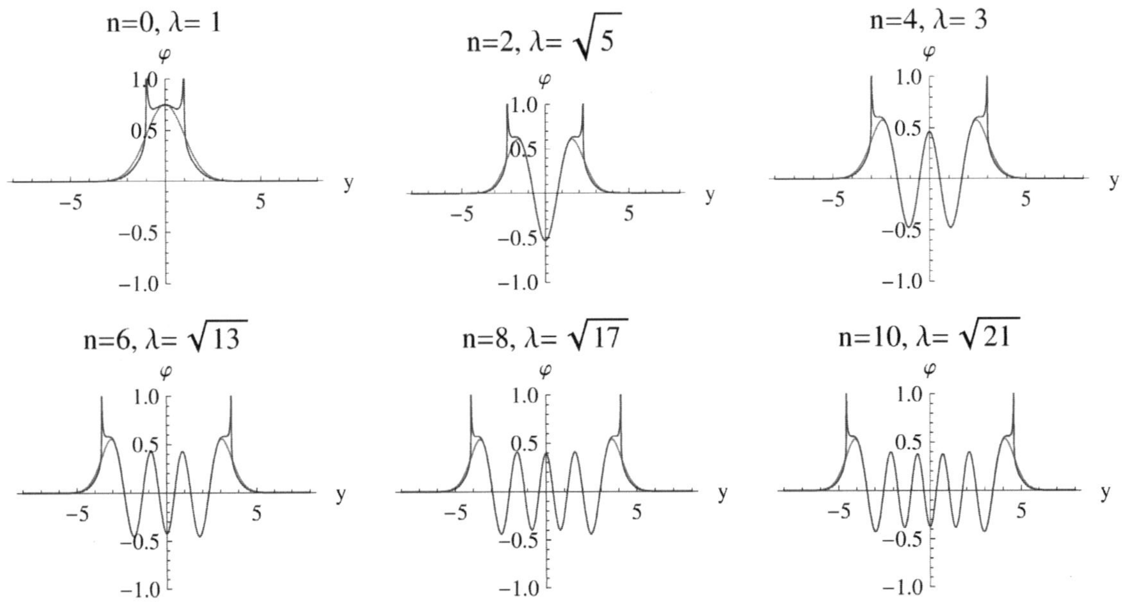


**Figure 4.** Exact and JWKB solutions of EP wavefunctions (for $p = 1$).

where $P_{(O.P.)}(\lambda_n)$ is for parity matching and defined by:

$$P_{(O.P.)}(\lambda_n) = (-1)^{[\lambda_n^2-1)/2-1]/2} \tag{19e}$$

As to the determination of the conversion factor, similar to our EP calculations, but we rather use (11b)-(12b) here for the OP wavefunctions as given in Fig. 5. We find the desired coefficient ($=B(\lambda_n)$) regarding the OP wavefunctions by using the initial values in (13) according to (11b). However, since the determination of constant $B$ requires the values of $\beta$ in (11b) and it is not convenient to apply (11b) for all $\beta$ values repeatedly (we might also not have found them yet), we similarly apply the followings:

$$\begin{aligned} &\text{our IVP: } \left.\frac{\partial\psi(\lambda,y)}{\partial y}\right|_{y=0} = \frac{\pm q}{\sqrt{\beta^3}} \Rightarrow B(\lambda_n) =? \\ &\text{modified IVP : } \left.\frac{\partial\psi(\lambda,y)}{\partial y}\right|_{y=0} = \pm q\Big|_{q=1} \Rightarrow B(\lambda_n) =? \end{aligned} \tag{20a}$$

Once we find $B(\lambda_n)$ from the modified IVP, then by using (11b) and (12b) we see that we should we replace it by $q\beta^{-3/2}B(\lambda_n)$ but the $\beta$ value does not change, namely:

$$B(\lambda_n) =: q\beta^{-3/2}B(\lambda_n)\ ,\ \beta =: \beta \tag{20b}$$

Related mathematica codes to find $B$ via this algorithm are given in Fig. 6 and the graphs of some exact and JWKB OP wavefunctions calculated according to the prescription discussed here are given for $q = 1$ in Fig. 7.

## 5. Conclusion

Although JWKB method fails at the classical turning points and hence we can not apply continuity relations to determine the JWKB wavefunction coefficients in the adjacent domains

```
g =.; s = {}; λ =.; β =.; Clear [β, φexact];
φexact[β_, n_, x_] = (Sqrt[β / (Sqrt[π] * 2^n * n!)]) * HermiteH[n, β * x] * Exp[-0.5 * (β * x)^2];
φexactderivative[β_, n_, x_] = D[φexact[β, n, x], x];
Print["n", "    λ=Sqrt[2*m+1]", "             β"];
Table[g = N[ ((φexactderivative[1, m, x])^(-2))^(1/3) /. {x → 0.} ]; s = Join[s, {g, 0}];
 Print[m, "    ", N[Sqrt[2*m+1], 7], "          ", N[s[[m]]], 7]; g =., {m, 1, 30, 2}]
```

| n | $\lambda=\sqrt{2*m+1}$ | $\beta$ |
|---|---|---|
| 1 | 1.732051 | 0.9605397 |
| 3 | 2.645751 | 0.8391087 |
| 5 | 3.316625 | 0.7789597 |
| 7 | 3.872983 | 0.7399447 |
| 9 | 4.358899 | 0.7114567 |
| 11 | 4.795832 | 0.6892087 |
| 13 | 5.196152 | 0.6710637 |
| 15 | 5.567764 | 0.6558067 |
| 17 | 5.916080 | 0.6426867 |
| 19 | 6.244998 | 0.6312077 |
| 21 | 6.557439 | 0.6210257 |
| 23 | 6.855655 | 0.6118917 |
| 25 | 7.141428 | 0.6036217 |
| 27 | 7.416198 | 0.5960757 |
| 29 | 7.681146 | 0.5891437 |

**Figure 5.** Mathematica codes to determine conversion factor $\beta$ for OP wavefunctions.

```
Print[
 "Determination of B(λ) for O.P. w.f. in Region II by odd parity condition at x=0:"];
B =.; λ =.; int[λ_, y_] = ∫_0^y k[λ, y] dy;
ψIIodd[λ_, y_] = Assuming[{λ, y} ∈ Re & 0 < y < λ, B / Sqrt[k[λ, y]] * Sin[int[λ, y]]];
ψIIoddder[λ_, y_] = D[ψIIodd[λ, y], y];
B = B /. B -> Quiet[B = B /. Solve[ψIIoddder[λ, 0] == 1, B]]
```

Determination of B(λ) for O.P. w.f. in Region II by odd parity condition at x=0:

$$\left\{\frac{1}{(\lambda^2)^{1/4}}\right\}$$

**Figure 6.** Mathematica codes to determine $B$ for OP wavefunctions.

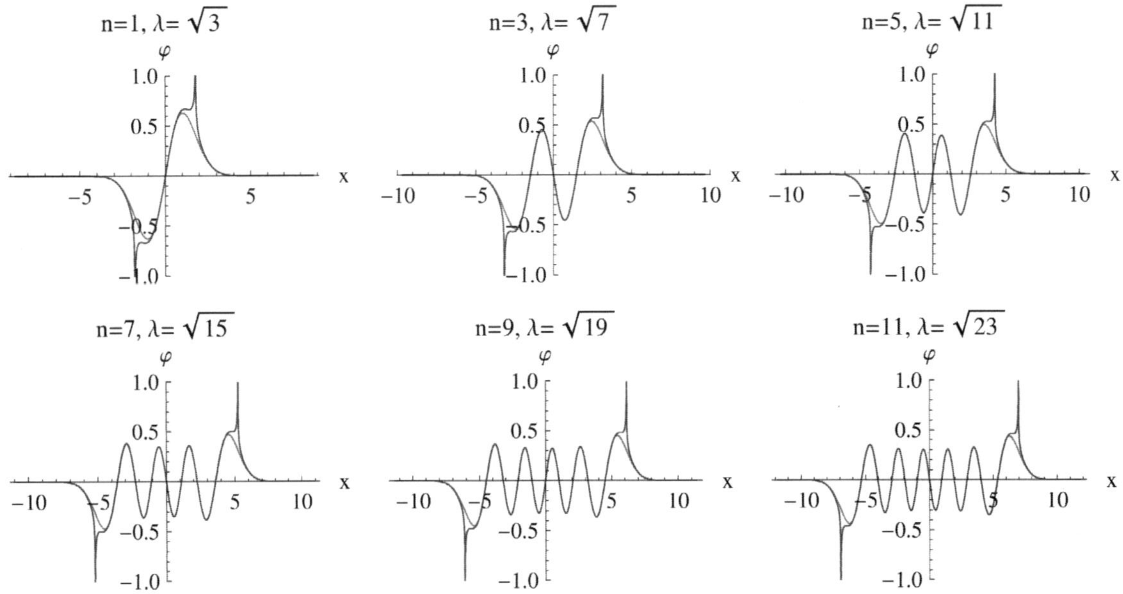


**Figure 7.** Exact and JWKB solutions of OP wavefunctions (for $q = 1$).

around these classical turning points, we find the JWKB solutions of the QHO successfully in the entire domain by applying the parity conditions of the wavefunctions. We have both exact JWKB eigenenergies and exact JWKB eigenfunctions (wavefunctions) in the entire domain except for a very narrow region around the classical turning points as typical to the JWKB method. The algorithm and the related mathematica codes presented here are very simple and can be used to find JWKB solutions of the QHO along with the conversion factor for the dimensionless form. Moreover, it is possible to adopt them to mathematica cdf (computable document file) player for interactive simulation and animation applications since EP and OP calculations here are simultaneous but in distinct variable and $\beta$-matrix representations.

## References

[1] Bender C M and Orszag S A 1999 *Advanced Mathematical Methods for Scientists and Engineers Asymptotic Methods and Perturbation Theory* (NY: Springer–Verlag)
[2] Deniz C 2011 *Ann. Phys.* **326** 8, 1816-1838.
[3] Hruska M, Keung W–Y, and Sukhatme U 1997 *Phys. Rev. A.* **55** 5, 3345–3350.
[4] Ghatak A K, Gallawa R L, andGoyal I C 1991 *Modified Airy Functions and WKB Solutions to the Wave Equation* (Washington: NIST).
[5] Landau L D and Lifshitz E M 1965 *Quantum Mechanics, Non-Relativistic Theory, second ed., Course of Theoretical Physics, vol. 3* (NY: Pergamon).
[6] Liboff R 2001 *Introductory Quantum Mechanics, 4/E* (Cornell University: Addison Wesley)
[7] Deniz C 2014 , *Proc. of the 19th International Conference on Applied Mathematics (AMATH'14)*, (Istanbul: Mathematical Applications in Modern Science, **38**), 22–31.
[8] Langer R E 1937 *Phys. Rev.* **51**, 669–676.

# Pinning properties of Y211 added cold top-seeded YBCO grown on $Y_2O_3$ layer

**Bakiye Çakır[1,2], Şeyda Duman[3], Alev Aydıner[1]**

[1]Department of Physics, Faculty of Sciences, Karadeniz Technical University, Trabzon 61080, Turkey,
[2]Department of Elementary Education, Faculty of Education, Artvin Çoruh University, Artvin 08000, Turkey
[3] Vocational School of Health Services, Artvin Çoruh University, Artvin 08000, Turkey

E-Mail: bakiye.cakir@gmail.com

**Abstract.** In this study, samples having different composition were prepared with the cold top seeding-melt-growth (TSMG) process by using Nd123 seed. $Y_2O_3$ buffer layer was placed to bottom of the pellets consist of Y123: Y211 powder mixtures. Two samples were fabricated in stoichiometric ratios of 1:0 and 1:0.4 labeled as Y0 and Y40, respectively. The $T_c$ onset values of Y0 and Y40 were found to be 93.4 and 93.6 K at 0 T, respectively. The dependence of the effective activation energy $U$ of the flux pinning on the magnetic field and temperature of the sample were determined using the Arrhenius activation energy law from the resistivity curves. The magnetization measurements were performed using a vibrating sample magnetometer (VSM) at 30, 50 and 77 K. The critical current densities ($J_c$) for Y0 and Y40 samples were determined to be $5.1\times10^3$ and $3.7\times10^3$ A/cm$^2$ at 77 K in 0 T, respectively. The normalized pinning force density versus the reduced field was examined at different temperatures to determine the pinning mechanism.

## 1. Introduction

Superconducting $YBa_2Cu_3O_{7-x}$ (Y123) bulks grown by TSMG method is a promising candidate for high performance superconducting samples. This is mainly due to their abilities of trapping magnetic flux and high critical current density ($J_c$), which allows practical applications such as permanent magnets, magnetic levitation, magnetic bearings, and flywheels. Because, for practical applications high temperature superconductors (HTS) need to carry a high critical current density under high magnetic fields. On the other hand, the supercurrent of YBCO bulk is limited by the weak links and flux creep, which is caused by grain boundaries and thermal activation, respectively [1].

On the other hand, it is well established that the presence of non-superconducting second phase Y211 particles of sizes ~1 μm embedded in the superconducting Y123 phase matrix correlates with enhanced flux pinning in YBCO. Reducing the size of these particles, ideally to dimensions of the order of double the coherence length of the system, increases their magnetic flux pinning potential further [2,3].

The unusual small coherence length leads to an unusual low pinning energy. However, due to vortex motion, the capability of YBCO to carry currents is significantly reduced at high value of the applied magnetic fields. To understand the complex pinning behavior in HTS, an important aspect is to study the dependence of the effective activation energy $U$ of the flux pinning on the magnetic field and temperature. One accepted method for probing this dependence is to measure the resistive transition in various applied magnetic fields and temperatures. It is commonly suggested that the resistivity of HTS in the region of low resistivity can be described by the Arrhenius equations [4].

There have been a lot of interests in understanding the mechanism of pinning in YBCO based superconducting nanocomposites. The non-superconducting nanoparticles such as Y211 particles trapped in YBCO superconducting matrix can act as effective pinning centers leading to enhanced

Journal of Physics: Conference Series **707** (2016) 012034 doi:10.1088/1742-6596/707/1/012034

critical current density. However, $Y_2O_3$ layer was used to absorb the unreacted Y211 liquid phase during the crystal growth process.

The critical current density $J_c$ at an induction $B$ is that which allows the Lorentz force to be just balanced by a flux pinning force ($F_p$). The flux-line lattice in mixed state is assumed to be pinned by the interaction with pinning centers, which are the features of the microstructure of the superconductors [5]. Hence, it is very important to correctly understand the flux pinning mechanism. Flux pinning in superconductors may be caused by spatial fluctuations of the transition temperature $T_c$ ("$\delta T_c$–pinning") or of the charge carrier mean free path ("$\delta l$–pinning") near a lattice defect [6]. $\delta l$ pinning centers behave like normal conducting particles and are very effective to provide pinning at low fields and low temperatures. However, $\delta T_c$ pinning centers behave like the local oxygen deficient regions and are only effective at the intermediate fields and temperatures [7].

In this study, due to the use of $Y_2O_3$ nanoparticles showed a potential as flux-pinning centers in YBCO thin films and the layer was effective on crystal growth mechanism for bulk YBCO [8], a TSMG YBCO sample was grown on the $Y_2O_3$ layer with Y211 addition and without any addition. Also, activation energies and the dominant flux pinning mechanism of strong pinning centers of the samples were investigated to obtain effect of the Y211 addition and $Y_2O_3$ layer.

## 2. Material and methods

YBCO bulk samples with nominal composition $YBa_2Cu_3O_{7-x}$ (Y123) with $YBaCuO_5$ (Y211) addition were fabricated by TSMG process by using Nd123 seed. The starting $Y_2O_3$, $BaCO_3$, and CuO powders were mixed with a molar ratio 1:2:3 to synthesize Y123 composition, which was then calcined within a box furnace at 900°C for 24h. The calcination process was performed three times with intermediate grindings for 15 minutes. The Y211 composition was made in a similar way to Y123, but ground for 45 minutes. After the calcination process the fine Y123 powders were then placed into a platinum crucible and put into the muffle furnace at room temperature. The furnace temperature was increased to 1450°C and held at that temperature for 5 min. to get molten material. The molten material was then poured onto a cupper plate, and afterwards sandwiched with another cupper plate to protect phase condition during melt. The obtained thin platelets were ground again for 15 min. to achieve fine powder. These Y123 powders were then mixed with Y211 powders in molar ratio Y123:Y211 = 1:0 and 1:0.4 and fabricated samples were labeled as Y0 and Y40, respectively. The mixed precursor powder was pressed into a pellet of diameter 20 mm and height 8 mm. An Nd123 seed was centrally placed on top of the pellet. Crystal growth process was performed on an alumina crucible with $Y_2O_3$ layer as heat treatment was applied as shown in Figure 1. Finally, the grown sample were annealed at 500 °C for 200 h in flowing oxygen and then cooled to room temperature at a rate of 1 °C.min$^{-1}$ in oxygen.

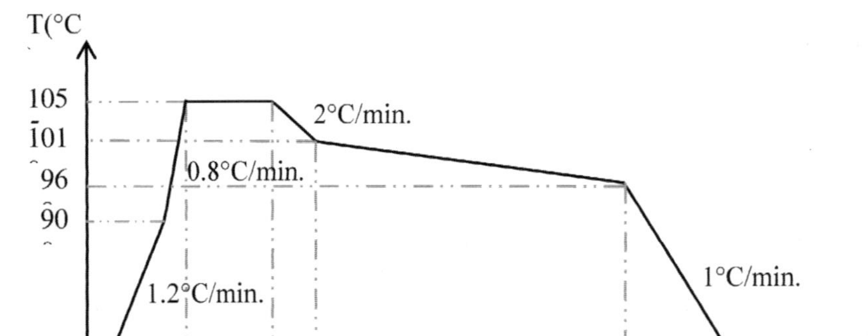


**Figure 1.** Schematic diagram of thermal process for YBCO samples

For determining characteristic peaks of Y123 sample was obtained after crystal growth, X-ray diffraction data were collected by using a Rikagu D/Max III C diffractometer, employing CuKα (λ = 1.5418 Å, 40 kV, 30 mA) radiation. 2θ range 20 to 60° was scanned in steps of 0.02° in the atmosphere with a counting time 3°/min.

Resistivity measurements were performed by a standard four-probe method at temperatures between 40 and 120 K at a rate of 4 $Kmin^{-1}$ using a Quantum Design physical properties measurement system (PPMS) under various constant magnetic fields from 1 to 5 T with the magnetic field parallel to the *c*-axis, which is the direction of pressing into the zero-field cooling regime (ZFC). Having completed the measurement at a given magnetic field for each cycle, the field was set to zero, and the temperature was increased to 120 K to completely remove the trapped field inside the sample. Effective activation energy $U$ of the flux pinning on the magnetic field and temperature of the sample were determined from the resistivity curves.

The magnetization properties were measured using a vibrating sample magnetometer (VSM) for the respective constant temperatures such as 30, 50, and 77 K with the magnetic field parallel to the *c*-axis, which is the direction of compaction. The M-H measurements were performed using the sweep rate of 100 $mTs^{-1}$. All the magnetization measurements were done in the ZFC regime. After completing the measurement at a given temperature for an each cycle, the field was set to zero and the temperature was warmed to 120 K to completely remove the trapped field inside the sample. The critical current density, $J_c$ ($A/cm^2$), was calculated from the magnetization hysteresis loops based on the extended Bean model [9]. The normalized pinning force density ($f_p$) derived from the $J_c$ versus the reduced field ($b$) was examined at different temperatures to determine the pinning mechanism.

## 3. Results and discussion

X-ray diffraction pattern of the Y0 and Y40 could be seen in Figure 2. It was observed in Figure 2(a) that only a dominant characteristic superconducting (103) peak was seen for Y0, indicating highly superconducting structure.

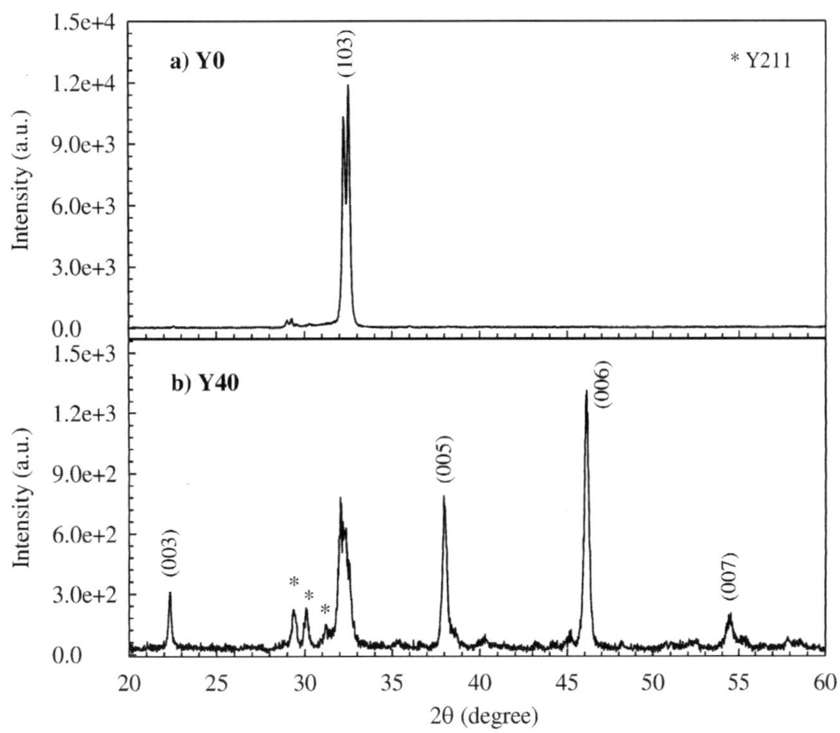


**Figure 2.** X-ray diffraction patterns of the a) Y0 and b) Y40 samples.

At the same time, there are small peaks of the Y211 phases revealing the presence of the Y211 particles in the sample and characteristic superconducting (103) peak in the spectrum at around $2\theta \sim 32.5°$ for Y40 as shown in Figure 2(b). Moreover the (00*l*) peaks were clearly observed, representing that these specimens are highly oriented by the *c*-axis perpendicular to the top surface. It was known that if the seed does not dissolve, this leads to the formation of the single-domain structure [10]. Although, both two samples were fabricated under the same condition, the samples had different XRD pattern. It was known that geometrical shape and single crystal structure of the seed are important parameters for fabrication of the bulk single crystal rather than the thermal process [11].

Figure 3 shows the normalized resistive superconducting transitions up to 5 T as a function of temperature for the Y0 and Y40 samples. Table 1 lists the onset point ($T_{c,onset}$), zero point ($T_{c,zero}$) and superconducting transition width (ΔT) of the samples estimated in the resistivity–temperature plots taken under different applied magnetic fields. Additionally, sharp superconducting transition were achieved for Y0 and Y40 samples and the $T_{c,onset}$ values obtained to be 93.4 and 93.6 K at 0 T, respectively. Also, it can be observed that the zero resistance temperature decreases with further increase of the magnetic field. However, the magnetic dispersion of Y40 was very little, indicating the strength of the sample against the applied magnetic field. For the samples, it was seen that the superconducting transitions was quite small compared with other results [4, 12]. The large ΔT values imply very sluggish oxygen diffusion into the large sized single Y123 grains in spite of the presence of the Y211/Y123 interfaces [13].

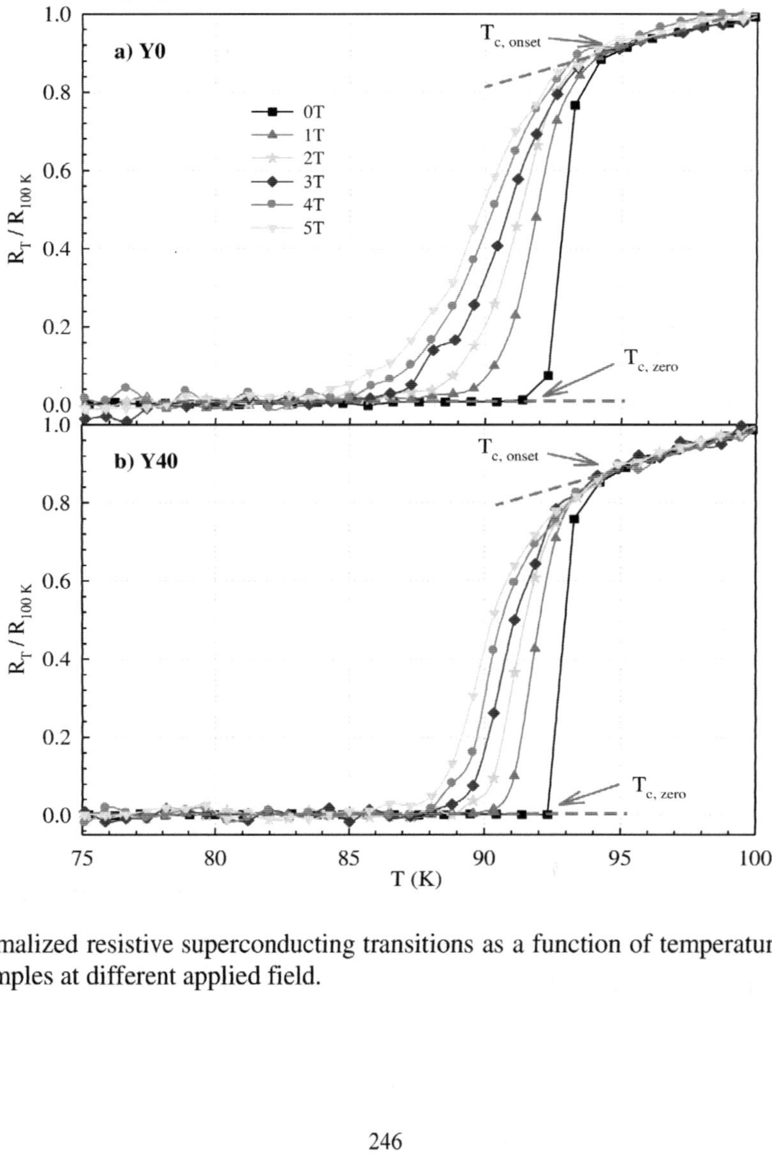


**Figure 3**. Normalized resistive superconducting transitions as a function of temperature for the a) Y0 and b) Y40 samples at different applied field.

**Table 1.** The values of critical transition temperatures for the sample Y0 and Y40 under different magnetic fields.

| | | | 0T | 1T | 2T | 3T | 4T | 5T |
|---|---|---|---|---|---|---|---|---|
| Y0 | $T_{c,onset}$ | (K) | 93.4 | 93.1 | 92.8 | 92.6 | 92.4 | 91.9 |
| | $T_{c,zero}$ | (K) | 92.2 | 90.5 | 89.4 | 87.9 | 87.2 | 85.4 |
| | ΔT | (K) | 1.2 | 2.6 | 3.4 | 4.7 | 5.2 | 6.6 |
| Y40 | $T_{c,onset}$ | (K) | 93.6 | 92.6 | 92.6 | 92.3 | 92.1 | 91.9 |
| | $T_{c,zero}$ | (K) | 91.9 | 89.9 | 89.0 | 88.1 | 87.8 | 87.3 |
| | ΔT | (K) | 1.7 | 2.8 | 3.6 | 4.2 | 4.3 | 4.6 |

When the Lorentz force per unit volume exceeds the pinning force per unit volume ($F_p$), electrical resistance is introduced because of flux flow. It is well established that lowering the temperature increases both the pinning strength and the pinning force. However, at high temperatures, the effect of pinning strength is suppressed by thermal fluctuations [14]. Because of the thermally activated flux flow (TAFF) of vortices, the resistive transition suppressed to lower temperature with the increase of applied magnetic field and it can be well described by Arrhenius dependence [15]

$$\rho(B,T) = \rho_0 \exp[-U(B,T)/k_B T] \qquad (1)$$

where $U$ (B,T) is activation energy for flux creep that depends on the temperature and magnetic field, $\rho$ (B,T) is resistivity in a magnetic field, $\rho_0$ is maximum resistivity, $T$ is temperature and $k_B$ is the Boltzmann constant. For the case of classical flux creep [16], where $U >> k_B T$, the activation energy should not depend on temperature. According to the Eq. 2, the activation energy ($U$) value can be directly deduced from the slope of the plot of $ln$ ($\rho/\rho_0$) versus $1/T$.

$$U = -k_B T \ln[\rho/\rho_0] \qquad (2)$$

Figure 4 illustrates the plots of $ln$ ($\rho/\rho_0$) versus $1/T$ of the Y0 and Y40 samples under different external magnetic fields from 0 to 5T and insets the plots of $U$ versus magnetic fields. As one can see for each curve the activation energy decreases with the increase of the applied magnetic field. However, the activation energies observed for the sample Y0 and Y40 in applied magnetic fields were 0.69 and 1.78 in zero magnetic field, 0.13 and 0.28 *eV* in 5 *T* magnetic field, respectively. Also, the activation energies of the Y40 were higher than the Y0 because of flux pinned by the Y211particles acting as pinning centers in Y40 is shown in insets plot of the Figure 4. The calculated values of the activation energies for Y0 and Y40 showed significantly higher than the values reported in literature for BSCCO and YBCO bulk produced by solid state reaction method [17–20]. Decrease of the activation energy depending on the increasing magnetic field causes vortex motion easily. That is, it can be said that the number of pinning centers are reduced and the superconducting properties are declined with the increasing magnetic field.

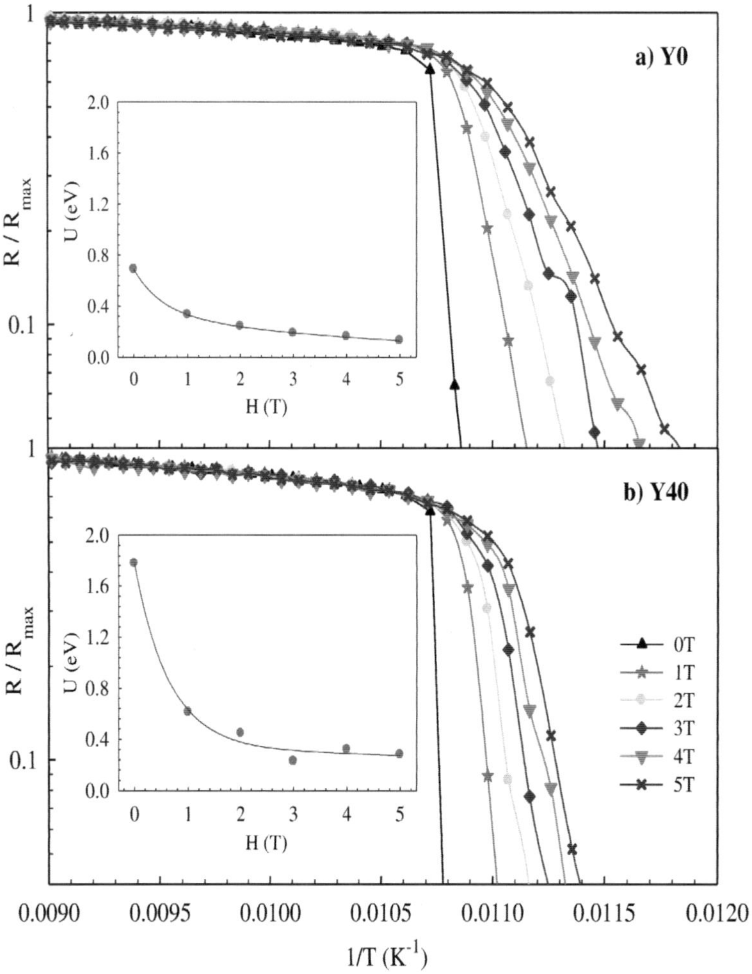


**Figure 4.** Resistive transitions as a function of temperature in semi-logarithmic scale and dependence of the activation energy on the magnetic field (*inset*) for a) Y0 and b) Y40 samples.

Figure 5 shows the variation of critical current density by the applied magnetic field for Y0 and Y40 samples at 30, 50 and 77 K. Critical current density ($J_c$) was calculated from the magnetization hysteresis based on the critical state model [9]. The $J_c$ values for Y0 and Y40 samples were determined to be $5.1\times10^3$ and $3.7\times10^3$ A/cm$^2$ at 77 K in 0 T, respectively and could be seen in Table 2 at different temperatures under magnetic field from 0 to 5T. Although any Y211 addition was not used for Y0, there was no explicit difference in terms of order between critical current densities for the samples, because, the $Y_2O_3$ layer prevent the losing of the liquid phase. However, $J_c$ of the Y0 samples higher than the Y40 at the lower temperatures in contrast to 77 K. In the low field and high temperature region, $J_c$ decreases sharply with temperature, which was attributed to the flux pinning created by the linear correlated disorders between Y211 and Y123 grains [21]. It is clearly seen from the figures that the $J_c$ varies almost linearly with the increasing magnetic field. This variation indicates that the samples are strong enough against a magnetic field. It is established that the Y211 particles are the source of strong pinning. It is also known that the linear behavior of $J_c$, attributes to the low values of flux creeps in samples.

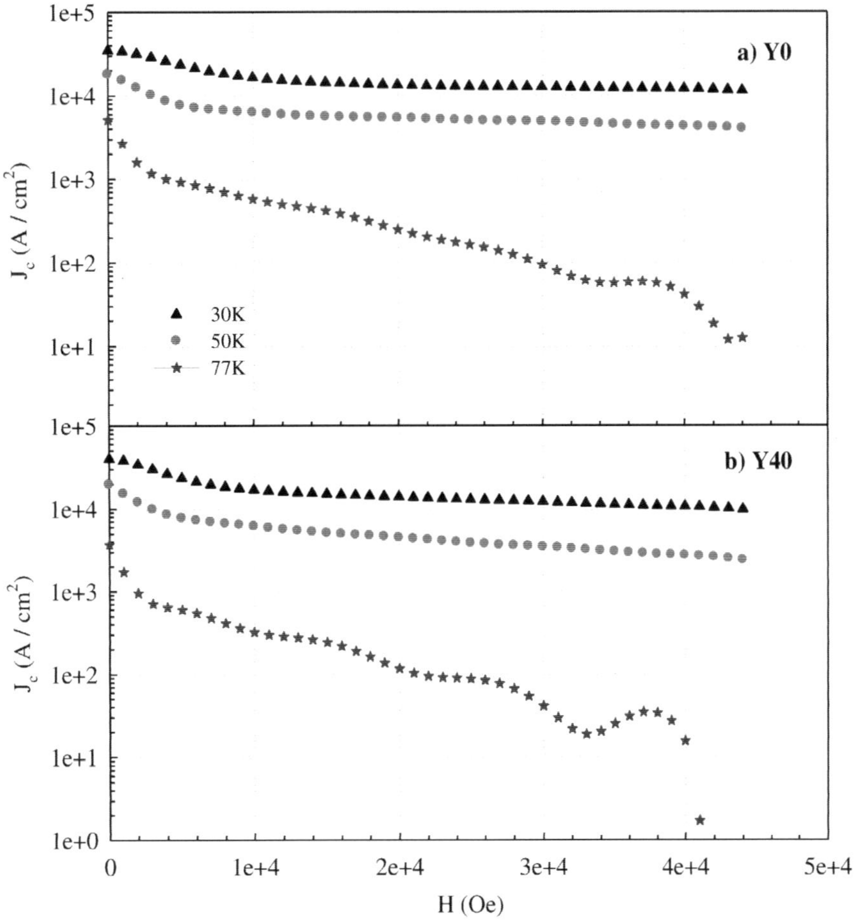


**Figure 5.** Critical current density as a function of the magnetic field at different temperatures for the a) Y0 and b) Y40.

**Table 2.** The values of critical current densities for the samples Y0 and Y40 under different magnetic fields.

| | | 0 T | 1 T | 2 T | 3 T | 4 T |
|---|---|---|---|---|---|---|
| Y0 | $J_c$ (A/cm$^2$) (30 K) | $3.48\times10^4$ | $1.64\times10^4$ | $1.34\times10^4$ | $1.26\times10^4$ | $1.21\times10^4$ |
| | $J_c$ (A/cm$^2$) (50 K) | $1.80\times10^4$ | $0.63\times10^4$ | $0.54\times10^4$ | $0.48\times10^4$ | $0.43\times10^4$ |
| | $J_c$ (A/cm$^2$) (77 K) | $5.13\times10^3$ | $5.73\times10^2$ | $2.48\times10^2$ | $0.94\times10^2$ | $0.41\times10^2$ |
| Y40 | $J_c$ (A/cm$^2$) (30 K) | $3.98\times10^4$ | $1.69\times10^4$ | $1.39\times10^4$ | $1.21\times10^4$ | $1.06\times10^4$ |
| | $J_c$ (A/cm$^2$) (50 K) | $1.96\times10^4$ | $0.61\times10^4$ | $0.45\times10^4$ | $0.34\times10^4$ | $0.27\times10^4$ |
| | $J_c$ (A/cm$^2$) (77 K) | $3.71\times10^3$ | $3.21\times10^2$ | $1.17\times10^2$ | $0.41\times10^2$ | $0.16\times10^2$ |

In order to give an explicit insight into the pinning mechanism an extended analysis of scaling laws was investigated because of the importance of the flux pinning force ($F_p$). According to the Dew-Hughes model [5], the $F_p$ can be written as follows;

$$F_p(b) = Ab^m(1-b)^n \tag{3}$$

where, $b = B/B_{max}$, A is a constant and *m*, *n* are parameters which is rely on the defect geometry (pointlike, linear, planar or surface, etc.), the type of interaction (core or magnetic) and the type of pinning center (normal or superconductor having different properties with respect to the bulk) that depend on the flux pinning mechanism. In order to obtain a deeper insight into the origin of the improved pinning properties, an extended analysis of the volume pinning force $F_p = J_c B$ is examined. Actually, in the high-temperature superconductors, $f_p(b) = F_p / F_{max}$ often scales with $b = B/B_{irr}$ but it is quite difficult to determine the correct $B_{irr}$ values using the DC magnetic measurement and several authors succeeded in scaling indicated that $B_{max}$ exhibits an identical temperature dependence with that of $B_{irr}$ and that $B_{max}$ can be used as a scaling field where $B_{max}$ is the field at which $F_p$ exhibits a maximum at $B = B_{max}$ [22, 23].

The scaling of $f_p - b$ for high–$T_c$ superconductors is often analyzed using the following Eq. 4(a–c) [22–24]:

| | | |
|---|---|---|
| $\Delta\kappa$ pinning | $f_p(b) = 3b^2(1-2b/3)$ | (4a) |
| Normal point pinning | $f_p(b) = (9b/4)(1-b/3)^2$ | (4b) |
| Surface pinning | $f_p(b) = (25\sqrt{b}/16)(1-b/5)^2$ | (4c) |

The normalized pinning force density $f_p(b) = F_p / F_{p\,max}$ versus reduced field $b = B/B_{max}$ are plotted in Figure 6, for the Y0 and Y40 samples at 30, 50 and 77K. Also Eq. 4(a–c) is represented in this figure to determine the pinning mechanism. At 30 K, the experimental results were appropriate for normal point pinning mechanism in a low field region (b < 0.2), the results are changed between $\Delta\kappa$ pinning and normal point pinning for both samples when b < 0.7 and b close to 1 then $\Delta\kappa$ pinning was good agreement for both samples. At 50 K, in low fields, normal point pinning was dominant for Y40 due to addition of the Y211 particles while the results for the Y0 between $\Delta\kappa$ pinning and normal point pinning. It has been proved that relatively large Y211 particles contribute to flux pinning in Y123 superconductors [25]. Y211 particles in Y123 superconductors have the proportionality of $F_p \propto N_p D^2$ where $N_p$ is number of the Y211 inclusions per unit volume and *D* their mean diameter. Also, contribution of the $\Delta\kappa$ pinning mechanism may be operating is the local variations in $T_c$ due to substitutions, doping, oxygen vacancy clusters [25, 26]. The normalized pinning force density results at 77 K disagree with a certain pinning mechanism (Figure 6(c)). The samples were convenient with surface pinning higher than the $B_{max}$ the mechanism was changed between $\Delta\kappa$ pinning and normal point pinning for b < 0.5 and the mechanism was varied between normal point pinning and surface pinning for b > 2.5. So, it was thought that there was some another contribution coming from the other magnetic interactions such as ferromagnetism or anti-ferromagnetism. Pinning performance for the samples in a low field region is closely related with the dispersion of non-superconducting particles, whose average diameter is larger than the coherence length ξ [23].

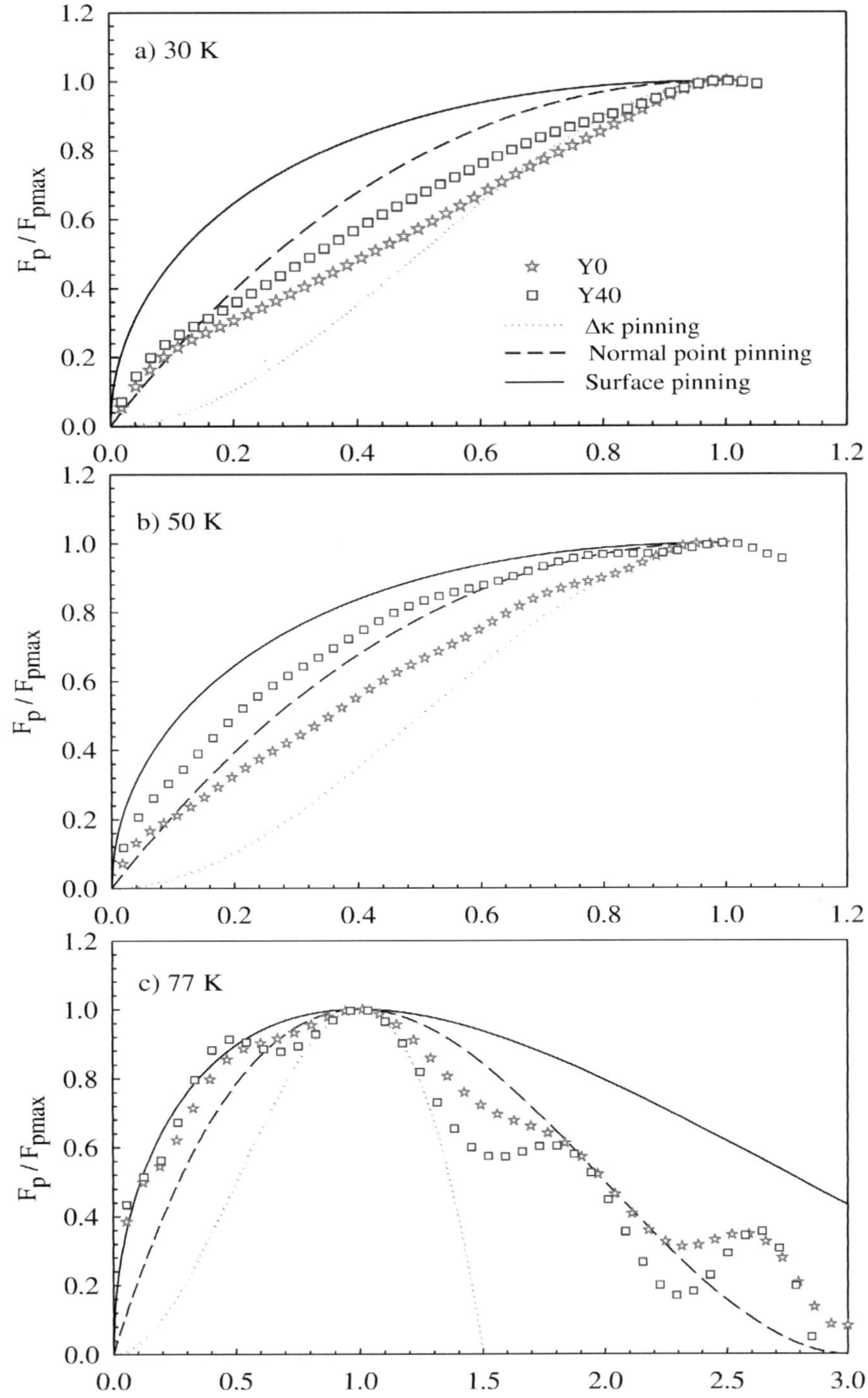


**Figure 6.** The normalized pinning force density $f_p(b)$ versus reduced field $b$ are plotted for the Y0 and Y40 samples at a) 30K, b) 50K and c) 77K.

## 4. Conclusion

Two samples were fabricated by TSMG methods with $Y_2O_3$ layer in stoichiometric ratios of Y123: Y211 as 1:0 and 1:0.4 named as Y0 and Y40, respectively. Results of the physical properties and pinning mechanism of the samples are shown in below;

1. The magnetic dispersion of Y40 was very little indicating strength of the sample against the applied magnetic field.
2. The activation energies of the Y40 were decreased more rapidly when it compared to the sample Y0. Because population of the non-superconducting Y211 particle of the Y40 sample was higher than the Y0 at low magnetic field flux pinned by the Y211 particles acting as pinning centers.
3. Although any Y211 addition was not used for Y0, there was no explicit difference between the critical current densities for the samples in terms of order. Because, $Y_2O_3$ layer prevent the losing of the liquid phase. However, $J_c$ of the Y0 samples higher than the Y40 at the lower temperatures in contrast to 77 K.
4. Both samples had almost same pinning behavior and normal cores type pinning and $\Delta\kappa$ pinning was observed dominantly at different temperatures. Although any Y211 adding did not used for Y0 pinning behavior not changed, because $Y_2O_3$ layer supports the Y211 phase during the crystal growth process. It was thought that the pinning is due to Y211 particles in low fields.

## References

[1] Li J, Yang W, Wang M, Guo Y and Feng Z 2015 *J Supercond. Nov. Magn.* 28 1725-1728.
[2] Hari Babu N, Reddy E S, Cardwell D A, Campbell A M, Tarrant C D and Schneider K R 2003 *Appl. Phys. Lett.* 83 23.
[3] Moutalbi N, Ouerghi A, Djurado E, Noudem J G and M'chirgui A 2011 *Physica C* 471, 97-103
[4] Zou X W, Wang Z H, Chen J L and Zhang H 2001 *Physica C* 356 31-38.
[5] Dew-Huges D 1974 *Philos. Mag.* 30 2 293-305.
[6] Giller D, Shaulov A and Yeshurun Y 2000 *Physica B* 288 687-688.
[7] Kim G C, Kim B J, Cheon M Y and Kim Y C 2003 *Physica C* 391 305-308.
[8] Çakır B and Aydıner A 2011 *J. Supercond. Nov. Magn.* 24 1577-1584.
[9] Ben Salem M K, Hannachi E, Slimani Y, Hamrita A, Zouaoui M, Bessais L, Ben Salem M and Ben Azzouz F 2014 *Ceramics International* 40 4953-4962.
[10] Cloots R, Koutzarova T, Mathieu J P and Ausloos M 2005 *Supercond. Sci. Technol.* 18 R9.
[11] Diko P, Zmorayova K, Babu N H and Cardwell D A, 2003 *Physica C* 398, 1-7.
[12] Liu S L 2008 *J. Supercond. Nov. Magn.* 21 199.
[13] Jun B-H, Jung S-A, Park S-D, Park B J, Han Y H and Kim C-J 2011 *Physica C* 471 876-879.
[14] Benkraouda M, Obaidat I M and Al Khawaja U 2006 *Physica C* 433 205-211.
[15] Anderson P W 1962 *Phys. Rev. Lett.* **9** 7 309-311.
[16] Balaev D A, Dubrovskiy A A, Popkov S I, Shaykhutdinov K A and Petrov M I 2008 *J. Supercond. Nov. Magn.* 21, 243.
[17] Balchev N, Nenkov K and Antonov V 2013 *J. Supercond. Nov. Magn.* 26 59-63.
[18] Çelik Ş, Öztürk K and Yanmaz E 2008 *J. Alloys and Compounds* 458 30-36.
[19] Kameli P, Salamati H, Abdolhosseini I and Sohrabi D 2008 *Physica C* 468 137-141.
[20] Kutuk S, Bolat S, Terzioğlu C and Altintas S P 2015 *J. Alloys and Compounds* 650 159-164.
[21] Okram G S, Aokia H and Nakamura K 1999 *Solid State Commun.* 110 327-331.
[22] Higuchi T, Yoo S I and Murakami M 1999 *Phys. Rev. B* 59 1514-1527.
[23] Goto T, Inagaki K and Watanabe K 2000 *Physica C* 330 51-57.
[24] Taylan Koparan E, Surdu A, Sidorenko A and Yanmaz E 2012 *Physica C* 473 1-5.
[25] Zablotskii V, Jirsa M and Petrenko P 2002 *Phys. Rev. B* 65 2245081.
[26] Koblischka M R, Muralidhar M and Murakami M 2000 *Physica C* 337 31-38.

# Comparison of Ge, InGaAs p-n junction solar cell

**M. Korun[1], T. S. Navruz[2]**
[1] Electrical and Electronics Engineering Department, Engineering Faculty, Gazi University, Ankara, Turkey,
[2] Electrical and Electronics Engineering Department, Engineering Faculty, Gazi University, Ankara, Turkey,

E-mail: muratkorun@gazi.edu.tr

**Abstract.** In this paper, the effect of material parameters on the efficiency of Ge and InGaAs p-n junction solar cells which are most commonly used as the sub-cell of multi-junction solar cells are investigated and the results due to these two cells are compared. The efficiency of Ge ($E_G$=0.67 eV) solar cell which is easy to manufacture and inexpensive in cost, is compared with the efficiency of InGaAs ($E_G$=0.74 eV) solar cell which is coming with drawback of high production difficulties and cost. The theoretical efficiency limit of Ge and InGaAs solar cells with optimum thickness were determined by using detailed balance model under one sun AM1.5 illumination. Since the band gap values of two cells are close to each other, approximate detailed balance efficiency limits of 16% for InGaAs and 14% for Ge are obtained. When drift- diffusion model is used and the thicknesses and doping concentrations are optimized, the maximum efficiency values are calculated as 13% for InGaAs and 9% for Ge solar cell. For each solar cell external quantum efficiency curves due to wavelength are also sketched and compared.

## 1. Introduction

The photovoltaic solar cell is becoming widespread and very important as a clean energy source for the earth. However, the efficiency of conventional and commercially available solar cells is still very low. To be competitive with the conventional energy source the efficiency of photovoltaic cell must be improved. Researchers are studying on both new technologies to improve the efficiency of single junction solar cells and new solar structures which are able to absorb a larger part of solar spectrum. Si is the most popular material for single junction solar cell production since it has a developed technology due to other electronic applications. The highest efficiency reported for Si single junction solar is 27.6% under 100 sun concentration [1]. GaAs single junction solar cell whose highest reported efficiency is 29.1% under 117 sun is a direct band gap material that is popular for especially space applications [2]. In order to use solar spectrum more efficiently, multijunction, intermediate and thermo photovoltaic solar cell structures are proposed [3]. In this study Ge and InGaAs single junction solar cells which are used as the bottom cell of triple junction solar cells are investigated [4].

Most common approaches to model a p-n junction solar cell are detailed balance model (DBM) and drift-diffusion model (DDM). DBM proposed by Shockley and Queisser in 1961 is used to calculate the efficiency limit of the cell if the bandgap is known [5]. In this model the one sun efficiency limit of a single junction cell is found as 33.7% for $E_G$=1.4 eV under some certain assumption. To have more realistic results DDM is preferred which includes the material parameter effects as doping concentrations, carrier mobility etc. In this study, both DBM and DDM calculations are carried out under AM1.5 illumination for Ge and InGaAs p-n junction solar cells whose bandgap values are close to each other. In DBM approach, minimum cell thickness to obtain maximum efficiency is determined and then in DDM approach the variation of efficiency due to the thicknesses and doping concentrations of p and n sides are obtained.

## 2. Methods

In this study, p-n junction solar cell is investigated using detailed balance and drift-diffusion approaches whose details are given in the following subsections.

### *2.1 Detailed balance model*

The detailed balance model, proposed by Shockley and Quiesser in 1961 [5], was used to calculate the potential maximum efficiency that can be provided by solar cells. In this approach, carrier mobilities are assumed to be infinite and each of the absorbed photon is assumed to produce one electron hole pair. Only radiative recombination is taken into account. Current voltage variation of a solar cell is given as below [6]:

$$J(V) = J_{SC} + J_{dark}(V) \tag{1}$$

Here $J_{SC}$ is short circuit current density obtained when the cell is illuminated under short circuit condition. $J_{dark}(V)$ is the radiative recombination current density changing with cell voltage under dark condition. $J_{SC}$ is calculated as follows:

$$J_{SC} = Xq\int(1-R(\varepsilon))(1-e^{-\alpha(\varepsilon)w}N_{ph}(\varepsilon))d\varepsilon \tag{2}$$

Where $X$ represents sun concentration, $q$ is electron charge, $R$ gives amount of reflection due to refraction index differences, $\varepsilon$ is energy, $\alpha$ is absorption coefficient, $w$ is the thickness of the cell and $N_{ph}$ is the number of photons incoming to cell surface per unit area per unit time. The dark current term in Eq. 1 is given as:

$$J_{dark} = q\int\frac{2\pi}{h^3c^2}(1-e^{-\alpha(\varepsilon)w})(\frac{\varepsilon^2}{e^{(\frac{\varepsilon-\Delta\mu}{k_BT_C})}})d(\varepsilon) \tag{3}$$

Here $h$ is Planck's constant, $c$ is the speed of light, $k_B$ is Boltzmann constant, $T_C$ is cell temperature and $\Delta\mu=qV$ is amount of quasi fermi level split where $V$ is output voltage of the cell. Using equations 1-3 current- voltage variation is obtained. Short circuit current, $J_{SC}$, open circuit voltage, $V_{oc}$, efficieny, $\eta$ and fill factor, $FF$ values are calculated from this variation.

$$\eta = \frac{P_m}{P_i} \tag{4}$$

$$FF = \frac{P_m}{J_{SC}V_{0C}} \tag{5}$$

Where $P_m$ is maximum output power and $P_i$ is the incident power density.

### *2.2 Drift-Diffusion model in semiconductors*

Continuity equations and drift-diffusion current equations are used to model p-n junction [6]. In this model the electric field outside the depletion region is assumed to be zero. Therefore diffusion current is dominant in n and p sides and drift current is dominant in the depletion region. Inserting the diffusion current equation into continuity equations [6] the following equations are obtained for p and n sides of the cell.

$$D_e \frac{d^2 \Delta n}{dx^2} - rg \frac{\Delta n}{n_{eq}} - \frac{\Delta n}{\tau_c} = -g_{eh} \tag{6}$$

$$D_h \frac{d^2 \Delta p}{dx^2} - rg \frac{\Delta p}{p_{eq}} - \frac{\Delta p}{\tau_\upsilon} = -g_{eh} \tag{7}$$

Where, $D_e$ , $D_h$ are the diffusion coefficients of electrons and holes respectively. $\tau_c, \tau_v$ are Shockley-Read Hall (SRH) recombination carrier lifetimes, $n_{eq}, p_{eq}$ represent carrier concentrations under thermal equilibrium and $r_g$ is radiative recombination coefficient given as below.

$$rg = \frac{2\pi}{h^3 c^2} \int_0^\infty \alpha_\varepsilon(\varepsilon) \varepsilon^2 \exp(-\frac{\varepsilon}{k_B T_C}) d\varepsilon \tag{8}$$

geh in equations 6 and 7 is photo generation rate given as below:

$$g_{eh}(x) = \int (1 - R(\varepsilon)) \alpha(\varepsilon) Nph(\varepsilon) \exp(-\alpha\ (\varepsilon) x) d\varepsilon \tag{9}$$

Equations 6 and 7 are solved first under short circuit condition with appropriate boundary conditions and electron and hole concentration variations are obtained. These variations are used to calculate diffusion current contribution of p and n sides ($J_n = D_n d\Delta n/dx$, $J_p = -D_p d\Delta p/dx$). Under dark condition dark current variation with the output voltage is also calculated. The photo and dark current contribution of the depletion region is given as below.

$$J_{photo\_dep} = -q \int_{-W_n}^{W_p} (g_{eh}) dx \tag{10}$$

$$J_{dark\_dep} = -q \int_{-W_n}^{W_p} (-u_{rad} - u_{nr}) dx \tag{11}$$

Here $u_{rad}$ and $u_{nr}$ are radiative and nonradiative recombination rates depending on output voltage. Summing the short circuit condition contributions of n side, p side and depletion region, total photo current of the cell is obtained. Similarly, total dark current is found by summing the dark current contributions of these three regions for varying values of output voltage. Finally, cell current-voltage variation hence *J-V* curve is obtained by subtracting the dark current from photo current and including the series ($R_S$) and shunt resistance ($R_{SH}$) effects [7]. Jsc, Voc, FF and η values are obtained similar to DBM.

*EQE* curve is found from short circuit calculations [8].

$$EQE = \frac{\text{electrons/sec}}{\text{photons/sec}} = \frac{J_{photo}(\lambda)}{qN_{ph}(\lambda)} \tag{12}$$

Where λ is the wavelength and $J_{photo}$ is the current density variation due to wavelength found by short circuit calculations.

## 3. Results and discussion

In this study, Ge and InGaAs p-n junction solar cells are investigated using the detailed balance and drift diffusion approaches under AM1.5 spectrum and the results are compared. The optical and electrical parameters of Ge and InGaAs are taken from [9].

### *3.1. Detailed Balance Approach Results*

The detailed balance efficiencies of Ge and InGaAs p-n junction solar cells are obtained using real absorption coefficient values and including the effect of reflection due to refractive index differences. As seen from Fig. 1, limiting efficiency values of Ge and InGaAs cells are close to each other but Ge cell should be thicker with respect to InGaAs cell to achieve maximum efficiency.

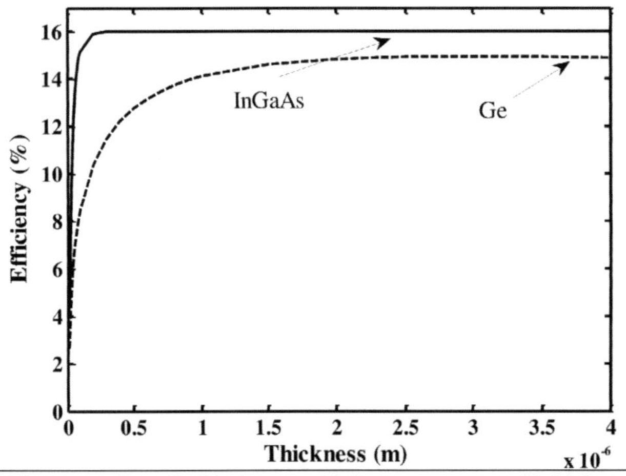


**Figure 1**. Detailed balance efficiency versus cell thickness for Ge and InGaAs cells

The output parameters of both cells; $J_{SC}$, $V_{OC}$, $FF$, $w_{OPT}$ and $\eta$ are summarized in Table 1 when their efficiencies are maximum.

**Table 1**. The detailed balance results of Ge and InGaAs solar cells

| Semiconductor | $J_{SC}$ (*mA*) | $V_{OC}$ (*V*) | FF (%) | $W_{OPT}$ (*μm*) | η (%) |
|---|---|---|---|---|---|
| Germanium | 37.71 | 0.4980 | 80.36 | 3 | 14.83 |
| InGaAs | 40.88 | 0.4970 | 80.37 | 0.6 | 16.014 |

### *3.2. Drift Diffusion Approach Results*

Using drift diffusion equations given above, the efficiency variation of Ge and InGaAs solar cells with p type emitter thickness ($w_p$) and doping concentration ($N_A$) and n-type base thickness ($w_n$) and doping concentration ($N_D$) are investigated. For each cell, one of these parameters is changed while the others are kept constant at each simulation. Finally the optimum cell structures achieving maximum efficiency are obtained as seen in Table 2. InGaAs solar cell's efficiency is slightly higher than Ge solar cell's efficiency. Doping concentrations are nearly same but InGaAs solar is thinner with respect to Ge one.

**Table 2**. Optimum Ge and InGaAs solar cell structures and their output parameters

| Semiconductor | $J_{SC}$ *(mA)* | $V_{OC}$ *(V)* | FF (%) | $W_N$ (μm) | $W_P$ (μm) | $N_A$ ($cm^{-3}$) | $N_D$ ($cm^{-3}$) | η (%) |
|---|---|---|---|---|---|---|---|---|
| Germanium | 35.75 | 0.3430 | 74.24 | 2.5 | 0.15 | $5x10^{24}$ | $9x10^{22}$ | 9.11 |
| InGaAs | 39.94 | 0.3940 | 76.61 | 0.7 | 0.14 | $2x10^{24}$ | $9x10^{22}$ | 12.06 |

The J-V curve of the optimum cell structures given in Table 2 is shown in Figure 2.

Finally, external quantum efficiency curves for Ge and InGaAs solar cells are given in the Fig.3. Since the bandgap values are close to each other the EQE curves are also similar to each other.

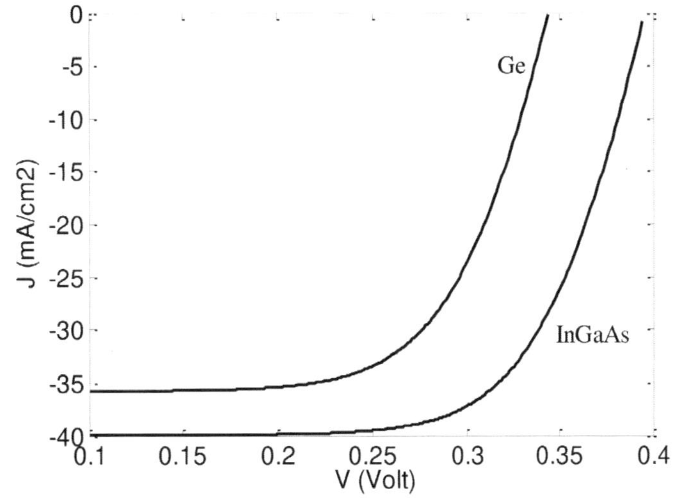


**Figure 2.** J-V curve of Ge and InGaAs solar cells

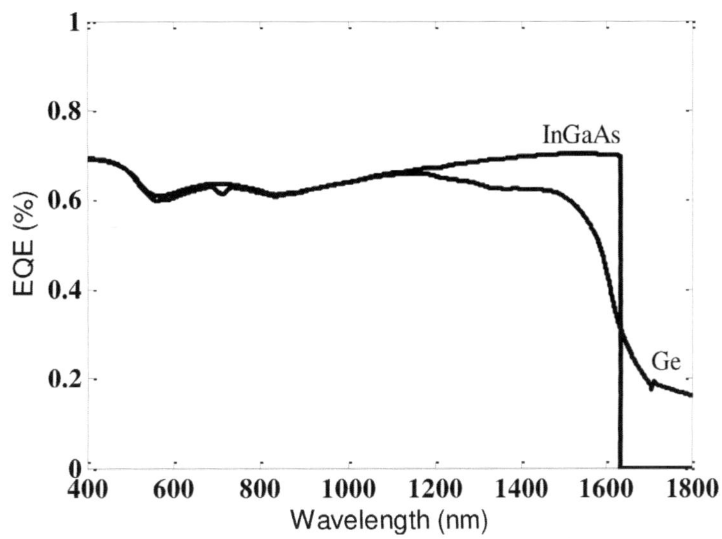


**Figure 3.** EQE curves of Ge and InGaAs solar cells

## 4. CONCLUSIONS

Despite the close $E_G$ values, the maximum efficiency for Ge (Eg=0.67) is 9.11 %, and maximum efficiency for InGaAs (Eg=0.74) is 12.26%. Ge and InGaAs do not give so high efficiencies when they are produced as p-n junction solar cells. But in multijunction solar cell structure they are used as bottom cell to absorb the low energy portion of the solar spectrum. In this study it is shown that these two materials show similar performance.

## References

[1] A Slade and V Garboushian,27.6% efficient silicon concentrator solar cells for mass production.technical Digest, *15th International Photovoltaic Science and Engineering Conference*, (Beijing. 2005)

[2] Green, Martin A., et al.Solar cell efficiency tables (Version 45).*Progress in photovoltaics: research and applications 23.1 (2015):* 1-9.

[3] King, R. R., et al. 40% efficient metamorphic GaInP/GaInAs/Ge multijunction solar cells. *Applied physics letters,* 90.18 (2007): 183516-183900.

[4] M Yamaguchi, III–V compound multi-junction solar cells: present and future. *Solar energy materials and solar cells* 75.1 (2003): 261-269.

[5] W Shockley and H J. Queisser, Detailed Balance Limit of Efficiency of p-n Junction Solar Cells, *Journal of Applied Physics,* Volume 32, pp. 510-519 (1961)

[6] J Nelson, *The physics of solar cells*. Vol. 1. London: Imperial college press, 2003.

[7] H Cotal, et al. III–V multijunction solar cells for concentrating photovoltaics.*Energy & Environmental Science* 2.2 (2009): 174-192.

[8] T J Nagle, Quantum efficiency as a device-physics interpretation tool for thin-film solar cells. *ProQuest*, 2007.

[9] http://www.ioffe.ru/SVA/NSM/Semicond/

# Comparative Investigation of Normal Modes and Molecular Dynamics of Hepatitis C NS5B Protein

**M S Asafi[1], A Yildirim[2] and M Tekpinar[1]**

[1]Yuzuncu Yil University, Van, Turkey
[2]Siirt University, Siirt, Turkey

E-mail: tekpinar@buffalo.edu

**Abstract**. Understanding dynamics of proteins has many practical implications in terms of finding a cure for many protein related diseases. Normal mode analysis and molecular dynamics methods are widely used physics-based computational methods for investigating dynamics of proteins. In this work, we studied dynamics of Hepatitis C NS5B protein with molecular dynamics and normal mode analysis. Principal components obtained from a 100 nanoseconds molecular dynamics simulation show good overlaps with normal modes calculated with a coarse-grained elastic network model. Coarse-grained normal mode analysis takes at least an order of magnitude shorter time. Encouraged by this good overlaps and short computation times, we analyzed further low frequency normal modes of Hepatitis C NS5B. Motion directions and average spatial fluctuations have been analyzed in detail. Finally, biological implications of these motions in drug design efforts against Hepatitis C infections have been elaborated.

## 1. Introduction

Hepatitis C virus (HCV) is an infection affecting millions of people worldwide. Unfortunately, there is not an effective treatment for this fatal viral disease. One of the drug targets in HCV infections is NS5B polymerase protein, which replicates the viral genome. NS5B polymerase protein performs conformational changes during genome replication. Therefore, understanding NS5B protein dynamics is vital for developing effective inhibitors against this protein.

Molecular dynamics and elastic network methods are widely used methods to investigate dynamics of proteins. There are a few molecular dynamics studies on HCV NS5B protein [1, 2]. However, there is not any study of HCV NS5B within framework of elastic network models. How do molecular dynamics simulation results are related with elastic network model data is one of answers we look for in this study. As a result, we employed both molecular dynamics and elastic network model to understand dynamics of HCV NS5B polymerase protein. We compared results of B-factors, normalized cross-correlations. Moreover, we compared principal components obtained from a molecular dynamics trajectory to low frequency normal modes from a coarse-grained elastic network model. Finally, we investigated low frequency normal modes obtained with elastic network model in detail and we elaborated on their biological implications.

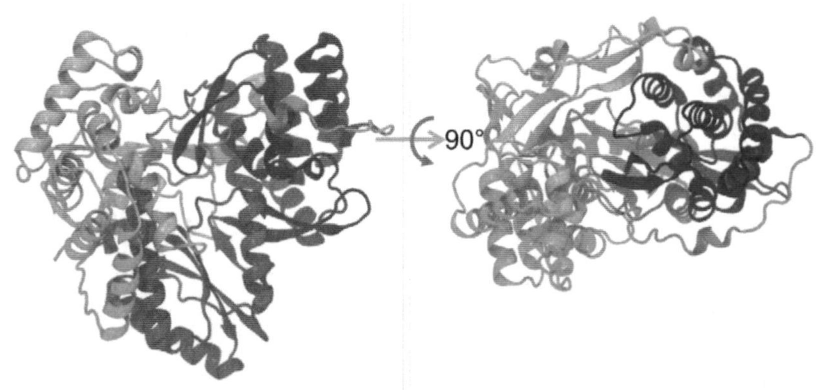


**Figure 1.** Side (left panel) and top (right panel) views of NS5B protein. Fingers domain is green (left group), palm domain is red (middle group) and thumb domain is blue (right group) new cartoon representation. Please check online version for colour figure.

## 2. Materials and methods

We utilized Visual Molecular Dynamics (VMD) program, Bio3D R package, and Prody package in addition to many custom made python scripts for visualization and analysis throughout the text [3-5].

### *2.1. Initial structure and its domains*

We used 4aep crystal structure of HCV NS5B deposited in Protein Data Bank for all calculations [6]. It has been suggested that HCV NS5B resembles to a right hand [7]. Therefore, this structure contains three domains (Figure 1). Residues 1 to 191 and 230 to 184 comprise fingers domain. Palm domain is made up of residues from 192 to 229 and residues 285 to 367. Finally, thumb domain is from residue 368 to 563. Thumb and fingers domains open or close during RNA replication process.

### *2.2. Molecular dynamics (MD) simulation and principle component analysis*

Only protein atoms and crystal waters of the crystal structure have been retained for simulations. 19607 solvent molecules of TIP3P model have been added to solvate the protein in a dodecahedral box[8]. 63 Na and 79 Cl ions have been added to the system to neutralize it. Particle Mesh Ewald method has been used for electrostatic energy calculations [9]. 50 picoseconds (ps) of constrained NVT simulations preceded 100 ps of unconstrained NPT simulations. All of the simulations have been performed under 300 K and 1 bar pressure. Gromacs simulation package with Charmm27 force field has been employed for the simulations [10, 11]. Production run was 100 nanoseconds with 2 femtoseconds time step. The trajectory frames were dumped out every 40 ps.

Normalized cross correlations ( $C_{ij}^{n}$ ) can be calculated from an MD trajectory by using

$$C_{ij}^{n} = \frac{\langle (R_i - \langle R_i \rangle)(R_j - \langle R_j \rangle) \rangle}{\sqrt{\langle R_i - \langle R_i \rangle \rangle^2 \langle R_j - \langle R_j \rangle \rangle^2}} \quad (1)$$

where <> denotes averaging over trajectory frames and $R_i$(or $R_j$) denotes position of Cα atom i (or j). This matrix can be diagonalized and eigenvectors of it are principal components of a trajectory. Low frequency principal components (PC) can indicate collective motions observed trajectories. Moreover, when MD trajectory frames are projected onto the first few PCs, major conformational changes can be visualized through clusters observed in 2D plots.

### *2.3. Elastic network model*

In an elastic network model, generally all Cα atoms of a protein within a cutoff radius (15 Å), which are connected with springs of uniform stiffness, interact through a simple Hookean potential given below [12, 13]:

$$V = \frac{\gamma}{2} \sum_{j(j \neq i)} \Gamma_{ij} \left(R_{ij} - R_{ij}^{0}\right)^2 \tag{2}$$

$\gamma$ denotes uniform spring constant, $R^0_{ij}$ is equilibrium distance between Cα atoms of residue i and j, while $R_{ij}$ is instantaneous distance between Cα atoms of residue i and j. $\Gamma_{ij}$ is ij$^{th}$ element of Kirchoff matrix $\Gamma$ and it is zero for interactions beyond the cutoff radius. A hessian matrix can be constructed from this potential. The hessian matrix has to be diagonalized and low frequency eigenvalues of the diagonalized matrix correspond to frequencies of collective motions. It has been shown that normal modes obtained with this type of potential can be used to predict observed conformational changes of proteins as well as fluctuations of proteins around native conformations [14, 15].

Cross-correlations ($C_{ij}$) can be calculated from a selected subset of L non-zero normal modes using

$$C_{ij} = \langle \Delta R_i \cdot \Delta R_j \rangle = \frac{3k_B T}{\gamma} \sum_{k}^{L} \left[ \frac{u_k u_k^T}{\lambda_k} \right]_{ij} \tag{3}$$

where $k_B$ is Boltzmann's constant, T is temperature, and $\gamma$ is uniform spring force constant. $\lambda_k$ denotes eigenvalue and $u_k$ is the corresponding eigenvector.

One can calculate normalized cross-correlations ($C^n_{ij}$) from the equation above as follows:

$$C^n_{ij} = \frac{\langle \Delta R_i \cdot \Delta R_j \rangle}{\sqrt{\langle \Delta R_i \cdot \Delta R_i \rangle \langle \Delta R_j \cdot \Delta R_j \rangle}} \tag{4}$$

## 3. Results and discussion

### *3.1. B-factor calculations with molecular dynamics and normal mode analysis*

We calculated theoretical B-factors using molecular dynamics trajectory and elastic network model. Correlation cosine (cc) agreement between experimental B-factors and the ones calculated with molecular dynamics is 0.812. B-factors calculated with elastic network model show a better overlap (0.954) with experimental data.

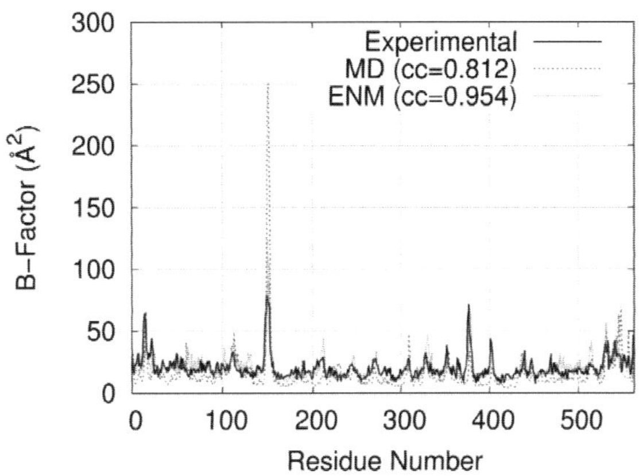


**Figure 2.** B-factors calculated with elastic network model (dotted) and molecular dynamics simulations (dashed) have been superimposed onto the experimental data (solid).

We can see that both methods overestimate fluctuation for residues around residue 150. However, degree of overestimation is better for elastic network model compared to MD. This excessively floppy region is located in fingers domain. Despite this overestimation, both methods can predict B-factors with a reasonable accuracy overall.

*3.2. Principal components of the molecular dynamics trajectory and their agreement with normal modes.*

We calculated the principal components (PCs) of our molecular dynamics trajectory. The first 5 principal components comprises of about ~35% of the motion observed in the trajectory. PC1 contributes 18% to the sum, PC2 contributes ~8%, and PC3 contributes ~6%. Contribution of the other principal components to the total motion decreases with principal component index. Contribution of the first 20 principal components reaches to 61%.

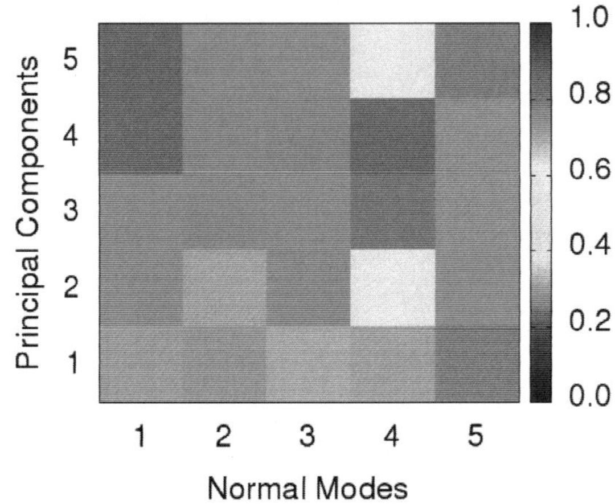


**Figure 3.** Overlaps of the first five principal components and normal modes. 1.0 indicates a perfect directional agreement while 0.0 denotes that the vectors are perpendicular or unrelated. Please check online version for colour figure.

Moreover, we wanted to observe if there is an agreement between principal components and normal modes calculated from elastic network model. So, we calculated the cosine angle between 3N dimensional principal components and normal modes. If angle is 0°, it means that two vectors are in same direction. If angle is 90°, it means that the vectors are perpendicular to each other and so on. Our data shows that the cosine correlation is greater than 0.6 and this indicates a good directional agreement between results of the two different methods. Finally, we should note that the order of low frequency motions between principal components and normal modes is not preserved.

*3.3. Normalized cross-correlations obtained with molecular dynamics and elastic network model.*

Normalized cross-correlations can give information about positive and negative cooperation within a protein. Normalized cross correlations have been calculated as explained above in Materials and Methods section (Figure 4). For normal modes, we considered only the first 100 low frequency modes. The first 10 residues have a high correlation with residue 280, 400 and 500. This pattern can be observed both in MD and normal mode cross correlations. Overall, we observe that positive correlation predictions are in better agreement between two methods. C terminal residues show a negative correlation with fingers domain residues (around residue 100). Despite methodical differences and some minor disagreements in the plots, general similarity of normalized cross-correlations is striking.

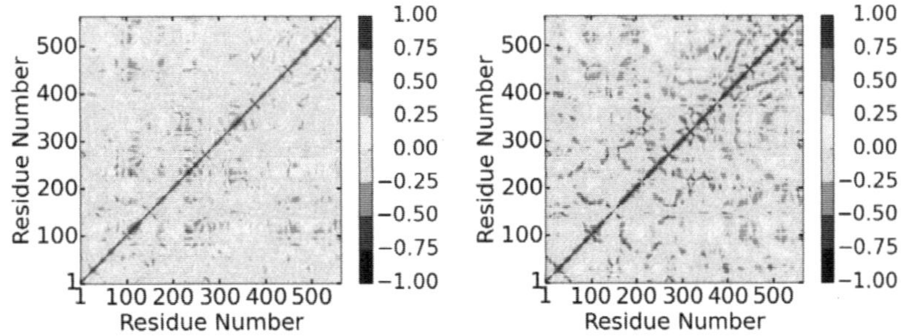


**Figure 4.** Normalized cross correlations obtained from molecular dynamics trajectory (left) and elastic network model (right). Please check online version for colour figure.

### *3.4. Low frequency normal modes and their biological meanings*
We visualized low frequency normal modes calculated with coarse-grained elastic network model (Figure 5). We observed that most of the low frequency modes are involved in fingers and thumb domain motions. The first mode is mainly about rotation of thumb domain and fingers in reverse directions. The width of RNA cavity is approximately same in this mode. However, the cavity opens up and gets closed in mode 2 and 3. It is highly possible that these two modes are functional modes corresponding to conformational states in RNA replication. Mode 4 is mostly an out of plane motion of a loop located in fingers domain. This motion is coupled with thumb domain. However, motion of the same loop is coupled with motions of all domains including the palm domain. Considering that major experimental conformational states are involved in changes of fingers-thumb domain distance variations, we conclude that low frequency normal modes can sample conformational states of NS5B observed in experimental studies [16].

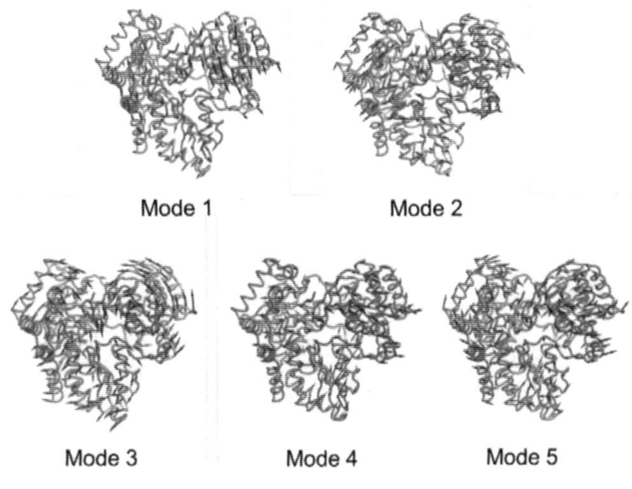


**Figure 5.** Side view of the first 5 low frequency normal modes. Colour indicates fluctuations. Blue means high fluctuations and red means low. Black arrows denote direction of motion and length of the arrow is proportional to amplitude of the motion. Please check online version for colour figure.

## 4. Conclusions
Molecular dynamics is an invaluable tool for understanding protein dynamics in atomistic detail. However, long computational times and big parallel computational resources are needed to perform MD simulations. On the other hand, elastic network models can provide a sufficiently accurate dynamics

information at a coarse-grained level. Our study also proves this point by showing the agreement between experimental, MD and coarse-grained results of B-factors for HCV NS5B. Moreover, when both methods are employed we obtain a similar picture for normalized cross-correlations of NS5B.

Low frequency motions of NS5B indicate that thumb and finger domains open up and get closed. Experimental studies show that these conformational states are required for RNA replication by HCV NS5B. Therefore, elastic network model can predict functional conformational states of NS5B protein. Moreover, inter-domain correlations may indicate possible allosteric interactions. These interactions can guide drug design efforts, which aim to stop transitions between conformational states of NS5B. Based on these results, we conclude that elastic network modes can provide invaluable information about dynamics of HCV NS5B protein without using lengthy MD simulations.

**References**

[1] Davis BC, Thorpe IF 2013 Molecular simulations illuminate the role of regulatory components of the RNA polymerase from the hepatitis C virus in influencing protein structure and dynamics. Biochemistry.52(26):4541-52.

[2] Brown JA, Thorpe IF 2015 Dual Allosteric Inhibitors Jointly Modulate Protein Structure and Dynamics in the Hepatitis C Virus Polymerase. Biochemistry.54(26):4131-41.

[3] Humphrey W, Dalke A, Schulten K 1996 VMD: visual molecular dynamics. J Mol Graph.14(1):33-8, 27-8.

[4] Grant BJ, Rodrigues AP, ElSawy KM, McCammon JA, Caves LS 2006 Bio3d: an R package for the comparative analysis of protein structures. Bioinformatics.22(21):2695-6.

[5] Bakan A, Meireles LM, Bahar I 2011 ProDy: Protein Dynamics Inferred from Theory and Experiments. Bioinformatics.27(11):1575-7.

[6] Scrima N, Caillet-Saguy C, Ventura M, Harrus D, Astier-Gin T, Bressanelli S 2012 Two crucial early steps in RNA synthesis by the hepatitis C virus polymerase involve a dual role of residue 405. J Virol.86(13):7107-17.

[7] Lesburg CA, Cable MB, Ferrari E, Hong Z, Mannarino AF, Weber PC 1999 Crystal structure of the RNA-dependent RNA polymerase from hepatitis C virus reveals a fully encircled active site. Nat Struct Biol.6(10):937-43.

[8] Jorgensen WL, Chandrasekhar J, Madura JD, Impey RW, Klein ML 1983 Comparison of Simple Potential Functions for Simulating Liquid Water. Journal of Chemical Physics.79(2):926-35.

[9] Darden T, York D, Pedersen L 1993 Particle Mesh Ewald - an N.Log(N) Method for Ewald Sums in Large Systems. Journal of Chemical Physics. 98(12):10089-92.

[10] Van Der Spoel D, Lindahl E, Hess B, Groenhof G, Mark AE, Berendsen HJ 2005 GROMACS: fast, flexible, and free. J Comput Chem. 26(16):1701-18.

[11] MacKerell AD, Brooks CL, Nilsson L, Roux B, Won Y, Karplus M 1998 CHARMM: The Energy Function and Its Parameterization with an Overview of the Program. In: Schleyer, editor. 1: John Wiley & Sons: Chichester. p. 271-7.

[12] Bahar I, Atilgan AR, Erman B 1997 Direct evaluation of thermal fluctuations in proteins using a single-parameter harmonic potential. Fold Des.2(3):173-81.

[13] Tirion MM 1996 Large Amplitude Elastic Motions in Proteins from a Single-Parameter, Atomic Analysis. Phys Rev Lett.77(9):1905-8.

[14] Atilgan AR, Durell SR, Jernigan RL, Demirel MC, Keskin O, Bahar I 2001 Anisotropy of fluctuation dynamics of proteins with an elastic network model. Biophys J. 80(1):505-15.

[15] Tama F, Sanejouand YH 2001 Conformational change of proteins arising from normal mode calculations. Protein Eng.14(1):1-6.

[16] Appleby TC, Perry JK, Murakami E, Barauskas O, Feng J, Cho A, et al 2015 Viral replication. Structural basis for RNA replication by the hepatitis C virus polymerase. Science.347(6223):771-5.

Journal of Physics: Conference Series **707** (2016) 012037 doi:10.1088/1742-6596/707/1/012037

# Comparison of dosimetry gels prepared by agar and bovine gelatine

**M E Sağsöz[1,7], Ö Korkut[2], N Alemdar[3], S Aktaş[4], E B Çalı[5] and M Kantarcı[6]**

[1]Atatürk University, Faculty of Medicine, Biophysics Dept., Erzurum, Turkey
[2]Atatürk University, Faculty of Engineering, Chem. Eng. Dept., Erzurum, Turkey
[3]Marmara University, Faculty of Engineering, Chem. Eng. Dept., İstanbul, Turkey
[4]Regional Directorate of Hygiene Laboratory Erzurum, Turkiye
[5]Erzurum Regional Research Hospital, Radiation Oncology Dept., Erzurum, Turkey
[6]Atatürk University, Faculty of Medicine, Radiology Dept., Erzurum, Turkey

E-mail: mesagsoz@atauni.edu.tr

**Abstract.** Gel dosimeters are unique materials capable of showing three dimensional (3D) dose distributions of therapeutic or diagnostic exposures. Fricke gel dosimeters can be considered as chemical dosimeters that rely on a radiation-induced chemical reaction. Dose distribution of Fricke solutions containing $Fe^{+2}$ ions determines the transformation of acidic, oxygen saturated $Fe^{+2}$ ions to $Fe^{+3}$ ions by the ionizing radiation in aqueous solutions. In this study we produced two different types of gel dosimeters using agar and bovine gelatin with similar fabrication methods. We compared the magnetic resonance (MR) T1 imaging responses of these two gel dosimeters to acquire a dose dependency of MR intensities. In conclusion agar gel dosimeters found to be produced easily and more consistent.

## 1. Introduction

Main purpose of treatment of cancer with radiation -briefly radiotherapy- is to expose previously diagnosed pathological volume of tissues with sufficient quantity of radiation dose to prevent the division or tumor cell proliferation and to realize this process with the least possible damage to normal tissues. During delivery of the prescribed dose to the patient it is critical to give radiation doses to specific organs (ie. eye, thyroid, spinal cord, adrenal glands, etc.) under standard threshold dose values for each organ.

Today, the determination of organ doses is measured during radiotherapy with point (ionization chambers, TLDs, diodes) or planar dosimeters (such as radiochromic films or 2D diode arrays). Therefore, the distribution of three-dimensional (3D) doses is needed to be measured with dosimetric systems [1].

The ferrous sulphate solution had been used for radiation dosimetry for a long time but firstly Gore [2] used magnetic resonance imaging to make possible three-dimensional radiation dosimetry. In a $FeSO_4$ doped gel the absorbed dose after exposure to ionizing radiation can be determined by optical absorption in the visible spectrum. Different reduction in the spin relaxation rates of hydrogen existing in aqueous solutions of ferrous and ferric ions can be used for the 3D determination of the absorbed dose [3].

The purpose of this study is to compare dose distribution of Fricke solutions containing $Fe^{+2}$ ions doped in different gel matrix containing agar agar or bovine gelatin by performing MR T1 relaxation time measurements.

## 2. Materials and Methods

In this study we produced two different types of gel dosimeters using agar or bovine gelatin with similar fabrication method described by Gambarini et al [4] . Gelling agent added to previously heated water at 70°C. Solution was regularly stirred under an air flow at $20 Lh^{-1}$ and heated to 100 °C. During the 20 min the solution was left to boil for saturation and then cooled down to 70 °C again. The acidic $FeSO_4$ solution was added and stirred for 5 minutes (Figure 1a). After that the gel was poured into the polyethylene tubes for MR (Figure 1b). The samples were maintained at 4 °C until their analyses were done.

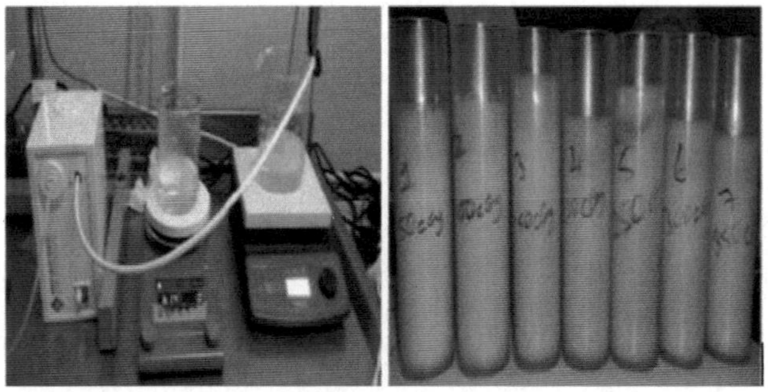

**Figure 1. a)** gel production assembly, **b)** the samples (after irradiation).

Prepared gels were irradiated with linear accelerator gamma radiation, dose rate of 300 cGy/min and energy 6MV, with doses between 50 and 350 cGy. The samples were maintained at least 30 min at room temperature before the MRI evaluations. The T1 weighted MRI signals were obtained by Siemens Symphony MR Scanner (1.5T), with quadrature head coil, using the TR= 367 ms, TI= 300ms and TE= 18 ms Inversion Recovery sequence parameters (Figure 2). Irradiation process and the MR measurements were performed within 24 hours after gel production for increasing the accuracy and sensitivity of the analysis.

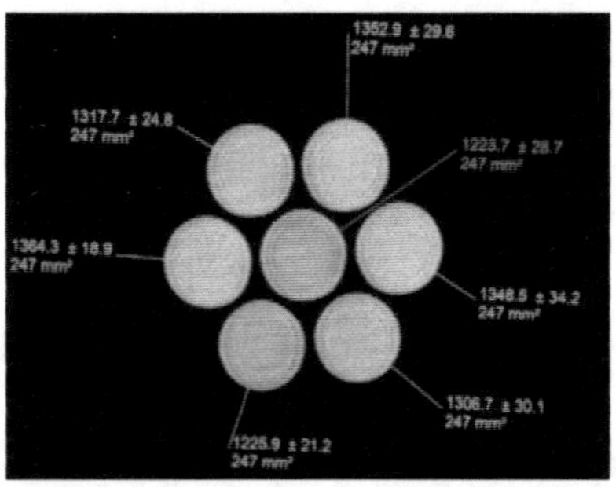


**Figure 2.** MR image of samples axial plane slice.

## 3. Results

Our initial results did not give a dose-intensity correlation with an expected linear slope. But at least there is a linear correlation for the agarose gel dosimeters we can say. A linear model can be shown between dose and MR intensity at Figure 3.

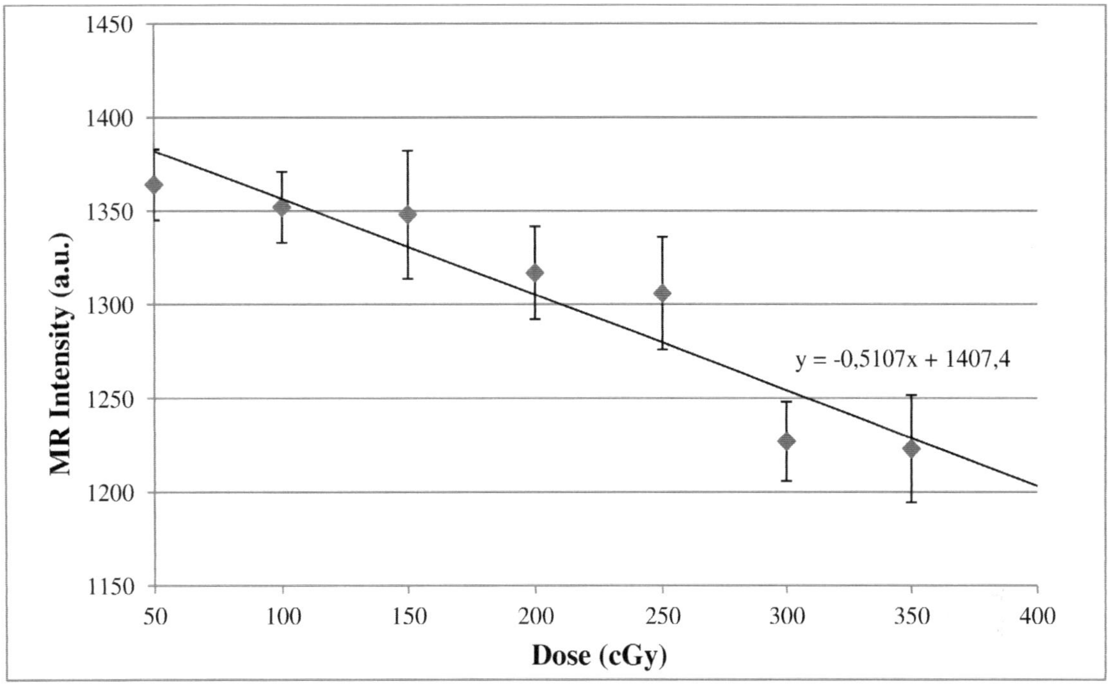


**Figure 3.** Dose-MR intensity response of agar gel dosimeters.

## References

[1] Simon Doran, Tina Gorjiara, Andrzej Kacperek, John Adamovics, Zdenka Kuncic and Clive Baldock Issues involved in the quantitative 3D imaging of proton doses using optical CT and chemical dosimeters, Phys. Med. Biol. 60 (2015) 709–726.

[2] Gore J C, Kang Y S and Schulz R J, The measurement of radiation dose distrubtions by nuclear magnetic resonans imaging, 1984 Magn. Reson. Imaging 2, 244.

[3] Schreiner L J, Review of Fricke gel dosimeters, Journal of Physics: Conference Series **3** (2004) 9–21.

[4] G. Gambarini, C. Birattari, D. Monti, M.L. Fumagalli, A. Vai, P. Salvadori, L. Facchielli and A.E. Sichirollo Fricke-Infused Agarose Gel Phantoms for NMR Dosimetry in Boron Neutron Capture Therapy and Proton Therapy Radiat Prot Dosimetry (1997) 70 (1-4): 571-575.

# Basic sciences agonize in Turkey!

**Fatma Akdemir[1], Aslı Araz[1], Ferdi Akman[2] and Rıdvan Durak[1]**

Sciences Faculty, Department of Physics, Atatürk University, Erzurum, Turkey
Vocational School of Technical Sciences, Department of Electronic Communication Technology/Bingöl University

**Abstract.** In this study, changes from past to present in the departments of physics, chemistry, biology and mathematics, which are considered as the basic sciences in Turkey, are shown. The importance of basic science for the country emphasized and the status of our country was discussed with a critical perspective. The number of academic staff, the number of students, opened quotas according to years for these four departments at universities were calculated and analysis of the resulting changes were made. In examined graphics changes to these four departments were similar. Especially a significant change was observed in the physics department.

Lack of jobs employing young people who have graduated from basic science is also an issue that must be discussed. There are also qualitative results of this study that we have discussed as quantitative. Psychological problems caused by unemployment have become a disease among young people. This study was focused on more quantitative results. We have tried to expain the causes of obtained results and propose solutions.

## 1. Introduction

In recent years, basic sciences have been seriously shaken in Turkey. In universities the situation of Physics, Chemistry, Biology and Mathematics departments is vexatious. In the study, we show that there are quotas for these four basic sciences in 2010-2015 and account for the amount of entered students and the changes according to years. Also, we accounted for the number of academic staff in these faculties and graphed according to years. This results are argued and solution suggested.

Basic sciences are basic of country. If there is no basic science in a country, the country is destroyed. There is need to take precautions to protect these basic sciences which are destroyed day by day. In the last years, most of the Faculty of Sciences, which is obligatory to open university by YÖK (Higher Education Institution), were closed because students do not choose these departments. The students do not find jobs when they are graduate. Those who are graduate these departments, face lack of jobs.

## 2. Quantitative datas

In 2010-2015 years, opened quota amounts and entered the amount of students are accounted and shown as graphically in Physics, Chemistry, Biology and Mathematics departments (Figure 1 and Figure 2).

When the Figure 1 is researched, seen that Physics is least but Mathematics is most. The amounts of opened quota to Chemistry and Biology departments are shown parallelism. It is seen that all departments are felt from 2010 to 2015, also in 2013 sudden falling is seen.
When the Figure 2 is researched, seen that the amount of entered students is least in Physics, but seen to the most in Mathematics. The amount of entered student in all departments are falling continuous unto 2015.
When the both graphic datas are researched, it is seen that none of opened quotas is full.

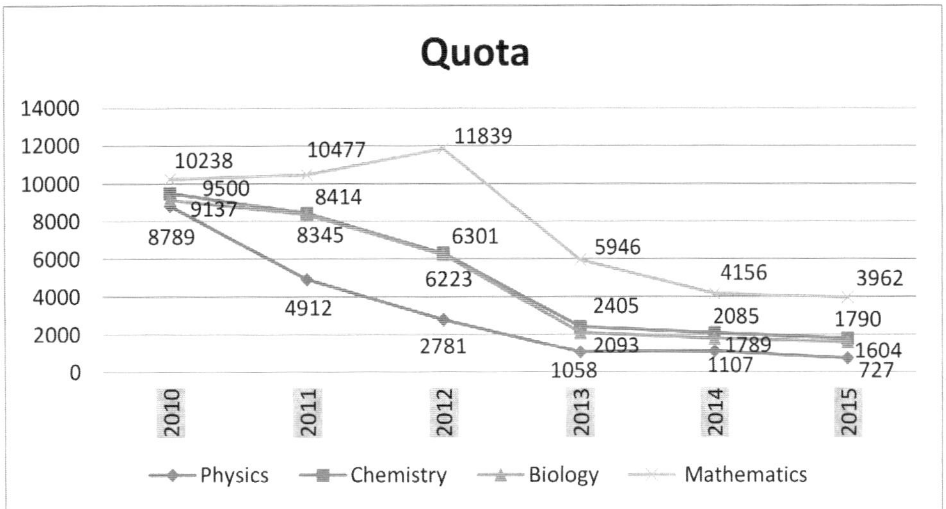


**Figure 1.** The number of quotas opened to year in departments of Physics, Chemistry, Biology and Mathematics [1, 2, 3].

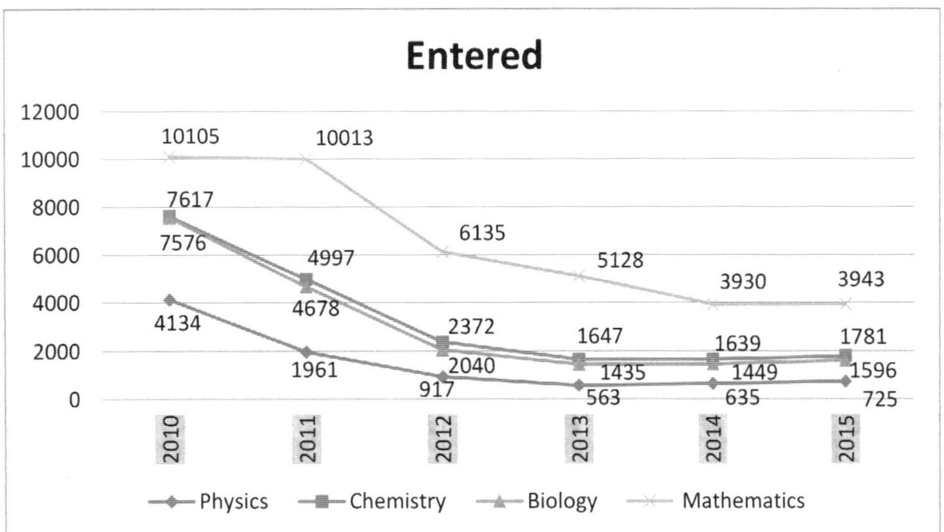


**Figure 2.** The number of entered students to year in departments of Physics, Chemistry, Biology and Mathematics [1, 2, 3].

Below, it is given that the departments of Physics, Chemistry, Biology and Mathematics are opened quotas by years and the amounts of entered students.

When the graphic is researched, the difference is big between the opened quotas in 2012 in these four departments and the amounts of entered students. Also, in 2015 the opened quotas to departments and the amounts of entered students are the least. It is seen to seriously falling from 2010 to 2015 years in each four departments. When the amounts of quota is looked in, Mathematics has the least falling in 5 years, Physics has the most falling. It is observed to same falling amount in Chemistry and Biology departments.

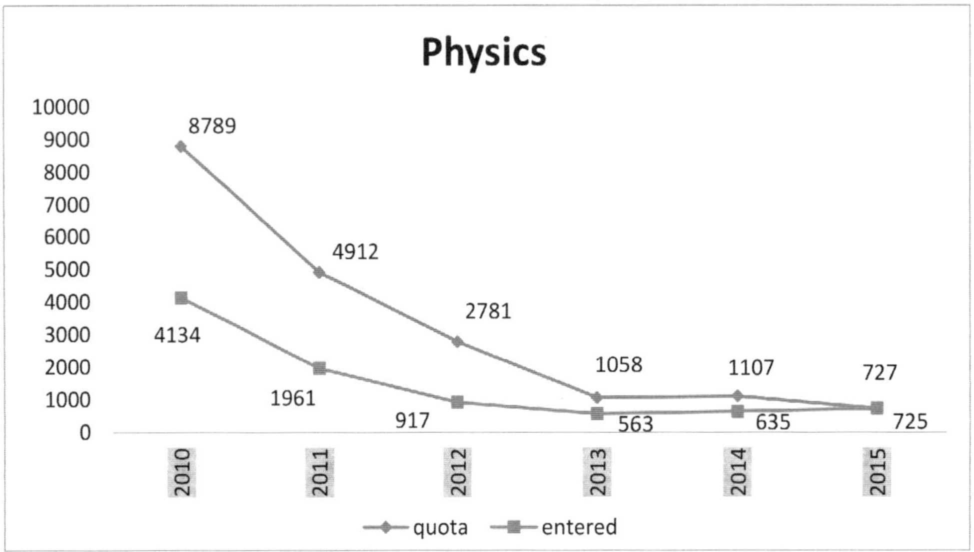


**Figure 3.** The opened quotas and the number of entered students to years in Physics Department [1, 2, 3].

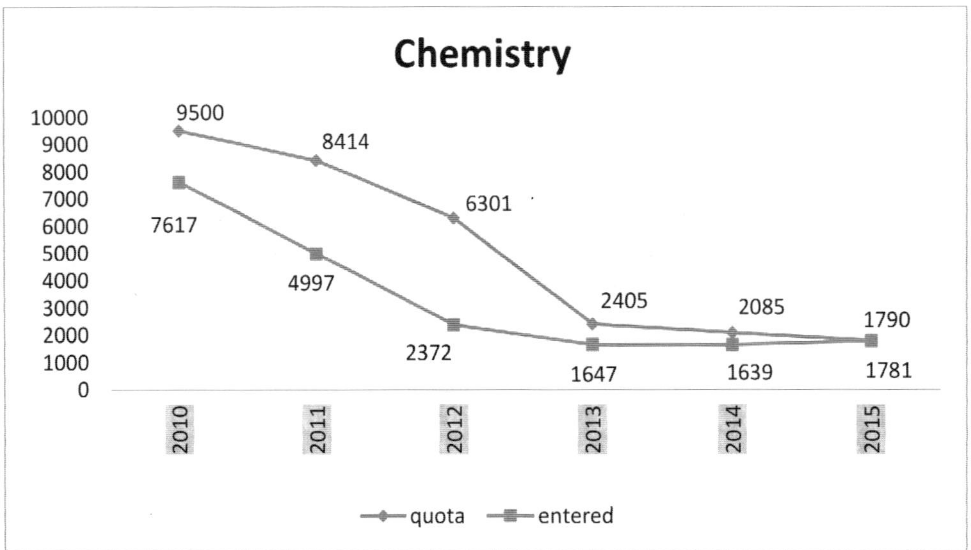


**Figure 4.** The opened quotas and the number of entered students to years in Chemistry Department [1, 2, 3].

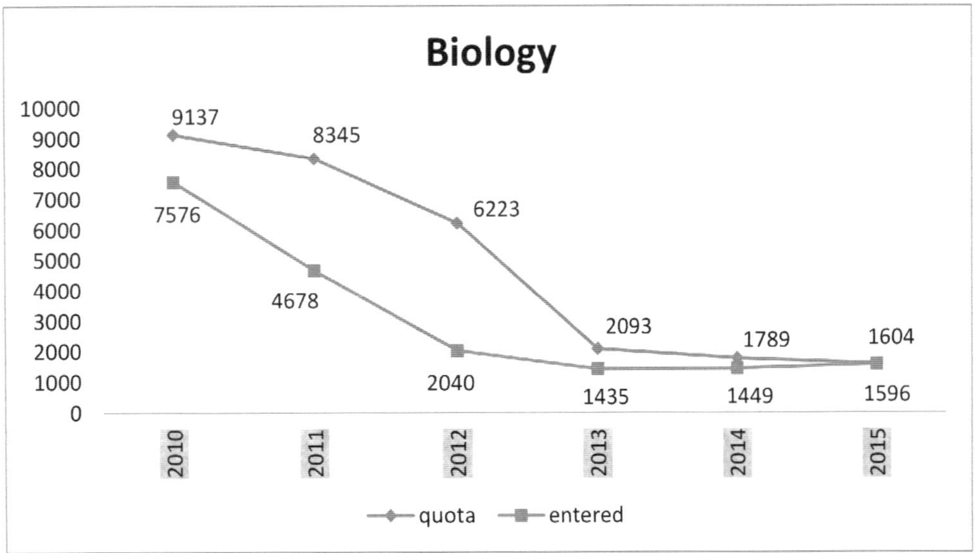


**Figure 5.** The opened quotas and the number of entered students to years in Biology Department [1, 2, 3].

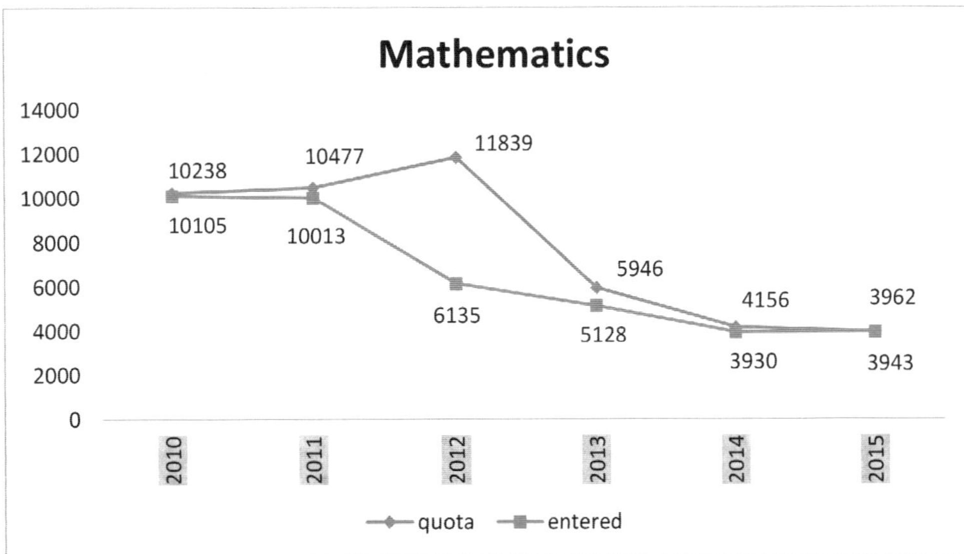


**Figure 6.** The opened quotas and the number of entered students to years in Mathematics Department [1, 2, 3].

Below, in Figure 7 it is given to the amounts of academic staff in Physics, Chemistry, Biology and Mathematics departments. When the graphic is researched, it is seen to Biology has the least academic staff, Mathematics has the most ones. It is seen to the amounts of academic staff are falling apart from Mathematics departments.

In our country, there are 193 universities with private university. If we look at the amount of universities, we can see to how the amounts of academic staff fallen in basic sciences.

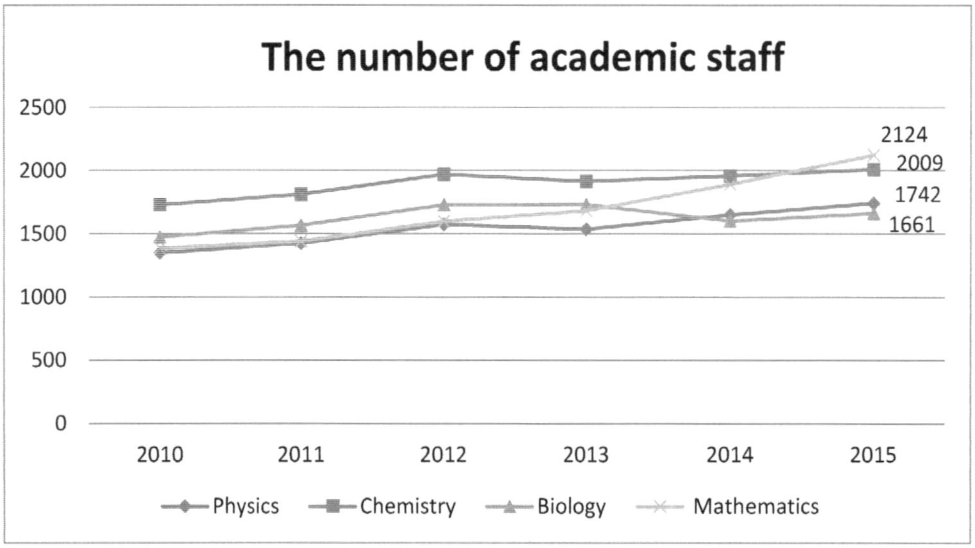


**Figure 7.** The number of academic staff according to years [4].

## 3. Conclusion and recommendation

We tried to make an analysis of our quantitative research which is calculated of opened quotas and number of entered students and also the numbers of academic staff in these departments between the 2010-2015 years in Physics, Chemistry, Biology and Mathematics departments of the state and private universities in Turkey. When the graphics are looked in, Mathematics Department is able to survive although it was difficult and destroyed department is Physics Department in these four departments. Especially after the year 2012, it is observed to a serious decline in the data values.

The result we obtained in this study, the basic sciences in Turkey is doomed to extinction. The sophistication of a country is directly proportional to the basic sciences. Without basic science is not scientific production. Without scientific production the country plays a role as a consumer, not producer and it becomes dependent on the outside. The problem is specific and clear. The life water should be given as soon as possible to the basic sciences. Therefore, the education system should be updated. Students should be encouraged to these departments when they are in secondary education. Above all, the youngs should not live in unemployment concern when they graduate from these departments. Particularly, job employment should be created for these departments. We hope, when is taken these precautions the basic sciences will find life and scientific production will begin to flourish.

## References

[1] Student Selection and Placement Center (ÖSYM) 2010-2015 Academic Year Higher Education Statistics Book
[2] Student Selection and Placement Center (ÖSYM) From 2010 until 2015 quantitative datas for the results of ÖSYS (Student Selection and Placement Examination)
[3] Student Selection and Placement Center (ÖSYM) From 2010 until 2015 ÖSYS (Student Selection and Placement Examination) higher education programs the smallest and the largest scores list
[4] Academic staff information on web page of universities.

# Adapting smart phone applications about physics education to blind students

**M. Ş. Bülbül**[1], **N. Yiğit**[2] **and B. Garip**[3]

[1]KafkasUniversity, Kars, Turkey,msahinbulbul@gmail.com
[2]Karadeniz Technical University, Trabzon, Turkey, nevzatyigit@yahoo.com
[3]Middle East Technical University, Ankara, Turkey, belkis@metu.edu.tr

E-mail: msahinbulbul@gmail.com

**Abstract.** Today, most of necessary equipment in a physics laboratory are available for smartphone users via applications. Physics teachers may measure from acceleration to sound volume with its internal sensors. These sensors collect data and smartphone applications make the raw data visible. Teachers who do not have well-equipped laboratories at their schools may have an opportunity to conduct experiments with the help of smart phones. In this study, we analyzed possible open source physics education applications in terms of blind users in inclusive learning environments. All apps are categorized as partially, full or non-supported. The roles of blind learner's friend during the application are categorized as reader, describer or user. Mentioned apps in the study are compared with additional opportunities like size and downloading rates. Out of using apps we may also get information about whether via internet and some other extra information for different experiments in physics lab. Q-codes reading or augmented reality are two other opportunity provided by smart phones for users in physics labs. We also summarized blind learner's smartphone experiences from literature and listed some suggestions for application designers about concepts in physics.

## 1. Introduction

The current way of communication is mostly based on smartphones which includes applications like software programs of personal computers. These applications vary from navigation to documentation. On the contrary to push-button telephones, smartphones have no button and this seems unusable feature for blind users. With appropriate model of smartphone which includes current operating system of Android, fast and accessible interface, some applications such as talkback help blind users to control their smartphone with different type of touching. Although there is no button on the screen, different type of touching on the screen are defined in the applications for blind users.

Application stores via internet like Google Play includes lots of applications for different educational purposes. For instance, you may download painting apps or animations about solar system or human body from any application stores and use them without internet connection. In this paper, we only focused on educational applications related with physics education. We selected applications from different content of physics which physics teachers or students may use to gather data and analyze it. The smartphones with sensors and applications are kind of laboratory tools for users; however, with this study we aim to analyze if there is enough physics education applications for blind users and how should we transform these applications for inclusive learning environments where both sighted and blind users enroll in.

### *1.1. Previous studies*

In this section, some projects about smartphone usages and other previous studies are given. Firstly N.E.T (New Media Experimental Tools) project which is located within the framework of situated learning theory is analyzed [1]. They provided an overview of possible usage of smartphones with example of production of an acoustic beats by Audacity or Cool Edit Pro, making infrared radiation visible and a free-falling smartphone emits a tone from a sound source by Spear. For gravitational acceleration, doing the experiment in a recreation park with free-fall tower may be more appropriate than small distance from the floor [2]. Secondly in the document from the "Science on Stage" project there are different smartphone applications not only for physics but also other science disciplines. In the guide about using smartphones in science education which is the product of the project and named "smartphones in science teaching istage 2" includes distance measuring and parallax method by stellarium, theodolite, angle meter and compass, measurement of the heavens by angle meter, solar walk, star chart, GPS Essentials, Theodolite and analyzing spectral sounds by iAnalyzer Lite and sound spectrum analyzer [3].Out of these projects, measuring acceleration with a turntable [4], measuring period of pendulum [5] or measuring Doppler effect and illuminance of a light source [6] are some other examples for using smartphones in physics education but none of these smartphone projects are appropriate for blind learners.

## 2. Methodology

:This study is a kind of survey for an instructional design and we investigated free smartphone applications with English key words from physics content in Google play application store according to their ratings from five star and number of downloads. These applications are compared with the possibility of usage by blind users and we tried to generate scenarios to to enable the usage of application for bothsighted and blinds users.

## 3. Findings

In this section, we aim to share a table including selected applications according to methodology and some additional explanations about how to adapt for inclusive usage.

### *3.1. Applications for physics education*

Although each day more and more applications are uploaded to application stores we collected some applications in Table 1 to give an overview about possible applications in physics education. There is no full or partially supported applications for blinds.

**Table 1.** Some of smartphone applications for physics education

| *Content in Physics* | *Logo of Application* | *Name of Application* | *Ratings* | *Number of Downloads* | *Sighted (Non-supported)* | *Blind (Full Supported)* | *Both (Partially Supported)* |
|---|---|---|---|---|---|---|---|
| Distance and area measurement | | GPS Fields Area Measure | 4.3 | 10000 | + | | |
| Velocity measurement | | Radar: Speed Gun | 3.5 | 50000 | + | | |
| Sound measurement | | Sound Meter | 4.2 | 200000 | + | | |
| Light measurement | | Light Meter | 4.0 | 50000 | + | | |
| Color Identification | | Color Meter | 4.2 | 10000 | + | | |
| Magnetic Field Measurement | | Gauss Meter | 4.2 | 100000 | | | + |

**Table 1.** (Cont.) Some of smartphone applications for physics education

| *Content in Physics* | *Logo of Application* | *Name of Application* | *Ratings* | *Number of Downloads* | *Sighted (Non-supported)* | *Blind (Full Supported)* | *Both (Partially Supported)* |
|---|---|---|---|---|---|---|---|
| Vibration Measurement | | Vibration Meter | 4.2 | 1000000 | + | | |
| Lens | | Smart Magnifier | 3.9 | 1000000 | + | | |
| Ruler | | Smart Ruler | 3.9 | 1000000 | + | | |
| Night Vision | | Night Vision Camera | 3.5 | 10000000 | + | | |
| Level | | Smart level | 4.3 | 100000 | + | | |
| Stopwatch and Timer | | Stopwatch and timer | 4.3 | 5000000 | + | | |
| Mirror | | Smart mirror | 3.7 | 500.000 | + | | |
| Hygrometer | | Physics Toolbox Hygrometer | 3.8 | 1000 | + | | |
| Converter | | Unit converter | 4.3 | 5000000 | + | | |
| Flashlight | | Flashlight | 4.4 | 1000000 | + | | |
| Thermometer | | Thermometer | 3.6 | 1000000 | + | | |
| Barometer | | DS Barometer - Air Pressure | 3.9 | 100000 | + | | |

| | | | | | | | |
|---|---|---|---|---|---|---|---|
| | 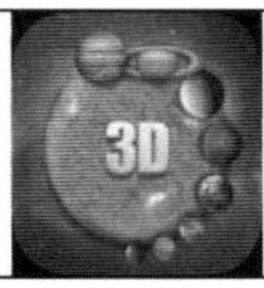 | Solar System 3D Viewer | 4.4 | 100000 | + | | |

### 3.2. *Adaptations for inclusive usage*

It is true that day after day a new sensor or/and applications are designed, using these smartphone applications in their lessons are helpful for physics teachers. With the help of these current applications and sensors in the smartphone teachers may transform anyplace to a laboratory. However, blind learners need some tactual and auditory support. In this section, two main adaptation are given; adaptations with a tool or without a tool. In both adaptations in inclusive environments, blind learners need sighted friend to vocalize data on the screen because none of these applications include screen readers.

*3.2.1. With just smartphone.* Each smartphone has frond and back sides which a blind learner may define and change. Moreover, turning smartphones parallel to the desk also is a possible action for blind learners. These tactile skills let them to attend an experiment. For instance, if you put a wire near a smartphone and take its position parallel to that wire, after giving current to the wire magnetic field will be changed and compass in the smartphone will also be changed (fig 1.). Sighted friend may support blind learner by changing the direction of smartphone according to new north pole (from A position to B position).

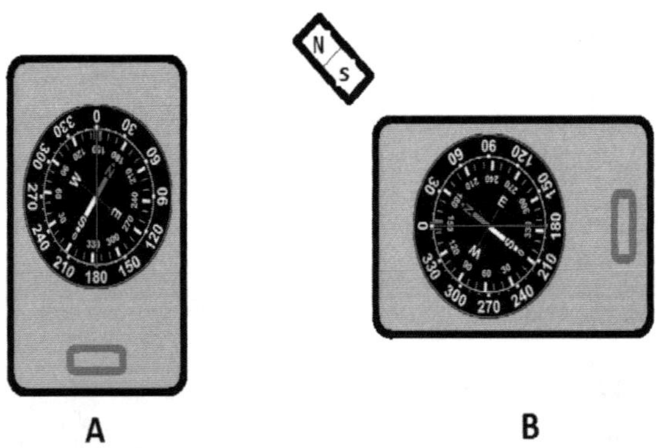


**Fig 1.** An adaptation for magnetic field perception of blind learner.

*With other tools.* Most popular tool among smartphone users is selfie stick. Users control smartphone with remote button on the stick. With the help of this button and moving the stick may be controlled by blind learner (fig 2.) and sighted friend may read what he or she sees. For instance, blind learner may take one toy car in left hand and take the stick in right hand and move together. Then, sighted friend help reading the speed.

It is possible to create a role and an experiment for inclusive usage with all listed applications, for instance, put a car toy on the front side and draw up the stick to vertical position. After opening "smart level" application they may decide whether the car will fall down or not due to slope of smartphone under car toy.

**Fig 2.** An adaptation for blind learner with selfie stick.

In all laser pointers there is a lens and you may put this lens with iron hair pins behind the smartphone on the inner lens of it to get an optic magnifier. In this activity, blind learners should be user with finger movements and sighted user should direct the user and describe what he or she watches.
If sighted learner cover flashlight with transparent tape and in each layer print the tape with blue before opening, sighted learner may read what blind learner draw on a paper with ultraviolet pen. In this scenario, blind learner is encoder and sighted learner is decoder.

## 4. Conclusion

Whether using selfie stick, other tools or without any tool, deciding the role of both sighted and blind users brings inclusive usage of smartphones. Generally sighted learner should read what he or she sees in the screen and transform graphs in the screen to tactual graphs [7]. However, blind learners may take other responsibilities and work cooperatively with their sighted peers during the application. Teachers should be careful to include blind learners to all activities and experiments. For instance, if a teacher want to measure the noise around, let sighted learner to read the screen and transform the graph to tactual version for his or her blind friend and let the blind create a noise. All more scenarios depend on teachers' creativity and believes that blind learners also should be included in all activities and they may do any appropriate duty with correct directions.

## References

[1] Vogt, Patrik, and Jochen Kuhn. "Analyzing free fall with a smartphone acceleration sensor." *The Physics Teacher* 50.3 (2012): 182-183.
[2] Kuhn, Jochen, and Patrik Vogt. "Smartphones as experimental tools: Different methods to determine the gravitational acceleration in classroom physics by using everyday devices." *European Journal of Physics Education* 4.1 (2013).
[3] Science on Stage Europe. 2014. Smartphones in science teaching. Berlin, Germany: Science on Stage Deutschland e.V. http://bit. ly/1MXWXTH
[4] Vieyra, Rebecca, et al. "Five challenges that use mobile devices to collect and analyze data in physics." http://static.nsta.org/files/tst1509_32.pdf
[5] Briggle, Justin. "Analysis of pendulum period with an iPod touch/iPhone."*Physics Education* 48.3 (2013): 285.
[6] Klein, P., M. Hirth, and J. Kuhn. "Classical experiments revisited: smartphones and tablet PCs as experimental tools in acoustics and optics." *Physics Education* 49.4 (2014): 412.
[7] Bülbül, M. Ş. "Görme engelli öğrenciler ile grafik çalışırken nasıl bir materyal kullanılmalıdır?." *Fen Eğitimi ve Araştırmaları Derneği Fen Bilimleri Öğretimi Dergisi* 1.1 (2013): 1-11.

# A discussion on velocity–speed and their instruction

**Ali Yıldız**
Atatürk University, Kazım Karabekir Faculty of Education, Erzurum, Turkey

E-mail: ayildiz@atauni.edu.tr

**Abstract.** This study was conducted to investigate how to teach velocity and speed effectively, with which activities and examples. Although they are different quantities, they are generally used in the same meaning. Study data and the quantities discussed were obtained from the examination of documents such as scientific articles and books about the instruction and they were examined by descriptive analysis approach. Expository instruction was supported with writing to learn activities and an approach actualized in seven stages was suggested so that velocity and speed could be understood at an anticipated level. At each stage, possible practices were explained; and especially at the fifth stage of the study, a detailed example on distance, displacement, velocity and speed promoted the understanding of presented quantities much more easily and correctly with their critical properties, thus students would be able to associate it with their prior knowledge. Moreover, it was anticipated based on these reasons that the example could be used as a tool to actualize permanent learning. At the last stage of the study, it was considered that having students write a letter and a summary to young respondents could support the practices in the previous stages, and also it would help promoting long-term retention of the learned concepts.

## 1. Introduction

Although speed and velocity are different quantities, they are used in the same meaning and this is something frequently encountered. It can be stated that among the individuals who receive higher education, those who use speed and velocity as synonyms or without knowing the difference or differences between them is far beyond the estimated numbers. What makes this observation more interesting is that this attitude has been identified among the university students who took mechanics course. In a qualitative study [1], science students who took mechanics course stated that "the speed of light" which is defined with the constant $c = 3x10^8\ m/s$ "is true because we were always taught like that and it is explained in that way in the sources". The percentage of the students who made this justification with that statement was 93.3% and it is really difficult to evaluate this explanation in another way. In the same study, the constant c ($c = 3x10^8\ m/s$) is called "the speed of light in vacuum" in the physics books studied in all secondary school classes, in the questions asked in the examinations (Undergraduate Placement Examination (LYS)) administered by the Measuring, Selection and Placement Center (OSYM), and in most of the national books written for the physics courses included in some higher education undergraduate programs. Similarly, some books of international level [2, 3, 4, 5] are used as a reference book for the physics course at undergraduate level. The constant c included as "Speed of light in vacuum" in the original form of these books is usually translated incorrectly into Turkish as "the speed of light in vacuum" or "the speed of light in free space" [1]. A similar misattribution is done for the speed of light. The course books, which incorrectly call the magnitude of speed of propagation of sound at 1 atmospheric pressure and $20^0$C or its velocity at 343 m/s value [5] as "the speed of sound in air" and also questions asked about sound in the exams, are frequently encountered.

It was found in a study [6] that pre-service science teachers had difficulties about understanding the main principles of scalar and vector quantity, scientific definition of speed, constant speed, and the relation between speed and velocity; and they hold the views that "on the condition that numerical value does not change, the speed will be constant and the change in direction will not change the speed". In another study conducted on the primary school pre-service mathematics teachers [7], it was found that the concepts of speed and velocity were not understood fully and speed was mostly used in place of velocity. It was determined in some studies that [8, 9, 10] generally all students at every level, but mainly the students studying in high school and at university, had difficulties in understanding the concept of speed. It was emphasized in another study [11] that even students in upper grades had difficulties in understanding some common quantities like speed.

Among the reasons why speed and velocity are used in the same meaning, we can point to the spoken language, wrong documents used, translation mistakes, and not understanding the main principles of scalar and vector quantity fully [1,7]. However, although pre-service teachers took mechanics course at university, it would be more realistic not to limit their evaluation of speed and velocity as different quantities to only abovementioned reasons.

It can be stated that the same condition is true for distance and displacement (change in position). We need to explain that distance and displacement are different quantities, and clear and suitable examples should be given for their explanations. Since, explaining the difference or differences between distance and displacement clearly with appropriate examples and describing average speed as distance travelled divided by the time elapsed (Total distance / Total time) and average velocity defined ($\Delta x/\Delta t$) as the displacement ($\Delta x$) over time interval may help these quantities to be understood much more easily.

It is predicted that using documents naming the constant c, the speed of light in vacuum or the magnitude of the propagation of the speed of light in free space, as "the speed of light", the value 343 m/s as "the velocity of sound" instead of "the speed of sound", explaining the differences between average velocity and average speed without identifying the differences between distance and displacement or not indicating that these quantities are different with suitable examples will cause problems in the teaching of these subjects. It is inevitable that students, pre-service teachers, teachers, researchers, and instructors who use such documents mentioned above will have wrong constructions in their minds about distance and displacement, and velocity and speed or experience misconceptions.

*The Purpose of the study*

This study was carried out to examine how to actualize the definitions of distance-displacement, speed-velocity quantities, their crucial properties and the instruction of the formulae more effectively with appropriate examples and activities.

## 2. Method

Study data were obtained by analysing the relevant documents to understand the definitions of distance and displacement and speed and velocity, their critical properties, formulae related to them and the differences between them. The findings of a qualitative study are usually composed of data gathered from interviews, observation and documents [12]. During the interviews carried out, some participants can feel uncomfortable due to the existence of the researcher or the tools used for recording or they may not consider appropriate being recorded. Therefore, analysis of documents provides advantages under such circumstances. Document analysis is preferred to observations and interviews, because the information can be obtained in a very short time, it is economical and easy. Any book, journal and article found in the libraries can become an available data and information source for every researcher who loves her/his job [12]. Many documents such as columns, course books, organizational documents, annual reports, scientific study reports, and articles, which are revised, checked for their originality, organized and arranged by the experts in the field, can become a source of data, and using these documents can increase the validity and reliability of the qualitative study [13]. Documents, which include the definitions, concepts, and formulae about distance, displacement, speed and velocity, were examined with regard to descriptive analysis approach in the study so that these quantities could be taught and understood easily.

## 3. Findings and interpretation

Scalars are quantities that are fully described by a magnitude or a numerical value only, and vectors are quantities that are fully described by both a magnitude and a direction as physical quantities.

Distance is a scalar quantity, but displacement is a vector quantity. Displacement is referred to as the simplest vector quantity [4]. Displacement vector is defined as both the distance and direction of an imaginary motion along a straight line from the initial position to the final position of the point. Distance is the ground an object covers linearly or curvilinearly during its motion from the initial point to the final point. Speed is a scalar quantity and velocity is a vector quantity. This is not the only difference between them. Average velocity and average speed are calculated in this way. Average velocity is displacement ($\Delta x$) over total time ($\Delta t$) and average speed is distance over time. Since displacement and numerical values of the distance are usually different, the magnitude of average velocity and average speed are different.

There are quasi-experimental studies [14, 15, 16, 17] in the literature conducted on examining pre-service science teachers' level of understanding of some subjects they learned in the introduction to modern physics course and the effect of writing activities on their academic achievement. Research findings revealed that pre-service teachers had low understanding levels about the subjects, which were examined. The views of the participating pre-service teachers about writing activities were gathered with the follow-up questions given with the post-test in written form. 87.5% of the pre-service teachers stated that they understood the Compton Effect, which they wrote letters about, 91.7% of them understood the photoelectric effect, 91.4% of them understood the uncertainty principle and 87.2% of them stated that they understood the special theory of relativity, about which they wrote a summary. They said that writing activities, which they used, were effective in learning the subjects, about which they wrote a letter and a summary.

Moreover, the use of writing activities, which facilitate materializing conceptual change within students [18], enable retention of knowledge [19, 20] and help designing student-centred learning, in which students discover and construct knowledge, can be beneficial. The use of these activities for the teaching of difficult subjects of the modern physics, which are difficult to understand and have effects on increasing academic achievement [14, 15, 16, 17], also reveal that they can be used for teaching quantities such as distance, displacement, velocity, and speed.

Teaching of distance, displacement, speed and velocity, the definitions of quantities, critical properties and equations can be actualized by following the stages suggested for the instruction of uncertainty principle [21], dimensional analysis, [22], and $E = mc^2$ paradigm.

**I)** At the first stage, students will be informed about the gains and operations that will be performed, so they will get prepared for learning [24]. As stated in a research study [21], students are informed about the subject to be presented, its importance, and the gains. Considering and discussing the student responses to the question "Suppose that there is an athlete who gets prepared for a race in a circular race track with a length of 600 m. He completes one lap around the track at the same pace in 300 s (returns to his starting point). In that condition, what is the athlete's distance travelled displacement, and average velocity and average speed?", finding the correct answer can raise students' interest and curiosity in the subject in the first stage.

**II)** Students can be given suitable and necessary explanations or suggestions about the quantities in the formulae related to average speed and average velocity, concepts and symbols ( $\Delta$, $\Delta x$, $\Delta t, v_{average}$).

**III)** "What are distance, displacement, speed, and velocity?" The names of these quantities are written on the board with their symbols in order and they are defined clearly and directly.

**IV)** The formulae for average speed and average velocity are written on the board and the quantities in each formulae and the relation between them are explained.

**V)** Appropriate examples and non-examples related to distance, average speed and average velocity must be given. It must be considered that appropriate examples must be used as well as using lecturing and talk in expository instruction [26].

*Example:*

Suppose that an athlete runs in a rectangular racetrack with a 60 m width and 80 m length. He begins running from the point A and he runs at the same pace (2 metres in 1 second) only one lap and he completes this one lap in 140 seconds.
What is the distance taken, displacement, average speed and velocity when
**a)** the athlete reaches the point B.
**b)** the athlete reaches the point C.
**c)** the athlete reaches the point D.
**d)** the athlete reaches the point A.
Please calculate.

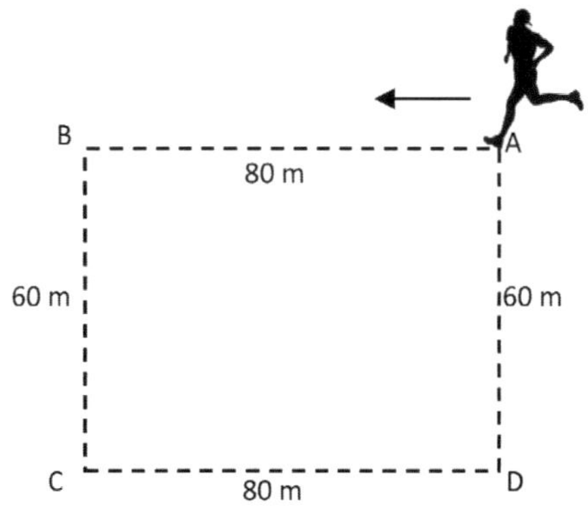


*Solution:*

a)

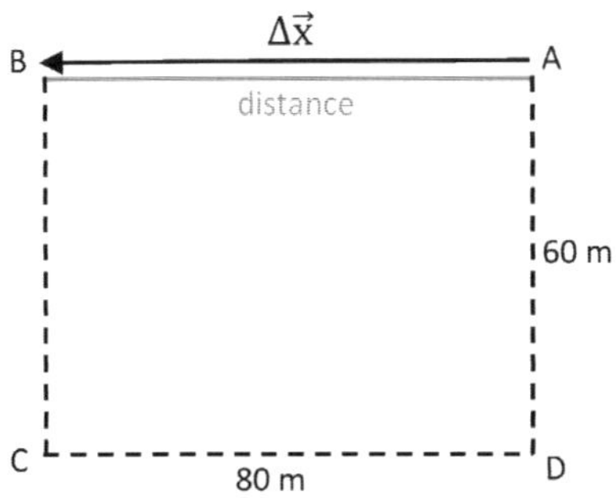


Distance travelled (length of the green line) = 80 m
The size of the displacement ($\Delta$x), $\Delta$x=80 m
Average speed= Total distance / Total time
$= \frac{80m}{40s} = 2\ m/s$
The value of average velocity is calculated as,
$v_{average} = \frac{\Delta x}{\Delta t} = \frac{80m}{40s} = 2\ m/s$

b)

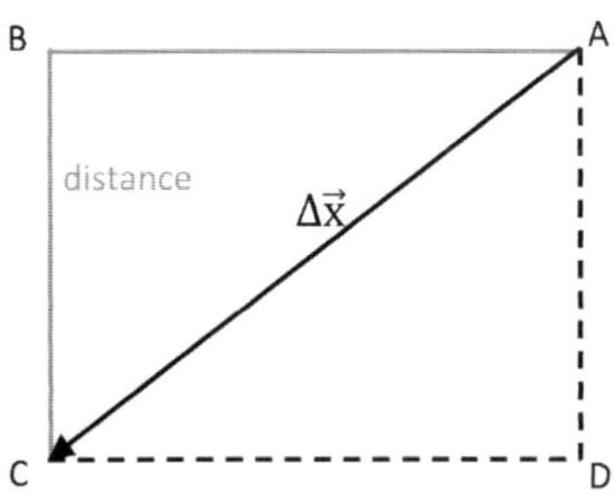


Distance travelled (length of the green line) = 80 m + 60 m = 140 m
The size of the displacement ($\Delta$x), $(\Delta x)^2 = (80)^2 + (60)^2 = 6400 + 3600$
$(\Delta x)^2 = 10000\ m^2 \rightarrow \Delta x = 100\ m$
Average speed= Total distance / Total time
$= \frac{140\ m}{70\ s} = 2\ m/s$ because he runs at the same pace.
The value of average velocity is calculated as
$v_{average} = \frac{\Delta x}{\Delta t} = \frac{100\ m}{70\ s} = 1.42\ m/s$

c)

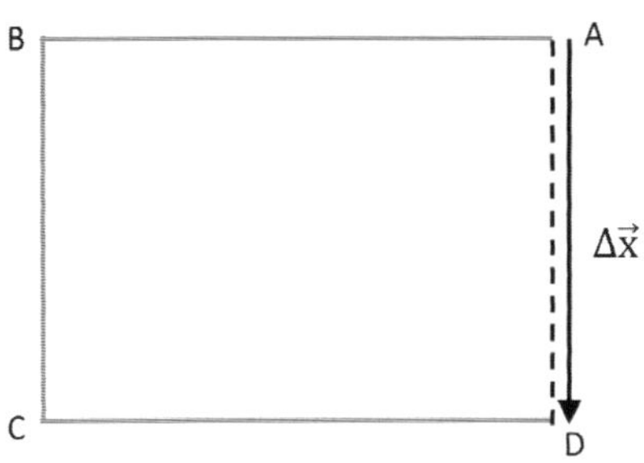


Distance travelled (length of the green line) = 80 m + 60 m + 80 m= 220 m
The size of the displacement ($\Delta$x), $\Delta$x= 60 m
Average speed= Total distance / Total time
$= \frac{220\ m}{110\ s} = 2\ m/s$ because he runs at the same pace.
The value of average velocity is calculated as
$v_{average} = \frac{\Delta x}{\Delta t} = \frac{60\ m}{110\ s} = 0.545\ m/s$

d)

| |
|---|
| Distance travelled (length of running track) = 80 m + 60 m + 80 m+ 60 m = 280 m<br>The size of the displacement ($\Delta x$) is zero ($\Delta x = 0$). Because the athlete is at point (A) where he begins to run, his first and last positions are at the same point ($x_1 = x_2$). $\Delta x = x_2 - x_1 = 0$<br>Average speed= Total distance / Total time = $\frac{280\ m}{140\ s} = 2\ m/s$<br>When the athlete reaches point A (he completed one lapse), the average velocity is zero for displacement ($\Delta x = 0$). Because there is no displacement, we cannot mention average velocity, in other words, average velocity is zero ($v_{average} = 0$). |

As it is viewed in the above example, the only difference between distance and displacement is not the fact that distance is a scalar quantity and displacement is a vector quantity. Only in the option a of the example, the magnitudes of distance and displacement are equal. It can be seen in the other options (b, c, and d) that the calculated values of distance and displacement are completely different. Especially in the last option, although displacement is zero, the distance travelled is 280 m., and this proves that distance and displacement are different quantities. Similarly, the only difference between speed and velocity is not the fact that speed is a scalar quantity and velocity is a vector quantity. Only in the option a, while average velocity and average speed have an equal value (2 m/s), in other options (b, c, and d), while the value for the average speed is the same, different values are calculated for average velocity. In the last three options, the calculated values demonstrate that speed and velocity are different quantities. It can be more useful if the examples solved are supported with appropriate visuals, then the concepts of distance and displacement, velocity and speed and the relation between the quantities related to them in the formulae would be understood much better.

**VI)** Appropriate examples and problems about distance, displacement, average speed, and average velocity are solved by the teacher clearly and concisely. Immediately after, students are asked to solve different examples and problems.

**VII)** In the final stage, writing activities (letter, summary, diary, journal, poster, banner, essay and so on) can be used. Summaries or letters written to younger respondents [26] can enable students to understand crucial properties of quantities and concepts such as distance, displacement, average speed and average velocity, and the formulae about them in a meaningful and permanent way.

## 4. Results and discussion

Regarding the documents examined [1, 7, 8, 27], it can be considered that although velocity and speed are different quantities or concepts, using them in the same meaning may be attributed to such reasons as the spoken language, wrong definitions in the documents, wrong translations, not understanding the fundamentals of scalar and vector quantities and the thought that distance and displacement are the same quantities. It would be effective to teach speed, velocity, and their crucial properties via inclusive, simple, and clear examples, which include the differences between distance and displacement.

According to the findings of the studies carried out about the writing activities [14-20, 26], after the lessons in which speed and velocity are explained, students can be asked to write a letter or a summary to the students who are younger as writing activities. These activities can help these students to construct the difference (s) between distance and displacement, and speed and velocity in their minds correctly and make their learning permanent.

If there are wrong descriptions in the course books or documents used as reference about speed and velocity such as $c = 3x10^8\ m/s$ constant is "speed of light" and "343 m/s" is "speed of sound in air", teachers or instructors must certainly point out that they are wrong. In other words, warnings or explanations must be made by the experts or people in charge on time about the wrong descriptions in the reference documents.

As stated in a study [27], sometimes it may not be enough for the person who is going to translate, speak and write the language, which the original book was written in. If the work which is going to be translated is about physics and the person who knows, speaks, and writes in both languages is not a physicist, he will not know the differences between speed and velocity, so his translation will be probably faulty. Ethically, although people translate, speak and write in a language very well, they should not make an attempt to translate a topic, which they are not experts at or they should work with and consult to an expert of the field.

**References**

[1] Yıldız A 2014b Views of pre-service teachers on speed constant "$c=3.10^8$ m/s" *The Journal Of Academic Social Science Studies* 28 Autumn II **28** 13-21 doi:10.9761/JASSS2474
[2] Beiser A 2008 Modern fizigin kavramları 6. Baskıdan Ceviri (cev.: Gulsen Onengut) Ankara: Akademi Yayıncılık
[3] Serway R A and Beichner R J 2011 Fen ve muhendislik icin fizik-3, 5. Baskıdan Ceviri (cev. ed.: Kemal Colakoglu) Ankara: Palme Yayıncılık
[4] Young H D and Freeman R A 2008 Sears and Zemansk's university physics with modern physics12th Edition, Pearson Addison Wesley
[5] Halliday D, Resnick R and Walker J 2014 Fizigin temelleri 9. Baskıdan Ceviri, 1. Kitap, Cev.: Bulent Akınoglu ve Murat Alev, Ankara: Palme Yayıncılık
[6] Yıldız A, Büyükkasap E, Erkol M and Dikel S 2007 Science students' understanding levels of velocity, stable velocity, speed and displacement concepts *Erzincan Egitim Fakultesi Dergisi* **9** (2), 1-12
[7] Yıldız A, Büyükkasap E and Gunel M 2011 Preservice teachers' comprehension levels of circular motion *Journal of Qafqaz University:Philology and pedagogy* **32**, 97-103
[8] Borghi L, De Ambrosis A and Massara C I 1993 Understanding average speed: a study on students aged 11 to 12 years *Physics Education* **28**, 33- 38
[9] Gagliardi M, Gallina G, Guidino P and Piscitelli S 1989 Forze, deformazioni, movimento. Emme Ed Torino
[10] Mcdermott L C 1990 Research and computer-based instruction: opportunity for interaction. *American Journal of Physics* **58**, 452
[11] Arons A B 1990 A Guide to Inroductory Physics Teaching John Wiley New York
[12] Merriam S B 2013 Nitel arastırma: Desen ve uygulama icin bir rehber (cev. ed.: S. Turan) Ankara: Nobel Yayıncılık
[13] Yıldırım A and Simsek H 2011 Sosyal bilimlerde nitel arastırma yöntemleri Ankara: Seckin Yayıncılık
[14] Yıldız A and Büyükkasap E 2011a Prospective teachers' levels of understanding compton effect and the impact of writing activities for learning purposes on academic success *Uluslararası İnsan Bilimleri Dergisi* **8** (1), 1643-1664
[15] Yıldız A and Büyükkasap E 2011b The Level of understanding of the photoelectric phenomenon in prospective teachers and the effects of writing with learning" on their success rates *Educational Sciences: Theory & Practice* **11** (4), 2259-2274
[16] Yıldız A and Büyükkasap E 2011c Prospective teachers' levels of understanding heisenberg uncertainty principle and the ımpact of writing activities for learning purposes on academic success *Journal of Turkish Science Education* **8**(4),134-148
[17] Yıldız A 2012 Prospective teachers' comprehension levels of special relativity theory and the effect of writing for learning on achievement. *Australian Journal of Teacher Education* **37** (12) 15-28 doi:10.14221/ajte.2012v37n12.1
[18] Mason L and Boscolo P 2000 Writing and conceptual change. What changes? *Instructional Science* **28**, 199–226, Printed in the Netherlands: Kluwer Academic Publishers
[19] Klein P D 2000 Elementary students' strategies for writing-to-learn science. *Cognition and Instruction* **18**, 317–348

[20] Rivard P L and Straw B S 2000 The Effect of talk and writing on learning science: an exploratory study *Science Education* **84**, 566-593
[21] Yıldız A 2014c Uncertainty principle and its instruction *Turkish Studies* **9** (11), 603-612 doi:10.7827/TurkishStudies.7359
[22] Yıldız A 2015a Dimensional analysis and its instruction *NWSA-Education Sciences* **10** (2), 100-112 doi:10.12739/NWSA.2015.10.2.1C0635
[23] Yıldız A 2015b The paradigm of $E= mc^2$ and its instruction *NWSA-Education Sciences* **10** (3), 225-236 doi:10.12739/NWSA.2015.10.3.1C0644
[24] Erden M and Akman Y 2012 Gelisim ve öğrenme, 20. Baskı. Ankara: Arkadas Yayınevi
[25] Yesilyaprak B and Ucar E 2014 Ogrenmeden ogretime B Yesilyaprak (ed) Egitim psikolojisi (309-368) 11. Baskı. Ankara: Pegem Akademi
[26] Yıldız A 2014a Letter as a writing to learn activity and its effective use *Turkish Studies* **9** (5) pp 2097-2104 doi:10.7827/TurkishStudies.6979
[27] Yıldız A 2015c A Discussion on an expression written about dimensional analysis in a physics textbook *European J of Physics Education* **6** (3) pp 1-7

# *STEREO* observations of HD 90386 (RX Sex): a $\delta$-Scuti or a hybrid star?

**D Ozuyar[1,2], I R Stevens[2], G Whittaker[3] and V Sangaralingam[4]**

[1] Ankara University, Deparment of Astronomy and Space Sciences, Ankara, Turkey
[2] The University of Birmingham, School of Physics and Astronomy, Birmingham, U.K.
[3] The University of Toronto, Toronto Ontario, Canada
[4] The University of Montreal, School of Physics, Montreal, Canada

E-mail: dozuyar@ankara.edu.tr, irs@star.sr.bham.ac.uk, whittaker@astro.utoronto.ca, vinothini.sangaralingam@asc-csa.gc.ca

**Abstract.**
HD 90386 is a rarely studied bright A2V type $\delta$ Scuti star (V = 6.66 mag). It displays short-term light curve variations which are originated due to either a beating phenomenon or a non-periodic variation. In this paper, we presented high-precision photometric data of HD 90386 taken by the *STEREO* satellite between 2007 and 2011 to shed light on its internal structure and evolution stage. From the frequency analysis of the four-year data, we detected that HD 90386 had at least six different frequencies between 1 and 15 c $d^{-1}$. The most dominant frequencies were found at around 10.25258 c $d^{-1}$ (A $\sim$ 1.92 mmag) and 12.40076 c $d^{-1}$ (A $\sim$ 0.61 mmag). Based on the ratio between these frequencies, the star was considered as an overtone pulsator. The variation in pulsation period over 35 years was calculated to be dP/Pdt = 5.39(2) x $10^{-3}$ $yr^{-1}$. Other variabilities at around 1.0 c $d^{-1}$ in the amplitude spectrum of HD90386 were also discussed. In order to explain these variabilities, possible rotational effects and $\gamma$ Dor type variations were focused. Consequently, depending on the rotation velocity of HD 90386, we speculated that these changes might be related to $\gamma$ Dor type high-order g-modes shifted to the higher frequencies and that HD 90386 might be a hybrid star.

## 1. Introduction

$\delta$ Scuti stars are located at the intersection of the classical Cepheid instability strip and the main sequence (MS). Their luminosity classes vary between V to III, since some massive samples evolve towards the giant region and across the strip at higher luminosities. Spectral types of these stars range from A2 to F0 on the MS and from A3 to F5 in the giant region [1]. Also, the mass values are between 1.5 and 2.5 $M_\odot$ for $\delta$ Scuties with solar metal abundance and 1.0 – 2.0 $M_\odot$ for metal-poor samples. Therefore, it is known that these stars are either at the core or shell hydrogen-burning stages. They are mostly non-radial pulsators, but some $\delta$ Scuties show radial oscillations. The periods of the pulsations are between 0.02 days and 0.25 days, with a typical amplitude of around 0.02 mag [2].

HD 90386 is a rarely studied A2V type, bright $\delta$ Scuti star ($B$ = 6.78 mag, $V$ = 6.66 mag). Its variability was accidentally discovered by Jerzykiewicz [3] while it was being used as a comparison star for 23 Sex. He detected short-term variations with the period of 0.0799 days (A $\sim$ 0.01 mag) in the light curve (LC), and reported that their origins were due to either a beating phenomenon or a non-periodic variation in LC.

## 2. *STEREO* Satellite and Data Analysis

In this paper, we present high-precision photometric data of HD 90386 taken by the *STEREO* satellite to shed light on its internal structure and evolution stage. During the seasonal observations, 684, 678, 651, and 660 data points were derived from the satellite in 2007, 2008, 2010, and 2011, respectively. This yielded a total of 2,673 points, corresponding to 1,809 hours of observation over four years. The data were provided only from the *HI-1A* instrument on-board the *STEREO-A* satellite. It has been capable of observing background stars with the magnitude of $12^m$ or brighter for a maximum of 20 days and a useful stellar photometer which has covered region around the ecliptic (20% of the sky) with the field of view (FOV) of $20° \times 20°$. A more detailed description of the data preparation and background information on *STEREO* can be found in Sangaralingam and Stevens [4] and Whittaker et al. [5]. The nominal exposure time of the camera is 40 seconds, and putting 30 exposures together on board, a 40-minute integrated cadence has been obtained to transmit for each *HI-1* image (For details of the *HI* instruments refer to Eyles et al. [6]). Therefore, the Nyquist frequency of the data is around 18 c $d^{-1}$.

Since the *STEREO* satellite has observed the region quite close to the solar disk, data of HD 90386 were mostly affected by solar activities. Therefore, the distorted effects were removed from the LCs by using a $3^{rd}$ order polynomial fit, and observation points greater than $3\sigma$ were clipped with the help of a freshly coded pipeline in *IDL*. Four years of seasonal and combined time series were then analysed with the Lomb-Scargle (LS) algorithm. The results found were compared to those of *Period04*. For each periodogram, regional noise levels were determined by averaging the noise values in every 0.5 c $d^{-1}$, and a specific noise characteristic was thus established. Based on this characteristic, a significance level was calculated with 99% probability. Frequencies whose amplitudes were greater than this level were then detected. The detection precision in four-year combined data given in Figure 1 was around $10^{-5}$ c $d^{-1}$ in frequency and $10^{-4}$ mag in amplitude.

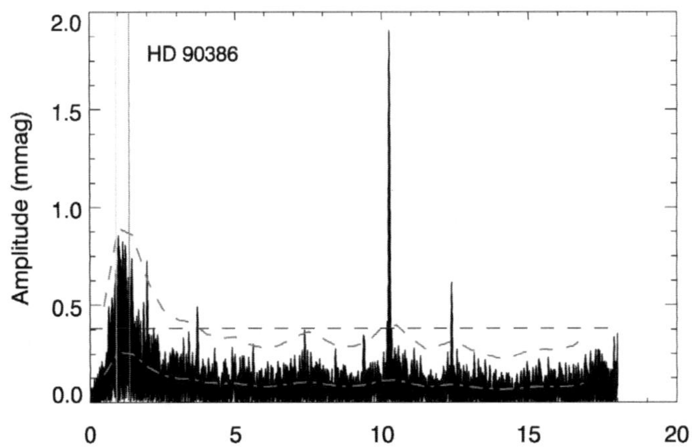


Figure 1: The amplitude spectrum of HD90386 obtained from four-year combined light curves. The mean noise level is plotted with red dash-dot line. Red dashed line is the significance level calculated based on the noise characteristics, and blue dashed line is the $4\sigma$ level. The estimated rotational periods are shown with green and red vertical lines. For more details, see the text.

Table 1: The frequencies of HD 90386, their amplitudes, SNR values, noise levels ($A_m$), and pulsation constants ($Q$).

| **HD 90386 Frequencies** | | | | | |
|---|---|---|---|---|---|
| No | Freq. (c d$^{-1}$) | Amp. (mmag) | SNR | $A_m$ (mmag) | $Q$ (days) |
| $f_1$ | 10.25258(2) | 1.92(10) | 16.88 | 0.11 | 0.031(6) |
| $f_2$ | 1.99122(5) | 0.74(10) | 4.00 | 0.19 | |
| $f_3$ | 12.40076(6) | 0.61(10) | 6.76 | 0.09 | 0.026(5) |
| $f_4$ | 3.67474(8) | 0.50(10) | 4.24 | 0.12 | |
| $f_5$ | 7.36338(11) | 0.36(10) | 3.43 | 0.10 | 0.043(8) |
| $f_6$ | 9.36921(11) | 0.34(10) | 3.92 | 0.09 | 0.034(6) |
| **2007** | | | | | |
| $f_1$ | 10.255(2) | 1.94(16) | 9.18 | 0.21 | |
| $f_2$ | 12.397(6) | 0.77(16) | 3.58 | 0.21 | |
| $f_3$ | 9.371(8) | 0.59(16) | 3.74 | 0.16 | |
| **2008** | | | | | |
| $f_1$ | 10.254(3) | 1.98(17) | 8.17 | 0.24 | |
| $f_2$ | 12.413(8) | 0.65(17) | 3.59 | 0.18 | |
| **2010** | | | | | |
| $f_1$ | 10.250(3) | 1.94(20) | 8.34 | 0.23 | |
| $f_2$ | 12.399(1) | 0.55(20) | 2.93 | 0.19 | |
| **2011** | | | | | |
| $f_1$ | 10.253(4) | 1.77(22) | 7.15 | 0.25 | |
| $f_2$ | 3.666(10) | 0.69(22) | 3.21 | 0.21 | |
| $f_3$ | 12.402(12) | 0.55(22) | 3.06 | 0.18 | |

The noise and related significance levels are presented with red dash-dot and red dashed lines in the figure, respectively. For a comparison, the constant significance level calculated from the mean noise value of the periodogram is shown with a blue dashed line. Additionally, the extracted frequencies, their amplitudes, signal-to-noise ratios (SNR) and noise ($A_m$) values are given in Table 1.

## 3. Results and Discussion

The frequency analysis of the combined data set showed that there were six frequencies having amplitude greater than the specific significance level in the amplitude spectrum (Table 1). As seen in Figure 1, four of these peaks were located between the frequencies of 5 and 15 c d$^{-1}$. Among them, the most dominant frequency was found at around 10.25 c d$^{-1}$ (0.0975 days) with the amplitude of 1.92 mmag. Also, the second powerful frequency showed up at 12.40 c d$^{-1}$ (0.0806 days) in our amplitude spectrum, and it was quite close to the result (0.0799 days) obtained by Jerzykiewicz [3]. Based on this alteration in the main frequency, it might be suggested that the star gradually switched pulsation modes from 12.50 to 10.25 c d$^{-1}$ over time. Furthermore, the intensity of the 12.40 c d$^{-1}$ in the seasonal LCs seemed to decrease between 2007 and 2011. As a result of this decrement, its amplitude observed in 2010 and 2011 fell below the significance level (Figure 2). However, their SNR values determined from *Period04* were over 4.0, and they were hence included in the analyses.

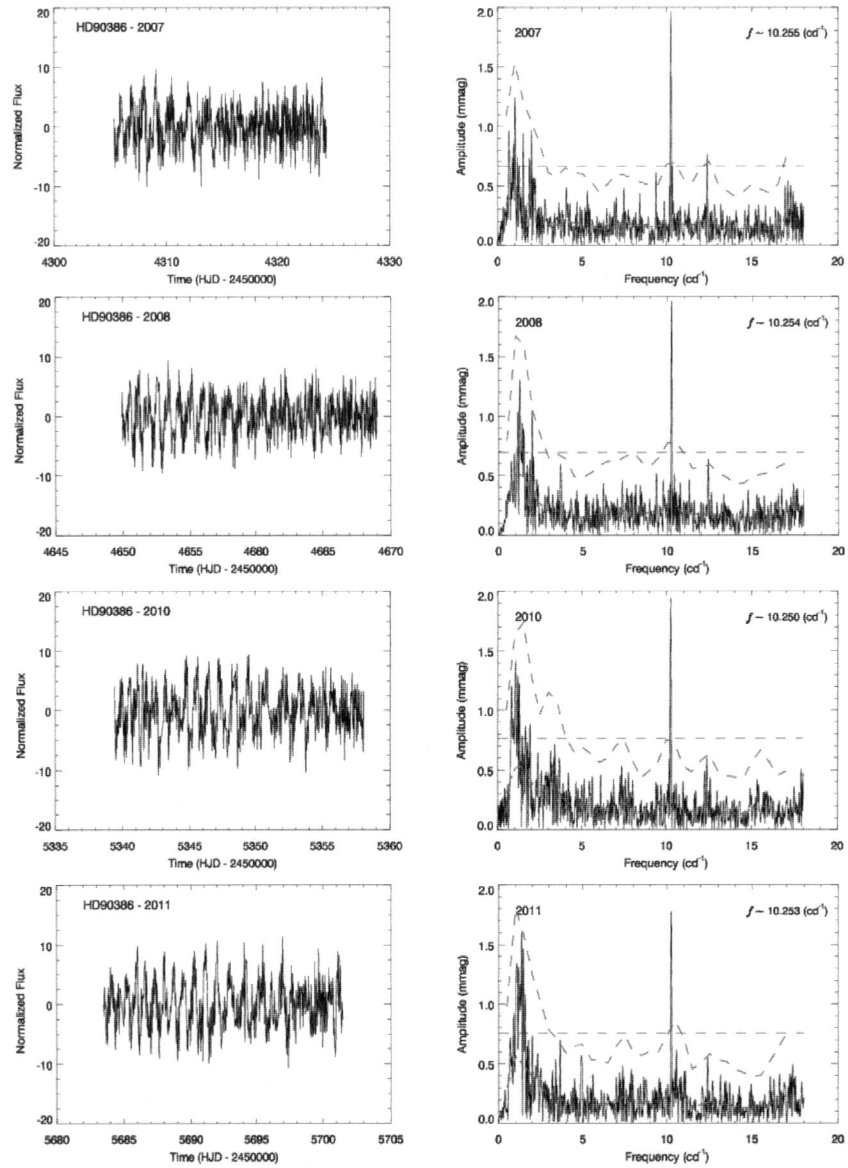


Figure 2: Seasonal light curves of HD 90386 and their amplitude spectra.

As seen in Figure 2, all seasonal LCs exhibit variabilities around the region of 1.00 c $d^{-1}$. Since there was not a sufficient number of studies on the physical, chemical, or orbital parameters of the star, interpretation of these variations might not be possible. However, Niemczura et al. [7] demonstrated that the projected rotational velocities of typical A and F stars ranged from 8 to about 280 km $s^{-1}$, with a mean of 134 km $s^{-1}$ in the analysis of high-resolution spectroscopic data of A/F type stars from the Kepler field. Moreover, they found that the maximum of the velocity distribution for A and F stars having an effective temperature between 8,000 and 9,000 K was around 200 km $s^{-1}$. Assuming that the temperature and radius of the star were around $logT = 3.914$ ($\sim$ 8200 K; approximated from the $B - V$ colour index) and $R = 2.85$ $R_\odot$ (calculated from the luminosity equation $L/L_\odot = [R/R_\odot]^2\ [T/T_\odot]^4$), its rotational period was estimated to be 0.93(7) and 1.39(11) c $d^{-1}$ for an inclination angle of $i = 90°$, and rotational velocities of 134 and 200 km $s^{-1}$, respectively. These values shown with green and red vertical lines in Figure 1 were considerably compatible with the frequency accumulation at around 1.00 c $d^{-1}$.

We also computed the pulsation constants ($Q$) of the $\delta$ Scuti type frequencies by using the relation provided by Breger and Bregman [8]. Consequently, $Q$ values of $f_1$ and $f_6$ were around 0.031(6) and 0.034(6) days, which were within the theoretical range for the fundamental mode given by Fitch [9]. If the main frequency was the fundamental mode, its first overtone would be around 13.47 c $d^{-1}$, which was not observed in our amplitude spectrum. In contrast, if 9.37 c $d^{-1}$ was assumed to be the fundamental mode, the first overtone was found as 12.31 c $d^{-1}$ based on the period ratio $P_1/P_0 = 0.761$. This value slightly matched with $f_3$, and its $Q$ constant was also within the theoretical range given for the first overtone.

Additionally, seasonal frequency and amplitude variations were investigated to gain better understanding of the short-term pulsation behaviour of RX Sex. To do this, we determined consistently observed frequencies in seasonal LCs (Figure 2) and surmised that only two frequencies potentially varied as seen in Figure 3. However, all variations were within the error limits, and no significant frequency or amplitude changes were found for both peaks over four years. On the other hand, combining the frequency value (12.5156 c $d^{-1}$) given by Jerzykiewicz [3] with our *STEREO* result, a long-term period variation of $2.49(22) \times 10^{-4}$ $yr^{-1}$ was calculated for the frequency at around 12.40 c $d^{-1}$. It should also be noted that if there was a gradual change between the frequencies (from 12.50 to 10.25 c $d^{-1}$), the variation rate of the period would be obtained as $dP/Pdt = 5.39(2) \times 10^{-3}$ $yr^{-1}$ over 35 years. According to Breger and Pamyatnykh [10], this suggested that the period change observed was not caused by stellar evolutionary changes, but produced by another factor such as non-linear mode interactions.

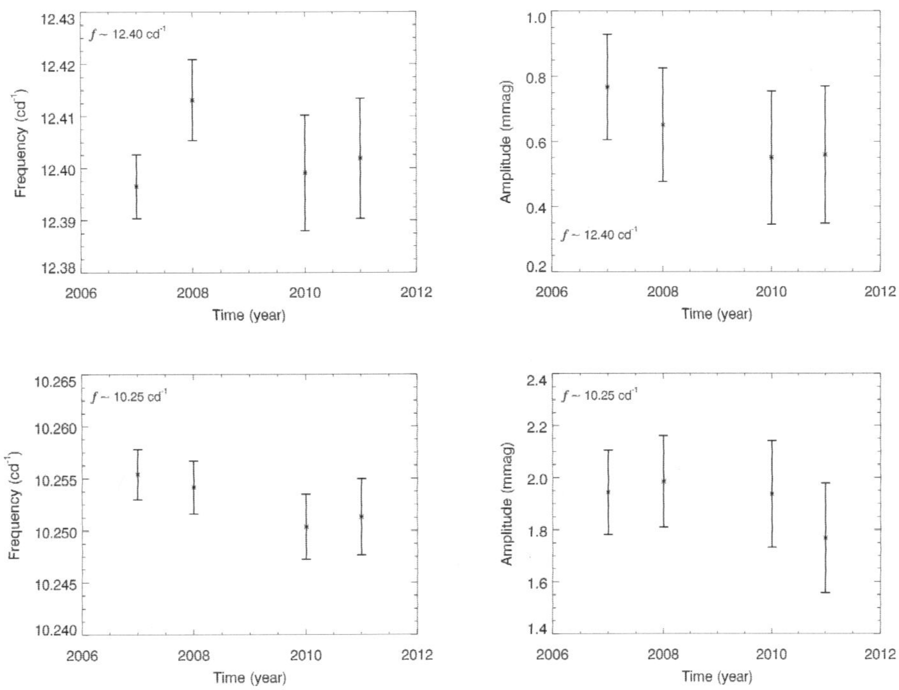


Figure 3: Two frequencies of HD 90386 and their amplitude variations between 2007 and 2011.

In order to have a better understanding of the evolutionary stage of HD 90386, we attempted to estimate some of the fundamental physical parameters. Its effective temperature and luminosity were approximated to be around $logT = 3.914(4)$ from the $B - V$ colour index and $logL/L_{\odot} = 1.517(67)$ from the parallax given in *Simbad* Database, respectively. Using these parameters, the radius of the star was derived as 2.85(23) $R_{\odot}$. The surface gravity ($logg = 3.887(159)$) was obtained from the relation given by Tsvetkov [11] and the mass ($M/M_{\odot} = 2.28(9)$) was computed by means of the equation taken from Cox and Pilachowski [12].

The luminosity and temperature values indicate that HD 90386 is quite close to the blue edge of the instability strip and to the end of the MS stage where central hydrogen is consumed (Figure 4). Its metallicity index ($m_1 = 0.173$ mag) is low [13]. So, this might result in higher luminosity and temperature values, which specify its position on the H-R diagram. Its mass value is also compatible with the theoretical evolution track. Further, since the inclination angle is not known and the rotational velocity is ambiguous, the effects of the rotation on the amplitude spectrum and the evolution status are not clear.

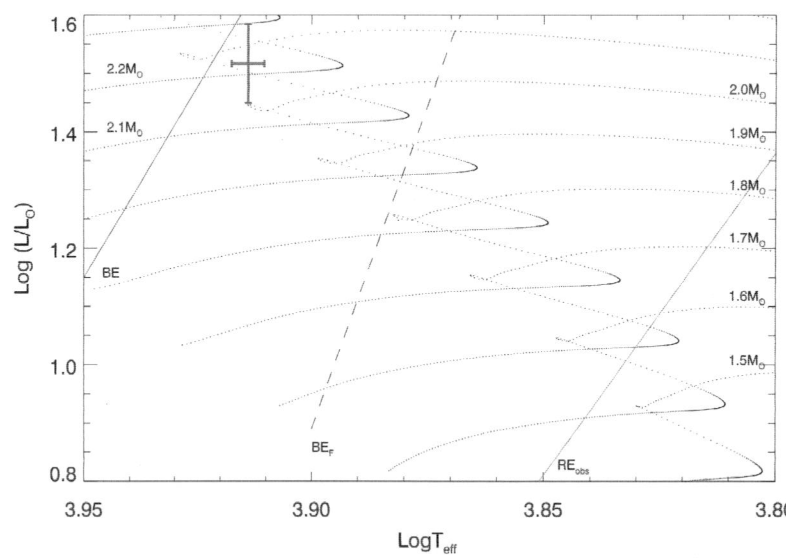


Figure 4: The evolutionary stage of HD 90386. The blue and red lines are the hot and cold edges of the instability strip, respectively. The vertical dashed line is the hot edge for the fundamental mode pulsators (Breger & Pamyatnykh, 1998b). The theoretical values in the MS and the post-MS phases are taken from Christensen-Dalsgaard [14].

HD 90386, a double-mode pulsator, has a simple amplitude spectrum. The period ratio between $f_1$ and $f_3$ is around 0.827, which is quite consistent with the ratio given by Breger [2] for the first and second overtones ($P_2/P_1 = 0.810$). Based on Breger and Bregman [8], HD 90386 might be an overtone pulsator since $\delta$ Scuties pulsating with the fundamental mode lie in the cool part of the instability strip whereas the overtone pulsators are found in the hotter part.

As previously mentioned, the variabilities around 1.00 c d$^{-1}$ in the amplitude spectrum are consistent with the rotation period. However, there is a possibility that these variations are related to $\gamma$ Dor type high-order g-modes and that HD 90386 might be a $\delta$ Scuti/$\gamma$ Dor type hybrid star. Although the location of the star on the H-R diagram suggests that it is a $\delta$ Scuti type star, in the studies related *Kepler* A and F type stars Bradley et al. [15] and Uytterhoeven et al. [16] showed that the locations of $\delta$ Scuti and $\gamma$ Dor classes on the H-R diagram have been extended, and that $\gamma$ Dor type pulsations have been seen in hotter stars than those known so far. They observed a great number of hybrid stars which spread between the blue edge of the $\delta$ Scuti instability strip and the red edge of the $\gamma$ Dor instability strip (two-third of the hybrids have been located in the hot side of the $\delta$ Scuti instability strip [15]. In this context, the frequency accumulation at around 1.00 c $d^{-1}$ might in fact be related to $\gamma$ Dor type variations and HD 90386 might be thought as a hybrid star based on the results of these studies.

Bear in mind that the stellar rotation might also affect the frequency distribution of the star. According to Bouabid et al. [17], if the star is a fast rotator, $\gamma$ Dor type pulsations have been shifted to higher frequencies and filled the gap between $\gamma$ Dor type high-order g-modes and $\delta$ Scuti type modes, which is compatible with the frequency range of 5 and 15 c $d^{-1}$. On this basis, such a situation might explain the low-amplitude variations seen between 5 and 15 c $d^{-1}$ in the periodogram of the star.

Even though HD 90386 is classified as a $\delta$ Scuti type pulsator in the literature, it is quite difficult to conclude the actual class of the star by using only the *STEREO* data. Therefore, long-term follow-ups similar to those of *Kepler* satellite, multi-colour photometry, and spectroscopic observations are needed in order to accurately determine physical parameters and the nature of the star.

## References

[1] Kurtz D W 2000 *ASPC* **210** 287
[2] Breger M 1979 *PASP* **91** 5
[3] Jerzykiewicz M 1972 *PASP* **84** 529
[4] Sangaralingam V and Stevens I R 2011 *MNRAS* **418** 1325
[5] Whittaker G N, Stevens I R and Sangaralingam V 2013 *MNRAS* **431**, 3456
[6] Eyles C J, et al. 2009 *SoPh* **254** 387
[7] Niemczura E, et al. 2015 *MNRAS* **450** 2764
[8] Breger M and Bregman J N 1975 *ApJ* **200** 343
[9] Fitch W S 1981 *ApJ* **249** 218
[10] Breger M and Pamyatnykh A A 1998 *A&A* **332** 958
[11] Tsvetkov T G 1988 *Ap&SS* **150** 357
[12] Cox A N and Pilachowski C A 2000 *PhT* **53** 77
[13] Rodríguez E, López-González M J and López de Coca P 2000 *A&AS* **144** 469
[14] Christensen-Dalsgaard J 1993 *ASPC* **40** 483
[15] Bradley P A, Guzik J A , Miles L F, Uytterhoeven K, Jackiewicz J and Kinemuchi K 2015 *AJ* **149** 68
[16] Uytterhoeven K, et al. 2011 *A&A* **534** A125
[17] Bouabid M P, Dupret M A, Salmon S, Montalbán J, Miglio A and Noels A 2013 *MNRAS* **429** 2500

# Preparation of Spectra for Surface Mapping with Doppler Imaging of a Peculiar Star V776 Her

**Hande Gürsoytrak[1] and Birol Gürol[1]**

[1]Astronomy and Space Sciences Department, Faculty of Science, Ankara University, Tandoğan, Ankara, TURKEY

hande.gursoytrak@science.ankara.edu.tr

**Abstract.** Our Sun is the only star that we can see thoroughly the surface of it. Chemically peculiar stars show anomalous abundances of some elements such as silicium, chromium and strontium. Doppler imaging is a useful technique for reconstructing the several structures on the surfaces of peculiar stars. We analyze our spectral observations as a preparation to Doppler imaging and compare V776 Her's observational spectra with the synthetic spectra that we produce with using the SPECTRUM code. The differences between the observational spectra and the synthetic spectra is discussed in this work.

## 1. Introduction

The Chemically peculiar (CP) stars show anomalously weak or strong absorption lines of some elements in their spectra. They are hot ($7000 \leq T_{eff} \leq 30000$) upper main sequence stars (1). However you can divide them into lots of subclasses basically there are four main groups. These are; CP1; metallic line (Am) stars, CP2; magnetic Ap stars, CP3; Hg-Mn stars and CP4; He-weak stars (1). It is believed that the chemical peculiarities of these stars is based on radiative diffusion of elements in the atmosphere that is influenced by the strength and the orientation of the magnetic field of the star (2).

V776 Her is a magnetic Ap star. According to Preston (1) CP2 star's effective temperatures generally are between 8000-15000 K and they show enhanced lines of Si, Sr, Cr or Eu in their spectra. Babcock (3) emphasized CP2 stars possess magnetic fields between $10^3 \sim 10^4$ Gauss. They show periodic magnetic field variations, this periodicity varies from a few days to years (1). All these variations can clearly be seen both photometric and spectroscopic investigations of this type stars.

## 2. V776 Her

V776 Her (l Her, HD 151525, HR 6234) is a bright (V=5.24 mag) and well known chemically peculiar (CP2) star. The first radial velocity observations of the star were made in 1909 and 1910 at Allegheny Observatory of The University of Pittsburgh and published by Jordan (4). It is classified as an $\alpha^2$ Canum Venaticorum star by Morgan (5). Babcock (6) emphasized a varying magnetic field is certain in this star and he indicated the presence and width variations of Si II, Sr II, Cr II and Fe II lines in the spectra of V776 Her. Jaschek & Jaschek (7) represented the peculiarity type of the star as "Eu-Cr". The effective temperature of star estimated from parameter Q is 9450 K and from multicolor photometry as 9050 K (8) The projected rotational velocity of the star derived 44 km $s^{-1}$ by Royer et al (9) and 35 $kms^{-1}$ by Abt et al (10).

There was contradiction between photometric and spectral period of the star. According to Burke and Barr (11) photometric period of this star is 1.31159 d. But Hatzes (12) emphasized that when phased his spectral data with this photometric period no spectral variations have been seen and he determined the spectral period of the star as 4.1164 d. This lastly determined period is entered to Renson and Catalano's last period catalogue too (13).

Hatzes (12) investigated the distribution of chromium on V776 Her with two other Ap stars, ε UMa and ω Her. He obtained chrome maps of these stars with using Doppler imaging method.

## 3. Observations and data reduction

The spectra used in this work were obtained using RTT150 telescope with Coude Echelle Spectrograph at the TUBİTAK National Observatory in Antalya. The observations were obtained at 10 different nights in May, July and August 2014. Unfortunately some parts of these observations can't be used for analysis because of bad weather conditions and/or technical problems. Coude Echelle type spectra cover a wavelength range from 3690 Å to 10275 Å and have a spectral resolution of about 40000. All data sets consist of bias frames, halogen and thorium-argon lamp spectra.

Bias subtraction, bad pixel and flat-field correction were made with standard IRAF procedures. The spectra were extracted using procedures from the IRAF (Image Reduction and Analysis Facility) package noao.imred.echelle. Wavelength calibration was performed using Th-Ar comparison lamp obtained during the same observing run. The heliocentric correction was applied with the RVCORR and DOPCOR task of the IRAF packages ASTUTIL and ECHELLE. After reduction of the observations we normalized the spectra to the continuum using areas free of absorption lines.

Typical exposure time varies 3000~4200 seconds. According to Iliev (14) for a successful Doppler imaging you need to have time series spectra at least 40 individual rotational phases which is considered in the preparation of nightly observing programs. A signal to noise ratio between 150 and 250 was achieved for central parts of all spectra. Rotational phases of V776 Her (see Table 1) were calculated according to the light elements derived by Hatzes (12):

$$HJD = 2446162.9090 + 4.1164 \times E.$$

Table 1. Log of V776 Her observations

| HJD (2400000+) | Phase | HJD (2400000+) | Phase | HJD (2400000+) | Phase |
|---|---|---|---|---|---|
| 56783.5649 | 0.0835 | 56789.4809 | 0.5207 | 56879.4058 | 0.3662 |
| 56785.4453 | 0.5404 | 56789.5299 | 0.5326 | 56880.2619 | 0.5742 |
| 56785.4957 | 0.5526 | 56789.5757 | 0.5437 | 56880.3126 | 0.5865 |
| 56849.2733 | 0.0461 | 56851.2833 | 0.5344 | 56880.3606 | 0.5982 |
| 56849.3213 | 0.0578 | 56851.3318 | 0.5462 | 56880.4080 | 0.6097 |
| 56849.3729 | 0.0703 | 56851.3802 | 0.5580 | 56881.2590 | 0.8164 |
| 56849.4209 | 0.0820 | 56851.4282 | 0.5696 | 56881.3070 | 0.8281 |
| 56849.4657 | 0.0929 | 56851.4767 | 0.5814 | 56881.3560 | 0.8400 |
| 56789.4317 | 0.5088 | 56879.3569 | 0.3544 | 56881.4027 | 0.8513 |

## 4. Analysis

The purpose of this study is testifying and preparing data (spectra) for obtaining the surface map of V776 Her using Doppler imaging method. We investigate our data for proving the line profile variations of especially for Cr II lines. Because of this we choose the spectral region between 4820 Å and 4850 Å that covers Cr II 4824, 4836 and 4848 lines. We compared lots of different models for these spectral region using SPECTRUM code published by Gray (15) with Kurucz atmosfer mod's (16). According to Wolff (17) the temperature and log g values given as 9333 K and 3.0 respectively for V776 Her. Because of the *vsini* =35 $kms^{-1}$ (10, 12) was given we obtained synthetic spectra's with those initial parameters and compared with our observations. Because of the published mod's are organized according the temperature differences of 250 K we take the mode prepared for $T_{eff}$=9250 K which is different only 83 K from the published temperature by Wolff (17). As can be seen from Fig.1 while the Cr II 4824 line is consistent with the model the other absorption lines are not.

Journal of Physics: Conference Series **707** (2016) 012042 doi:10.1088/1742-6596/707/1/012042

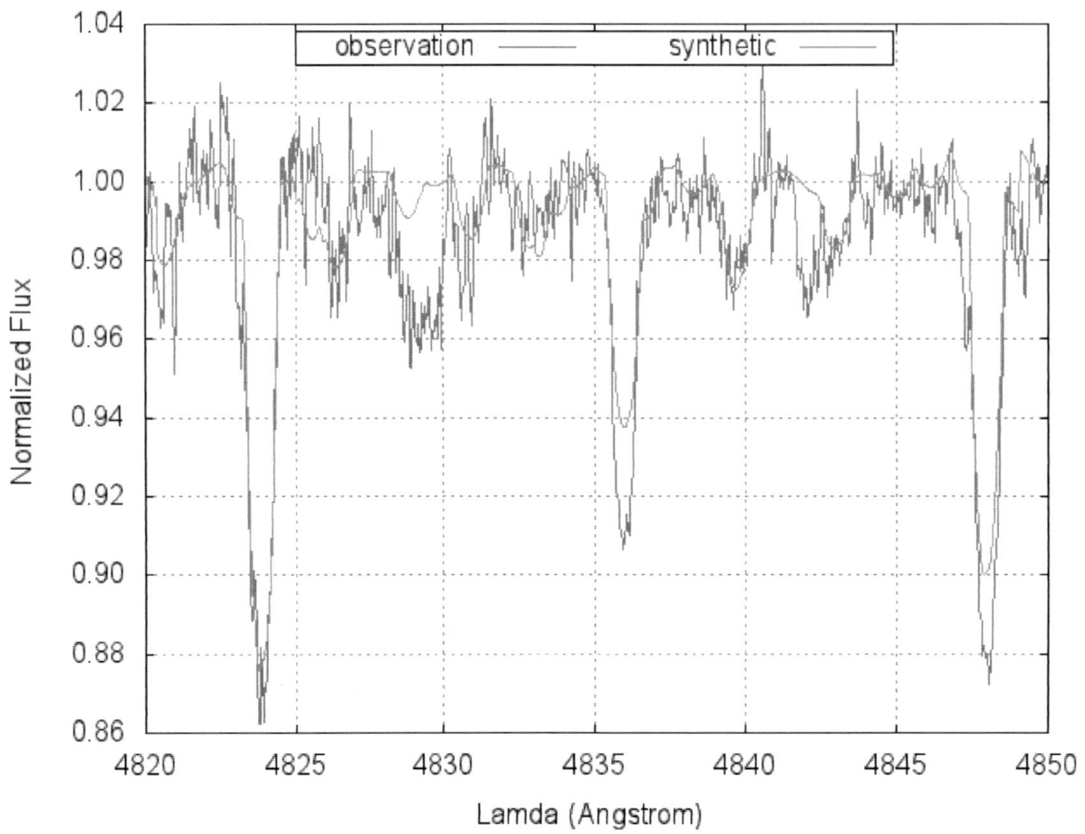


**Figure 1.** Comparison of the synthetic and observed spectra (phase=0.509) obtained for $T_{eff}$=9250 K and log g=3.0.

Because of the inconsistency of the synthetic and observed spectra we searched to find best synthetic spectra using different log g values, $2.0 \leq \log g \leq 4.5$, including same temperature. According to this search best consistency obtained with lover log g values as shown in Fig 2 and 3. Additionally we compared synthetic spectra for temperatures of lower and upper limits (9000 K and 9500 K) but can't obtained better consistency than $T_{eff}$=9250 K and log g=2.0 models. The comparison of the synthetic spectra with our observed spectra is given in Fig. 2 for log g= 2.0.

Because of the rotational velocity affect the observed spectral lines we obtained synthetic spectra with using different velocity values near the value given in the literature. After some trials we found that the best consistency is obtained when we use log g=2.5 with *vsini* as 32 kms$^{-1}$. The comparison of the observations with the synthetic spectra is given in Fig. 3. Essentially we don't expect perfect consistency for this type comparison since we used temperature restricted models. Another reason that causes this inconsistency is the spectral line variations which can be seen from observations obtained for different phases of V776 Her.

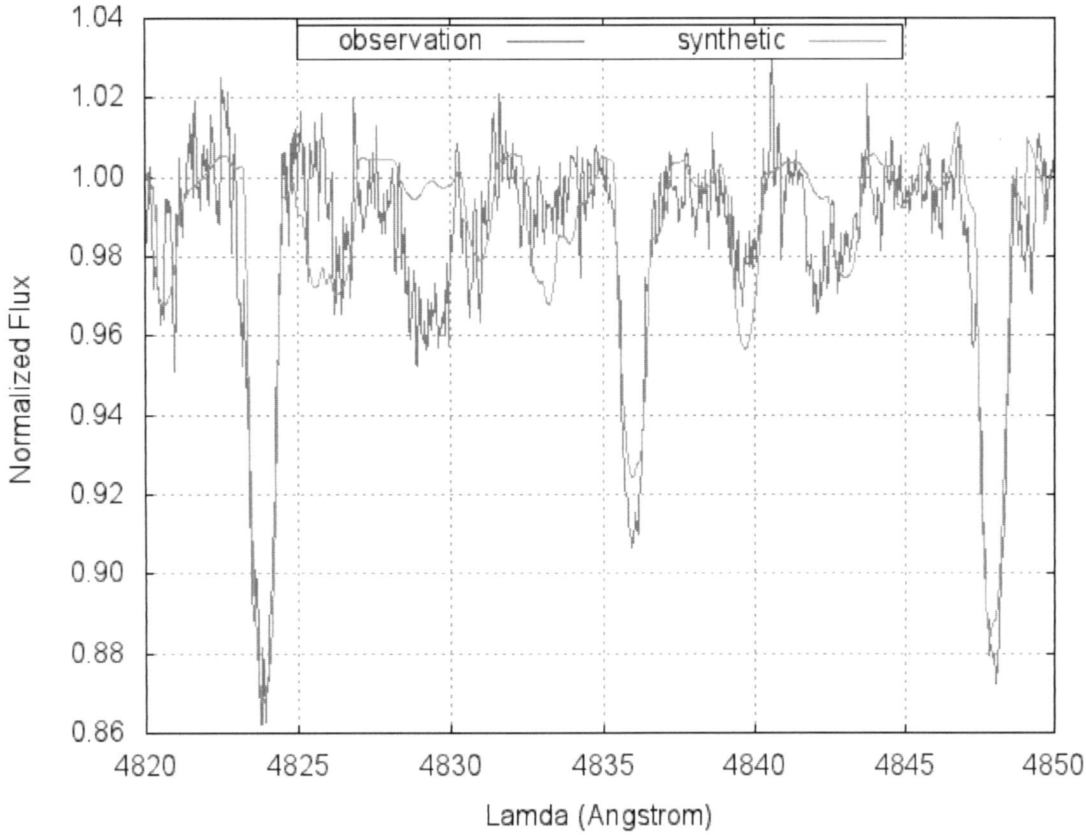


**Figure 2.** Same as Fig. 1 but for log g=2.0

After comparison of the synthetic spectra with the observations we conclude that the best parameters for V776 Her are $T_{eff}$=9250 K, log g=2.5 and vsin i=32 kms$^{-1}$ which are a bit different given in the literature. In Fig. 4 we compared synthetic spectra with our observations for 4 different phases. It is clearly seen that on some of the observations, the line center is splitting. This type of splitting can be seen mostly for stars where the atmosphere is influenced by the strength and orientation of the magnetic field (2).

Additionally we obtained the radial velocity of V776 Her comparing observed spectral lines with the laboratory lines and found that Vr=-13.6 km s$^{-1}$ which is in agreement published by Jordan, Gontcharov, Wielen et al (4, 18, 19).

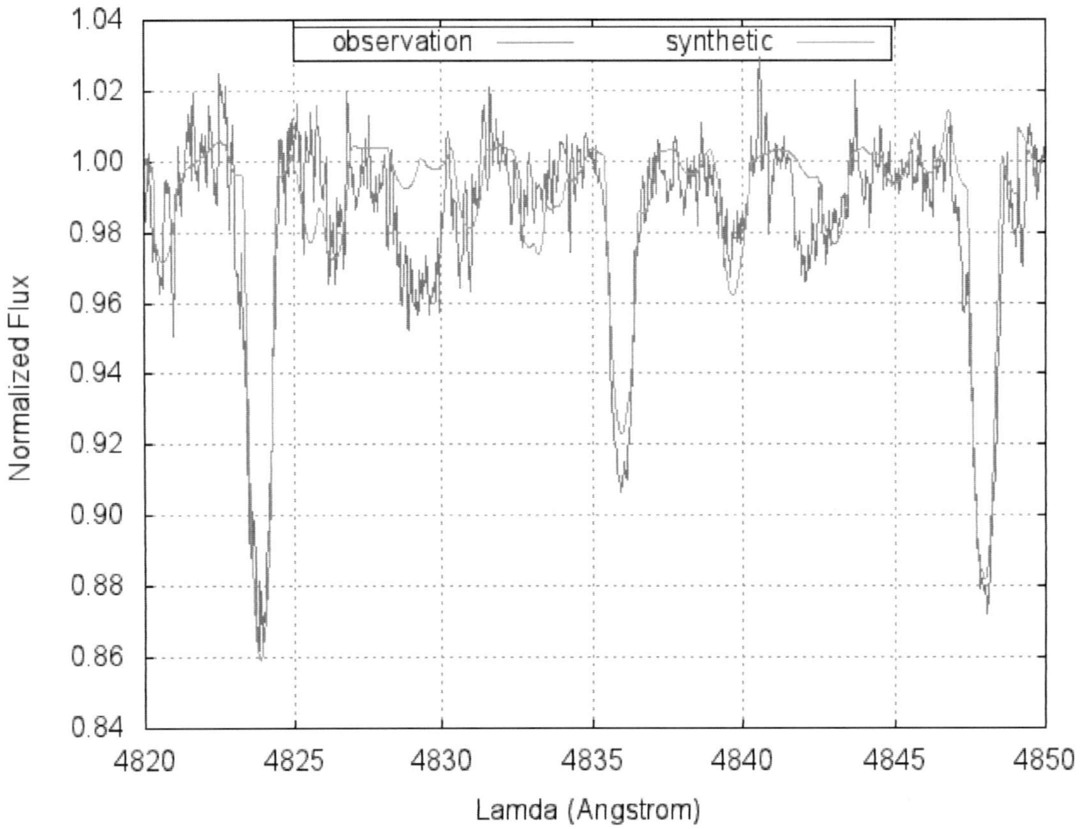


**Figure 3.** Same as Fig. 1 but for log g=2.5 with vsini=32 $kms^{-1}$.

## 5. Conclusions

In this work we showed that the parameters ($T_{eff}$, log g and vsini) obtained in this work are a bit different for V776 Her. The real parameters can be obtained with the detailed analysis of the observed spectra's which is planned to do in the near future. We showed that the observed spectra were changing with time as the star revolving around the rotational axis. The line profiles show us that the variation is real and can be used to map of the surface of this type stars.

Although the exposure time used in the observations are a bit above one hour, the obtained Signal/Noise values not enough as expected. It appears that observations with the RTT150 Coude Echelle Spectrograph are suitable for Doppler imaging. We see that the SPECTRUM code published by Gray (16) is very useful for computing synthetic spectra for different stellar atmospheric parameters.

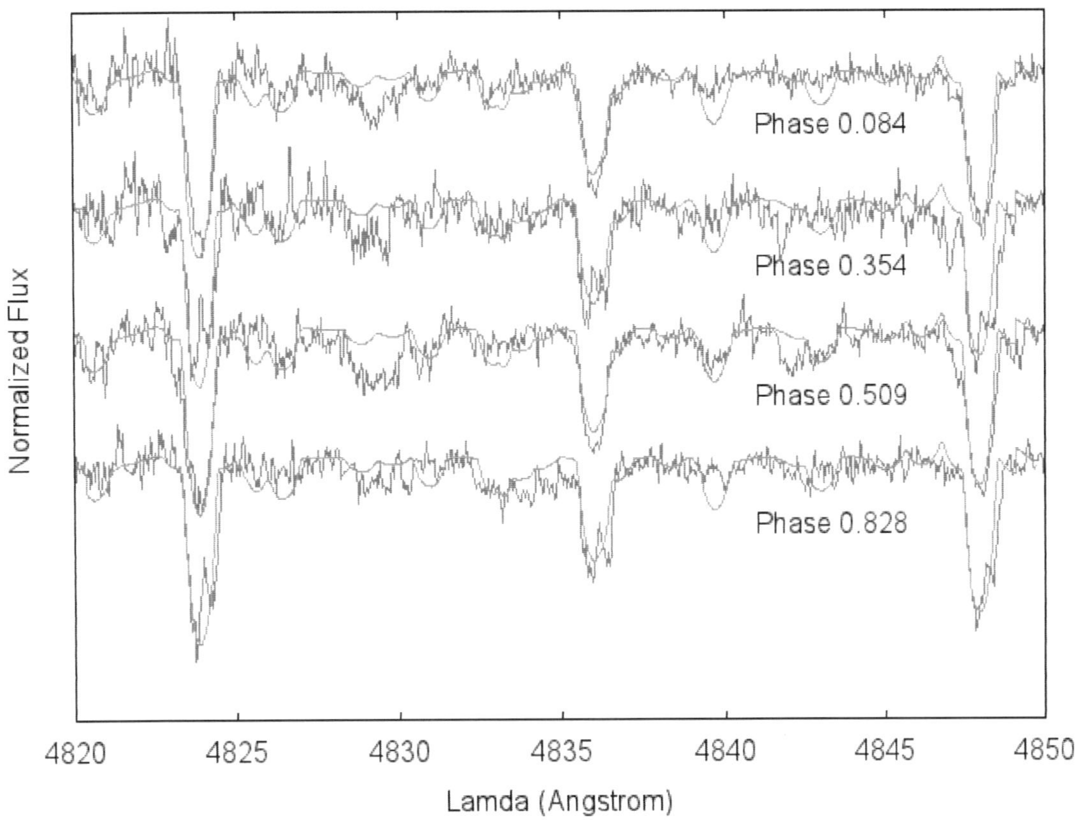


**Figure 4.** Change of the spectral line profiles for different phases because of rotation. It can be clearly seen that Zeeman splitting causes line widening and line splitting.

**Acknowledgments**
We thank to TUBITAK for a partial support in using RTT150 (Russian-Turkish 1.5-m telescope in Antalya) with project number 13BRTT150-499. This research has made use of the SIMBAD database, operated at CDS, Strasbourg, France, and NASA's Astrophysics Data System Abstract Service. We also thank to Richard Gray for accessing his SPECTRUM code and Francesco Castelli and Robert Kurucz for grid models.

**References**
[1] Preston G V 1974 *ARA&A* **12** 257
[2] Michaud G 1970 *ApJ* **160** 640
[3] Babcock H W 1958 *IAUS* **6** 161
[4] Jordan F C 1911 *Publ. Allegheny* **2** 121
[5] Morgan W W 1933 *ApJ* **77** 330
[6] Babcock H W 1958 *ApJS* **3** 141
[7] Jaschek M and Jaschek C 1958 *ZA* **45** 35
[8] Glagolevskij Y V 1994 *BSAO* **38** 152
[9] Royer F, Grenier S, Baylac M -O, Gomez A E and Zorec J 2002 *A&A* **393** 897
[10] Abt H A, Levato H and Grosso M 2002 *ApJ* **573** 359
[11] Burke E W and Barr T H 1981 *PASP* **93** 344
[12] Hatzes A P 1991 *MNRAS* **253** 89
[13] Renson P and Catalano F A 2001 *A&A* **378** 113

[14] Iliev Kh I 2010 *Publ. Astron. Obs. Belgrade* **90** 91
[15] Gray R O 1999 *Astrophysics Source Code Library, record ascl:*9910.002
[16] Castelli F, Kurucz R L 2003 *IAUS* **210** A20
[17] Wolff S C 1967 *ApJS* **15** 21
[18] Gontcharov G A *Astron. Lett.* **32** 759
[19] Wielen R, Schwan H, Dettbarn C, Lenhardt H, Jahreiss H, Jahrling R, Khalisi E 2000 *Veroeff. Astron. Rechen-Inst. Heidelberg* **37** 1

# Low luminosity AGNs in the local universe

**Tuba Ikiz[1,2], Reynier F Peletier[2], Cahit Yesilyaprak[1,3]**

[1] Atatürk University, Science Faculty, Department of Astronomy and Astrophysics, Erzurum, Turkey
[2] Kapteyn Astronomical Institute, University of Groningen, Groningen, the Netherlands
[3] Atatürk University, Research and Application Center for Astrophysics (ATASAM), Erzurum, Turkey

E-mail: tubaaaikiz@gmail.com

**Abstract.** Galaxies are known to contain black holes (e.g. Ferrarese & Merritt 2000), whose mass correlates with the mass of their bulge. A fraction of them also has an Active Galactic Nucleus (AGN), showing excess emission thought to be due to accretion of mass by the supermassive black hole at the center of the galaxy. It is thought that AGNs play a very important role during the formation of galaxies by creating large outflows that stop star formation in the galaxy (see e.g. Kormendy & Ho 2013). The aim is to detect the fraction of Low Luminosity Active Galactic Nucleus (LLAGN) in the nearby Universe. At present, they are typically found using optical spectroscopy (e.g. Kauffmann, Heckman et al. 2003), who discuss the influence of the AGN on the host galaxy and vice versa. However, optical spectra are seriously affected by extinction in these generally very dusty objects, and therefore can only give us partial information about the AGN. I used a newly-found method, and apply it to the S4G sample, a large, complete, sample of nearby galaxies, which I am studying in detail with a large collaboration, to detect the fraction of low luminosity AGNs, and to better understand the relation between AGNs and their host galaxy which is thought to be crucial for their formation.

## 1. Introduction

Galaxies are the basic building blocks of the Universe, and understanding their formation and evolution is crucial to many areas of current astrophysical research. Nearby galaxies contain the 'fossil record' of the evolution of galaxies and provide a wealth of detail to extensively test the current models of galaxy formation and evolution. A galaxy's structure is linked to both its mass and evolutionary history. Probing galactic structure requires understanding the distribution of stars among galaxies of all types luminosities across full range of environments. Understanding this history is to obtain census of stellar structure in galaxies in the local volume.

Galaxies are known to contain black holes (e.g. Ferrarese & Merritt 2000), whose mass correlates with the mass of their bulge. The tight correlation between nuclear black hole and bulge mass (e.g., Magorrian et al. 1998; Tremaine et al. 2002) implies the processes are intimately connected. These black holes are located in the nuclei of almost all galaxies. Interestingly, their mass scales with the mass of their surrounding spheroids. This fundamental connection shows that the super massive black hole and its host system must have formed in concert. A small fraction of galaxies also has a very bright nucleus called as an Active Galactic Nucleus (AGN), showing excess emission thought to be due to accretion of mass by the super massive black hole at the center of the galaxy. It is thought that AGNs play a very important role during the formation of galaxies by creating large outflows that stop star formation in the galaxy (see e.g. Kormendy & Ho 2013). The complex central regions of active and inactive galaxies are thus an ideal laboratory for study of the evolution and formation of galaxies.

Here we propose to study the presence of Low Luminosity Active Galactic Nucleus (LLAGN). At present, they are typically found using optical spectroscopy (e.g. Kauffmann, Heckman et al. 2003), who discuss the influence of the AGN on the host galaxy and vice versa. However, optical spectra are seriously affected by extinction in these generally very dusty objects, and therefore can only give us partial information about the AGN. Here we propose to use a newly-found method, and apply it to a large, complete, sample of nearby galaxies, which we are studying in detail with a large collaboration, to detect the fraction of low luminosity AGNs, and to better understand the relation between AGNs and their host galaxy.

## 2. Samples

In this study we took the Spitzer Survey of Stellar Structures in Galaxies ($S^4G$; Sheth et al. 2010) sample of more than 2500 galaxies, for which deep [3.6] and [4.5] micron images have been reduced and archived deep [3.6] and [4.5] micron images in collaboration with the DAGAL EU RTN network (http://www.dagalnetwork.eu/), and $S^4G$ Extension sample of more than 300 galaxies (http://sha.ipac.caltech.edu/applications/Spitzer/SHA/) both taken with the Spitzer Space Telescope.

The $S^4G$ is a volume, magnitude, and size-limited survey of more than 2500 nearby galaxies using the IRAC Infrared Array Camera (Fazio et al. 2004) with deep imaging at 3.6 and 4.5 μm on board the Spitzer Space Telescope which one of the Spitzer Legacy Program of late type galaxies (Figure 1).

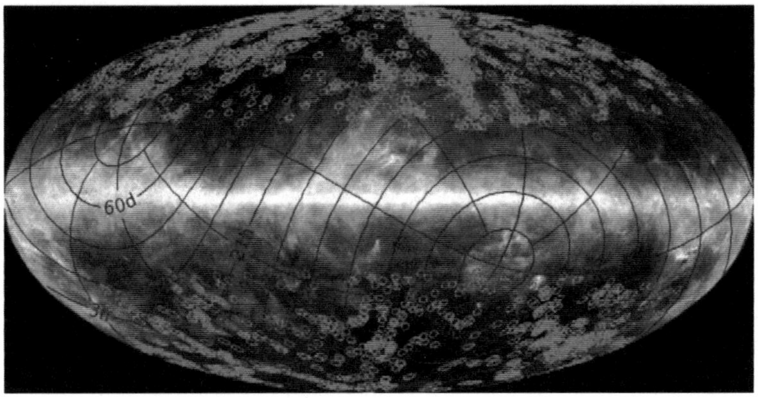


**Figure 1.** The 2,352 S4G galaxies are represented as overlays in red on the all-sky ISSA (IRAS Sky Survey Atlas) image above *(http://irsa.ipac.caltech.edu/data/SPITZER/S4G/)*.

This survey contains over 2300 galaxies within 40 Mpc (v < 3000 km/s), away from galactic plane (|b| < 30°), with corrected for internal extinction B band Vega magnitude brighter than 15.5 and B-band diameter larger than 1 arcmin with distances determined from HI radial velocities. Since S4G selection based on radial velocities from HI, it lacks early-type galaxies. This has been corrected for in the S4G Extension, a sample containing 465 galaxies. We also include them in this study.

## 3. Method

We used a newly-found method, to discover LLAGNs. The method is based on the work of Stern et al. (2005), who show that these objects fill a particular area in the Spitzer-IRAC [3.6]-[4.5] vs. [5.8]-[8.0] color-color diagram. We determined the central colors for all these galaxies, and detected LLAGNs by just studying the central [3.6]-[4.5] color, as has been shown by Van Der Wolk et al. (2011), and reference therein (Figure 2).

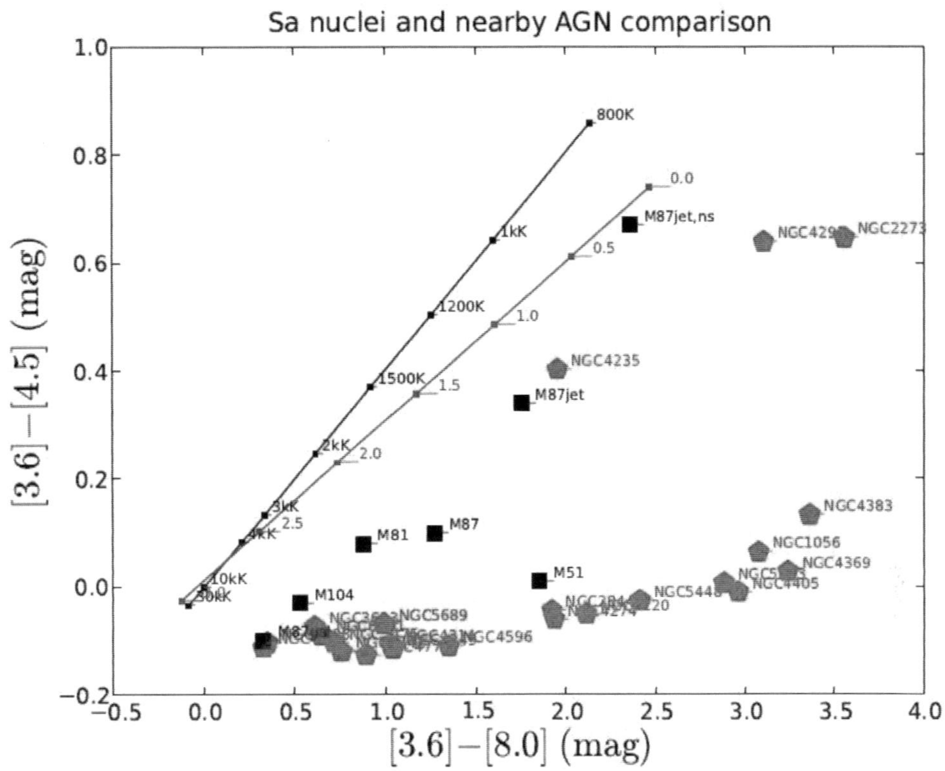


**Figure 2.** Spitzer colour-colour diagram. SAURON Sa sample nuclei are plotted in red and nearby active galaxy nuclei in black. For M 87 we also indicate the colours of the large-scale jet knots A/B indicates as M87 jet, the stellar emission at the same radius as the jet (M87gal) and the stellar-subtracted jet emission (M87jet,ns). Black bodies at various temperatures are plotted with a black line. Power-law models $F_\upsilon \alpha \upsilon^\alpha$ at various spectral indices $-1 \leq \alpha \leq 3$ are plotted with a green line (Van der Wolk thesis, 2011).

Van der Wolk (2011), working with Peletier and Barthel in Groningen, showed that for nearby galaxies the [3.6]-[4.5] colour by itself is also able to isolate LLAGNs. AGNs contain a hot core, which causes [3.6]-[4.5] to be larger than 0.4 (Vega system). Other emission in the center of galaxies, such as stars or PAH-dust emission never reach such red [3.6]-[4.5] color. The [3.6] micron images are almost unaffected by extinction by dust, and can be used to obtain accurate estimates of the stellar mass in these galaxies (e.g. Meidt et al. 2014). Also, the [3.6]-[4.5] color is a useful indicator of the fraction of AGB stars in these galaxies (Peletier et al. 2012).

We show that Spitzer Space Telescope is a powerful tool for studying AGN demographics. We investigate nature of galactic nuclei with a strong near infrared [3.6]-[4.5] color excess which provides a robust technique for identifying AGNs. We used this color criteria method which is one of the best way of distinguishing AGNs from normal galaxies and an efficient way of finding LLAGNs in a large sample. We measured the central colors of galaxies and detected AGNs with using near infrared excess. The nearby active galactic nuclei show infrared excess in comparison to normal nuclei which are dominated by stellar and star formation emission. This gives rise to the interperation that this excess is related to supermassive black hole acceration.

## 4. Analyses and Results

We measured the central colors of galaxies and detected AGNs with using near infrared excess. The nearby active galactic nuclei show infrared excess in comparison to normal nuclei which are dominated by stellar and star formation emission.

Color-color diagram of the galaxies in $S^4G$ sample is shown in Figure 3. In the diagram, Aperture 3 is versus Aperture 30. Aperture 3 shows the central color of the galaxy and Aperture 30 shows approximately the color of the whole galaxy, which is not affected by the central color.

Galaxies in both samples whose [3.6]-[4.5] colors ≥ 0.1 mag are represented by the yellow area. These galaxies show redder central colors indicating (Stern et al. 2005) that most of the galaxies in yellow area have an active galactic nucleus.

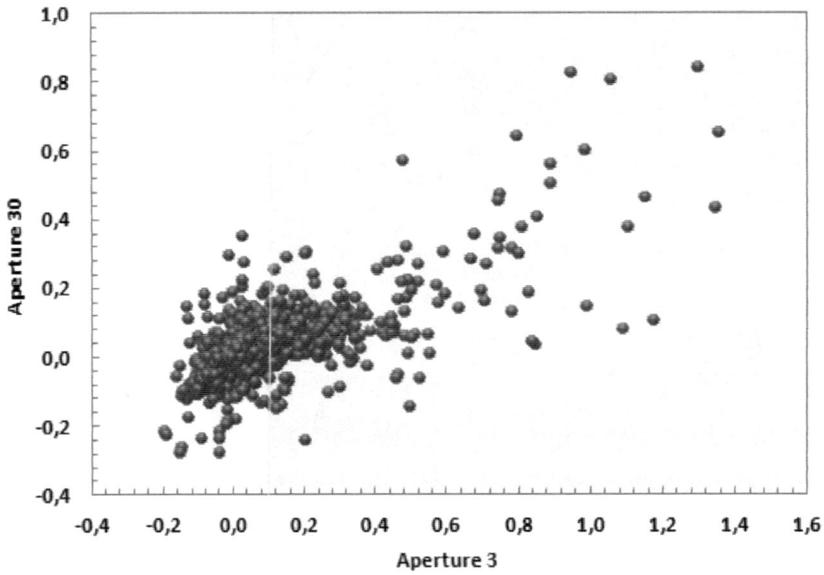


**Figure 3**. Central [3.6]-[4.5] color- outer [3.6]-[4.5] color for the S4G sample.

Once the AGN candidates have been detected, the next step is to understand what kinds of object this method gives. The first thing to do is to see whether these objects have radio and x-ray continuum counterparts, expected for AGNs. I searced galaxies whose [3.6]-[4.5] colors ≥ 0.1 magnitudes as defined by AGN in VLA FIRST and NVSS surveys for radio detections and CHANDRA and XMM Newton surveys for X-ray detections (Figure 4 and 5).

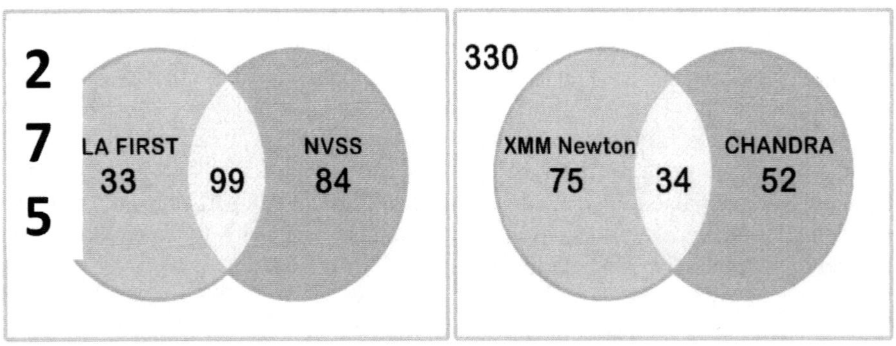


**Figure 4.** Number of AGNs detected in VLA FIRST, NVSS, CHANDRA and XMM Newton surveys.

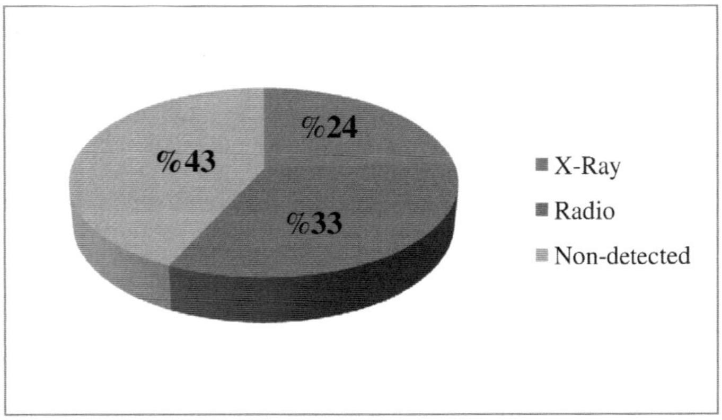


**Figure 5.** AGNs survey detections is given by percentages.

AGNs survey detections show in Figure 5 is given by percentages: %24 X-ray surveys, %33 radio surveys, %43 non-detected AGNs. These percentages represent the radio and X-ray measurements of AGNs in both samples.

## 5. Future Works

I will find the fraction of LLAGNs in both samples and represent relations of AGN fractions with galaxy mass, morphological type and environment, in order to understand the occurance of AGN.

## References

[1] Galbany L *et al* 2014 *A&A* ArXiv e-prints.
[2] Ferrarese L and Merritt D 2000 *ApJ* **539** L9
[3] Kauffmann G *et al* 2003 *MNRAS* **346** 1055.
[4] Kormendy J and Ho J C 2013 *ARAA* **51** 511.
[5] Magorrian J *et al* 1998 *AJ* **115** 2285.
[6] Pahre M A *et al* 2004 *ApJS* **154** 235.
[7] Peletier R F *et al* 2011 *MNRAS* ArXiv e-prints.
[8] Stern D *et al* 2005 *ApJ* **631** 163.
[9] Tremaine S *et al* 2002 *ApJ* **574** 740.
[10] van der Wolk G *et al* 2010 *A&A* **511** A64+.

# Colour relations for Mira and Semiregular (SR) type stars

**Yavuz Guney**[1], **Cahit Yesilyaprak**[2,3]

[1] Atatürk University, Science Faculty, Department of Physics, Erzurum, Turkey
[2] Atatürk University, Science Faculty, Department of Astronomy and Astrophysics, Erzurum, Turkey
[3] Atatürk University, Research and Application Center of Astrophysics (ATASAM), Erzurum, Turkey

E-mail: yguney@atauni.edu.tr

**Abstract.** In this study, the period-colour relations, the colour-colour relations and the effective temperature were examined for Semiregular (SR) and Mira type variable stars. SR variables show an obvious period-colour relations, especially in infrared (IR). There are differences between SR and Mira type variable stars with respect to their colour relations. It has been thought that these differencies are caused by their mass loss rates and their effective temperatures.

## 1. Introduction
Semiregular (SR) and Mira variable stars are pulsating stars. In the General Catalog of Variable Stars (GCVS4) , SR variable stars are defined as intermediate (F, G, K) or late spectral type (M, C, S) giants or supergiants. Mira variable stars are defined late spectral type. Long period variables (LPV) form an important class of red giant stars. They traditionally comprise Miras, SR and irregular (L) variables according to the amplitude and the regularity of their visual light curves. They are known to be either O-rich or C-rich, and comprise thus M, S and C stars. More recently OH-IR sources have been found from infrared and radio observations showing that they belong to the LPV population with periods up to 2000 days [1,2].

The periods known classically change from 20 to 2000 d or more for SR variables in GCVS4. Many SR variable stars have multiple periods [3,4,5]. The visual (V) light amplitude is, as a rule, less than 2.5 mag in SR variable stars [1]. The visual light amplitude is bigger than 2.5 mag in Mira variable stars.

## 2. Stars selection and data base
The stars were selected from: Hipparcos and Tycho Catalogue [6], General Catalog of Variable Stars (GCVS4) [1,7] and SIMBAD database [8] for SR and Mira type variable stars. GCVS4 is selected stars which parallax. Hipparcos and Tycho Catalogue is joinde by selecting the SR and Mira variable stars. The total number of stars are 1288. Data from different types of selected stars were taken the following catalog. The Guide Star Catalogue (GSC), The AC 2000.2 Catalogue, All-Sky Comiled Catalogue of 2.5 Million Stars, AKARI/HIP and AKARI/2MASS samples, The USNO-A2.0 Catalogue, NOMAD Catalog, UCAC3 Catalogue, The USNO-B1.0 Catalog, TASS Mark IV Patches Photometric Catalog, The PPMXL Catalog, IRAS Point Source Reject Catalog (IRAS), SKY2000 Catalog, Stellar Photometry in Johnson's 11 Color System, Orion Spiral Arm Catalogue (OSACA), Hipparcos Red Stars, The FON Astrographic Catalogue (FONAC), Tycho-2 Spectral Type Catalog.

## 3. Method
Colour values are not the stars of colours, known brightness values of different wavelengths is calculated by taking the difference. Fluxs value in infrared bands, are converted to magnitude, as shown in equations 1 below [9].

$$\begin{aligned}
[12] &= 3.63 - 2.5Log_{10}(F_{12}) \\
[25] &= 2.07 - 2.5Log_{10}(F_{25}) \\
[60] &= 0.19 - 2.5Log_{10}(F_{60}) \\
[100] &= -0.92 - 2.5Log_{10}(F_{100})
\end{aligned} \tag{1}$$

## 4. Results and Discussions

A two-colour (V – I / V – [12]) diagram for SR and Mira type variable stars is, as shown in Figure 1.

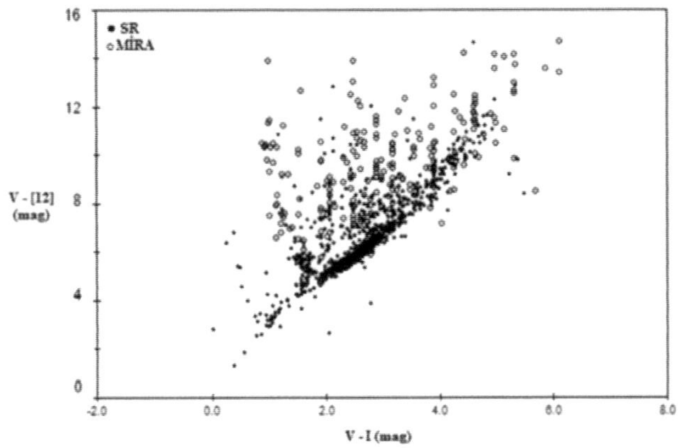


**Figure 1.** V – I versus V – [12] for SR and Mira variable stars. Mira variable stars are shown as open symbols while SR variable stars are closed symbols.

It was determined that Mira diverge more from this linear correlation, while SRs are gathered around a linear correlation. In SR type stars, this reason is either the different modes are pulsating red stars or SRa are the stars. Mira type variable stars has appeared that showed more unknown distributions. However, SR and Mira type stars have been found that it can be easily distinguished from the colour correlation.

Period – colour diagram for SR and Mira type variable stars is, as shown Figure 2.

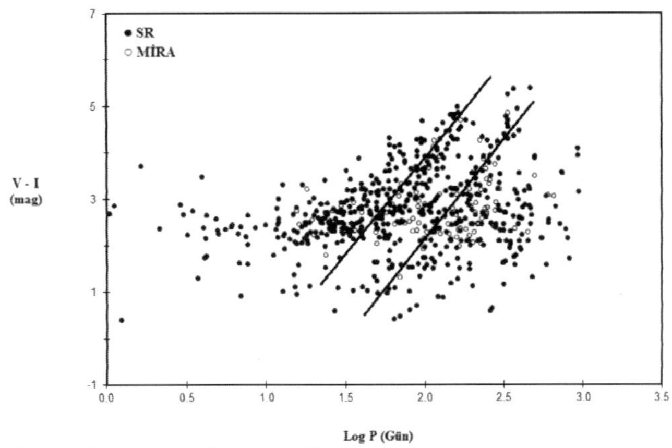


**Figure 2.** Period- colour relation for SR and Mira variables stars. Mira variable stars are shown as open symbols while SR variable stars are closed symbols.

Period – colour relation has two separate correlations for these two stars. So, Mira and SR stars are dispersed form two separate groups. Mira stars and SR stars with long periods have the same distribution. The main reason for this, some SR stars have pulsations features such both SR and Mira stars.

Period – colour diagram for SR and Mira type variable stars is, as shown Figure 3.

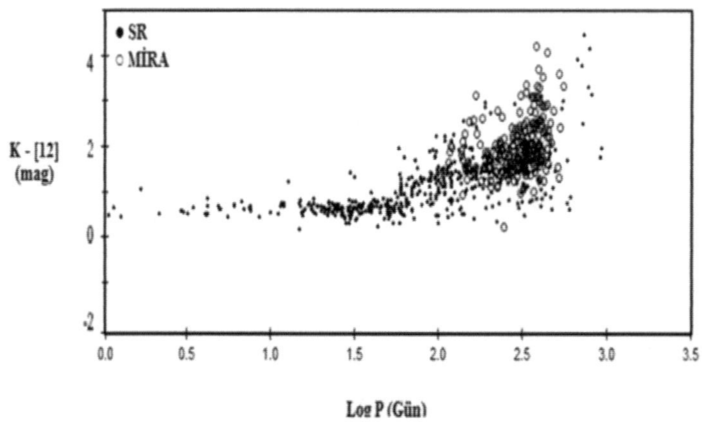


**Figure 3.** LogP versus K – [12] for SR and Mira variable stars. Mira variable stars are shown as open symbols while SR variable stars are closed symbols.

It is know that differences in colour between SR and Mira type variable stars. It is found that is show different period-colour and colour-colour relation in short and long period of SR and Mira type variable stars.

The more SR type stars is show sudden increase in K – [12] colour values that have K – [12] > 0.75 colour values, after 60-70 days. It was detected to be mass loss limit in 60-70 days period for SR variable stars [10].

Effective temperature – colour diagram for SR type variable stars is, as shown Figure 4.

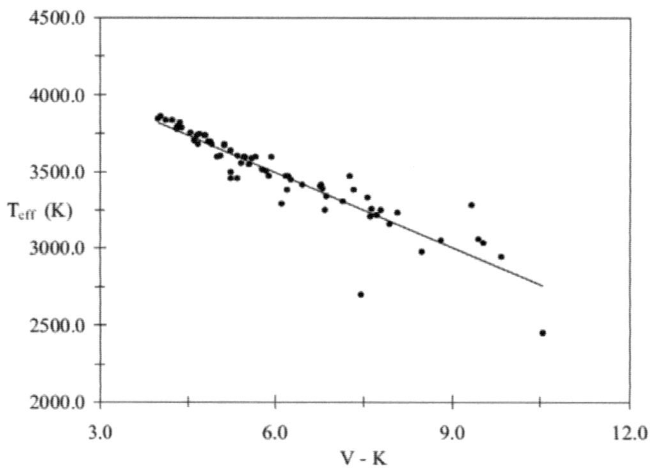


**Figure 4.** V – K versus $T_{eff}$ for SR variable stars.

The effective temperature of SR stars change about between 2500-4000 K. The effective temperature of SR stars can be expressed by more than one star in the red and infrared colour. The responsive colours to the changes in the atmosphere of these stars are J – K and H – K colours [11].

A linear correlation is found between V – K colour index and effective temperature for SR type variable stars. The deviations on the diagram are thought to be caused by the errors in the measurements of effective temperature or determination of colour index. Because of their atmospheres, there is no significant linear correlation between V – K colour index and effective temperature for Mira type variable stars.

A two-colour (V – I / B – V) diagram, was plotted for all SR variables with different spectral types (F, G, K, M, C, N, R and S) in the Hipparcos Catalogue [6]; the result is shown in Figure 5.

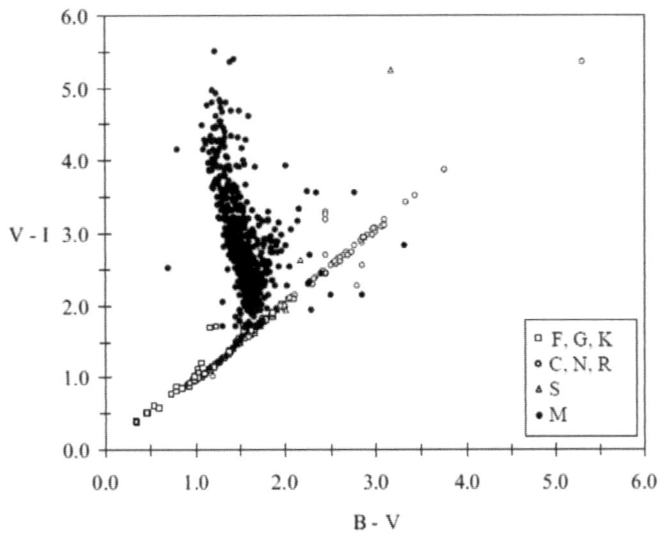


**Figure 5.** Two-colour diagram (V – I / B – V) of all SR variables in the Hipparcos Catalogue according to different spectral types.

Figure 5 shows that M-type SR variable in particular behave distinctly and are well separated from the linear relation formed by other spectral types of SR variables [10]. The distributions of both Mira and L-type variables are similar to that of the of M-type SR variables shown in Figure 5 [12].

## References

[1] Kholopov P N *et al* 1988 General catalogue of variable stars (GCVS and GCVS4) *Nauka Publishing House* **1,2,3,4**
[2] Mennesier M O *et al* 2001 Long period variable stars: Galactic populations and infrared luminosity calibrations *A&A* **374** 968
[3] Johnson H R and Querci F R 1986 The M type stars *NASA Publication* 502
[4] Mattei J A *et al* 1998 Classification of red variables *The ESA Symposium* 269
[5] Percy J R *et al* 2001 Periods of 25 pulsating red giants *IBVS* 5041
[6] ESA 1997 The hipparcos and tycho catalogues. Calestia (CD ROM) *ESA Publications Division*
[7] ADC 1997 Selected astronomical catalogs *NASA Publication* **4**
[8] CDS 2002 Centre de Donnees astronomiques de Strasbourg (CDS) *SIMBAD Astronomical Database* http://simbad.u-strasbg.fr/Simbad
[9] Walker H J and Cohen M 1988 The classification of stars from IRAS colours *AJ* **95** 1801
[10] Yesilyaprak C and Aslan Z 2004 Period-luminosity relation for M-type semiregular variables from Hipparcos parallaxes *MNRAS* **355** 601
[11] Barthes *et al* 1999 PLC distribution and classification of galactic O-rich LPVs I.Luminosity calibrations *A&AS* **140** 55
[12] Yesilyaprak C and Aslan Z 2008 Kinematics of M-type giant semi-regular variables from the Hipparcos catalogue *Publications of the Astronomical Society of Australia* **25** 63

**Acknowledgments**
This study is supported by Atatürk University and ATASAM.

# The effects of fuel type in synthesis of $NiFe_2O_4$ nanoparticles by microwave assisted combustion method

**Zeynep Karcıoğlu Karakaş,[1*] Recep Boncukçuoğlu [1] and İbrahim H. Karakaş [2]**

[1] *Environmental Engineering Department, Atatürk University, Erzurum 25240, Turkey*

[2] *Chemical Engineering Department, Bayburt University, Erzurum 69000, Turkey*

E-mail: kzeynep@atauni.edu.tr

**Abstract.** In this study, it was investigated the effects of the used fuels on structural, morphological and magnetic properties of nanoparticles in nanoparticle synthesis with microwave assisted combustion method with an important method in quick, simple and low cost at synthesis of the nanoparticles. In this aim, glycine, urea and citric acid were used as fuel, respectively. The synthesised nanoparticles were characterized by X-ray diffraction (XRD), scanning electron microscopy (SEM), Brunauer-Emmet-Teller surface area (BET), and vibrating sample magnetometry (VSM) techniques. We observed that fuel type is quite effective on magnetic properties and surface properties of the nanoparticles. X-ray difractograms of the obtained nanoparticles were compared with standard powder diffraction cards of $NiFe_2O_4$ (JCPDS Card Number 54-0964). The results demonstrated that difractograms are fully compatible with standard reflection peaks. According to the results of the XRD analysis, the highest crystallinity was observed at nanoparticles synthesized with glycine. The results demonstrated that the nanoparticles prepared with urea has the highest surface area. The micrographs of SEM showed that all of the nanoparticles have nano-crystalline behaviour and particles indication cubic shape. VSM analysis demonstrated that the type of fuel used for synthesis is highly effective a parameter on magnetic properties of nanoparticles.

## 1. Introduction

Nanotechnology is area of intense scientific research, due to a wide variety of potential applications in biomedical, optical, and electronic fields. Nanoparticles are of great scientific interest as they are effectively a bridge between bulk materials and atomic or molecular structures.

The properties of materials change as their size approaches the nanoscale and as the percentage of atoms at the surface of a material becomes significant. For bulk materials larger than one micrometre the percentage of atoms at the surface is minuscule relative to the total number of atoms of the material.

Nanoparticles have, so far, been used in many applications including ferrofluids, catalysts, microwave devices, gas sensors and magnetic materials [1-3]. Synthesis and applications of magnetic nanoparticles of spinel structured ferrite ($AB_2O_4$) are among the most important research topics due to their remarkable chemical, physical and magnetic properties. The type of nickel ferrite with spinel structure is an essential composite material for the production of electronic and magnetic components extensively used in advanced technological applications, the most remarkable of which are magnetic ferrofluids, dense information storage system and high frequency systems [4, 5].

The ferrite is an inverse spinel in which eight units of $NiFe_2O_4$ goes into one unit cell of the spinel structure. Half of the ferric ions preferentially fill the tetrahedral sites (A-sites), while the other occupies the octahedral sites (B-sites). Therefore, the compound can be represented by the formula $(Fe^{3+})_A[Ni^{2+}Fe^{3+}]_BO_4{}^{-2}$, where A and B represent tetrahedral and octahedral sites, respectively. Its remarkable electrical and magnetic properties are based on the nature of the ions, their charges and their distribution among the tetrahedral (A) and octahedral (B) sites [6]. Nickel ferrite shows ferromagnetism, which originates from magnetic moment of anti-parallel spins between $Fe^{3+}$ ions at tetrahedral sites and $Ni^{2+}$ ions at octahedral sites [7, 8].

Various chemical and physical techniques have been developed for the synthesis of $NiFe_2O_4$ nanoparticles such as sol-gel, hydrothermal, co-precipitation, aerosol, microwave, etc. that have the advantages for instance the low cost, low reaction time, large scale production. Common problems for many of these techniques are factors such as complex processes, expensive precursors and low production rates [9].

Alike the techniques, the combustion method is a promising chemical method for preparation of various nano-sized ferrites in laboratory and semi-pilot scale that is fast and safe furthermore enables the formation of nano-sized metal oxide particles and allows to produce the ferrites without need of sophisticated equipment [3, 10]. Organic compounds like glycine, urea, citric acid, alanine etc. are used as fuels in the combustion method, suitable that they can be mixed directly with the metal salts to enhance the efficiency of the synthesis. Among the organic compounds, urea seems to be the most commonly used compound that it appropriately is cheap, readily available, non-toxic and safe [3, 9, 10].

It is known that synthesis method of nanoparticles is highly effective on the structural, morphology and magnetic properties of nanoparticles. The organic materials used as fuel also has a significant impact on various properties of the nanoparticles in synthesis studies made with microwave assisted combustion method. In the literature, various chemicals have been proposed for use as fuel by various researchers. The most widely used fuels are compounds such as urea, glycine, citric acid, organic thio urea. Among these fuels, the most widely used substance are urea. Because, urea is readily and abundantly available, inexpensive, non-toxic and it can be produced in high purity, etc. Therefore, it finds wider usage area because of these important advantages [11, 12].

$NiFe_2O_4$ nanoparticles can synthesis with various methods such as combustion, sol-gel, hydrothermal etc... In this study, $NiFe_2O_4$ nanoparticles were synthesized with microwave assisted combustion method. In this method; some substances have to be used as fuel. The aim of this work was to investigate the effect of fuel type above structural, morphology and magnetic properties of $NiFe_2O_4$ nanoparticles synthesized with microwave assisted combustion method.

## 2. Materials and method

***The preparation of materials***

All chemicals used in this study have analytical purity, and these materials commercially produced is used without any additional purification. Nickel nitrate hexahydrate ($Ni(NO_3)_2.6H_2O$,>99%), iron (II) nitrate nonahydrate ($Fe(NO_3)_2.9H_2O$, >99%), urea ($CO(NH_2)_2$,> 99%), glycine ($NH_2CH_2COOH$,> 99%), citric acid ($C_6H_8O_7$,> 99%) and all other necessary chemicals were purchased from Sigma-Aldrich company.

***Synthesis***

Nickel ferrite was prepared from an exothermic reaction of a mixture of metallic nitrates and urea. Nickel nitrate, ferric nitrate and fuel reagent were mixed in certain stoichiometric ratios in a baker. The mixture was formed due to the hygroscopicity of metal nitrates, which absorbs air humidity [13]. Therefore, this method does not require the use of water or any solvent. Then, the mixture was placed in a laboratory-type microwave oven at a maximum power of 800 W for 10 min. When the solution reached the point of spontaneous combustion, it started burning by releasing a very dense gas and heat, and then the sample instantaneously became a solid [9].

In the experiments made to investigate the effects of fuel type in the synthesis of the $NiFe_2O_4$ nanoparticles with microwave assisted combustion method were used urea, glycine and citric acid as fuel reagent, respectively. The reactions occured between nickel nitrate, iron nitrate and fuel reagents were given with Eq.1, Eq. 2 and Eq. 3 for urea, glycine and citric acid, respectively.

$$Ni(NO_3)_2.\ 6H_2O + 2Fe(NO_3)_3.\ 9H_2O + 6.66(NH_2)_2\,CO \longrightarrow NiFe_2O_4 + 37.33H_2O + 6.66CO_2 + 10.66N_2 \quad (1)$$

$$Ni(NO_3)_2.\ 6H_2O + 2Fe(NO_3)_3.\ 9H_2O + 4.44C_2H_5NO_2 \longrightarrow NiFe_2O_4 + 35.11H_2O + 8.88CO_2 + 6.22N_2 \quad (2)$$

$$Ni(NO_3)_2.\ 6H_2O + 2Fe(NO_3)_3.\ 9H_2O + 2.22C_6H_{10}NO_6 \longrightarrow NiFe_2O_4 + 35.11H_2O + 13.33CO_2 + 4N_2 \quad (3)$$

According to the reactions given above, nickel ferrite nanoparticles were obtained as a single product in the all of the reactions. However, $H_2O$, $CO_2$ and $N_2$ gases were evolved as gases in the products.

***Characterization***

In this study, XRD analysis were performed for the structural characterization of the synthesised nanoparticles. X-ray diffraction (XRD) analysis were performed with CuKα irradiation (λ=1.5418 Å) in the 2θ range10–80°. The microscopic and morphological properties of the nanoparticles were characterized by Nova NS450 scanning electron microscopy (FEI, USA). The surface areas and the other surface characteristics were measured by the nitrogen adsorption principle the particle surfaces at low temperature. These measurements were made with Gemini VI model surface analyzer (Micromeritics Instruments, USA). Magnetic hysteresis loops were determined at room temperature, for fields up to 3 T.

## 3. Results

The synthesised nanoparticles should to have small grains, high surface area and adequate level of saturation magnetization values. It is known that these characteristics significantly depend on the type of the fuel used as reagent. In this study, glycine, urea and citric acid were used as fuel, respectively. Then the obtained samples were characterized by XRD, SEM, BET and VSM techniques. The results are presented below, respectively.

***XRD Analysis***

X-ray analysis performs to identify the atomic and molecular structure of a crystal. XRD is one of the key analysis in the determining of the structure. The synthesis mixture prepared in stoichiometric rates was put in to the kitchen type microwave oven. In the end of reaction time was obtained a brown-black solid and this solid was milled in a balling mill. Then obtained this powder were characterized with X-Ray powder diffraction system. The X-ray difractograms of the obtained particles were given in Fig. 1.

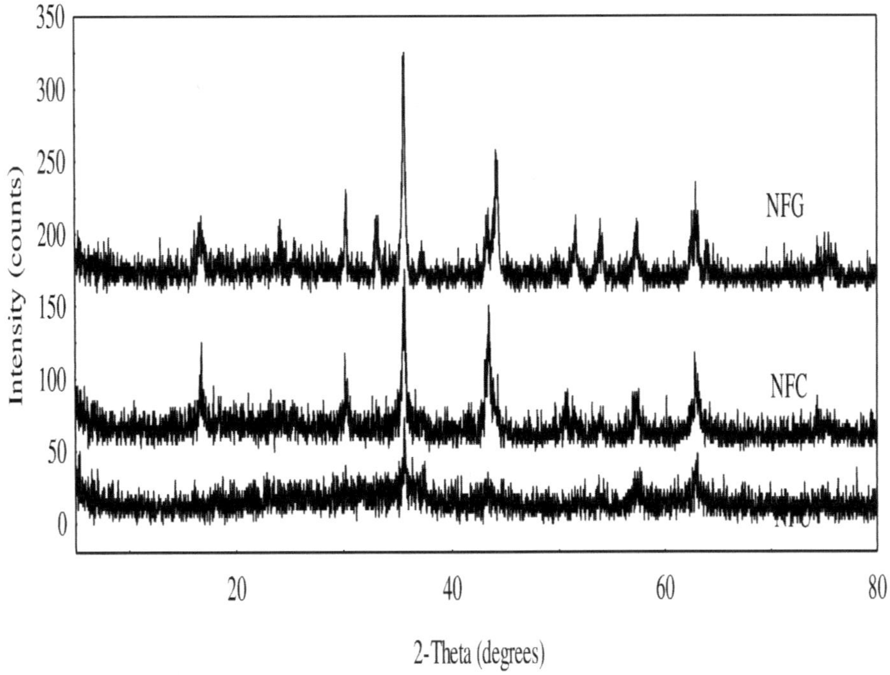


**Figure 1.** The X-Ray difractograms of the nanoparticles prepared with various fuels

According to the XRD patterns given in Fig. 1., narrow and sharp reflection peaks were observed in the measured diffraction pattern for three fuel types. Moreover, it can be said that the glycine between used fuels have the most significant positive effect on the crystal structure and the conversion. However, all the diffraction patterns of the obtained nanoparticles were matched with standard powder diffraction cards of $NiFe_2O_4$ (JCPDS Card Number 54-0964) and determined to be fully compatible with standard reflection peaks.

***Electron Microscopy (SEM) Analysis***

The $NiFe_2O_4$ nanoparticles prepared with microwave assisted combustion method used different fuel type were characterized with scanning electron microscopy. The SEM images of the $NiFe_2O_4$ nanoparticles synthesized were given in Fig. 2.

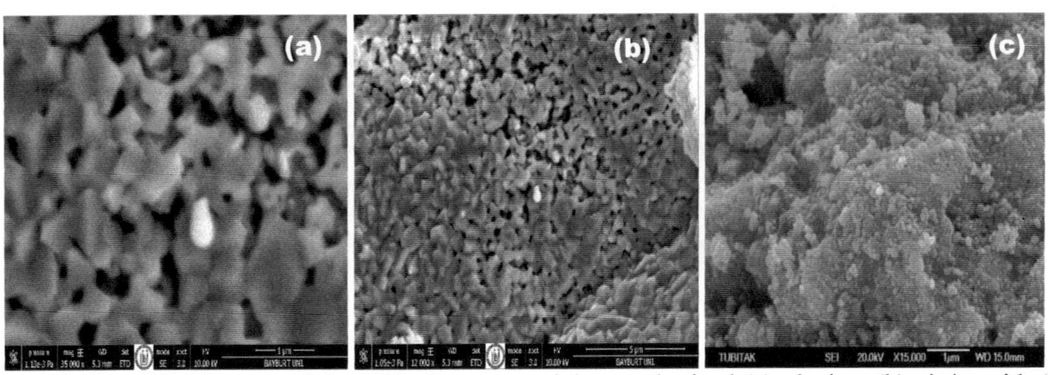


**Figure 2.** The SEM images of the $NiFe_2O_4$ nanoparticles synthesized (a) glycine, (b) citric acid, (c) urea

According to the results of SEM analysis, all nanoparticles prepared with three different fuel types consists of nanoscale particles. In addition, it was appeared that particles form large agglomerates. It was thought that formation of agglomerates originated from the electrostatic interactions between particles. Moreover, sintering may have occurred due from the happened high temperature during chemical reaction.

***BET Analysis***

To investigate the effects onto the surface properties of nanoparticles prepared with different fuels, BET analysis were made by using a surface analyzer. A sample contained in an evacuated sample tube is cooled (typically) to cryogenic temperature, and then is exposed to an analysis gas at a series of precisely controlled pressures. With each incremental pressure increase, the number of gas molecules adsorbed on the surface increases. The equilibrated pressure (P) is compared to the saturation pressure ($P_o$) and their relative pressure ratio ($P/P_o$) is recorded along with the quantity of gas adsorbed by the sample at each equilibrated pressure.

As adsorption proceeds, the thickness of the adsorbed film increases. Any micropores in the surface are filled first, then the free surface becomes completely covered, and finally the larger pores are filled by capillary condensation. The process may continue to the point of bulk condensation of the analysis gas. Then, the desorption process may begin in which pressure systematically is reduced resulting in liberation of the adsorbed molecules. As with the adsorption process, the changing quantity of gas on the solid surface at each decreasing equilibrium pressure is quantified. These two sets of data describe the adsorption and desorption isotherms. Analysis of the shape of the isotherms yields information about the surface and internal pore characteristics of the material. The obtained results in the end of measurement are shown in Figure 3. Various surface properties of the $NiFe_2O_4$ nanoparticles prepared with different fuels were presented in Table 1.

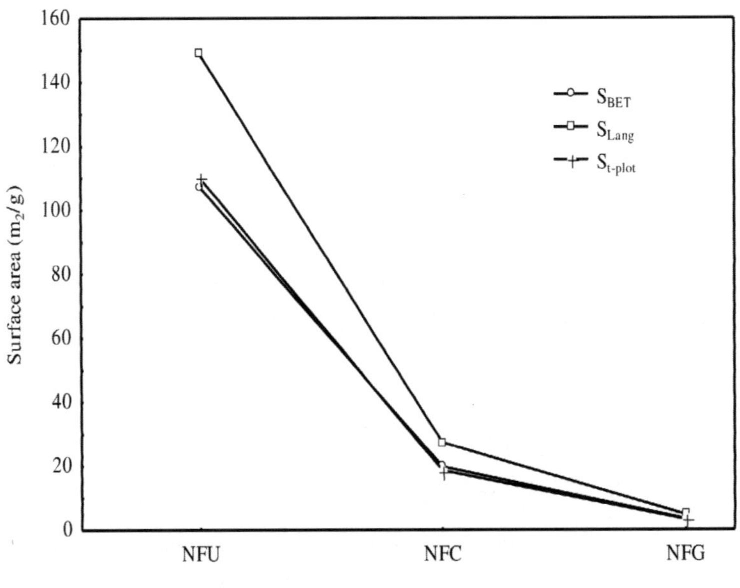


**Figure. 3**. Effect of type of fuel used over the surface area of the $NiFe_2O_4$ nanoparticles

**Table 1.** Various surface properties of the $NiFe_2O_4$ nanoparticles prepared with different fuels

| $NiFe_2O_4$ Samples | Fuel type | $S_{BET}$ ($m^2/g$) | $S_{Lang.}$ ($m^2/g$) | $S_{t\text{-}plot}$ ($m^2/g$) | Pore Volume ($cm^3/g$) | Pore Diameter (nm) |
|---|---|---|---|---|---|---|
| NFU | urea | 107.094 | 148.8080 | 110.082 | 0.187014 | 6.8648 |
| NFC | citric acid | 19.0784 | 27.3815 | 18.6425 | 0.052319 | 10.2866 |
| NFG | glycine | 3.4268 | 4.7601 | 3.2934 | 0.016057 | 15.1929 |

According to the results given in Fig. 3., the highest surface area was measured for the $NiFe_2O_4$ nanoparticles prepared with urea. The surface area properties of $NiFe_2O_4$ nanoparticles depend on the fuel type used in the synthesis. According to Table 1, the BET surface area of nanocyristalline $NiFe_2O_4$ was determined as ~108 $m^2/g$ for urea, ~19 $m^2/g$ for citric acid, ~4 $m^2/g$ for glycine. The pore volume per mass was determined to be 0.1870 $cm^3/g$ for urea, 0.0523 $cm^3/g$ for citric acid, 0.0160 $cm^3/g$ for glycine. The effect of fuel type used on the total pore volume and average pore diameter of the $NiFe_2O_4$ nanoparticles is given in Fig. 4.

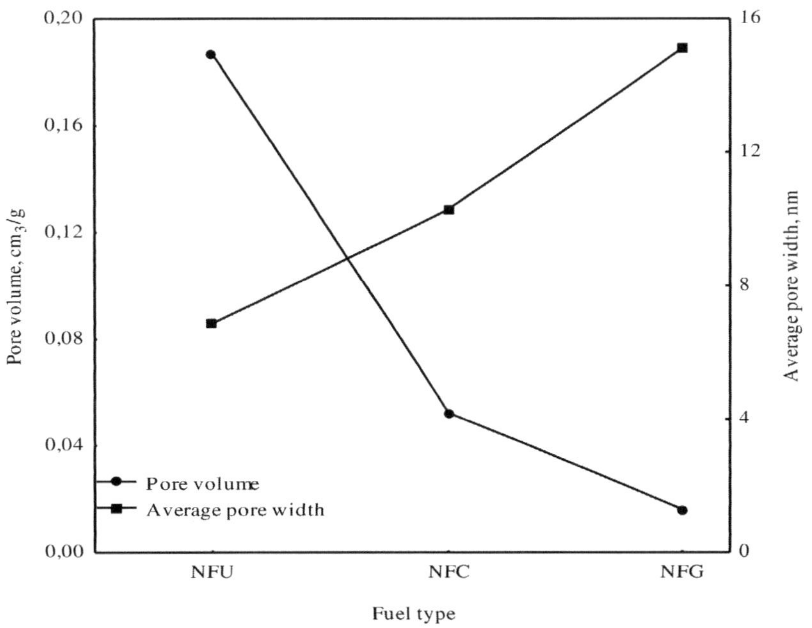


**Figure 4.** The effect of fuel type used on the total pore volume and average pore diameter of the $NiFe_2O_4$ nanoparticles

The highest value of average pore diameter was measured for the $NiFe_2O_4$ nanoparticles prepared with glycine. However, the highest value of total pore volume was measured for the $NiFe_2O_4$ nanoparticles synthesized using urea as fuel. Despite synthesized $NiFe_2O_4$ nanoparticles by using urea have the smallest pore diameter, they have the highest pore number. The smallest value of total pore was measured for sample prepared with glycine. Adsorption-desorption isotherms of the $NiFe_2O_4$ nanoparticles prepared with different fuels are showed Fig. 5. Hysteresis loop of adsorption and desorption isotherms were associated with a characteristic type IV and with mesoporous materials.

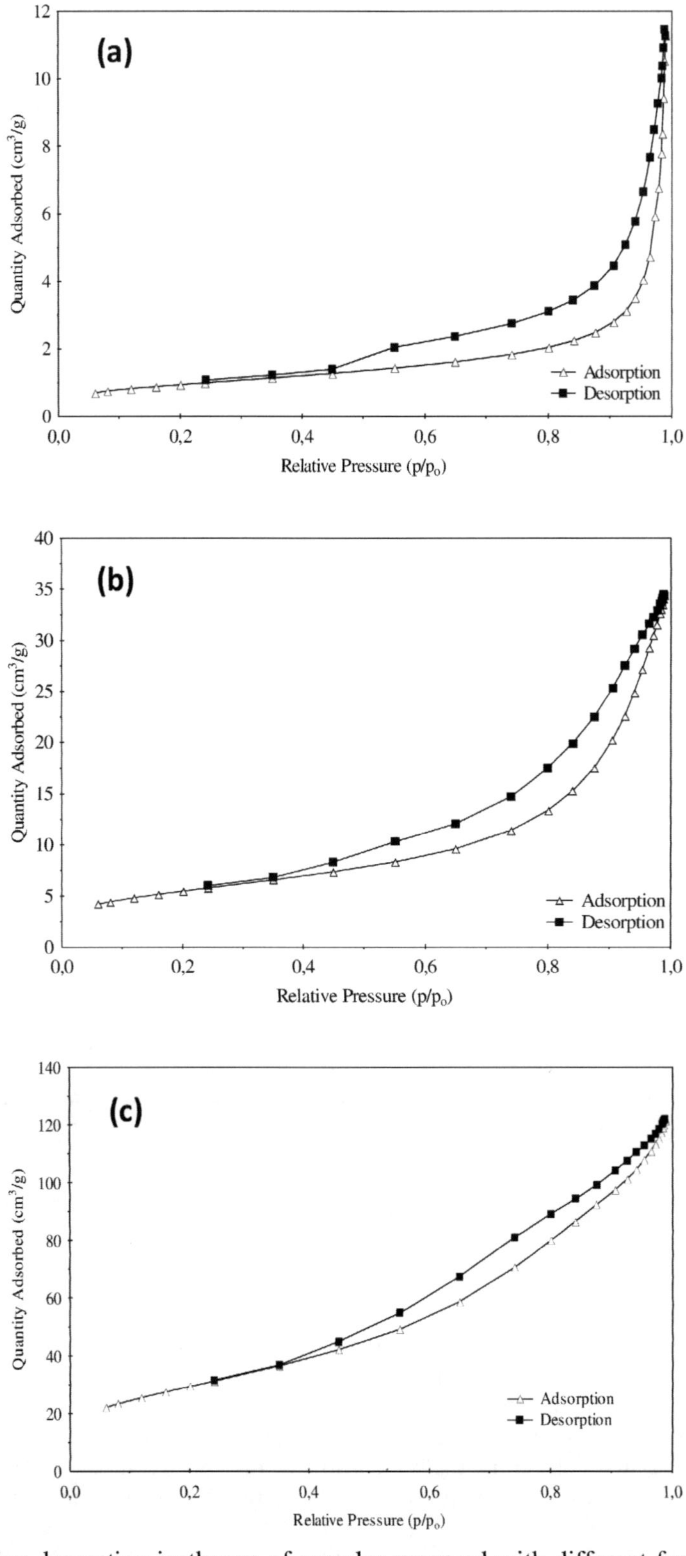


**Figure 5.** Adsorption-desorption isotherms of samples prepared with different fuels, **(a)** glycine, **(b)** citric acid, (c) urea

***VSM analysis***

In this study, it was investigated the effect of fuel type above magnetic properties of $NiFe_2O_4$ nanoparticles synthesized with microwave assisted combustion method. The fuel type used in synthesis of nanoparticles is quite effective above the crystal structure of nanoparticles. The crystal structure is the most important parameter on the magnetic properties of the nanoparticles. The magnetic properties of the materials are dependent on the sample shape, crystallinity, magnetization direction, etc. [14]. Magnetic characterization of the nickel ferrite nanoparticles is performed using a vibrating sample magnetometer (VSM) at room temperature with a maximum of 10kOe. The hysteresis loop curves of the nanoparticles were formed with the measured experimental data. The obtained results were given graphically in Fig.6.

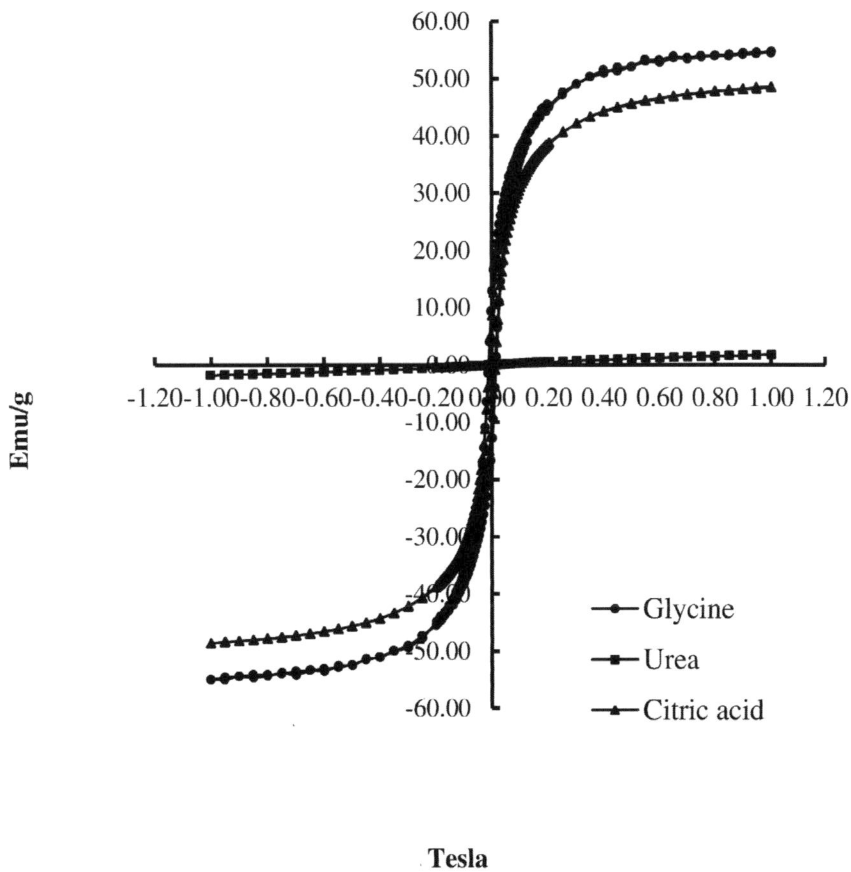


**Figure 6.** The hysteresis loop curves of the nanoparticles prepared with various fuels

VSM measurements performed at room temperature show that all of the nanoparticles synthesized with different fuel type exhibited ferromagnetic behaviour. The magnetic properties of an substance alike saturation magnetization ($M_s$), remanent magnetization ($M_r$) and coercivity ($H_c$) can be calculated from the hysteresis curve. Accordingly, magnetic properties measured at room temperature for nickel ferrite nanoparticles prepared with various fuels were presented in Table 2.

**Table 2.** The magnetic properties measured at room temperature for nickel ferrite nanoparticles prepared with various fuels.

| $NiFe_2O_4$ examples | Fuel Type | Saturation magnetization $M_s$ (emu/g) | Remanent magnetization $M_r$ (emu/g) | $M_r / M_s$ | Coercivity field $H_c$ (Oe) |
|---|---|---|---|---|---|
| NFU | Urea | 1.73 | 0.0084 | 0.00490 | 23 |
| NFC | Citric Acid | 48,43 | 5.1950 | 0.10727 | 112,4 |
| NFG | Glycine | 54.55 | 9,3259 | 0.17096 | 128,6 |

The expanded graphs of origin region on the hysteresiz curve in the range of approximately -400 to +400 Oe for the clear examination of hysteresiz are given in Fig. 7. Accordingly, the largest saturation magnetization value was measured as 54,55 emu/g for the nanoparticle prepared with glycine and the smallest value was measured as 1.73 emu/g for the nanoparticles produced with urea.

The saturation magnetization values of the nickel ferrite nanoparticles are observed to change with different fuel type. The changing in saturation magnetization values is associated with the cyristallinity of particles. According to graph given in Fig. 1, depending on the improvement in the crystal structure, the magnetization values of the nanoparticles were observed to increase.

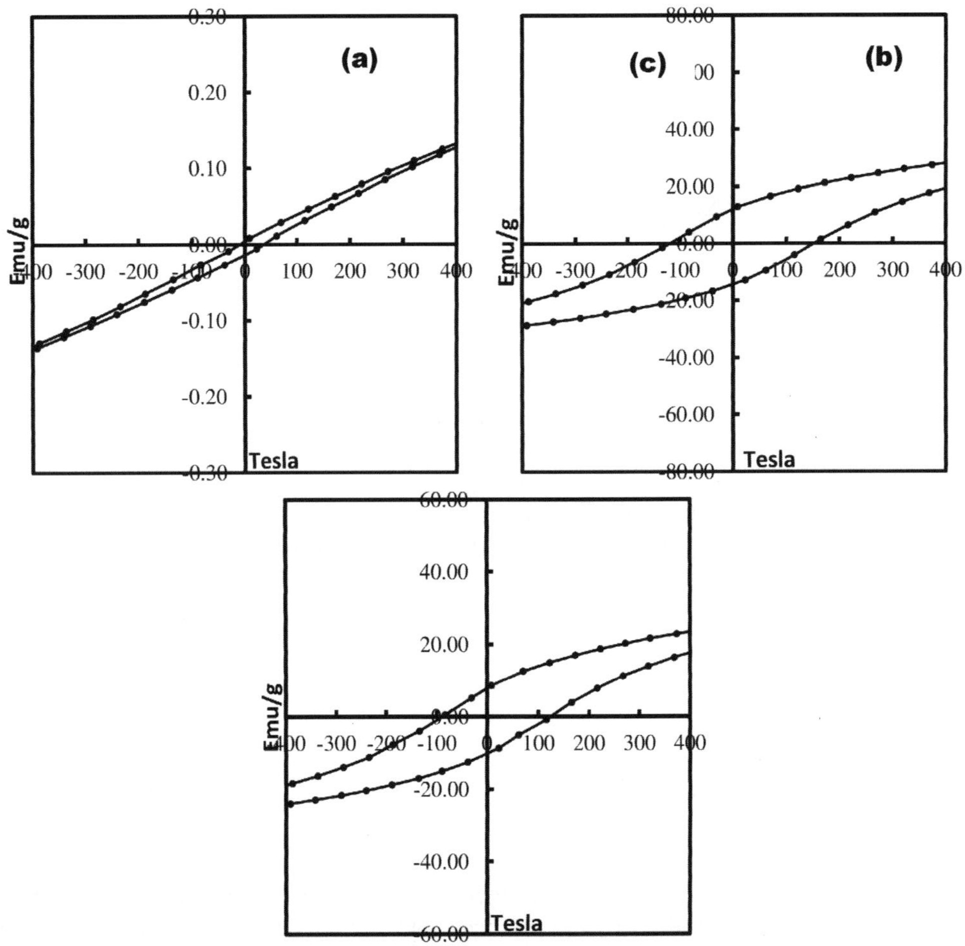


**Figure 7.** The expanded graphs of origin region on hysteresiz curve in the range of nearly -400 to +400 Oe. (a) urea, (b) glycine, (c) citric acid

## 4. Conclusion

$NiFe_2O_4$ nanoparticles can synthesis with various methods such as combustion, sol gel, hydrothermal etc... In this study, $NiFe_2O_4$ nanoparticles were synthesized with microwave assisted combustion method. In this method; some substances have to be used as fuel. The aim of this work was to investigate the effect of fuel type above structural, surface and magnetic properties of $NiFe_2O_4$ nanoparticles synthesized with microwave assisted combustion method. We observed that fuel type is quite effective on structural, surface and magnetic properties of the particles. The nanoparticles synthesized by using urea as fuel have got the highest surface area. But they have quite poor cyristallinity. On the other hand, the nanoparticles synthesized by using glycine and citric acid have got more than better cyristallinty. But their surface areas are smaller than the particles produced with urea.

According to the hysteresis loop curves of the nanoparticles, the magnetic properties of the $NiFe_2O_4$ nanoparticles changed with fuel type. The all of the samples showed ferromagnetic behaviour. But, the nanoparticles prepared by using glycine as fuel have got the more better saturation magnetization than the others.

## References

[1] M. Sugimoto, J. Am. Ceram. Soc., 82 [2] 269–80 (1999), 82 (1999) 269-280.
[2] V. Sepelaka, D. Baabeb, D. Mienertb, D. Schultzec, F. Krumeich, F.J. Litterstb, K.D. Becker, Journal of Magnetism and Magnetic Materials, 257 (2003) 377-386.
[3] A. Alarifi, N.M. Deraz, S. Shaban, Journal of Alloys and Compounds, 486 (2009) 501-506.
[4] A. Goldman, Modern Ferrite Technology, second ed., Springer, New York, 2006.
[5] M. Salavati-Niasari, F. Davar, T. Mahmoudi, Polyhedron, 28 (2009) 1455-1458.
[6] C.V. Gopal Reddy, S.V. Manorama, R. V.J., Sensors and Actuators B 55 (1999) 90-95.
[7] Y. Kinemuchi, K. Ishizaka, H. Suematsu, W. Jiang, K. Yatsui, Thin Solid Films 407 (2002) 109–113.
[8] A. Goldman, Modern Ferrite Technology, Marcel Dekker, NewYork, 1993.
[9] Z. Karcıoğlu Karakaş, R. Boncukcuoğlu, İ.H. Karakaş, M. Ertuğrul, Journal of Magnetism and Magnetic Materials, 374 (2015) 298-306.
[10] Y. Köseoğlu, A. Baykal, F. Gözüak, H. Kavas, Polyhedron, 28 (2009) 2887-2892.
[11] T. Mimani, K.C. Patil, Mater.Phys.Mech. , 4 (2001) 134-137.
[12] B.M. Abu-Zied, Colloids and Surfaces A: Physicochem. Eng. Aspects 211 (2002) 27-42.
[13] M.R. Phadatare, V.M. Khot, A.B. Salunkhe, N.D. Thorat, S.H. Pawar, Journal of Magnetism and Magnetic Materials, 324 (2012) 770-772.
[14] M.-z. Wu, G.-y. Quan, Y.-m. Liu, Y.-q. Ma, P. Dai, L.-d. Zhang, Transactions of Nonferrous Metals Society of China, 19 (2009) 1562-1566.

# The comparison of performance by using alternative refrigerant R152a in automobile climate system with differant artificial neural network models

**A T Kalkisim[1], A S Hasiloglu[2], K Bilen[3]**

[1]MYO Mechatronic, Gumushane University, Baglarbasi, 29100, Gumushane,Turkey,
[2]Computer Engineering, Ataturk University, 25100, Erzurum, Turkey
[3]MechanicalEngineering, Ataturk University,25100, Erzurum, Turkey,

E-mail: atkalkisim@gumushane.edu.tr

**Abstract.** Due to the refrigerant gas R134a which is used in automobile air conditioning systems and has greater global warming impact will be phased out gradually, an alternative gas is being desired to be used without much change on existing air conditioning systems. It is aimed to obtain the easier solution for intermediate values on the performance by creating a Neural Network Model in case of using a fluid (R152a) in automobile air conditioning systems that has the thermodynamic properties close to each other and near-zero global warming impact. In this instance, a network structure giving the most accurate result has been established by identifying which model provides the best education with which network structure and makes the most accurate predictions in the light of the data obtained after five different ANN models was trained with three different network structures. During training of Artificial Neural Network, Quick Propagation, Quasi-Newton, Levenberg-Marquardt and Conjugate Gradient Descent Batch Back Propagation methodsincluding five inputs and one output were trained with various network structures. Over 1500 iterations have been evaluated and the most appropriate model was identified by determining minimum error rates. The accuracy of the determined ANN model was revealed by comparing with estimates made by the Multi-Regression method.

## 1. Introduction

Today, cooling systems stand out as increasingly important systems and they are needed in many areas. The advancement of technology and the development of faster systems make the need for better cooling capacity. As well as cooling systems are meeting the technological needs, they meet an important need to improve important factors of living standards in terms of climate and comfort. Among them, air conditioning systems that provide in-cab comfort become even more important with also considering more than '1 billion vehicles' [1] on the world.

In a world where the number of vehicles exceeded 1 billion, while refrigerants used in air-conditioning systems in these vehicles are providing comfort, these refrigerants need to be used by considering environmental factors of them. In recent years, fluorinated gases damaging the ozone layer are banned and gases which have less decay time in the atmosphere are preferred with considering the global warming factor [2]. These gases or gas mixtures must be compatible with over 1 billion vehicles already used and work efficiently, safely and environmentally friendly without causing significant changes.

In this regard, many gases are being tested and recommended as alternative refrigerants. One of these gases is the refrigerant R152a. An artificial neural network model has been designed by using thermodynamic properties of this fluid and estimates close to real were obtained in wide ranges.

## 2. Artificial neural networks

Artificial neural networks (ANN) are implementation technique as a computable model based on the human brain by imitating the biological nervous system. In general, neurons are the basic elements of an artificial neural network model. These neurons are connected by synapses with specific weights [3]. ANN models are composed of 3 main layers. Although the task and function of each layer are different, they are associated with each other.

### *2.1. Input Layer*

There are entries in this layer in accordance with the principle of human brain. These entries consist of raw data waiting to be processed like our data that we get in communication with the outside world by 5 sense. These entries will be processed in ANN and they will have lots of weights that take us to an output value in the algorithm. Artificial neural network model input values should only be introduced as mathematical values and not contradict with learning of the model. These values in the input layer are transmitted to the intermediate layer without being subjected to a process in the input layer.

### *2.2. Hidden Layer (Intermediate Layer)*

This layer is the layer where the input values which came connected with certain weights are processed and evaluated. The number of nodes within this layer affects the relationship condition and the weight value between inputs and nodes. In this regard, it is being tried to produce estimates with minimal error by increasing or decreasing the number of nodes for an optimal network model.

### *2.3. Output Layer*

Values entered in the input layer are carried out to the output layer with another weight value after processed in the intermediate layer. The desired output value is given by ANN in this layer [4].

Many important issues pose while ANN modeling. It is aimed to build a network model with minimum error distribution for ANN models. As a result of access to the local values of this stage, it can be detected that a better distribution of errors is blocked. Important factors emerge for the best estimate of the model, such as the teaching network is starting from which value, how many iterations are performed and at which stage of learning the system would stop. At this stage, the working of artificial neural network model is not known exactly so which method and model are the most appropriate must be found with experiments. In this context, different models are used for different neural network analysis.

## 3. ANN modeling

ANN models which will allow for the best estimate of the thermodynamic properties of R152a refrigerant for automobile air conditioning systems was trained and the different models of ANN were compared. Created ANN models were given in Table 1.

The importance of the generated data sets for ANN models affects the estimate accuracy of the model. In this context, an adequate number of data collection in a wide range and then the desired estimating values must be in the solution space of the input value.

The number of data in the modeling phase in the input layer of ANN is 5 for all modeling and the output layer is designed to be 1. Network was trained at every turn with 3, 5 and 7 nodes in the intermediate layer, respectively. At this stage, the unused test data in the training stage given and the system was asked to estimate. The results of error rates after estimates were compared and each model which correctly estimated with minimum error was determined. Due to the use of different ANN structures, it was determined to have a different estimate error rates. Evaporator Temperature (Te),

Condenser Temperature (Tk), Condenser Heat Load (Qk), Evaporator Heat Load (Qe) and Mass Flow (m) were used as input values. That's why the input values in all methods and models are 5 pcs. Output values are Cooling Coefficient of Performance (COP), the Compressor Work (Wk) and Compression Ratio (Pr).

**Table 1.** Representation of ANN models used in this research (for a single output value)

| ANN No | Method Name | Naming of ANN Models | Number of Nodes in Layers | | |
|---|---|---|---|---|---|
| | | | *Input Layer* | *Inter Layer* | *Output Layer* |
| 1 | Quick Propegation | QP 531 | | 3 | |
| 2 | Quick Propegation | QP 551 | | 5 | |
| 3 | Quick Propegation | QP 571 | | 7 | |
| 4 | Back Propegation | BP 531 | | 3 | |
| 5 | Back Propegation | BP 551 | | 5 | |
| 6 | Back Propegation | BP 571 | | 7 | |
| 7 | Levenberg Marquerdt | LM 531 | | 3 | |
| 8 | Levenberg Marquerdt | LM 551 | 5 | 5 | 1 |
| 9 | Levenberg Marquerdt | LM 571 | | 7 | |
| 10 | Quasi-Newton | QN 531 | | 3 | |
| 11 | Quasi-Newton | QN 551 | | 5 | |
| 12 | Quasi-Newton | QN 571 | | 7 | |
| 13 | Conjugate Gradient Descent | CG 531 | | 3 | |
| 14 | Conjugate Gradient Descent | CG 551 | | 5 | |
| 15 | Conjugate Gradient Descent | CG 571 | | 7 | |

3 different individual ANN models were developed for 3 different outputs. So ANN was trained by asking only one output value with 5 input values every time. Therefore, ANN models were created separately for the COP, Wk and Pr values. The total number of training was done by 5 inputs and 3 individual outputs for each. it was identified as 45, as a result of the calculation with 15 different ANN models. 45 ANN modeling and training were made and the same training set and the same testing set were used for each. All ANN models were limited with 1500 iterations and training was completed in 1500 iterations. Besides training of 45 ANN models, estimates were made by the Multi-Regression method separately for COP, Pr and Wk values and it has participated in the evaluation to make comparisons.

## 4. Calculations of cooling system

The knowledge of thermodynamic properties of the fluid in a cooling cycle gives an idea in terms of where the fluid can be used. In calculations made by using COOLPAC program, variations were calculated and simulated by including compressor constant speed n = 1000 rpm, n = 0.7 for a volumetric efficient system, the fluid pressed to system between Tk 40 °C to 60 °C and Ta -20 °C to -6 °C with a 4-cylinder compressor. Pressure Ratio (Pr), the compressor work (Wk), and Cooling Efficacy Coefficient (COP) values were determined by the results of calculations and modeling and calculations were completed.

R152a which may be used as an alternative gas in automobiles instead of R134a is also required to have high value up to R134a in terms of COP. In this regard, the load on the compressor and the compression ratio are also required to have close to the values of R134a gas. Because a system in a vehicle is established and being used according to the compression ratio and compressor work of currently used R134a gas. Pr and Wk values are essential to be close to R134a to charge of R152a gas without making major structural changes in the car. In this regard, COP, Wk and Pr values were estimated in our ANN model.

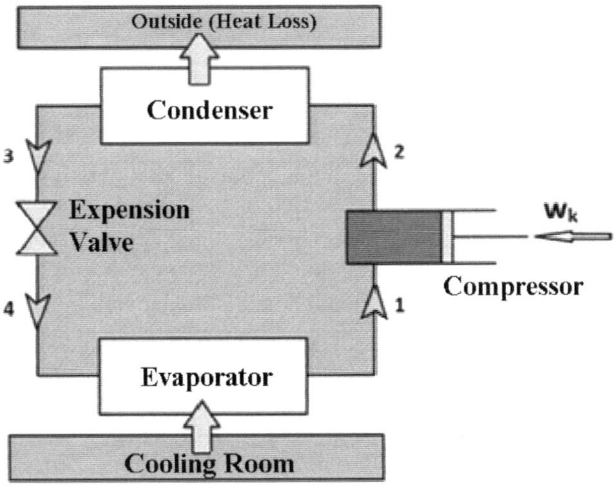


**Figure 1.** The General Design of a Cooling System [5]

## 5. Creation of ANN model

89 data were used in models. 49 of these data were separated for training data, 40 of them were used as test data to be used for control purpose subsequently in trained ANN models. The obtained data were mixed in completely mixed form and randomly selected without being subjected to a particular sequence. Determined ANN models and structures were trained with obtained data respectively and finally a suitable model for this system was identified and the system was designed. In performed application, formulas were created by using the same values in the calculation of Multi-Regression method and these results were compared with results of ANN.

Applied ANN models are Conjugate Gradient Descent, Levenberg Marquardt, Quick Propagation, Batch Backpropagation and Quasi Newton. They were trained with a maximum of 1500 iterations. The system was trained repeatedly and best results were obtained by using structures with the number of 3, 5 and 7 nodes individually in the intermediate layer in general methods.

## 6. Operating Data

Indications of the completion of training for ANN models are Correlation and $R^2$. The case of these values are close to 1 shows that the ANN model is well trained.

As shown in Table 2, correlation and $R^2$ values are close to 1 and it is understood from these values that our networks are very well trained and tested. After the training phase accuracy of trained and tested networks was evaluated with never trained 20 non-tested data. Root Mean Square Error (RMSE) value is a value that indicates the estimated error rate for the test data made by each trained structure. The case that this value is close to zero shows that the closest estimate was made to the real value. RMSE (obtained from Equation 1). In Equation 1; N: the estimate number, t: the estimate value and y: the actual value [6].

$$RMSE = \sqrt{\frac{1}{N}\sum_{i=1}^{N}(t_i - y_i)^2} \tag{1}$$

Regression was performed by using Multi Regression (MR) method with the same training data to perform the comparison of the ANN models. COP, Wk and Pr values were found for the test data as a result of multiplying the input values with related coefficients. Related formula and training data for MR developed for each output value are as follows.

**Table 2.** Correlation and $R^2$ values of the trained ANN methods and models

| Name of ANN Method | Correlation | | | R-Square | | |
|---|---|---|---|---|---|---|
| | COP | Wk | Pr | COP | Wk | Pr |
| QP 531 | 0,995511 | 0,994809 | 0,998055 | 0,990534 | 0,988507 | 0,995887 |
| QP 551 | 0,998361 | 0,99901 | 0,997789 | 0,996444 | 0,997898 | 0,995294 |
| QP 571 | 0,996581 | 0,995502 | 0,998354 | 0,992887 | 0,989918 | 0,996589 |
| BP 531 | 0,964123 | 0,97711 | 0,975913 | 0,784525 | 0,90767 | 0,922974 |
| BP 551 | 0,973479 | 0,986111 | 0,978356 | 0,891928 | 0,958919 | 0,932567 |
| BP 571 | 0,989038 | 0,988618 | 0,978116 | 0,976656 | 0,967273 | 0,934104 |
| LM 531 | 0,99133 | 0,997268 | 0,992393 | 0,982422 | 0,993948 | 0,984604 |
| LM 551 | 0,984679 | 0,993045 | 0,992754 | 0,96551 | 0,981259 | 0,98543 |
| LM 571 | 0,985546 | 0,992786 | 0,999995 | 0,965478 | 0,985459 | 0,999988 |
| QN 531 | 0,996561 | 0,999897 | 0,998921 | 0,992956 | 0,999784 | 0,99779 |
| QN 551 | 0,996426 | 0,999983 | 0,998912 | 0,99267 | 0,999965 | 0,997802 |
| QN 571 | 0,997125 | 0,999723 | 0,998823 | 0,994184 | 0,999432 | 0,997555 |
| CG 531 | 0,996586 | 0,998863 | 0,998749 | 0,99301 | 0,997557 | 0,997435 |
| CG 551 | 0,996798 | 0,99992 | 0,998835 | 0,993379 | 0,999832 | 0,997613 |
| CG 571 | 0,997205 | 0,999916 | 0,998721 | 0,994379 | 0,999822 | 0,997344 |

**Table 3.** Multi Regression Training Data Set

| *Regression Statistics* | | | *Coefficients* | *P-value* |
|---|---|---|---|---|
| Multiple R | 0,995777752 | Intercept | 11,55384695 | 7,17E-05 |
| R Square | 0,991573331 | Tk | -0,011123381 | 0,147668 |
| Adjusted R Square | 0,990893761 | Te | 0,15328854 | 0,002431 |
| Standard Error | 0,061958923 | Qk | -23,73707892 | 0,003187 |
| Observations | 68 | Qe | 19,3153009 | 1,78E-05 |
| | | M | 1412,804823 | 0,145967 |

In tables 4, 5 and 6, RMSE values of estimates done by ANN models created with different models and structures for COP, Wk and Pr output values were calculated and given below respectively. To make a comparison for RMSE values of created ANN models, RMSE values of estimates made by MR method were also added to the tables and the comparison was made with performed ANN models. Structures which have the best RMSE values were determined for each structure from the ANN models indicated in the table below. Then, the structure and the model which estimated closest to the actual were determined by evaluating determined models with the best structure among themselves.

When we examine RMSE values of each model in the tables, they are marked in bold. The best value is shown underlined among these values. The real-estimate graphic is given in test data of a model that has the best results and structures were shown.

As seen in Table 4, an optimal ANN model is 7 knotted and intermediate layered CG model structure. For the COP output values, 7 knotted intermediate layers showed good results for all other methods except QN. As it can be seen in Chart 1, the estimation curve decomposes for 1 value. All other estimates are consistent with the actual values and it was seen that a good learning was provided. While RMSE value obtained with the MR method was 2.99, RMSE value in the CG model decreases to 1.46. This shows that the ANN models are very successful. The model structure is as shown in Figure 2.

**Table 4.** COP estimate RMSE values for ANN models

| | COP | | |
|---|---|---|---|
| Model | QP 531 | QP 551 | QP 571 |
| RMSE | *1,952573* | *2,966885* | ***1,510595*** |
| Model | BP 531 | BP 551 | BP 571 |
| RMSE | *4,143737* | *2,759268* | ***2,154733*** |
| Model | LM 531 | LM 551 | LM 571 |
| RMSE | *2,965355* | *2,991904* | ***2,368911*** |
| Model | QN 531 | QN 551 | QN 571 |
| RMSE | *2,025418* | ***1,673851*** | *1,944325* |
| Model | CG 531 | CG 551 | **CG 571** |
| RMSE | *1,813621* | *1,504902* | ***1,462868*** |
| Model | MR | | |
| RMSE | ***2,990578*** | | |

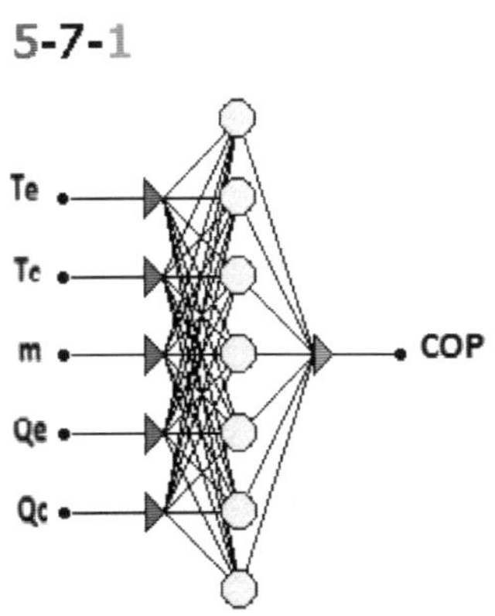


**Figure 2.** 5-7-1 ANN structure with Conjugate Gradient Descent method

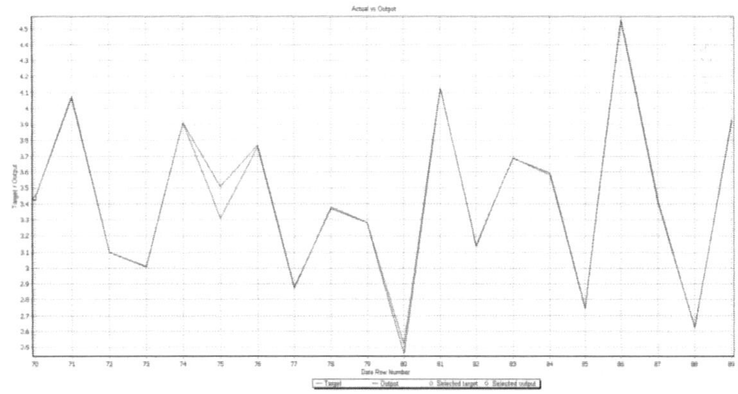

**Chart 1.** "The Real-Estimate Values" Chart for Test Data

As seen in Table 5, an optimal ANN model is 5 knotted and intermediate layered QN model structure. Number of nodes in the intermediate layers were found to vary among the best-estimate models for Wk output values. What number of nodes will give the better results were not determined for the intermediate layer. As it can be seen in Chart 2, estimated values were found to be very close to the actual results for all values. RMSE value in the QN model was calculated as low as 0,15 while RMSE value obtained with MR method was 0,19. In this respect, the ANN models was shown to be much more successful. Model structure is as shown in Figure 3.

**Table 5.** Wk estimate RMSE values for ANN models

| | Wk | | |
|---|---|---|---|
| Model | QP 531 | QP 551 | QP 571 |
| RMSE | *1,021062* | ***0,50382*** | *0,840244* |
| Model | BP 531 | BP 551 | BP 571 |
| RMSE | *2,305302* | *1,371869* | ***1,229686*** |
| Model | LM 531 | LM 551 | LM 571 |
| RMSE | ***0,818307*** | *0,956933* | *1,184893* |
| Model | QN 531 | **QN 551** | QN 571 |
| RMSE | *0,180927* | ***0,158108*** | *0,561539* |
| Model | CG 531 | CG 551 | CG 571 |
| RMSE | *0,376303* | *0,183277* | ***0,176379*** |
| Model | MR | | |
| RMSE | ***0,199599*** | | |

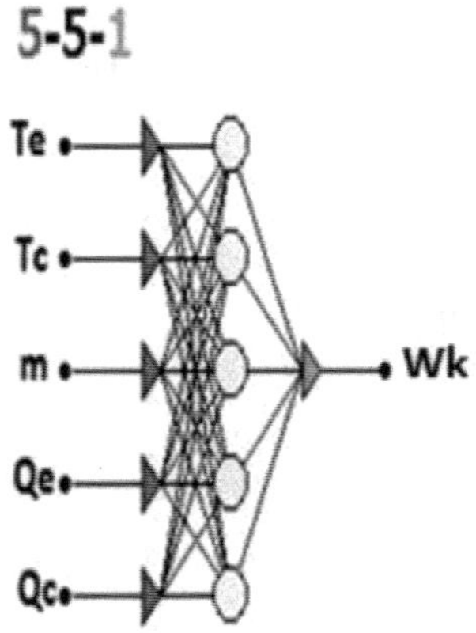


**Figure 3.** Created ANN structure with 5-5-1 structure with Quasi-Newton method.

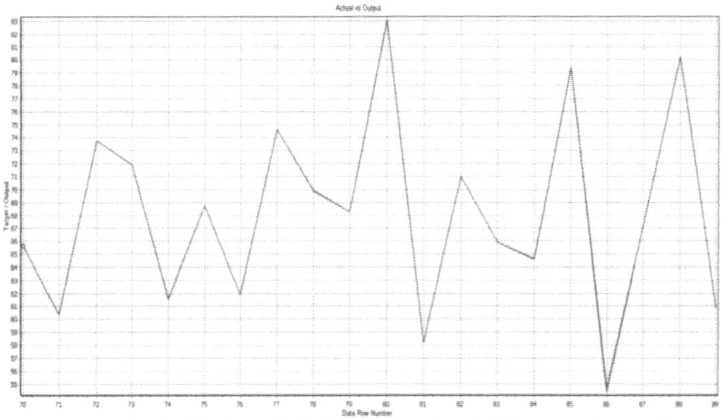

**Chart 2.** "The Real-Estimate Values" Chart for Test Data

**Table 6.** Pr estimate RMSE values for ANN models

| | Pr | | |
|---|---|---|---|
| Model | QP 531 | QP 551 | QP 571 |
| RMSE | *0,77135* | *0,987961* | ***0,442178*** |
| Model | BP 531 | BP 551 | BP 571 |
| RMSE | *3,221279* | ***3,06141*** | *3,180418* |
| Model | LM 531 | LM 551 | LM 571 |
| RMSE | *2,625556* | *2,738073* | ***0,692605*** |
| Model | QN 531 | **QN 551** | QN 571 |
| RMSE | *0,607295* | ***0,403461*** | *1,003959* |
| Model | CG 531 | CG 551 | CG 571 |
| RMSE | **0,515271** | 0,574139 | 0,52654 |
| Model | MR | | |
| RMSE | ***2,531808*** | | |

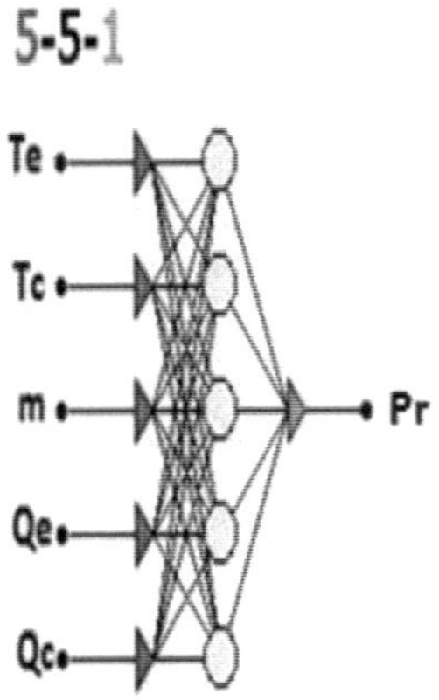


**Figure 4.** Created ANN structure with 5-5-1 structure with Quasi-Newton method.

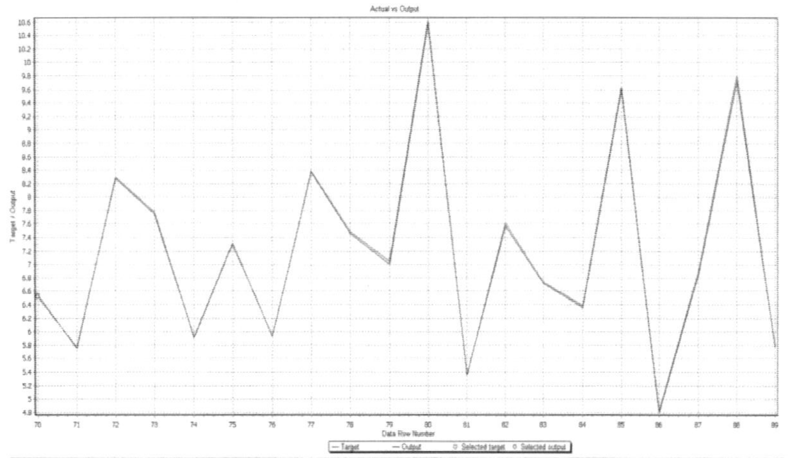


**Chart 3.** "The Real-Estimate Values" Chart for Test Data

As seen in Table 6, an optimal ANN model is 5 knotted and intermediate layered QN model structure. Number of nodes in the intermediate layers were found to vary among the best-estimate models for Pr output values. What number of nodes will give the better results were not determined for the intermediate layer. As it can be seen in Chart 3, estimated values were found to be very close to the actual results for all values. RMSE value in the QN model was calculated as low as 0,40 while RMSE value obtained with MR method was 2,53. In this respect, ANN models were shown to be much more successful. Model structure is as shown in Figure 4.

## 7. Conclusion

Because network models are made separately for *COP, Wk* and *Pr,* it was found that different models gave good results for each output value. For this reason, when determining the best ANN models, the difference between the percentages of RMSE error rates was taken into account. The same training data, the same validate data and the same test data were used when determining the network design which made the best estimation for 3 different output values. Networks with 3 different structures were created to determine the best model. Results were obtained by also making comparisons with values calculated with Multi Regression for 45 different network models.

It was observed that ANNs estimated unseen data with very minor differences with the success of our training set. The most suitable ANN model was identified as QN model in our system for the alternative refrigerant R152a for car cooling systems. When looking at the RMSE value, it was determined that it is 27% more successful at Pr value estimation and 11% more successful at Wk value estimation. It was seen that it made 14% poor estimation in the COP value according to the CG model. Also, QN 551 model yielded the average of 210.6% better results compared to estimations made by MR method. Second model providing the best results is CG and it yielded the average of 169.33% better results compared to estimations made by MR method.

It was determined that the QN model within trained 15 network models with 5 intermediate nodes was the model providing best average results across all models for 3 different output values with the number of iterations limited to 1500.

## References

[1] http://www.worldometers.info/cars/
[2] Talley E 2010 *R1234yf Refrigerant* Fall 2010 ICAIA Conference.

[3] Kamar H M, Ahmad R, Kamsah N B and Mustafa A F M 2013 *Artificial Neural Networks for automotive air-conditioning systems performance prediction* Applied Thermal Engineering pp 63-70

[4] Kuzucuoglu A E 2010 *İklimlendirme Deney tesisatının Neuro-Fuzzy Yöntemi ile Kontrolü* (İstanbul: Marmara Üniversitesi Press) pp 20-52

[5] Kalkisim A T 2015 *Otomobil Klima Sistemleri için Alternatif Soğutkan R152a ve ANFIS Uygulaması*, (Erzurum : Ataturk University Press) p 5

[6] Kuo Y M, Liu C W 2004 *Evaluation of the ability of an artificial neural network model to assess the variation of groundwater quality in an area of blackfoot disease in Taiwan*, Water Research pp 148-154

# Artificial Intelligence Techniques for the Estimation of Direct Methanol Fuel Cell Performance

**Abdulsamet Hasiloglu[1*], Ömür Aras[2], Mahmut Bayramoglu[2]**

[1]Department of Computer Engineering, Ataturk University, Erzurum 25240, TURKEY
[2]Department of Chemical Engineering, Gebze Technical University, Kocaeli 41400, TURKEY

asamet@atauni.edu.tr

**Abstract.** Artificial neural networks and neuro-fuzzy inference systems are well known artificial intelligence techniques used for black-box modelling of complex systems. In this study, Feed-forward artificial neural networks (ANN) and adaptive neuro-fuzzy inference system (ANFIS) are used for modelling the performance of direct methanol fuel cell (DMFC). Current density (I), fuel cell temperature (T), methanol concentration (C), liquid flow-rate (q) and air flow-rate (Q) are selected as input variables to predict the cell voltage. Polarization curves are obtained for 35 different operating conditions according to a statistically designed experimental plan. In modelling study, various subsets of input variables and various types of membership function are considered. A feed -forward architecture with one hidden layer is used in ANN modelling. The optimum performance is obtained with the input set (I, T, C, q) using twelve hidden neurons and sigmoidal activation function. On the other hand, first order Sugeno inference system is applied in ANFIS modelling and the optimum performance is obtained with the input set (I, T, C, q) using sixteen fuzzy rules and triangular membership function. The test results show that ANN model estimates the polarization curve of DMFC more accurately than ANFIS model.

**Keywords:** Direct Methanol Fuel Cell, Artificial Neural Networks, Neuro-Fuzzy, Modelling, Performance

## 1. Introduction

In recent years, the depletion of fossil fuel reserves, global warming and other environmental concerns have accelerated researches on fuel cell technology. Among various types of fuel cells, polymer electrolyte (PEM) hydrogen and direct methanol fuel cells (DMFC) are the most intensively searched as principal candidates for stationary and portable electrical energy sources. Liquid methanol has advantages of easy fuel storage and higher energy density compared to hydrogen. But, slow anode kinetics and methanol crossover are still major barriers to its commercialization.

Researches on fuel cells are mainly focused on the development of the various components of the membrane-electrode assembly (MEA), especially on the anode catalyst and PEM. On the other hand, the development of elaborated mathematical models is another active research area, as they are valuable tools to analyse ultimate details of electrochemical and physical phenomena occurring in various parts of the fuel cell [1- 7]. In this context, 3D models have been developed in order to detect the effects of operation conditions as well as the other structural-spatial and material properties on the complex mass transport, electrode kinetics, and two-phase flow in various parts of the fuel cell [8, 9]. These computational fluid dynamics (CFD) models which need commercial softwares and also a huge amount of thermo-physical parameters values to be estimated are usually used for single cell analysis but they are not suitable for analysis, design, optimization and control tasks at system level. Thus, semi-analytical or empirical models find still applications in these areas.

Empirical models can be divided as parametric and non-parametric ones. Parametric empirical models are based on pre-determined relations derived from scientific knowledge on activation, ohmic and mass transfer losses [10-13]. They correlate well fuel cell output as a function of current density, temperature, and fuel-oxidant concentrations, but they usually lack of other process variables affecting fuel cell performance. Furthermore, the model parameters have to be determined under different operating conditions for different systems.

In recent years, fuzzy logic and artificial neural networks (ANN) have been extensively used as effective non-parametric modelling tools capable of estimating non-linear nature of complex systems by means of sufficiently representative data sets. ANN offer exciting advantages, such as learning, adaptation, fault-tolerance, parallelism and generalization. Highly nonlinear complex relationships are implicitly put down in the weights of the network without the need of an explicit mathematical relation. On the other hand, fuzzy logic provides an inference morphology that enables to approximate human reasoning capabilities to be applied to knowledge-based systems. Fuzzy modelling is concerned with the construction of fuzzy inference systems that can predict and explain the behaviour of ill-defined or complex systems which requires otherwise an excessive number of parameter data and sophisticated numerical software for model solving. In recent years, various advantages of these two modelling approaches are beneficiated by hybrid approaches such as adaptive neuro-fuzzy inference system (ANFIS). Besides the modelling purposes, these techniques are successfully applied in process control for various engineering applications.

Despite the broad applications of these artificial intelligence techniques, a few studies have been conducted on the modelling of various types of single fuel cells [14-19] and cell stacks [20, 21]. In this study, feed-forward ANN and ANFIS models have been used to predict the polarization curve of DMFC, using current density, fuel cell temperature, methanol concentration, gas and liquid flow-rates as operating variables for a given MEA, channel geometry and dimensions. Various aspects of these modelling techniques have been investigated and comparative results on the test performance have been given.

## 2. Theory

### *2.1 Feed -Forward ANN*

Inspired from the human brain, ANN is a layered organization of neurons. Among various types of organization (architecture), the most popular is multilayer feed-forward ANN consisting of an input layer, a series of hidden layers and an output layer, as exemplified in Fig.1 . The weighted sum of the input signals fed to a neuron is transformed to a normalized output signal by means of an activation function and transferred to the subsequent layer. The most popular activation functions are “tansig” and “logsig” function defined mathematically as

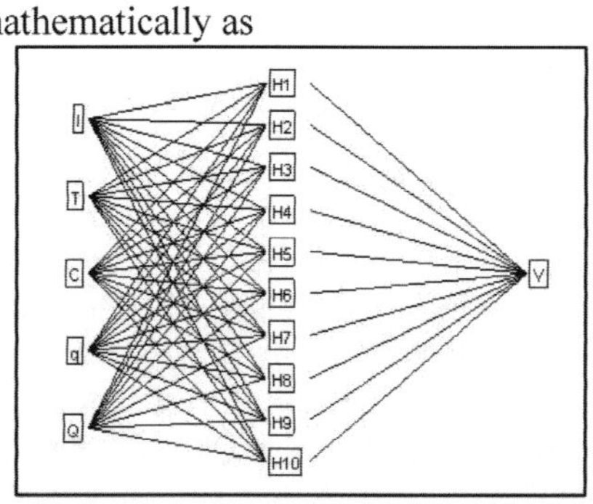


**Figure 1.** One hidden layered feed-forward ANN

$$\textit{Logsig activation function}: \; y = 1/(1+e^{-x}) \tag{1}$$

$$\textit{Tansig activation function}: \; y = 2/(1+e^{-2x}) - 1 \tag{2}$$

Mathematically, ANN is a black box containing a weight matrix optimized by a training process. For prediction or forecasting purposes, ANN model is trained usually by supervised learning using a data set organized usually in matrix form with rows corresponding to cases (experimental results) and columns corresponding to variables. The most widely used training algorithm for multilayer feed-forward ANN is the back-propagation. During the training process, model outputs are recalculated until some function of the network errors (the differences between model and experimental outputs) is minimized iteratively to yield finally the optimum weight matrix. "Overtraining" which falls the forecasting capability of the network due to the structure excessively adopted to the training data is avoided by using an independent check data; the training process is terminated when checking error tend to increase even if the training error continues to decrease. The ultimate success of the model is always assessed by an independent test data.

### 2.2 *Fuzzy inference system*

Fuzzy inference is a logical method that interprets the values in the input vector (x), and based on some set of fuzzy "*if-then*" rules, assigns values to the output vector (y):

$$\textit{IF } x \text{ is } A \quad \textit{THEN } \; y \text{ is } B \tag{3}$$

Where $A$ and $B$ are labels of fuzzy sets, e.g., "low", "high". Each fuzzy set is characterized by appropriate membership functions that map each element to a membership value between $0$ and $1$. Triangular, trapezoidal, Gaussian and disigmoidal types are examples of widely used membership functions.

A fuzzy inference system consists of three conceptual components namely, a *rule base* comprising fuzzy rules, a *database* defining the membership functions of the fuzzy sets used in the fuzzy rules and a *reasoning mechanism* which performs the inference procedure. A fuzzy inference process is realized in five steps:

1) Fuzzification of crisp inputs by means of membership functions
2) Application of fuzzy operations to antecedent parts
3) Application of implication to consequent parts
4) Aggregation of rules outputs into a single fuzzy set
5) Defuzzyfication of the final fuzzy set into a single crisp output.

Tagaki-Sugeno fuzzy inference system is widely used in data based modelling. The $i$.th rule for a first order linear Sugeno model with two input variables ($x_1$, $x_2$) is given as ;

$$\text{IF } (x_1 \textit{ is } X_{1,i}) \text{ AND } (x_2 \textit{ is } X_{2,i}) \text{ THEN } y_i = p_{i,0} + p_{i,1}x_1 + p_{i,2}x_2 \tag{4}$$

where $p$ constitutes a tunable parameter set . The crisp output, $y$, is the weighted sum of the rule outputs , $y_i$

$$y = \sum W_i y_i \tag{5}$$

Where $W_i$ is the weight of the $i$.th rule.

### 2.3 *Fuzzy and Neuro-Fuzzy Modelling*

Fuzzy modeling involves two phases namely, structure identification and parameter optimization. Structure identification encompasses the selection of input variables, the specification of the fuzzy inference system, the rule base set, the type and number of the membership functions. The parameter optimization is the fine-tuning of model parameters to best fit the input-output data set. Neural networks learning techniques are successfully applied in this phase.

Adaptive neuro-fuzzy inference system (ANFIS) is proposed to integrate fuzzy logic and ANN. ANFIS incorporates a Sugeno type fuzzy inference system into adaptive feed forward ANN structure, as illustrated in Figure 2. For training ANFIS, a hybrid algorithm has been developed by combining the gradient method and linear regression, which is faster than the classical back-propagation method [22].

The functionality of neurons in different layers of ANFIS is summarized as follows:

Layer 1: Neurons are adaptive; membership functions of input variables are used as activation functions, and parameters in this layer are referred to as antecedent (usually non-linear) parameters.

Layer 2: Neurons are fixed with outputs representing the firing strengths of the rules.

Layer 3: Neurons are fixed with outputs representing normalized firing strengths.

Layer 4: Neurons are adaptive with parameters referred to as consequent (constant or linear) parameters.

Layer 5: The single neuron is fixed with output equal to the sum of all the rules outputs (eq. (3)).

## 3.0 Experimental

### *3.1 Experimental Set-up*

A commercial MEA provided by Fuel Cell Store was used in experimental study, with the following specifications:

Membrane type: Nafion 117
Anode catalyst: % 20 (1:1 Pt-Ru) /C with 2 mg/cm$^2$ loading
Cathode catalyst: Pt black with 2 mg/ cm$^2$ loading
Gas diffusion and backing layer: Carbon cloth (for both electrodes)
Effective surface area: 25 cm$^2$

The single cell unit has flow field of serpentine geometry: the flow channels are 1.5 mm wide, 1 mm deep and 38 cm total length. The rib between the channels is 1 mm. A computer controlled fuel cell test station (Fideris, USA) was used to obtain the polarization curves.

### *3.2 Experimental design*

Fuel cell temperature, gas and liquid flow-rates and methanol concentration were selected as operating variables. The operating ranges of these variables are given in Table 1 and chosen as broad as possible to encompass practical ranges considered in the literature survey. The inlet air temperature and pressure were kept constant at 25 °C and 5 bar.

Polarization curves were obtained for 35 different operating conditions according to the second order experimental plan given in Table 2 where each operating variable is examined at six different levels. Furthermore, 3 central replicates were included in the plan to estimate the experimental error.

**Table 1**. Operating variables and ranges

| Variable | | range |
|---|---|---|
| Methanol concentration | C (M) | 0.25- 3.25 |
| Liquid flowrate | q ( ml/min) | 2- 14 |
| Air flowrate | Q (ml/min) | 450- 1200 |
| Temperature | T (°C) | 25- 65 |

## 4.0 Results and discussion

MATLAB neural networks and fuzzy toolboxes were used in the modelling study. A polarization data comprising 460 data points collected according to the experimental plan was partitioned in three subsets in the following proportions; 60 % for training, 20 % for check and 20 % for test. The same datasets were used for both modelling studies. The data points from the activation, ohmic and mass-transfer limiting regions of polarization curves were distributed as evenly as possible to three data subsets. Five input variables were considered for the modelling; current density I, Temperature T, methanol concentration C, liquid flow-rate q, and gas flow-rate Q. The polarization data of three center replicate experiments were used to assess the experimental error on voltage values. The mean standard deviation was calculated as 9.8 mV, corresponding to percent relative error of 4.1 %.

**Table 2.** The experimental plan

| Run no. | q (ml/min) | Q(ml/min) | C (M) | T (°C) |
|---|---|---|---|---|
| 1 | 5 | 450 | 1 | 35 |
| 2 | 5 | 450 | 1 | 55 |
| 3 | 5 | 450 | 2.5 | 35 |
| 4 | 5 | 450 | 2.5 | 55 |
| 5 | 11 | 450 | 1 | 35 |
| 6 | 11 | 450 | 1 | 55 |
| 7 | 11 | 450 | 2.5 | 35 |
| 8 | 11 | 450 | 2.5 | 55 |
| 9 | 5 | 950 | 1 | 35 |
| 10 | 5 | 950 | 1 | 55 |
| 11 | 5 | 950 | 2.5 | 35 |
| 12 | 5 | 950 | 2.5 | 55 |
| 13 | 11 | 950 | 1 | 35 |
| 14 | 11 | 950 | 1 | 55 |
| 15 | 11 | 950 | 2.5 | 35 |
| 16 | 11 | 950 | 2.5 | 55 |
| 17 | 8 | 700 | 1.75 | 25 |
| 18 | 8 | 700 | 1.75 | 65 |
| 19 | 8 | 700 | 0.25 | 45 |
| 20 | 8 | 700 | 3.25 | 45 |
| 21 | 2 | 700 | 1.75 | 45 |
| 22 | 14 | 700 | 1.75 | 45 |
| 23 | 8 | 200 | 1.75 | 45 |
| 24 | 8 | 1200 | 1.75 | 45 |
| 25 | 8 | 700 | 1.75 | 45 |
| 26 | 8 | 700 | 1.75 | 45 |
| 27 | 8 | 700 | 1.75 | 45 |
| 28 | 8 | 700 | 1.75 | 30 |
| 29 | 8 | 700 | 1.75 | 60 |
| 30 | 8 | 700 | 0.59 | 45 |
| 31 | 8 | 700 | 2.91 | 45 |
| 32 | 3 | 700 | 1.75 | 45 |
| 33 | 13 | 700 | 1.75 | 45 |
| 34 | 8 | 313 | 1.75 | 45 |
| 35 | 8 | 1087 | 1.75 | 45 |

The maximum number of training epochs was set to 300 for both ANN and ANFIS modelling. Meanwhile, all the training stopped early by validation control. The root of the mean sum of squares of the networks errors (RMSE) was used as the termination criteria.

*4.1 ANN modelling*

A feed-forward ANN architecture with one hidden layer, and one linear neuron in the output layer was used in the modelling study. The input data was scaled between {-1 1} for an efficient training. "Levenberg-Marquardt optimization method was used in the training process. The training process was started from several different initial conditions and the ANN with the minimum RMSE was accepted as the best.

Various subsets of input variables and number of hidden neurons were investigated to determine the optimum ANN structure. On the hidden layer, "tansig" and "logsig" activation functions were comparatively used from performance of view. Generally, tansig function performed better than logsig function. Table 3 gives the cumulative results of RMSE for different input subsets. As seen, by increasing the number of neurons, the generalization performance of ANN increased up to an optimum value and then it decreased due to the over-fitting caused by the excessive number of parameters. Furthermore, during the selection of most relevant input set, the initial examination showed that the base input set (I, T, C) was insufficient to estimate accurately the cell voltage. This is the situation of most empirical models where liquid and gas flow dynamics are usually neglected. By the inclusion of liquid flow rate into input set, the model performance was increased. But, further addition of the air flow-rate, the model performance is not increased as expected; the optimum number of neurons was less but the RMSE value was higher. In conclusion, minimum RMSE was obtained with the input set (I, T, C, q) and twelve hidden neurons; the total number of network parameters (weights and biases) is 73. The minimum value of RMSE is 10.8 mV, very similar to the mean standard deviation of the experimental error (9.8 mV) ; this supports the statistical validity of the model. The model fit is shown in Figure 2.

**Table 3.** Test RMSE values for various input sets

| Number of neuron | Set (I,T,C) | Set (I,T,C,q) | Set (I,T,C,q,Q) |
|---|---|---|---|
| 4 | 20.2 | 18.9 | 16.9 |
| 6 | 16.9 | 15.8 | 15.0 |
| 8 | 16.3 | 14.6 | 15.1 |
| 10 | 16.5 | 11.4 | **11.6** |
| 12 | 13.8 | **10.8** | 11.7 |
| 14 | 13.3 | 15.4 | **13.1** |
| 16 | 17.6 | 15.8 | 14.4 |

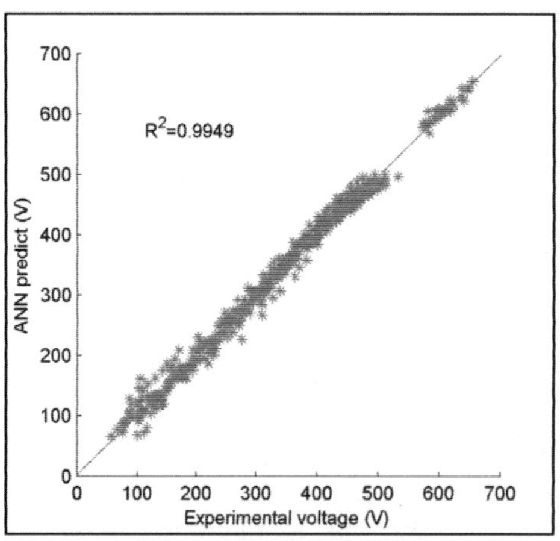


**Figure 2.** The fit of ANN model with experimental voltage values

*4.2 ANFIS modelling*

The rule base set was formed by means of "grid partition" method. Two membership functions were defined for each input, thus the number of fuzzy rule was equal to $2^N$ in which gived also the number of output membership functions. Zero and first order Sugeno inference systems were tried, first order system giving much higher performance.

Various subsets of input variables and various types of transfer functions namely triangular, trapezoidal, gaussian and disigmoidal transfer functions, were tried in the modelling study. RMSE values of various input sets and using different membership functions are given in Table 4. The highest test performance was obtain again with input set (I, T, C, q) and with triangular membership function. The optimum ANFIS architecture with 16 rules and 104 parameters is shown in Fig.3. The fit between model outputs and experimental values is plotted in Fig. 4.

**Table 4**. Test MSSE values for various input sets and membership function types

| | Membership function type | | | |
|---|---|---|---|---|
| Input set | Trimf | Trapmf | Gaussmf | Dsigmf |
| I, T, C | 25.2 | 81.9 | 65.5 | 44.2 |
| I, T, C, q | **15.5** | 92.4 | 112.8 | 43.8 |
| I, T, C, q, Q | 21.2 | 183.0 | 92.4 | 49.6 |

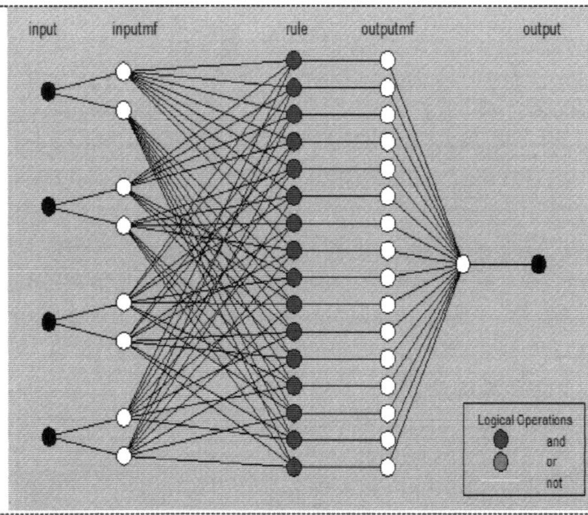


**Figure 3.** ANFIS structure with 4 inputs and 16 rules

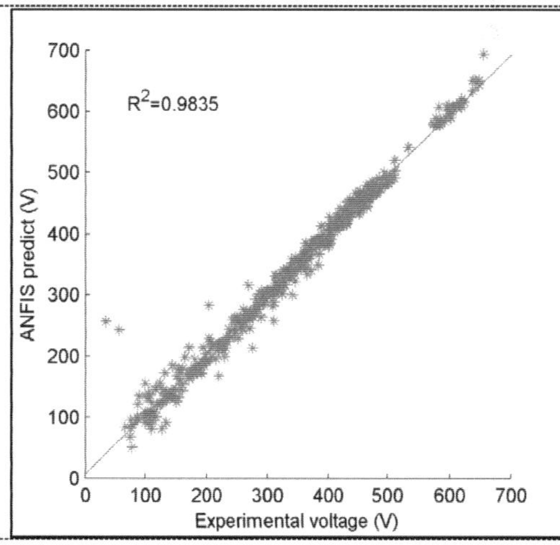


**Figure 4.** The fit of ANFIS model with experimental voltage values

Finally, when performance comparison between the two modelling techniques was made according to $R^2$ value of the model fit (0.9949 for ANN against 0.9835 for ANFIS) it was seen that feed-forward ANN model predicted more successfully the voltage values of DMFC. Simulated I-V curve shown in Fig. 5a and the corresponding power curve depicted in Fig. 5b approximate very well the experimental values.

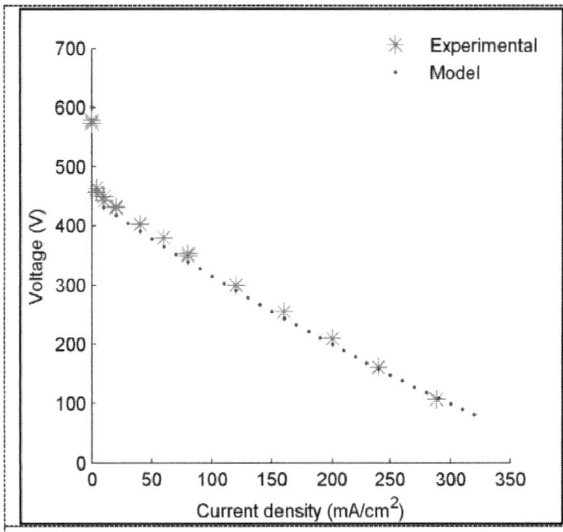


**Figure 5-a.** ANN model simulation I-V curves

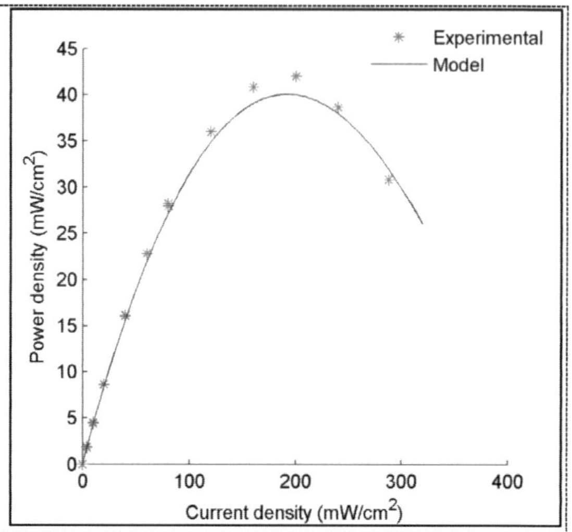


**Figure 5-b.** ANN model simulation I-W curves (Operating conditions: T = 55 °C, C = 2.5 M, q = 11 ml/ min, Q = 950 ml/min)

**5.0 Conclusion**

In both modelling approaches, in addition to principal inputs (current density, temperature, methanol concentration) liquid flow rate was detected as an effective input influencing the cell voltage; liquid dynamics affects the mass transfer rates in flow channels through the backing and diffusional rates to catalyst particles surfaces. Meanwhile, fuel cell voltage is not so sensitive to air flow rate. For the optimum feed-forward ANN model with twelve hidden tansig neurons, minimum test RMSE was found as 10.8 mV which is the estimate of the mean standard deviation of the experimental error (9.8 mV); this result supports further the statistical validity of the model. In the case of ANFIS model, the minimum RMSE value (15.5 mV) was obtained with an architecture using triangular membership function and 16 fuzzy rules. Finally, further statistical tests demonstrated that feed-forward ANN model predicted more successfully the voltage values of DMFC.

**References**

[1] H Guo C-F Ma 2004 *Electrochem Commun* **6** 306
[2] H Liu J Ge 2006 *J Power Sources* 160 413
[3] A A Kulikovsky 2003 *Electrochem Commun* **5** 530
[4] K Sundmacher T Schultz S Zhou K Scott M Ginkel E D Gilles 2001 *Chem Eng Sci* **56** 333
[5] Xu C Follmann P M Biegler L T Jhon M S 2005 *Computuers and Chem Eng* **29** 1849
[6] Vidakovic T Christov M Sundmacher K J 2005 *Electroanal Chem* **580** 105
[7] T Schultz K Sundmacher 2005 *J Power Sources* **145** 435
[8] ZHWang CYWang 2003 *J Electrochem Soc* **150** A508
[9] VADanilov JLim IMoon HChang 2006 *J Power Sources* **162** 992
[10] Dohle H Wippermann K 2004 *J Power* Sources **135** 152
[11] Squadrito G Maggio G Passalacqua E Lufrano F Patti A 1999 *J Appl Electrochem* **29** 1449
[12] Argyropoulos P Scott K Shukla AK Jackson C 2003 *J Power Sources* **123** 190
[13] Tu H-C Wang Y-Y Wan C-C Hsueh KL 2006 *J Power Sources* **159** 1105
[14] SOu LEK Achenie 2005 *J Power Sources* **140** 319
[15] WYLee GPark TYang YYoon C Kim 2004 *Int J Hydrogen Energy* **29** 961
[16] SOT Ogaji RSingh P Pilidis M Diacakis 2006 *J Power Sources* **154** 192
[17] EEntchev LYang 2007 *J Power Sources* **170** 122
[18] K K Justesen S J Andreasen H R Shaker M P Ehmsen J Andersen **2013** *Int J Hydrogen Energy* 38 10577

[19] S Bozorgmehri and M Hamedi 2012 *Fuel Cells* **12** 1
[20] SJamei DHisselMC Pera JMKauffmann 2003 *J Power Sources* **124** 479
[21] R Wang L Qi X Xie Q Ding C Li C M Ma 2008 *J Power Sources* **185** 1201
[22] JSR Jang CT Sun E Mizutani 1997 *Neuro-Fuzzy and Soft Computing :A Computational Approach to Learning and Machine Intelligence* London: Prentice Hall NY

Journal of Physics: Conference Series **707** (2016) 012049 doi:10.1088/1742-6596/707/1/012049

# Calculation of Solar Radiation by Using Regression Methods

**Ö Kızıltan and M Şahin**

Electrical and Electronic Engineering, Engineering Faculty, Siirt University, 56100, Siirt, Turkey

**Abstract.** In this study, solar radiation was estimated at 53 location over Turkey with varying climatic conditions using the Linear, Ridge, Lasso, Smoother, Partial least, KNN and Gaussian process regression methods. The data of 2002 and 2003 years were used to obtain regression coefficients of relevant methods. The coefficients were obtained based on the input parameters. Input parameters were month, altitude, latitude, longitude and land-surface temperature (LST).The values for LST were obtained from the data of the National Oceanic and Atmospheric Administration Advanced Very High Resolution Radiometer (NOAA-AVHRR) satellite. Solar radiation was calculated using obtained coefficients in regression methods for 2004 year. The results were compared statistically. The most successful method was Gaussian process regression method. The most unsuccessful method was lasso regression method. While means bias error (MBE) value of Gaussian process regression method was 0,274 MJ/m$^2$, root mean square error (RMSE) value of method was calculated as 2,260 MJ/m$^2$. The correlation coefficient of related method was calculated as 0,941. Statistical results are consistent with the literature. Used the Gaussian process regression method is recommended for other studies.

## 1. Introduction

Solar radiation (SR) predictions are performed in a geographical region, operating conditions of many solar energy systems, which are in design and development stage, can be simulated in the related region [1]. More clearly, long-term SR predictions are an essential parameter for engineering applications such as modelling solar power plants, modelling photovoltaic cells, and modelling solar heating systems [2]. For this reason, it is necessary to accurately prediction SR, which is excessive valuable information. In the previous years, SR prediction models have been developed in harmony with parameters such as air temperature, humidity, sunshine duration, and cloud coverage, which were measured from conventional meteorological stations and which were evaluated indirectly as a function of SR [3]. These models can be categorized into two groups, namely parametric methods such as Angstrom [4] and nonparametric methods that are based on artificial intelligence [5, 6]. In the literature, it has been observed that SR information in a specific location can be estimated using these models. However, it may not be possible to obtain accurate and continuous data from every station because maintenance and calibration of measuring devices, which are installed in meteorological stations for SR measurement, are difficult and installation costs are high [7]. In addition, the fact that the number of meteorological stations are limited especially in developing countries and inefficient recording of data due to device malfunctions constitute another limitation in obtaining SR data [8]. Thus, these limitations forced researchers toward a tendency to develop alternative estimation methods and find more reliable data sources for the regions where SR data cannot be directly measured or stations are insufficient. In recent years, satellite-based remote sensing (RS) techniques are widely used as an alternative method and as a data source for SR estimations [7, 9]. One of the most especial advantages of RS is that it is a reliable and rapid method for obtaining up-to-date and continuous information about wide geographical areas. In addition to this, satellite-based RS provides opportunity to perform SR estimations in rural, mountainous, and remote places where meteorological stations are insufficient. Calculations of solar radiation can be made by with various methods. It has been made with regression models using data from ground stations. Regression analysis, to explain the relationship between variables as a functional and this relationship is described by a model. It used to measure the magnitude of the relationship between variables. And it enables us to find the cause-effect relationships between variables. To sum up; Regression analysis is to find the proper function of the data table.

## 2. The Machine Learning

The learning machine, which was developed by Huang et al. [10], is a novel learning algorithm for single hidden layer feed forward networks (SLFNs). It has been widely used for the solution of estimation problems in many different fields [11, 12]. There are some advantages of the ELM algorithm. (i) It is easy to use, and no parameters need to be tuned except predefined network architecture and thus avoid many difficulties faced by gradient- based algorithms such as learning rate, learning epochs, and local minima. (ii) It is proven to be a faster learning algorithm compared with other conventional learning algorithms such as back propagation (BP) algorithm. Most training can be accomplished in seconds and minutes (for large-scale complex applications), which might not be easily obtained using other traditional learning methods. (iii) It possesses similar high generalization performance as BP [13] whatever the application area, the method of analyzing a lot of data available in machine learning methods to make predictions about the future

## 3. Methodology

In this study, Solar Radiation was predicted for 53 locations by using Linear, Ridge, Lasso, Smoother, Partial least, KNN and Gaussian process regression methods. These methods were used of data 2002 and 2003 years and accuracy of the mentioned methods was tested yearly. The data set of each year was consisting of 12 months. Whereas the values of month, altitude, latitude, longitude and land surface temperature were used as input for developing models, SR was obtained as output. Land surface temperature (LST) was taken from radiometry of NOAA-AVHRR satellite. Estimated solar radiation data were compared with actual data that were obtained from satellite data.

Learning Machine has a very valuable method of modelling tools, many of them with a proven track record in applications. K-NN is a non-parametric method used for categorization and regression. [14].LR measures the relationship between the categorical dependent variable and one or more independent variables by prediction probabilities using a logistic function, which is the cumulative logistic distribution.

### *3.1. Linear regression*

LR is a machine learning method which finds the relation between the dependent variable and the independent variable or variables [15]. This methods depends on the assumption of the relationship of this variables are polynomial as in Eq. 1.

$$y_i = \beta_1 x_{i_1} + \cdots + \beta_p x_{i_p} + \varepsilon_i \tag{1}$$

Where, equation 1 can also be shown as a format of equation 2.

$$y = X\beta + \varepsilon \tag{2}$$

Where $y$ is dependent variable, $X$ is independent variable, $\beta$ is parameter vector and $\varepsilon$ is error term shown below.

$$y = \begin{bmatrix} y_1 \\ y_2 \\ \vdots \\ y_p \end{bmatrix}, X = \begin{bmatrix} x_1^T \\ x_2^T \\ \vdots \\ x_n^T \end{bmatrix} = \begin{bmatrix} x_{11} & \cdots & x_{1p} \\ \vdots & \ddots & \vdots \\ x_{n1} & \cdots & x_{np} \end{bmatrix}, \beta = \begin{bmatrix} \beta_1 \\ \beta_2 \\ \vdots \\ \beta_p \end{bmatrix}, \varepsilon = \begin{bmatrix} \varepsilon_1 \\ \varepsilon_2 \\ \vdots \\ \varepsilon_p \end{bmatrix} \quad (3)$$

*3.2. Ridge Regression*
For the ease of presentation, we rewrite regression Eq. (1) into the following matrix form:

$$y = X\beta + e \quad (4)$$

where **y** is a (n×1) vector of the observation values of the dependent variable, X is a (n×p) matrix of the p explanatory variables or p−1 explanatory variables with the first column being a column vector of 1s as the regression intercept, ˇ is a (p×1) vector of the unknown regression coefficients, and **e** is a (n×1) vector of normally distributed errors with E(e) = 0 and Var (e) $\sigma^2$In where In is a (n×1) vector with every element equals to (1).

The ordinary least square (OLS) regression is the fundamental method in the regression family. Though OLS regression is one of the oldest methods for cost estimation, it is still being continuously improved or frequently applied by many researchers. [16]

$$\hat{\beta} = (X'X)^{-1}X'y \quad (5)$$

The key properties of $\hat{\beta}$ are (7) it is unbiased, which is E ( $\hat{\beta}$ ) = β and it has the minimum variance among all linear unbiased estimators, which is assured by Gauss–Markov theorem

Mathematically, ridge regression parameter estimation has the following form:

$$\hat{\beta}(k) = (X'X + kIn)^{-1} X'y, k > 0 \quad (6)$$

$\hat{\beta}$ (k) in (9) is obtained by minimizing the combination of both the sum of square error and the norm of the $\hat{\beta}$ vector

$$\min(y - \hat{\beta})'(y - x\hat{\beta}) + k\hat{\beta}'\hat{\beta} \quad (7)$$

The norm $\hat{\beta}'\hat{\beta}$ is the length of the vector $\hat{\beta}$ in a vector space. The purpose of adding one norm term to the sum of square errors is to restrict the values of $\hat{\beta}$ into a certain boundary. It is noted that the methodology of minimizing.

*3.3. Lasso Regression*
In statistics and statistical and machine learning, lasso (least absolute shrinkage and selection operator) is a regression analysis method that performs both variable selection and regularization in order to enhance the prediction accuracy and interpretability of the statistical model it produces. It was introduced by Robert Tibshirani in 1996 based on Leo Breiman's Nonnegative Garrote.[16][17] lasso was originally formulated for least squares models and this simple case reveals a substantial amount about the behavior of the estimator, including its relationship to ridge regression and best subset selection and the connections between lasso coefficient estimates and so-called soft thresholding.

Lasso regularization can be extended to a wide variety of objective functions such as those for generalized linear models, generalized estimating equations, proportional hazards models, and M-estimators in general, in the obvious way. [16] Given the objective function

$$\frac{1}{N}\sum_{i=1}^{N} f(x_i, y_i, \alpha, \beta) \tag{8}$$

the lasso regularized version of the estimator will be the solution to

$$\min_{\alpha,\beta} \frac{1}{N}^{2} \sum_{i=1}^{N} f(x_{i,} y_{i,} \alpha, \beta) \text{ subject to } \lVert \beta \rVert_1 \leq t \tag{9}$$

Where only is penalized while is free to take any allowed value, just as was not penalized in the basic case.

### 3.4. Kernel Smoother regression

Kernel-based methods are very popular in non-parametric estimation methods. The goal is, to find a non-linear functional relationship between random variables and [18]. Suppose that and are random variables defined in, the non-parametric regression model in equation 7.

$$Y(X) = \Psi(X) + \varepsilon \tag{10}$$

Nadaraya - Watson kernel method smoothed regression function provides estimation based on the weighted average of local values of in equation (8)

$$\Psi^{n}(X) = \frac{\sum_{i=1}^{n} K\left(\frac{x - x_i}{h}\right) y_i}{\sum_{i=1}^{n} K\left(\frac{x - x_i}{h}\right)} \tag{11}$$

Where is a kernel, usually chosen from a positive value of the symmetric functions such as Epanechnikov, Tri-cube and Gaussian kernels and is zero outside of defined inferable area [19]. h is the bandwidth and also scaling factor parameter and controlling the smoothness of regression function $\Psi(X)$.

### 3.5 Partial least squares regression

PLS is a bilinear calibration method which compresses by reducing a large number of measured collinear spectral data to a few non-correlated principal components (PCs) [20]. The aim of linear PLC is to reduce the data a few latent variables, such as $t_j$ and $u_j$ ($j = 1, \ldots, A$; where is the number of hidden variables) and then to develop a regression model to reveal the relationship between $t_j$ and $u_j$ [21] is in equation 4

$$u_j = b_j t_j + e_j \tag{12}$$

where $e_j$ is the error vector, and $b_j$ is represented by the $\hat{b}_j = \left(t_j^T t_j\right)^{-1} t_j^T u_j$ and Latent variables are obtained by $t_j = X_j w_j$ and $u_j = Y_j q_j$, where $w_j$ and $q_j$ of unit length and $t_j$ and $u_j$ must have a maximum of the covariance. $X_j$ is defined

$$X_{j+1} = X_j - t_j p_j^T \tag{13}$$

Where $X_1 = X$ and $p_j = X_j^T t_j / (t_j^T t_j)$ and $Y_{j+1} = Y_j - b_j t_j q_j^T$ where $Y_1 = Y$.

If $\hat{u}_j = \hat{b}_j t_j$ then matrices $X$ and $Y$ can be written ;

$$X = \sum_{j=1}^{A} t_j p_j^T + E \quad \text{and} \quad Y = \sum_{j=1}^{A} \hat{u}_j q_j^T + F \tag{14}$$

where $E$ and $F$ are residuals of $X$ and $Y$

*3.6 K Nearest Neighbors Regression*

k-NN, which converges the output to the mean of k nearest neighbors of input at feature space, is in equation 3[22].

$$f_{knn}(x) = \frac{1}{K} \sum_{i \in N_{K(x)}} y_i \tag{15}$$

where; $f : x \to y$ is regression function, $x$ is input, $y_i$ is known output values and $N_{K(x)}$ is the indeces of k nearest neighbor of $x$ .

*3.7 Gaussian Process regression*

GPR is a machine learning process which depends on Bayesian probability model. This model, which is shown in equation 9, assumes that the random variables are normal distributed and therefore is based on mean and covariance matrix of random variable for regression [23].

$$E\{y \backslash L, x\} = \int y p(f \backslash L, x) dy \tag{16}$$

where; $E\{y \backslash L, x\}$ is expectation of output conditioned to $L, x$; $y$ is best estimation output, $x$ is input, $L$ is quantity condition, $p(f \backslash L, x)$ is predictive distribution and is assumed that predictive distribution is normal (Gaussian) distribution.

So;

$$p(f \backslash L, x) \sim N(\mu_x, \sigma_x^2) \tag{17}$$

where; $\mu_x = k_x^T [K + \sigma_n^2 I]^{-1} . y + b$ and $\sigma_x^2 = k(x, x') - k_x^T [K + \sigma_n^2 I]^{-1} k_x$ (18)

$k(x, x')$ is covariance matrix of training samples, $K$ is covariance vector between the training samples and $k_x$ is covariance vector of sample$x$. $b$ is bias factor, $\sigma_n$ is the noise variance and $I$ is identity matrix.

## 4. Results

In this study, SR was calculated as related to 53 locations. Regression methods were used to make the calculations. Linear, Ridge, Lasso, Smoother, Partial least, KNN and Gaussian process regression methods were used to calculate SR. Locations of the 2002 and 2003 values were used to obtain the regression coefficients. Input parameters were month, altitude, latitude, longitude and land-surface temperature (LST).The values for LST were obtained from the data of the National Oceanic and Atmospheric Administration Advanced Very High Resolution Radiometer (NOAA-AVHRR) satellite. After obtaining the coefficients of the regression methods were used for 2004 of locations. The obtained SR values are evaluated with actual values statistically. R, MBE and RMSE methods were used statistical methods. These methods depending on the values are shown in Table1.

**Table 1.** Regression of Methods MBE, RMSE and R values.

| METHOD | MBE(MJ/m$^2$) | RMSE(MJ/m$^2$) | R |
|---|---|---|---|
| Linear Regression | -2,732 | 5,742 | 0,660 |
| Lasso Regression | -1,970 | 5,532 | 0,601 |
| K.SmootherRegression | 0,952 | 6,079 | 0,711 |
| Partial Least squares reg. | -2,079 | 5,679 | 0,639 |
| k-NN Regression | -0,998 | 3,500 | 0,864 |
| Gaussian Process Regression | 0,274 | 2, 260 | 0,941 |
| Ridge Regression | -3,358 | 5,908 | 0,648 |

When table 1 examined, the smallest correlation coefficient value have been found 0,601 MJ/m$^2$.This result have been obtained by using Lasso Regression. The biggest correlation coefficient have been evaluated 0,901 MJ/m$^2$ . The obtained result have been got using Gaussian Process regression. On the other side, correlation coefficient values of other methods is changeable values between 0,639 MJ/m$^2$ and 0,864 MJ/m$^2$ .The best MBE value have been calculated 0,274 MJ/m$^2$.This result have been got by Gaussian Process regression methods. The worst MBE value is 358 MJ/m$^2$. Related Value is provided from the ridge regression method. Other statistical methods are used the RMSE method. Accordingly largest RMSE value of 6.079 MJ / m$^2$ is calculated. This value is obtained from the kernel regression smoothing method. This result is very important because the highest error of solar radiation calculations have been obtained by smooth Kernel regression method. The smallest RMSE value is 2,260 MJ/m$^2$, which have been got by using Gaussian Process regression method. So, it proved that this is the most successful method. This results shows that, Linear, Ridge, Lasso, Smoother, Partial least, KNN and Gaussian process regression methods are enough for correctly calculation. Depending on the location of this method, MBE, RMSE and R values are shown in table 2

**Table 2.** Statistical results of locations

| LOCATION | MBE(MJ/m$^2$) | RMSE(MJ/m$^2$) | R |
|---|---|---|---|
| Amasya | -0,600 | 1,411 | 0,984 |
| Antakya | -1,091 | 1,372 | 0,990 |
| Antalya | -1,916 | 2,244 | 0,986 |
| B.Kesir-Gönen | -0,555 | 1,857 | 0,960 |
| Bingöl | 2,631 | 3,541 | 0,939 |
| Bitlis | -1,393 | 2,392 | 0,953 |
| Denizli | -1,655 | 2,476 | 0,980 |
| Edirne | -0,094 | 2,787 | 0,959 |
| G.antep | -1,201 | 2,081 | 0,967 |
| Gümüşhane | -0,812 | 1,344 | 0,988 |
| Hakkari | 2,421 | 4,220 | 0,803 |
| Iğdır | 0,270 | 2,418 | 0,917 |
| İst-Göztepe | 1,190 | 2,179 | 0,962 |
| | | | 0,962 |
| İzmir | -0,339 | 2,034 | |
| Konya | 1,139 | 1,824 | 0,974 |
| Malatya | 0,165 | 1,182 | 0,988 |
| Mersin | 0,296 | 0,972 | 0,989 |
| Muş | 0,722 | 2,463 | 0,944 |
| Niğde | 0,943 | 1,651 | 0,978 |
| Ordu | 0,959 | 2,693 | 0,891 |

| | | | |
|---|---|---|---|
| Rize | 1,493 | 2,417 | 0,904 |
| Siirt | -1,256 | 1,960 | 0,967 |
| Silifke | 0,068 | 0,693 | 0,994 |
| Sinop | 3,520 | 4,358 | 0,941 |
| Ş.Urfa | 2,602 | 3,163 | 0,963 |
| Yalova | 0,480 | 2,369 | 0,941 |
| Yozgat | -0,485 | 1,873 | 0,977 |
| Zonguldak | 0,804 | 1,652 | 0,979 |
| Adana | -1,125 | 1,930 | 0,966 |
| Adıyaman | -1,215 | 2,097 | 0,977 |
| Ağrı | -0,880 | 1,922 | 0,963 |
| Aksaray | 1,510 | 2,033 | 0,988 |
| Akşehir | -0,198 | 1,533 | 0,974 |
| Ankara | 0,020 | 1,082 | 0,986 |
| Artvin | -0,331 | 1,885 | 0,947 |
| Aydın | -1,910 | 2,430 | 0,980 |
| Batman | 1,279 | 2,637 | 0,950 |
| Bilecik | 0,584 | 1,546 | 0,983 |
| Birecik | 1,442 | 1,874 | 0,983 |
| Burdur | 2,278 | 2,452 | 0,991 |
| Çanakkale | 0,253 | 1,878 | 0,958 |
| Çorum | 0,358 | 1,719 | 0,979 |
| Elazıg-Bölge | -0,947 | 1,947 | 0,973 |
| Erzincan | 0,052 | 1,845 | 0,987 |
| Isparta | 2,049 | 2,964 | 0,947 |
| K.maras | 0,326 | 1,473 | 0,977 |
| Karaman | 1,506 | 1,884 | 0,987 |
| Karataş | -0,613 | 1,452 | 0,972 |
| Kars | -0,607 | 1,406 | 0,985 |
| Kastamonu | -0,874 | 2,676 | 0,965 |
| Kayseri | -0,708 | 1,549 | 0,988 |

When table 2 is investigated, the lowest correlation coefficient value have been evaluated 0,803 MJ/m$^2$. This result has been provided from Hakkari location. The highest correlation coefficient value has been found 0,994 MJ/m$^2$. The related value has been obtained from Silifke location. In this study, MBE values also calculated. The best MBE value is is 0,020 MJ/m$^2$ which have been got from Ankara location. The worst MBE value has been found 3,520 MJ/m$^2$. This value has been provided from Sinop location. The maximum value for the calculation of RMSE is calculated 4,358 MJ / m$^2$. This calculation is performed at the location of Sinop. It proved that Sinop location is the most unsuccessful according to other calculations of SR. The smallest RMSE value has been calculated to as 0,693 MJ / m$^2$ .The related value is belonging to Silifke location. According to Gaussian Process regression method, most successful calculation of SR was provided from Silifke location. SR variation of Silifke location monthly was shown in Figure 1

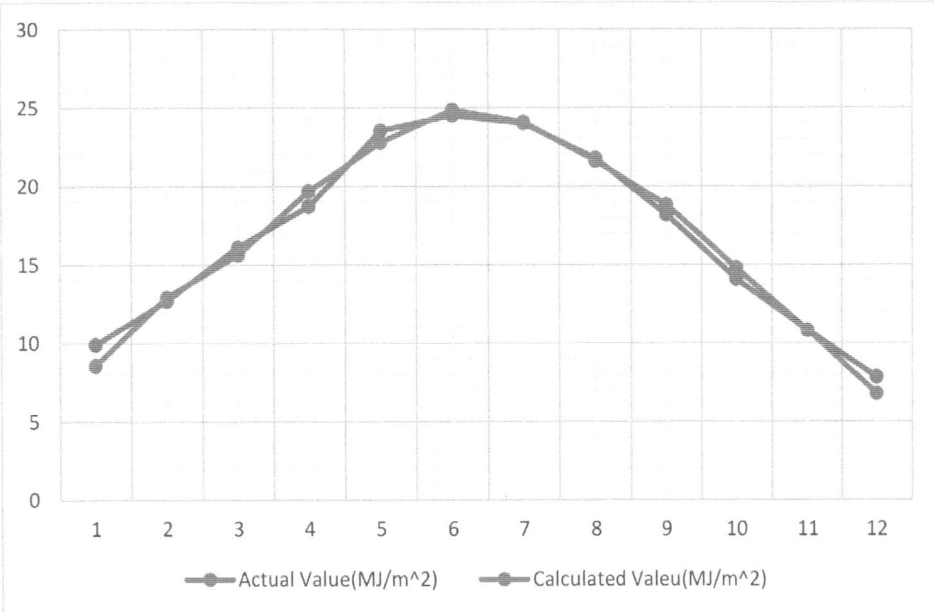


Figure 1.GSR change of Silifke location monthly

When we examined the GSR variation of location, we can observed that actual value and calculated value is very close to each other. The second, sixth, seventh, eight and eleventh month's value are more conformable that it has been observed. We cannot say the same for the other months.

## 5. Conclusion

In this study, calculation of SR was the different used regression methods. These methods are Linear, Ridge, Lasso, Smoother, Partial least, KNN and Gaussian process regression methods. It used to calculate SR. Gaussian process regression method has been the most successful method calculation of SR. On the other hand Linear Regression method was found to be the most unsuccessful method. The other methods gave agreeable results. If researcher studies calculation of SR we may suggest them to use Gaussian process regression method

## References

[1] Marti P Gasque M, Improvement of temperature based solar radiation estimation through exogenous data assistance. *Energy Conversion and Management* 2011; **52** pp: 990– 1003.

[2] Koca A Oztop HF Varol Y Koca GO, Estimation of solar radiation using artificial neural networks with different input parameters for Mediterranean region of Anatolia in Turkey. Expert Systems with Applications 2011; **38** pp: 8756–8762.

[3] Ehnberg JSG, Bollen MHJ. Simulation of global solar radiation based on cloud observations. Solar Energy 2005; **78** pp: 157–62.

[4] Akinoglu BG, Ecevit A. Construction of a quadratic model using modified Angstrom cofficients to estimate global solar radiation. Solar Energy 1990; **45** pp: 85–92

[5] Sözen A, Arcaklıoğlu E, Özalp M, Kanıt EG. Use of artificial neural networks for mapping of solar potential in Turkey. Applied Energy 2004; 77 pp: 273-86

[6] Mellit A. Artificial Intelligence technique for modelling and forecasting of solar radiation data: a review. International Journal Artificial Intelligence and Soft Computing 2008; **1** pp: 52– 76.

[7] Qin J, Chen Z, Yang K, Liang S, Tang W. Estimation of monthly-mean daily global solar radiation based on MODIS and TRMM products. Applied Energy 2011; **88** pp: 2480–2489

[8] Chen JL, Liu HB, Wu W, Xie DT. Estimation of monthly solar radiation from measured temperatures using support vector machines: a case study. Renewable Energy 2011; **36** pp: 413–420.

[9] Kandırmaz HM, Yeğingil I, Peştemalcı V, Emrahoğlu N. Daily global solar radiation mapping of Turkey using Meteosat satellite data. International Journal of Remote Sensing 2004; **25** pp: 2159–2168.

[10] Huang GB, Zhu QY, Siew CK. Extreme learning machine: theory and applications. Neurocomputing 2006; **70** pp: 489–501...

[11] Malathi V, Marimuthu NS, Baskar S, Ramar K. Application of extreme learning machine for series compensated transmission line protection. Engineering Application of Artificial Intelligence 2011; **24** pp: 880–887

[12] Yang H Zhao J Wang D and Dong Z, Predicting the probability of ice storm damages to electricity transmission facilities based on ELM and Copula function. Neurocomputing 2011; **74** pp: 2573–2581.

[13] Suresh S Saraswathi S Sundararajan N, Performance enhancement of extreme learningmachine for multi category sparse data classification problems. Engineering Application of artificial Intelligence 2010; **23** pp: 1149–1157.

[14] Altman N S, An introduction to kernel and nearest-neighbour nonparametric regression The American Statistician 1992; **46 (3)** pp: 175–185.

[15] Almorox J Hontoria C, Global Solar Radiation estimating using sunshine duration in Spain 2004; **45** pp: 1529-1535

[16] Tibshirani R, Regression Shrinkage and Selection via the lasso. Journal of the Royal Statistical Society. Series B (methodological) **58 (1)** pp Wiley: 267 88.

[17] Breiman, L, 1995 Better Subset Regression Using the Nonnegative Garrote. Techno metrics **37 (4)** pp. Taylor & Francis, Ltd.: 373–84. Doi: 10.2307/1269730.

[18] Bacher P Madsen H Aalborg H, Online short-term solar power forecasting 2009; **83** pp: 1772-1783.

[19] Şenkal Ö, Modeling of Solar Radiation Using Remote sensing and artificial neural network in Turkey December 2010; **35** pp: 4795-4801.

[20] Seshu D V Cady F B, Response of Rice to Solar Radiation and Estimated from International Yield Trials 1983; **40** pp: 649-654

[21] Guillermo P P Liliana N Carlos A V Maria A S, Estimating daily Solar Radiation in the Argentine Pampas 2004; **123** pp: 41-53.

[22] Paoli C Voyant C Muselli M Nivet M L, Forecasting of preprocess daily solar radiation time series using neural networks. December 2010; **84** pp.2146-2160

[23] Dinçer İ Dilmac Ş Ture E Edin M, A simple technique for estimating solar radiation parameters and its application for Gebze 1996; **37** pp: 183-198

# The heat recovery with heat transfer methods from solar photovoltaic systems

**A. N. Özakın[1], S. Karsli[1], F. Kaya[1] and H. Güllüce[1]**

[1]Atatürk University, Erzurum, Turkey

E-mail: ahmet.ozakin@atauni.edu.tr

**Abstract**. Although there are many fluctuations in energy prices, they seems like rising day by day. Thus energy recovery systems have increasingly trend. Photovoltaic systems converts solar radiation directly into electrical energy thanks to semiconductors. But due to the nature of semiconductors, whole of solar energy cannot turn into electrical energy and the remaining energy turns into waste heat. The aim of this research is evaluate this waste heat energy by air cooling system. So, the energy efficiency of the system will be increased using appropriate heat transfer technologies such as fin, turbulator etc.

## 1. Introduction

Since the humankind began to show his presence in the earth, every newborn has brought an energy burden and this burden is rising day by day [1]. Both growing world population and rising energy demand were caused an insistent uptrend. This increase in energy demand is providing with fossils as much as possible. However the global warming truth that can't be repaired has caused a prejudice to the fossils. For this reason, various government and non-governmental organizations welcomes to replace the fossils with renewables.

Solar energy has the highest energy density in renewables and has a limitless energy[2]. Solar energy source as well as being directly a source, the trigger of renewable resources such as wind energy. There are many areas in which the solar energy is used. These can mainly be divided as thermal and photovoltaic. Photovoltaic systems produced with using 4A group element of the periodic table. While scientific studies on this topic continues, theoretically efficiency reached 45% s by German researchers at the Fraunhofer Institute.

Industrial photovoltaic cells are used today with large scales. However, the efficiency obtained theoretically is impossible to achieve in practice. The efficiency of the cells used in solar power plants are only around 17%.[3] There are many reasons underlying this value. The subject of this research is cell temperature that one of the reasons of reducing the cell efficiency. The photovoltaic cell temperature rises because of wavelength of the incident photons' being less or more than remove an electron from silicon element.

When the photovoltaic cell manufacturers determining the specifications of the product, the ideal conditions are considered. When it is going out of this ideal conditions, it is impossible that reach this values given with products data sheet. With the increasing photovoltaic cell temperature, the cell efficiency drops 0.4% with 1°C temperature rising[4]. To prevent this decline the surface temperature of photovoltaics must be about 25°C which is optimum value.

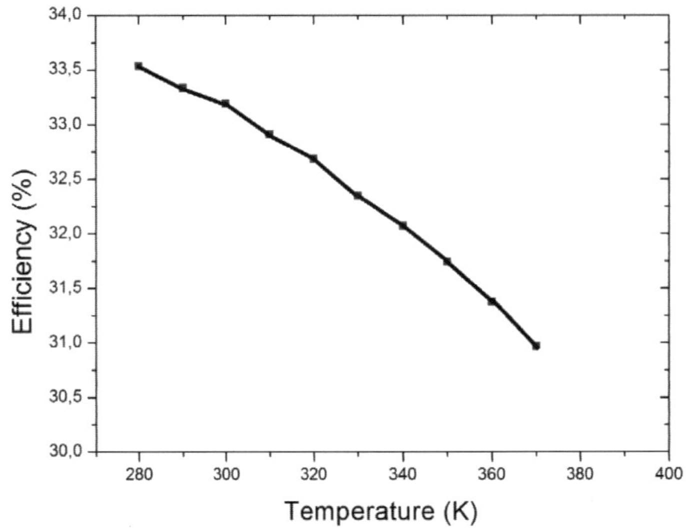


2. **Figure 1:** The temperature dependence of the photovoltaic efficiency.[5]

It is possible to use this surface temperature rising for contribute to photovoltaic cell efficiency.[6] Both for keep the temperature at an optimal level and for use the waste heat for benefit, the heat transfer methods used in engineering will be used as a practical way. The optimum one of these methods will be determined and energy that occurred waste heat recovery period will be recommended for various industrial applications.

## 2. Heat Transfer

Heat transfer is one of the subjects of scientific activity, gained importance with the progress of the industry. The laws of nature provides the perfect functioning for the heat transfer mechanism. This heat transfer mechanisms actually working even in the human body. Human skin and lung can be example to heat transfer device. [7]

There are three types of heat transfer mechanisms; radiation, convection and conduction. For example to these mechanisms; solar energy is transferred to the world with radiation, human skin balances temperature with convection, a bar conducts heat from heated end to the other end with conduction. This heat transfer mechanisms mostly used in industry for heating, cooling and keep the thermal balance.

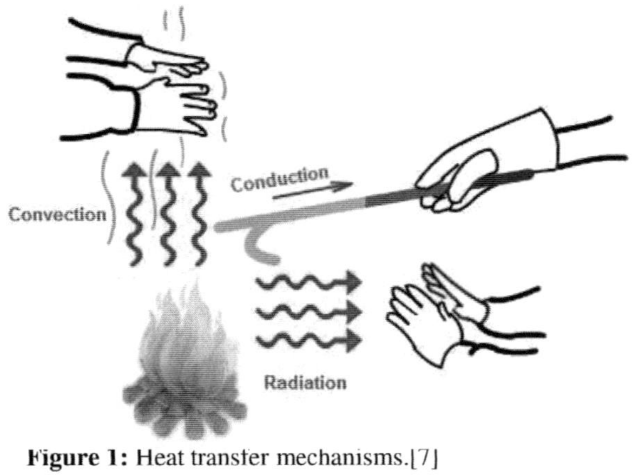


**Figure 1:** Heat transfer mechanisms.[7]

## 3. Heat Recovery

Heat recovery issues are available in every plants both large and small scales. According to the energy conversation, when the cooling fluid cool the system it will gain energy with a certain amount. In such cases if the heat releases into the environment, it will create an environmental pollution. In order to prevent this pollution and to improve the overall efficiency of the system, heat recovery unit are designed, these systems are often installed with a heat exchanger. With this heat exchangers, the energy is only to be transferred to the cold from the hot fluid without mixing[7]. While the hot fluid leaving the heat, the other one is heated for using energy in another cycle. The recovered heat from such systems mainly use in heating applications.

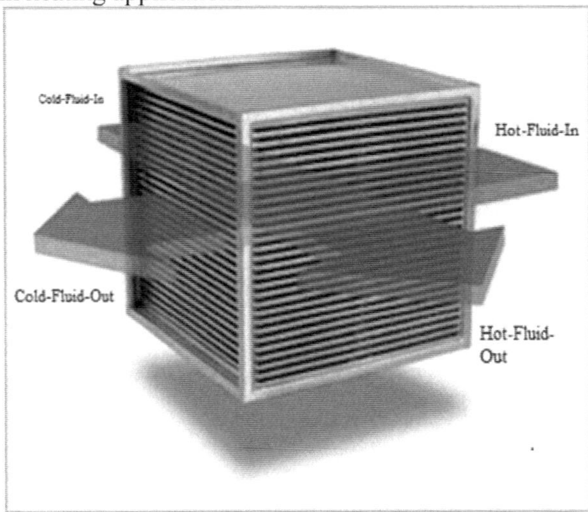


**Figure 2:** Heat exchanger.[8]

## 4. Waste Heat in Photovoltaics

Semiconductor materials are 4A group elements of periodic table that used for photovoltaic cell production. In the solar system the most commonly used semiconductor material is silicon which is a ceramic material. Silicon has 14 electrons and valence electron number is 4. The band gap energy of silicon is 1.12 Ev.[9]. When a photon with higher energy is absorbed by the cell, the electron will be excited and the remained energy will occurre heat. Similarly if a photon with smaller energy is absorbed by the cell, the energy of photon will be directly turned into heat. This absorbed heat causes irregularities by increasing cell temperature. Thus cell efficiency drops dramatically. So the heat remained in the system will create a negative situation.

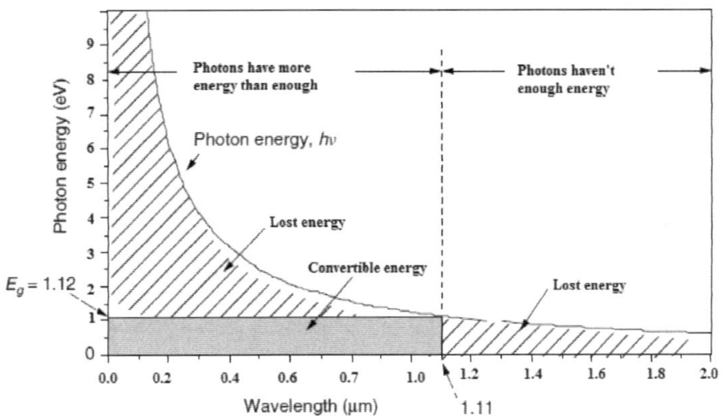


**Figure 3:** Photons with wavelengths above or shorter than needed to excite an electron.[9]

## 5. Results and Findings

It is known that the high surface temperature caused by solar radiation drops system efficiency drastically [10]. To prevent this decrease and waste heat recovery a secondary system used.

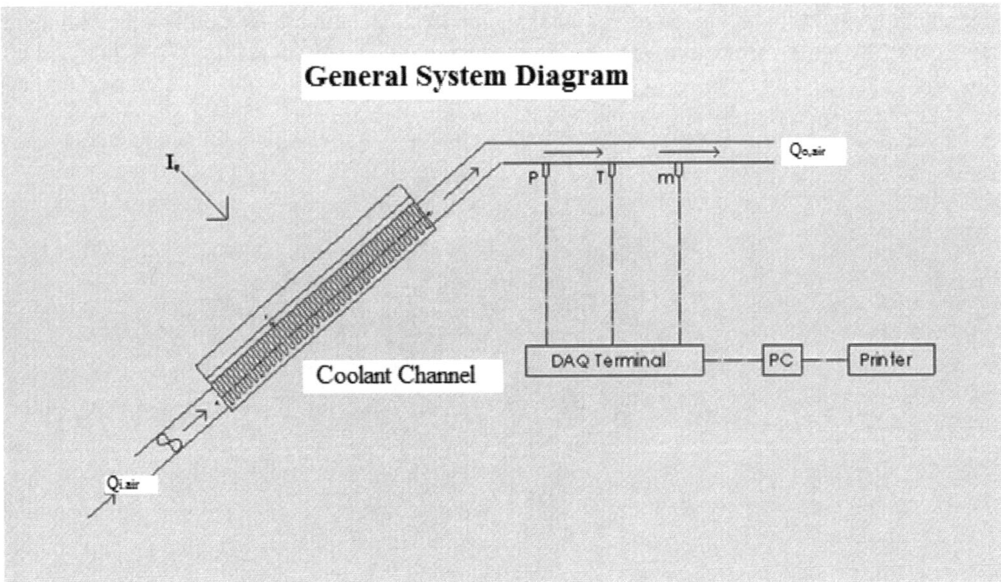


**Figure 4:** Designed heat recovery system.

In such systems, there are two types of efficiency [11]. Electrical and thermal efficiencies vary depending on operating conditions. System cooling is done by means of a working fluid and heat transfer calculations made to use waste heat.

$$Nu = C \times \mathrm{Re}^{m} \times \mathrm{Pr}^{n} \tag{1}$$

$$Q = h \times A_s \times \Delta T \tag{2}$$

The heat transfer coefficient of the system is determined in this equality and is written into the general heat transfer equation.

$$\bar{Q}_w = \dot{m} \times c_p \times \Delta T \tag{3}$$

With the help of general heat transfer equation, on july climate conditions with considering that the average surface temperature of our system is 45°C, it is calculated that every 0.1 ° C increase in fluid temperature provides about 4.85W power gain in thermal efficiency. At the same time it helps increase electrical efficiency.

### References

[1] Zimmermann, S., et al., A high-efficiency hybrid high-concentration photovoltaic system. International Journal of Heat and Mass Transfer, 2015. 89: p. 514-521.
[2] Cuce, E.C.a.P.M., Improving thermodynamic performance parameters of silicon photovoltaics

cells via air cooling. International Journal of Ambient Energy, 2014. 35.
[3] Othman, M.Y., et al., Performance analysis of PV/T Combi with water and air heating system: An experimental study. Renewable Energy, 2016. 86: p. 716-722.
[4] Rahman, M.M., M. Hasanuzzaman, and N.A. Rahim, Effects of various parameters on PV-module power and efficiency. Energy Conversion and Management, 2015. 103: p. 348-358.
[5] Messmer, E.R., Solar Cell Efficiency vs. Module Power Output: Simulation of a Solar Cell in a CPV Module. "Solar Cells - Research and Application Perspectives", 2013.
[6] Teo, H.G., P.S. Lee, and M.N.A. Hawlader, An active cooling system for photovoltaic modules. Applied Energy, 2012. 90(1): p. 309-315.
[7] Cengel, Y.A., Heat and Mass Transfer-A practical Approach.
[8] Company, U.K.E.L., Heat Excanger. Product Catalog, 2016.
[9] Masters, G.M., Renewable and Efficient Electric Power Systems. 2004.
[10] Reddy, S.R., M.A. Ebadian, and C.X. Lin, A review of PV-T systems: Thermal management and efficiency with single phase cooling. International Journal of Heat and Mass Transfer, 2015. 91: p. 861-871.
[11] Tonui, J.K. and Y. Tripanagnostopoulos, Performance improvement of PV/T solar collectors with natural air flow operation. Solar Energy, 2008. 82(1): p. 1-12.

# Determination and Fabrication of New Shield Super Alloys Materials for Nuclear Reactor Safety by Experiments and Cern-Fluka Monte Carlo Simulation Code, Geant4 and WinXCom

**Bünyamin Aygün[1], Turgay Korkut[2] and Abdulhalik Karabulut[3]**

[1]Atatürk University Faculty of Science, Department of Physics, Erzurum, Turkey,
[2]Sinop University, Faculty of Engineering, Department of Nuclear Engineering, Sinop, Turkey.
[3]Atatürk University Faculty of Science, Department of Physics, Erzurum, Turkey,

E-mail: akara@atauni.edu.tr

**Abstract.** Despite the possibility of depletion of fossil fuels increasing energy needs the use of radiation tends to increase. Recently the security-focused debate about planned nuclear power plants still continues. The objective of this thesis is to prevent the radiation spread from nuclear reactors into the environment. In order to do this, we produced higher performanced of new shielding materials which are high radiation holders in reactors operation. Some additives used in new shielding materials; some of iron (Fe), rhenium (Re), nickel (Ni), chromium (Cr), boron (B), copper (Cu), tungsten (W), tantalum (Ta), boron carbide (B4C). The results of this experiments indicated that these materials are good shields against gamma and neutrons. The powder metallurgy technique was used to produce new shielding materials. CERN - FLUKA Geant4 Monte Carlo simulation code and WinXCom were used for determination of the percentages of high temperature resistant and high-level fast neutron and gamma shielding materials participated components. Super alloys was produced and then the experimental fast neutron dose equivalent measurements and gamma radiation absorpsion of the new shielding materials were carried out. The produced products to be used safely reactors not only in nuclear medicine, in the treatment room, for the storage of nuclear waste, nuclear research laboratories, against cosmic radiation in space vehicles and has the qualities.

## 1. Introduction

Radiation is a natural part of our lives. When we breath the radioactive gases in the atmosphere (such as the actinonin 86 and radioactive isotopes of radon and thoron), as a result of this radioactive materials naturally occure in our bodies, our muscles, our bones, our cells. For example, potassium has a half-life of 1.42 million years in our body ($^{40}$K). The radiation, we are exposed to, contains cosmic radiation, comes from outside of the world, and also the natural radiation exist in our world naturally. Beside this, we are also subjected to the radiation (X-ray) used in medicine for diagnosis and treatment, nuclear tests, nuclear power centrals, nuclear accidents and radioactive waste. One of the most important problems of mankind is to meet the growing need for energy. Fossil fuels are used to meet the energy needs to a great extent. However, both are in line and came to a halt near the damage they cause to the environment.

Other energy sources are solar, wind and water as known renewable energy. However, the use of these resources is limited and insufficient. In this case, considering the huge damage caused as a result of an accident that occurs to provide energy, nuclear power plants are coming to the fore. Therefore the safety of nuclear power plant is essential. Ordinary materials cannot be used to prevent radiation leaks in nuclear power plants. The materials to be used specifically must have both a good ability to withstand radiation and high retention in addition to the feature temperatures. For these reasons to meet the growing needs in this field studies about new generation high temperature resistant shielding materials have gained speed. Work has been done on this subject in the literature. Alloy materials, such as CS-516, SS-403, SS-410, SS-316, SS-316, SS-304L, Incoloy-600, Monel-400 and Cupero-Nickel had determined the neutron and gamma shielding properties. Cupero-Nickel was found to be best shielding for gamma rays. For neutron, SS-316 was found to be the best shielding materials in energy 2–12 MeV [1]. Ferro-tungsten investigated the neutron shielding properties. Fast neutron shielding property of the ferro-tungsten steels has soft compare. They ultimately determine ferro-tungsten is doing better than steel shielding against fast neutrons [2]. 0.8% C, 0.05% Pb, 17% Cr, 11% Ni, 20% Mo + 1.92% W, 4.5 to 0.5% B, 0.05% Gd by using thermal corrosion resistant material they managed to get a patent by neutron shielding alloys [3]. To stop the fast neutron fission in the reactor tungsten (W) and rhenium with the compound (Re), tantalum (Ta), titanium (Ti) with such elements, alloys has done [4]. Medical applications and the of radio isotopes transportation during in nuclear reactors, gamma, neutron radiation in order to avoid leakage tungsten ratio of 90-98% (W), nickel (Ni), iron (Fe) such as metals combining have made heavy armor alloy. This heavy armor of material lead to commonly used gamma radiation of, to be said better than uranium [5]. They increased the percentage of armor materials in austenitic and ferritic stainless steel pipe and developed ductile and received patents against neutron radiation with good heat conduction [6]. They examined the scattering of neutrons at high pressure and to neutron shielding used in the tungsten carbide ceramic tubes [7]. Suitable for use in nuclear applications for control purposes, good neutron absorbent the contents of nickel, chromium, molybdenum, gadolinium alloy have developed armor material with about 5% by weight of boron [8]. Storage of nuclear waste for transport with high boron and Fe (49.7%), CR (17.7%), Mn (1.9%), Mo (7.4%), W (1.6%), B (15.2%), Cu (3.8%) and Si (2.4%) with the elements and stainless steel alloys, Ni-Cr-Mo high dose of neutron radiation in the desired diameter using alloys have made tubes to good shielding [9]. The base metal is iron, being more resistant to corrosion and neutrons, gamma radiation tungsten to its high absorption capacity, Al-boron carbide, borate, stainless steel, Ni, Cr, Mo, have developed a composite armor material containing Gd alloy [10].

The principal goals of this study are to acquire neutron cross section and neutron capture via Geant4 Monte Carlo code for samples.

## 2. Material and methods

The used materials to develop super alloys, of iron (Fe), rhenium (Re), nickel (Ni), chromium (Cr), boron (B), copper (Cu), tungsten (W), tantalum (Ta), boron carbide (B4C). Mixture ratios have identified with Geant4 Monte Carlo Simulation Code high fast neutron cross sections. Benefiting from international alloys standard powder metallurgy method we used for the production samples of super alloys. All of these samples have 2.5 g. weights and 2cm. diameter. These powder materials mixtured for 4 hours. After was pressed by SPECAC model hydraulic pellet press machine at 600MPa. Finally, disc samples were annealed at $1100^0$C during four hours. In this study 5types of super alloys were produced.

Radiation test was carried out by exposing to neutron source $^{241}$Am-Be (number of events processed 1000000) and $Ba^{133}$ 10mCi. Five types of super alloys have identified GEANT 4 code generated by quasi-experimental the macroscopic cross section of 7 MeV gamma and 4.5 MeV neutron radiation.

### 2.1. Monte carlo simulation code

Geant4 is a toolkit for the simulation of the passage of particles through matter. Its areas of application include high energy, nuclear and accelerator physics, as well as studies in medical and space science. In the GEANT4 code system, neutrons have been modeled by using high-precision (HP) data-driven hadronic physics models (G4Neutron HPElastic, G4Neutron HPInelastic, G4Neutron HPCapture, and G4Neutron HPFission). So neutrons are simulated from thermal neutron energies (0.025 eV) up to 20 MeV. The neutron cross-section data are taken from the GEANT4's G4NDL4.0 library, in which data comes largely from the ENDF/B-VII evaluated data library which explicitly includes all nuclear resonances in the form of point-like cross-sections rather than in the form of parameterizations. In the present study 4.9.4.p01 released version was used to simulate interactions between neutrons and shield samples. As outputs absorbed doses and macroscopic cross sections were given.

We obtained cross section and neutron capture via GEANT4 Monte Carlo code. In the first place, atomic stoichiometry and densities of the samples have been entered. In second place simulation has been started for 1000000 primary neutron particles. Then in the practical section absorbed dose by the detector is obtained.

## 3. Results and discussion

Macroscopic cross-section shows the probability of collisions of atomic radiation target substance. This value higher can be said that the material well absorbs radiation. The macroscopic cross section with neutron capture is influential factors to determine neutron and gamma radiation shielding characteristics of the sample. Monte Carlo simulations, five sample were produced and the stainless steel 316LN reference sample, then equivalent dose rate measurements were performed. Table.3.1. shows experimental results.

**Table 1.** Percentage composition of manufactured samples by mass.

| **Sample code** | **Material** | **Geant4 4.5 MeV neutron total macroscopic cross sections ($cm^{-1}$)** | **7 MeV gamma mass attenuation coefficient ($cm^2/g$)** | **Equivalent dose rates 4.5 MeV neutron (μSv/h)** | **Melting point range** (°C) |
|---|---|---|---|---|---|
| - | **Emp.** | - | - | 1.38556 | - |
| **[a]RNS** 316LN (reference samples) | Armor steel is used in nuclear reactors | 0.269224 | 0.03008 | 0.97812 | 1345-1440 |
| **SA1[b]** | 40%Ni-20%Cr-30%Fe-1%$B_4C$-4%Re-5%W | 0.337380 | 0.31794 | 0.590105 | 1480-1500 |
| **SA2** | 50%Ni-20%Cr-25W-5%Cu | 0.354598 | 0.30594 | 0.391248 | 1480-1500 |
| **SA3** | 50%Ni-25%Cr-15%Fe-5%Re-5%W | 0.371489 | 0.32393 | 0.525218 | 1480-1500 |
| **SA4** | 45%Re-40%B-10%Cr-5%Fe | 0.418609 | 0.39377 | 0.876505 | 1480-1500 |
| **SA5** | 55%Re-38%B-2%Cr-5%Fe | 0.446621 | 0.46391 | 0.733460 | 1480-1500 |

[a]**RNS:** References nuclear stainless steel
[b]**SA** : New super alloys material

Figure 1 and Figure 2 shows five types of super alloys have identified GEANT 4 code generated by quasi-experimental the macroscopic cross section of 7 MeV gamma and 4.5 MeV neutron radiation.

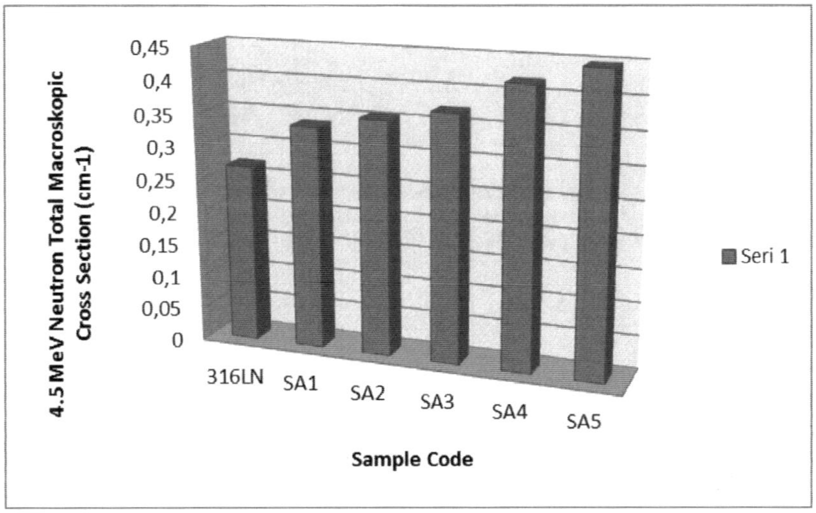


**Figure 2.** Total macroscopic cross sections by GEANT4 (4.5MeV neutron) of RNS 316LN.SA1.SA2.SA3.SA4.SA5

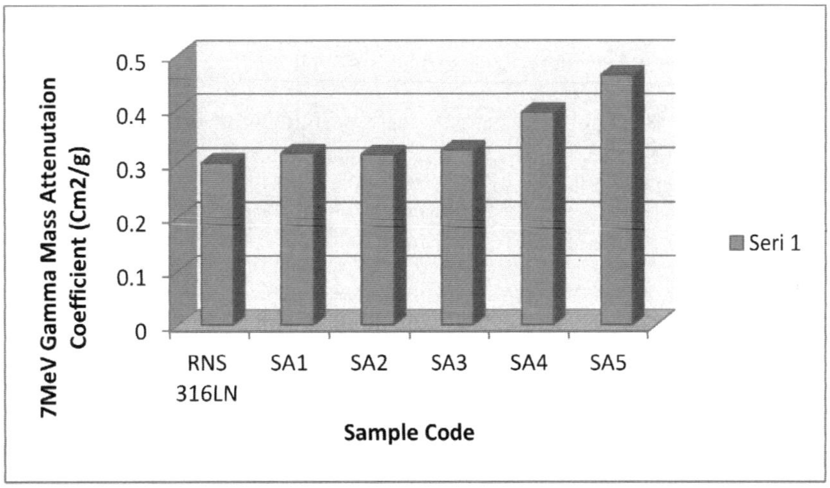


**Figure 2.** Gamma linear absorption coefficient by GEANT4 (7 MeV gamma) of RNS 316LN.SA1.SA2.SA3.SA4.SA5

These dose values mean that equivalent dose rates (μSv/h) by the detector. Decreased equivalent dose rate value indicates that this sample has high neutron shielding performance. As can be seen from Table 1 cross section is increasing with the using of boroncarpide ($B_4C$), wolfram (W) in the samples and decreasing absorbed dose by a detector of stainless steel. From table 3.1. that SA1, SA2, SA3, SA4, SA5 sample have high cross section value and low absorbed dose rate. The dose from the example source 1.38556 (μsv/ h), while the second sample supply SA2 front titled these values 0,391248 (μsv/h) decreased value. However, in the reactor, the commonly nuclear steels used dose value for the 0.97812 (μsv/h) decreased. Figure 3 shows 4.5 MeV neutrons for experimental dose measurement results. So SA2 has high neutron shielding properties in comparison to other samples.

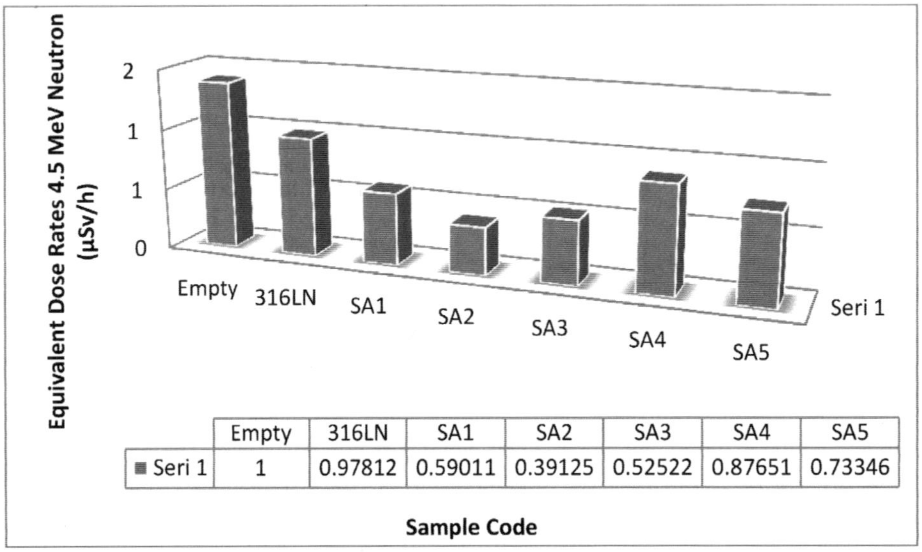


**Figure 3.** 4.5 MeV neutrons for experimental dose measurement results.

Thus, as can be seen from table.1, is more effective shielding material because it has a high cross section, high neutron capture values and low absorbed dose rate.

SA1, SA2, SA3, SA4, SA5 WinXcom theoretical and experimental results for samples 276.389 (keV) 302.853 (keV) 356.017 (keV) 383.851 (keV) gamma radiation energy mass attenuation coefficient ($cm^2$/ g). This result is shown in Table 2.

**Table 2** Experimental and theoretical (WinXcom) mass attenuation coefficients.

| | **Mass Attenuation Coefficient ($cm^2$/g)** | | | | | | | |
|---|---|---|---|---|---|---|---|---|
| | **276.389 keV Gamma** | | **302.853 keV Gamma** | | **356.017 keV Gamma** | | **383.851 keV Gamma** | |
| | ***Exp.** | **Theo.** | **Exp.** | **Theo.** | **Exp.** | **Theo.** | **Exp.** | **Theo.** |
| **SA1** | 0.088 ±0.0013 | 0.119 | 0.081 ±0.0054 | 0.112 | 0.073 ±0.0067 | 0.102 | 0.071 ±0.0010 | 0.098 |
| **SA2** | 0.090 ±0.0037 | 0.133 | 0.082 ±0.0010 | 0.123 | 0.071 ±0.0010 | 0.109 | 0.068 ±0.0017 | 0.104 |
| **SA3** | 0.109 ±0.0032 | 0.164 | 0.097 ±0.0010 | 0.147 | 0.083 ±0.0010 | 0.124 | 0.075 ±0.0020 | 0.116 |
| **SA4** | 0.099 ±0.0037 | 0.143 | 0.092 ±0.0010 | 0.133 | 0.081 ±0.0010 | 0.119 | 0.063 ±0.0017 | 0.114 |
| **SA5** | 0.121 ±0.0032 | 0.182 | 0.0107 ±0.0010 | 0.153 | 0.093 ±0.0010 | 0.134 | 0.085 ±0.0020 | 0.126 |

*Exp: experimental, Theo: Theoretical

Accordingly SA2 has greater value than standard steel and other samples from the prepared sample. As a result, samples prepared photon radiation can be used as an effective protective material.

## 4. Conclusions

We produced new super alloys samples having properties superior to the commonly used steel armor in this work in particular in nuclear power plants. In this production, we made largely dependent on international standards. We have identified materials ratio of combination of code using the Geant4 Monte Carlo Simulation mode. Produced made experimental measurements of the new super alloys which have proved to be better than the reference samples of materials by neutron shielding 316LN steel. These materials can be used for building walls of nuclear energy centrals, as moderator for nuclear reactors, in nuclear medicine departments and nuclear investigation centers, etc., to protect damages from neutron particle.

## References

[1] Singh. V.P. and Badiger. N.M., 2014 Gamma ray and neutron shielding properties of some alloy materials.*AnnalsofNuclearEnergy,* 301-310
[2] Mawutorli Nyarku. Ramanathapura S. Venkata D., 2013 Experimental neutron attenuation measurements in possible fast reactor shield materials. *Annals of Nuclear,* **53**, 135-139.
[3] Schmidt. E. Schmieden, 2013, Search for long-lived stopped R-hadrons decaying out of time with pp collisions using the *ATLAS detector. Physical Review D* 88, 112003.
[4] Gilbert. MR. Sublet. JC., 2011. Neutron-induced transmutation effects in W and W-alloys in a fusion environment. *Nuclear Fusion,* **6**.
[5] Caldwell. Steven.. GWM Symposia.. 2013.. A Review of Tungsten Heavy Alloy Utilization in Isotope Transport Containers-13380 *Material science Waste Menagemnet.* United States.
[6] Kibata. M.. Saito.Y.. Tsubota. M.. Doken. Y.. Sato.M.. 2013. Neutron shielding material. method of manufacturing the same. and cask for spent fuel. US**8481986** B2 US 13/360.
[7] Zhao. L. Robinson. L. Paul. RL.. 2007. Application of cold neutron prompt-gamma activation analysis in environmental studies of aquatic plants. *Journal of Radioanalytical and Nuclear Chemistry*. Volume **271**. Number 3.
[8] Mizia. RE. Lister. TE. Pinhero. PJ. Swank. WD.2007. Neutron absorbing coating for nuclear criticality control. US Patent **6919576**.
[9] Blink. J. Choi. J. Farmer. J. 2007. Applications in the Nuclear Industry for Thermal Spray Amorphous Metal and Ceramic Coatings. Materials Science & Technology 2007 *Conference and Exhibition Detroit.* MI. United States September 16.Volume **40.**
[10] Choi. J-S. Farmer. J.C. Lee. C. Fischer. L. M. July 42007. Neutron-Absorbing Coatings for Safe. *Chemistry Materials and Life Sciences Directorate Lawrence Livermore National Laboratory* UCRL-TR-232463 Volume **1**.

# Schottky Diode Applications of the Fast Green FCF Organic Material and the Analyze of Solar Cell Characteristics

**Z. Çaldıran[1], Ş. Aydoğan[1] and Ü. İncekara[2]**

[1]Department of Physics, Faculty of Science, Atatürk University, Erzurum, Turkey
[2]Department of Biology, Faculty of Science, Atatürk University, Erzurum, Turkey

E-mail: saydogan@atauni.edu.tr

**Abstract.** In this study, a device applications of organic material Fast Green FCF ($C_{37}H_{34}N_2Na_2O_{10}S_3Na_2$) has been investigated. After chemical cleaning process of boron doped *n*-Si crystals, Al metal was coated on the one surface of crystals by thermal evaporation and fast green organic materials were coated on other surface of crystals with spin coating method (coating parameters; 800 rpm for 60 s). Finally, Ni metal was coated on Fast Green by sputtering and we obtained the Ni/Fast Green FCF/n-Si/Al Schottky type diode. And then we calculated the basic diode parameters of device with current-voltage (*I-V*) and capacitance-voltage (*C-V*) measurements at the room temperature. We calculated the ideality factory (n), barrier height ($\Phi_b$) of rectifing contact from *I-V* measurements using thermionic emission methods. Furthermore, we calculated ideality factory (n), barrier height ($\Phi_b$) and series resistance ($R_s$) of device using Cheung and Norde functions too. The diffusion potential, barrier height, Fermi energy level and donor concentration have been determined from the linear $1/C^2$-$V$ curves at reverse bias, at room temperature and various frequencies. Besides we measured the current-voltage (*I-V*) at under light and analyzed the characteristics of the solar cell device.

## 1. Introduction

It is rapidly developing electronics industry and organic electronics has been started to make significant contribution to the commercial World of industry [1]. Organic materials are inexpensive and easy to prepare. Therefore, there is an increasing application of organic materials such as organic photovoltaic devices (OPV) and organic light emitting Diode (OLED) in electronics technology [1]. Nowadays, the developments in organic electronics have raised and therefore there is an increasing interest to the organic materials and their research.

Fast Green FCF is a progressive dye and it is soluble in either alcohol or water and it is used counterstain to safranin O. It can be used for tinned green peas, sauces, fish, vegetables, jellies, desserts. It may be used instead of light green SF yellowish in Masson's trichrome since it is less likely to fade and it has more brilliant color. Fast Green FCF is used for the preparation of a staining solution that can be used for modified connective tissue. Furthermore, at alkaline pH, it is used as a quantitative stain for histones after acid extraction of DNA. It is also used as a protein stain in electrophoresis. In addition, the absorbance of organic Fast Green FCF by the intestines is poor [2].

## 2. Experimental procedures

In this work, n-Si wafer with phosphorus doped, (100) orientation, 400 μm thickness and 1–10 Ω-cm resistivity was used and the n-Si wafer was chemically cleaned using the RCA cleaning procedure before fabrication device. The RCA cleaning procedure is**:** 10 min. boil in $NH_3 + H_2O_2 + 6H_2O$ followed by a 10 min $HCl + H_2O_2 + 6H_2O$ at 60 °C. We completed the three stage of the device fabrication. Firstly, the ohmic contact was made by thermal evaporating Al metal on the one surface (back) of the substrate and then was annealed at 450 °C for 10 min in $N_2$ atmosphere. Then, the native oxide on the front surface of the n-Si substrate was removed in $HF + 10H_2O$ solution at before forming a Fast Green FCF layer on the Si substrate. Secondly, Fast Green FCF was dissolved in deionized water to make 0.5 mg/1 ml solution and an organic Fast Green FCF layer formed on other surface of n-Si by spin coating method with 800 rpm, 60 s in clean rooms (class 1000). Furthermore, Fast Green FCF was coated on cleaned glass for the optical absorption and transmittance measurements. Finally, Ni contact was coated by DC sputtering method on Fast Green FCF film at $10^{-5}$ torr pressure and hence, we obtained device shown in Fig.1 a and the chemical structure of Fast Green FCF is depicted in Fig. 1b. The *I–V* and *C–V* measurements of the device have been performed with KEITHLEY 487 Picoammeter/Voltage Source and HP 4192A (50 Hz to13 MHz) LF IMPEDENCE ANALYZER, respectively. All electrical measurements have been performed at room temperature in dark and under illumination.

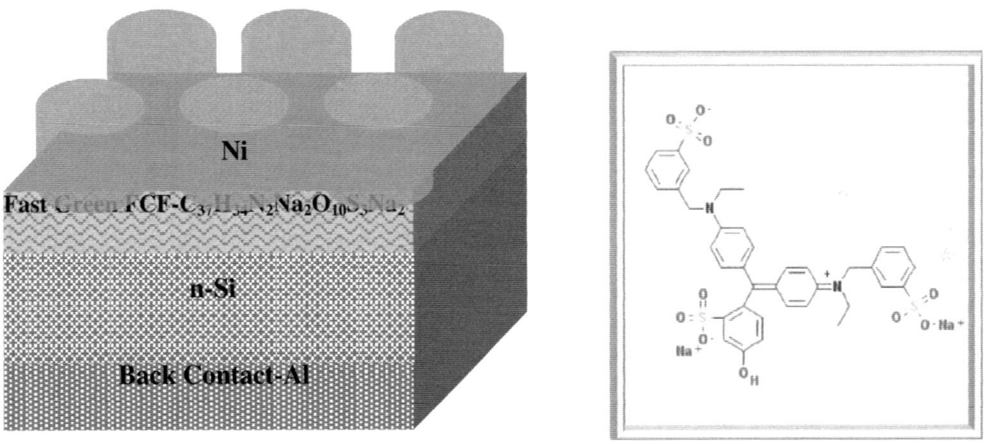


**Fig. 1.** (a) Schematic diagram of Ni/Fast Green FCF/n-Si/Al Schottky diode and (b) the chemical structure of Fast Green FCF

## 3. Results and discussion

Fast Green FCF organic film was coated on n-Si and the electrical measurements were performed of Ni/Fast Green FCF/n-Si/Al Schottky diode. Furthermore, Fast Green FCF film coated on glass substrate for optical measurements. The optical transmittance and spectra absorption of Fast Green FCF organic film is shown in Fig. 2. Also, plotting $(\alpha)^2$ versus hν is shown in this figure. The optical band gap of the Fast Green FCF film can be determined by linear region of plotting $(\alpha)^2$ versus hν. The band gap of the Fast Green FCF film has been calculated as 1.79 eV using the optical measurements. The typical absorption peak is at $\lambda_{max}$= 693 nm obtained absorption spectra of the Fast Green FCF organic film.

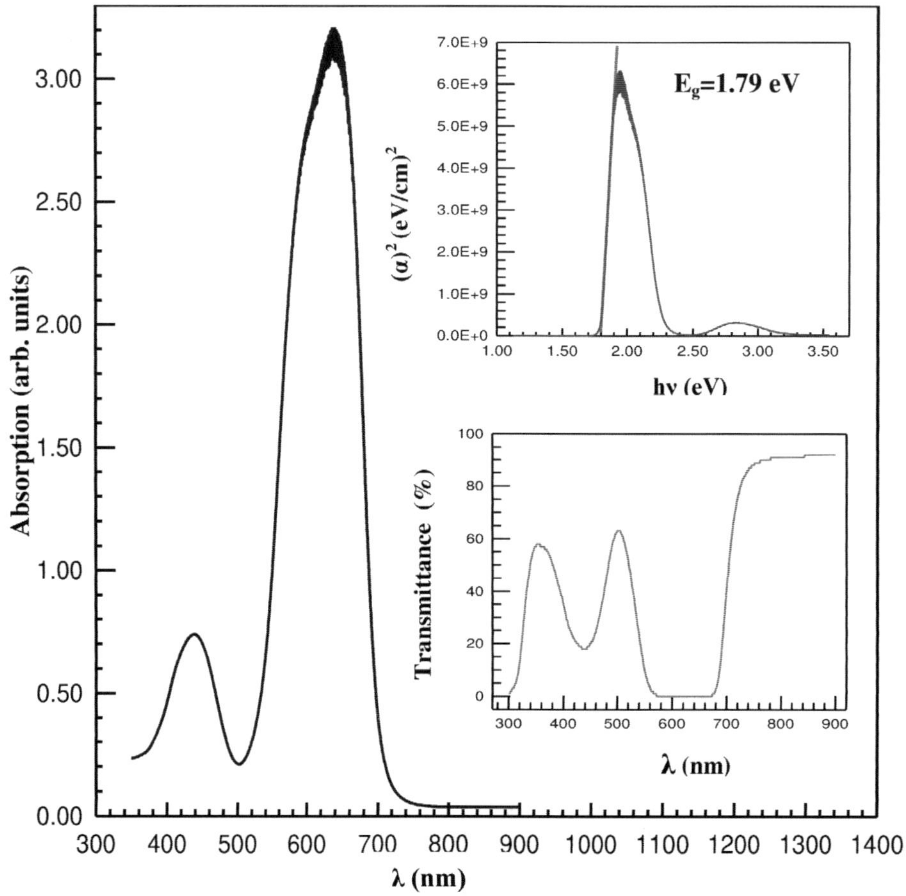


**Fig. 2.** Absorption spectra and optical transmittance of Fast Green FCF organic film

*3.1. The ideality factory (n) and barrier height ($\Phi_b$) parameters by the conventional (TE) method:*

The current-voltage relation of Schottky barrier diodes (SBDs) can be expressed as:

$$I = \left[ AA^{*}T^{2}\exp\left(\frac{-e\Phi_{b}}{kT}\right)\right]\left[\exp\left(\frac{eV_{o}}{nkT}\right)\right] \tag{1}$$

$I_o$ is the saturation current and it is given by;

$$I_{0} = AA^{*}T^{2}\exp\left(-\frac{e\Phi_{b}}{kT}\right) \tag{2}$$

$$I = I_{0}\left[\exp\left(\frac{eV}{nkT}\right)\right] \tag{3}$$

$$n = \frac{e}{kT}\frac{dV}{d(\ln I)} \tag{4}$$

Where $A$ is effective diode area, $A^*$ is Richardson constant of semiconductor. Eq.(4) equals to "unity" for an ideal Schottky diode. However, $n$ has usually a value greater than unity for non-ideal diode. High values of $n$ can be attributed to the presence of the interfacial thin native oxide layer between junction materials and to a wide distribution of low-Schottky barrier height, to higher series resistance values at higher forward biases, to barrier inhomogeneities, and to the bias dependence of the Schottky barrier height (SBH) [3]. The barrier height of a Schottky diode $\Phi_b$ can be given as:

$$e\Phi_b = kT\ln(AA^*T^2/I_0) \tag{5}$$

The dark current–voltage ($I$–$V$) characteristics of Ni/Fast Green FCF/$n$-Si/Al Schottky barrier diode is shown in Fig.3. This plot gives significant information about the rectifying device such as the leakage current and the series resistance. It is clear from Fig. 3. that a linear relationship exists between $I$ and $V$ for small applied voltages. This linearity deviated at higher biases due to the higher series resistance value.

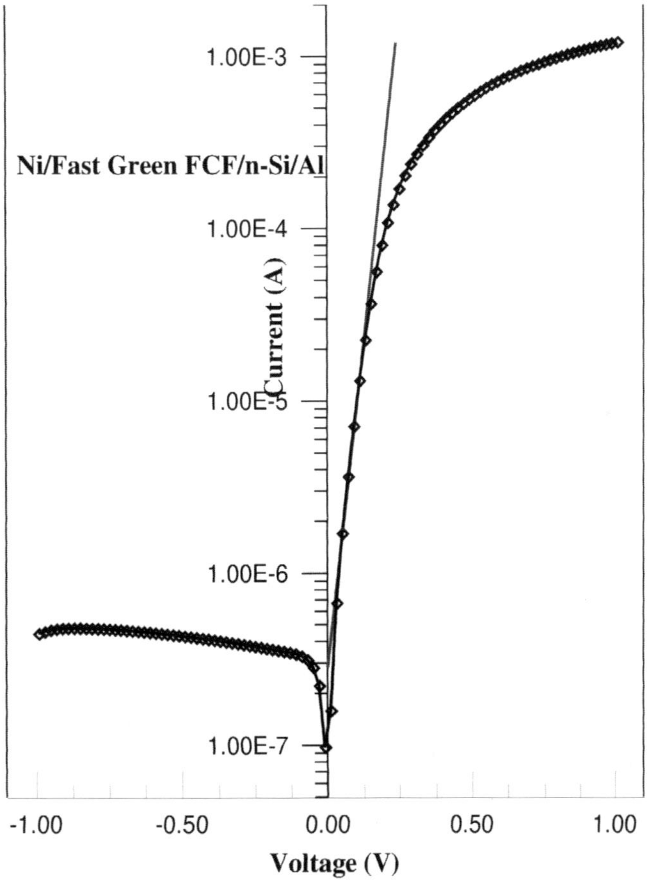


**Fig. 3.** The log *(I–V)* characteristics of the Ni/Fast Green FCF/n-Si/Al Schottky barrier diode.

The dark current–voltage ($I$–$V$) characteristics of Ni/n-Si/Al and Ni/Fast Green FCF/n-Si/Al devices are shown in Fig.4. The ideality factor $n$ and barrier height $\Phi_b$ values of the Ni/Fast Green FCF/n-Si/Al Schottky barrier diode were determined as 1.11 and 0.69 eV, respectively using the $I$–$V$ characteristics from Eqs .4 and 5. The higher values than unity of the ideality factor can be attributed to the inhomogeneities of the Fast Green FCF film thickness, to non-uniformity of the interface states and to the effect of the series resistance $R_s$ [4]. The ideality factor and the barrier height of Ni/Fast Green FCF/n-Si/Al device has been compared with the Ni/*n*-Si/Al reference diode. The ideality factor and barrier height of Ni/n-Si/Al were calculated as 1.18 and 0.58 eV. As seen that Fast Green FCF increased the barrier height and decrased the ideality factor. Namely, the organic film has improved the quality of device with respect to the Ni/n-Si/Al reference diode.

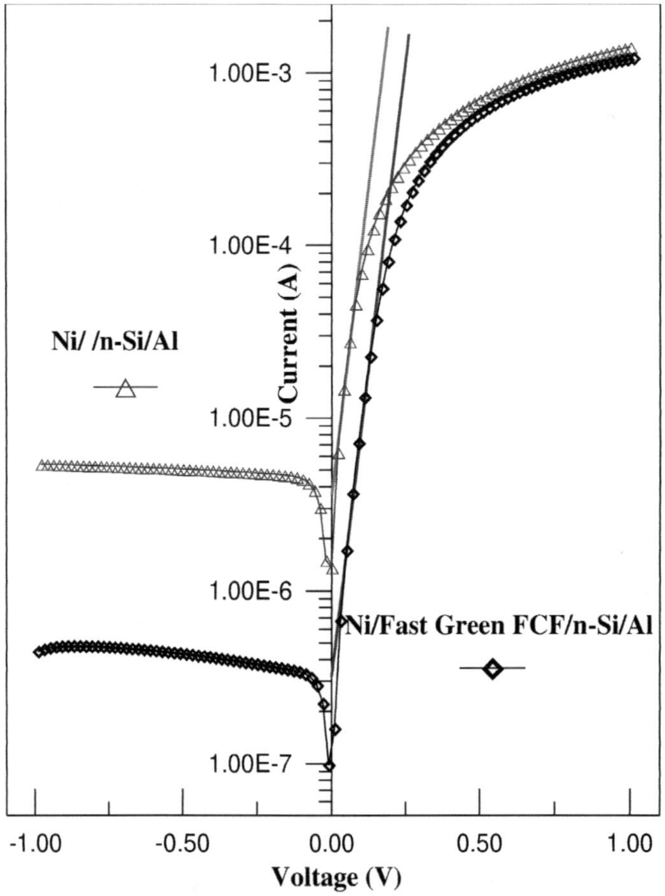


**Fig. 4.** The log *(I–V)* characteristics of reference the Ni/n-Si/Al the Ni/Fast Green FCF/n-Si/Al Schottky barrier diode.

Figure 5 shows the dark and under illumination the both linear and logarithmic current–voltage (*I–V*) characteristics of Ni/Fast Green FCF/n-Si/Al Schottky barrier diode, at room temperature. It is clearly seen that the device has a photosensitive behavior. At higher forward voltages of device, the dark and illuminated currents do not differ by appreciable amounts. But there are changes in the reverse bias voltages [5]. This result suggests that the photodiodes are a photosensitive and show a photodiode behavior [6].

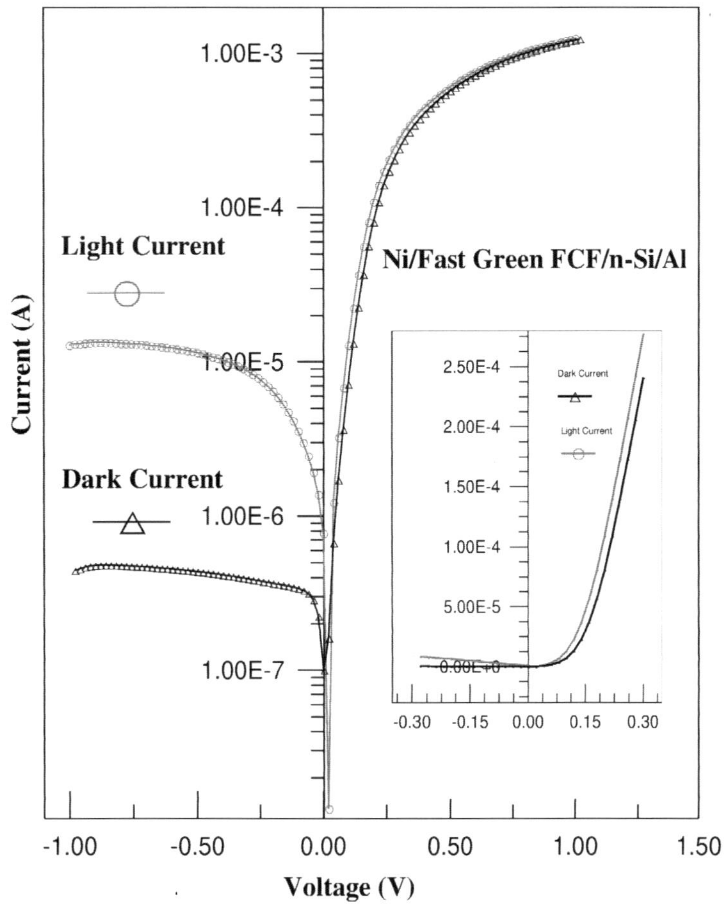


**Fig. 5.** *I–V* characteristics of the Ni/Fast Green FCF/n-Si/Al device in dark and under illumination

Figure 6 shows the logarithmic plots of the *I–V* characteristics obtained under forward bias of Ni/Fast Green FCF/n-Si/Al Schottky barrier diode. As seen in the forward bias characteristics three different regions can be identified. This indicates that three distinct current conduction mechanisms: Region I, these plot suggests ohmic conduction at low voltages. Namely, ohmic currents increase linearly with the voltage. The occurrence of space charge limited current (SCLC) requires that at least one contact has good injecting properties to provide an inexhaustible carrier source. In the region II, the slope was found to be 3.14. In the region III, the slope was found to be 1.14 indicating the trap filled SCLC process. The interface states between n-Si and Fest Green FCF are effective in electrophysical behavior of device.

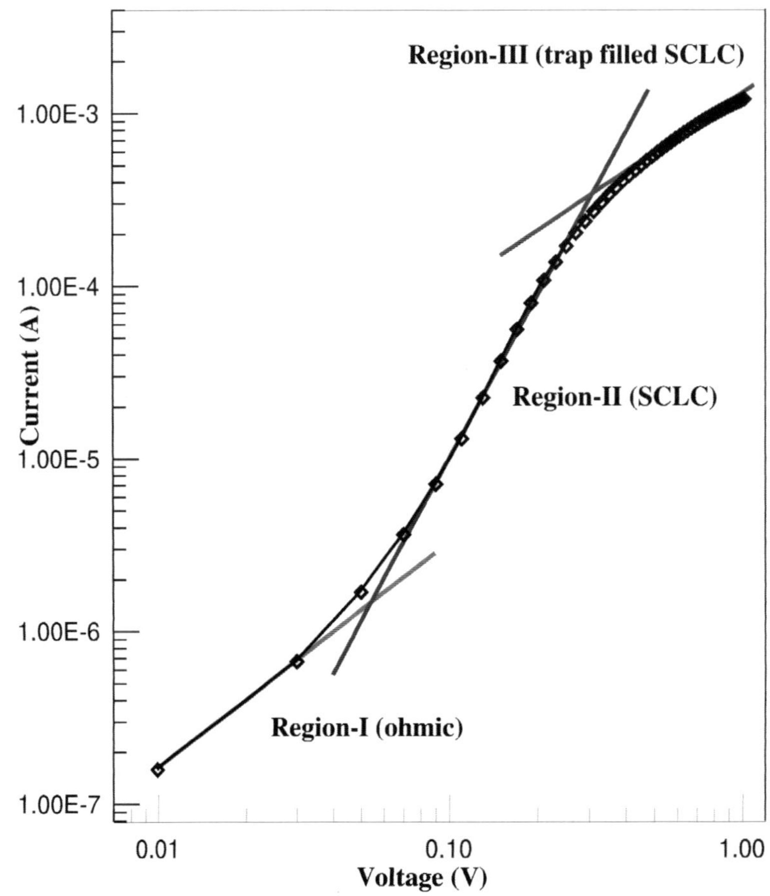


**Fig. 6.** The forward bias log(I) versus log(V) plot of the Ni/Fast Green FCF/n-Si/Al Schottky barrier diode

### *3.2. The ideality factory (n), barrier height ($\Phi_b$) and series resistance ($R_s$) parameters by the Cheung functions:*

For Ni/Fast Green FCF/n-Si/Al Schottky barrier diode, at the lower forward bias, both current is increasing as linear with voltage. However, series resistance effect is observed at higher voltages. The series resistance values are calculated with Cheung functions as follows:

$$I = A.J = \left[ AA^{*}T^{2} \exp\left(\frac{-e\Phi_b}{kT}\right)\right]\left[\exp\left(\frac{eV}{nkT}\right) - 1\right] \text{ if } V{=}(V)\text{-} IR_s \tag{6}$$

$$I = A.J = \left[ AA^{*}T^{2} \exp\left(\frac{-e\Phi_b}{kT}\right)\right]\left[\exp\left(\frac{e(V - IR_s)}{nkT}\right) - 1\right] \tag{7}$$

$$V = \left(\frac{nkT}{e}\right) \ln\left(\frac{I}{AA^{*}T^{2}}\right) + n\Phi_b + IR_s \tag{8}$$

$$\frac{dV}{d(\ln I)} = \frac{nkT}{e} + IR_s \tag{9}$$

$$H(I) = V - \left(\frac{nkT}{e}\right)\ln\left(\frac{I}{AA^{*}T^{2}}\right) \tag{10}$$

$$H(I) = n\Phi_b + IR_s \tag{11}$$

Figure 7 shows the $dV/d(lnI)$ versus $I$ and $H(I)$ versus $I$ obtained from forward bias. The values of n and $R_s$ have been calculated as $n = 2.47$ and $R_s = 0.76$ kΩ using Eq. 9, respectively. Similarly, the values of $\Phi_b$ and $R_s$ have been founded as $\Phi_b = 0.51$ eV and $R_s = 0.75$ kΩ, respectively with help of $H(I)=n\Phi_b+ IR_s$ . As seen, there is a significiant difference for the ideality factor values obtained from the forward-bias $lnI–V$ and from the $dV/d(ln\ I)–I$ plotes. This can be explained by the existence of a high series resistance $R_s$ and the interface states and to the voltage drop across the interfacial layer of the device [3].

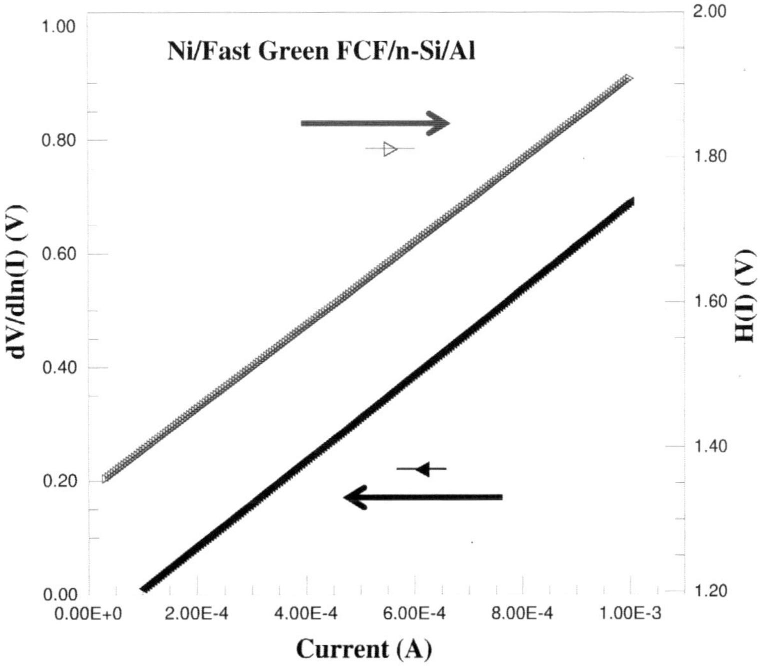


**Fig. 7.** The plots of $dV/d(lnI)$ versus $I$ and $H(I)$ versus $I$ obtained from forward bias current–voltage characteristics of the Ni/ Fast Green FCF/n-Si/Al Schottky barrier diode.

### *3.3. The barrier height ($\Phi_b$) and series resistance ($R_s$) parameters by the Norde methods:*

Norde's method is an alternative method for calculation the barrier height ($\Phi_b$) and series resistance ($R_s$) parameters of Schottky barrier diodes. Eq. (12) is defined in the modified Norde's method.

$$F(V) = \frac{V}{\gamma} - \left(\frac{kT}{2}\right) In\left(\frac{I(V)}{AA^{*}T^{2}}\right) \tag{12}$$

$$\varphi_b = F(V_0) + \frac{V_0}{\gamma} - \frac{kT}{q} \tag{13}$$

$$R_s = \frac{kT(\gamma - n)}{qI_o} \tag{14}$$

where γ is an arbitrary integer (dimensionless) greater than ideality factor. *I(V)* is current which obtained from the *I–V* curve. In determining the junction parameters, firstly, the F against V curve should be plotted, and then the value of barrier height can be obtained from it. where $F(V_0)$ is the minimum point of F(V), and $V_0$ is the corresponding voltage [7].

Figure 8 shows a plot of F(V) versus V for the device. From the plot, the some parameters of the structure have been obtained as $\Phi_b$ = 0.79 eV, $R_s$ = 0.41 kΩ using $F(V_0)$=0.75 V, $V_0$ = 0.16 V values.

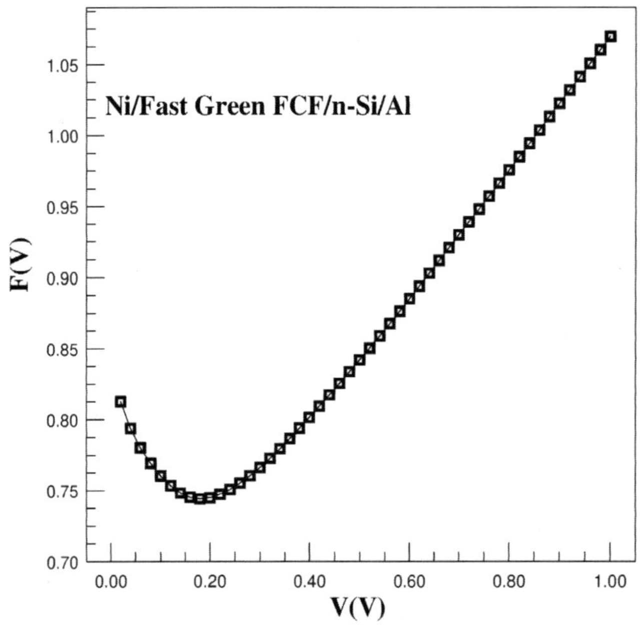


**Fig. 8.** The F(V) versus V plot of the Ni/ Fast Green FCF/n-Si/Al Schottky barrier diode

Ideality factor and barrier height value can be controlled by the interfacial energy states. In an n-type semiconductor, the energy of the interface states $E_{ss}$ with respect to the bottom of the conduction band at the surface of the semiconductor is given with Eq. 15 [3]. For our device, the energy of the interface states $E_{ss}$ with respect to the bottom of the conduction band $E_c$ at the surface of the Si is shown in Fig.9. It is seen that the interface state density Nss has an exponential decreasing toward higher Ess -Ec values for Ni/ Fast Green FCF/n-Si/Al Schottky barrier diode. The interface state density $N_{ss}$ obtained from the forward bias *I–V* ranges from $1\times10^{15}$ $cm^{-2}$ $eV^{-1}$ to $6 \times 10^{15}$ $cm^{-2}$ $eV^{-1}$.

$$Ess - Ec = q\Phi_e - qV \tag{15}$$

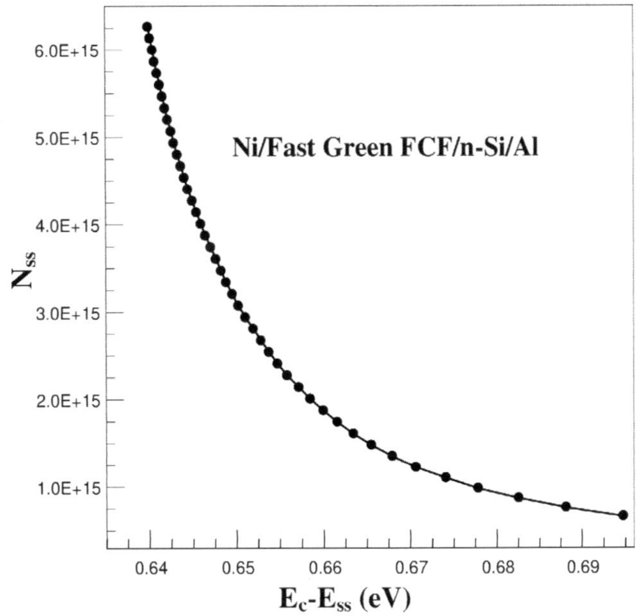


**Fig. 9.** $N_{ss}$ versus $E_{ss}$–$E_c$ plots of the Ni/ Fast Green FCF/n-Si/Schottky barrier diode

### *3.4. Calculated of the device parameters with the capacitance-voltage (C-V) measurements:*

Capacitance–voltage (*C–V*) measurement is another common electrical measurement technique for calculate the parameters of the Schottky barrier diode. We can use the following equations for calculate the basic diode parameters as the diffusion potential $V_d$, barrier height, Fermi energy level $E_f$ and donor concentration $N_d$:

$$C = A\left(\frac{\varepsilon_s \varepsilon_0 e N_d}{2}\right)^{\frac{1}{2}}\left(V_d - \frac{kT}{e}\right)^{\frac{1}{2}} \tag{16}$$

$$C^{-2} = \frac{2(V_d + V)}{\varepsilon_s \varepsilon_0 e A^2 N_d} \tag{17}$$

$$\frac{d(C^{-2})}{dV} = \frac{2}{\varepsilon_s \varepsilon_0 e A^2 N_d} \tag{18}$$

$$n_0 = N_c \exp\left(\frac{E_f - E_c}{kT}\right) \qquad n_o \cong N_d \; for \; N_d >> n_i \text{ and } N_c = 2.8\text{x}10^{19} \text{ cm}^{-3} \tag{19}$$

$$N_d = N_c \exp\left(\frac{E_f - E_c}{kT}\right) \tag{20}$$

$$E_f = kT\ln\left(\frac{N_d}{N_c}\right) \tag{21}$$

$$\Phi_b = \frac{V_d}{n} + E_f \tag{22}$$

Fig. 10 show the forward and the reverse bias *C–V* characteristics of the Ni/ Fast Green FCF/n-Si/Al device at 100, 200 and 500 kHz frequencies. It is clearly seen in Fig.10 the capacitance increased with voltage. Basic diode parameters can calculate from the reverse bias $1/C^2$–$V$ characteristics. Fig. 11 show the reverse bias $1/C^2$–V characteristics of the Ni/ Fast Green FCF/n-Si/Al structure at various frequencies. In the $1/C^2$–V plots the dashed lines shows the linear fits.

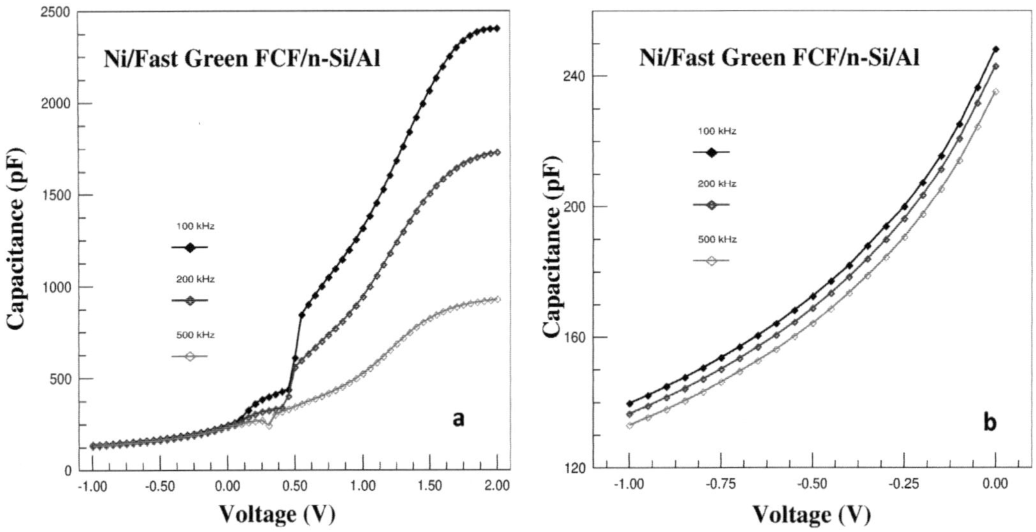


**Fig. 10.** (**a**)The forward and reverse bias *C–V* characteristics, (**b**) the reverse bias *C–V* characteristics of the Ni/Fast Green FCF/n-Si/Al device at various frequencies.

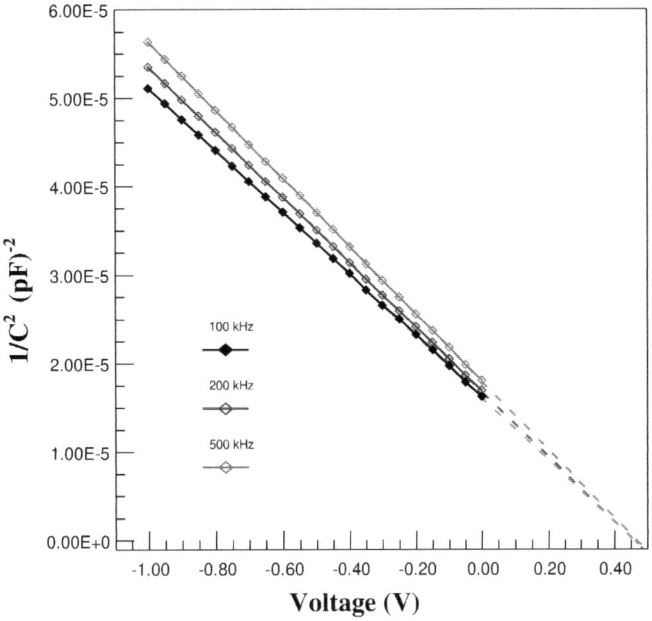


**Fig. 11.** The reverse bias $1/C^2$–$V$ characteristics of the Ni/ Fast Green FCF/n-Si/Al structure at various frequencies.

The junction parameters of Ni/ Fast Green FCF/n-Si/Al Schottky barrier diode obtained from the reverse bias $1/C^2$–$V$ characteristics are given in Table 1. As seen, there is no significant variations with frequency.

**Table 1:** The junction parameters of Ni/ Fast Green FCF/n-Si/Al Schottky barrier diode obtained from the reverse bias $1/C^2$–$V$ characteristics

| Frequencies (kHz) | $V_d$ (V) | $N_d$ (cm$^{-3}$)x10$^{15}$ | $E_f$ (eV) | $\Phi$ (eV) |
|---|---|---|---|---|
| **100** | 0.46 | 5.5 | 0.22 | 0.63 |
| **200** | 0.46 | 5.3 | 0.22 | 0.63 |
| **500** | 0.47 | 5.1 | 0.22 | 0.65 |

## 4- Conclusions

In summary, a Ni/Fast Green FCF/n-Si/Al device was successfully fabricated and it was investigated the electronic properties of the device using the current–voltage *(I–V)* and capacitance–voltage *(C–V)* characteristics at room temperature. It has been shown that thin films of the Fast Green FCF between Ni and n-Si may be employed to increase the barrier heights of the Ni/n-Si Schottky contacts. At The forward bias log *(I–V)* characteristics, three separate regions; as ohmic (region-I), space charge limited currents (SCLC) (region-II) and trap filled SCLC(region-III) has been observed for Ni/Fast Green FCF n-Si/Al Schottky barrier diode . These characteristics show ohmic conductivity at low voltages, space charge limited currents and trap filled SCLC at higher voltages. The difference in apparent barrier height as obtained from the capacitance–voltage and current–voltage measurements on Ni/ Fast Green FCF Schottky barrier has been attributed to

*I–V* and *C–V* techniques to have a different nature. Similarly, the difference in apparent barrier height from the thermionic emission method and Cheung, Norde methods has been attributed to the difference of the methods and non-ideal behavior of the Ni/Fast Green FCF/n-Si/Al device. The response of the device with a Fast Green FCF interlayer investigated and it has been thought that Fast Green FCF may be a novel material for solar cell applications in the future.

## References

[1] Yilmaz M., Caldiran Z., Deniz A. R., Aydogan S., Gunturkun R., Turut A., 2015. Preparation and characterization of sol–gel-derived n-ZnO thin film for Schottky diode application Appl. Phys. A (2015) 119:547–552.
[2] https://en.wikipedia.org/wiki/Fast-Green-FCF.
[3] Çaldıran Z., Deniz A.R., Aydoğan Ş., Yesildağ A., Ekinci D., 2013. The barrier height enhancement of the Au/n-Si/Al Schottky barrier diode by electrochemically formed an organic Anthracene layer on n-Si Superlattices and Microstructures 56 (2013) 45–54.
[4] Orak I., Ejderha K., Sönmez E., Alanyalıoğlu M., Turut A., 2015. The effect of annealing temperature on the electrical characterization of Co/n type GaP Schottky diode. Materials Research Bulletin 61(2015) 463–468.
[5] Aydogan S., Turut A., 2011. Influence of 12 MeV electron irradiation on the electrical and photovoltaic properties of Schottky type solar cell based on Carmine .Radiation Physics and Chemistry 80 (2011) 869–875.
[6] Khusayfan Najla M., 2016. Electrical and photoresponse properties of Al/graphene oxide doped NiO nanocomposite/p-Si/Al photodiodes. Journal of Alloys and Compounds 666 (2016) 501-506.
[7] Norde H., 1979. A modified forward I-V plot for Schottky diodes with high series resistance. J. Appl. Phys. 50 (1979) 5052.

**Acknowledgement:**
*The financial support from Atatürk University Scientific Research Project Council (Project no: 2014/181) is gratefully acknowledged*

Journal of Physics: Conference Series **707** (2016) 012053 doi:10.1088/1742-6596/707/1/012053

# A Novel Numerical Method for Fuzzy Boundary Value Problems

**E Can[1], M A Bayrak[2] and B Hicdurmaz[3]**

[1] Department of Physical Engineering, Istanbul Medeniyet University, Istanbul, Turkey
[2] Department of Mathematics, Kocaeli University, Kocaeli, Turkey
[3] Department of Mathematics, Istanbul Medeniyet University, Istanbul, Turkey

E-mail: emine.can@medeniyet.edu.tr

**Abstract.** In the present paper, a new numerical method is proposed for solving fuzzy differential equations which are utilized for the modeling problems in science and engineering. Fuzzy approach is selected due to its important applications on processing uncertainty or subjective information for mathematical models of physical problems. A second-order fuzzy linear boundary value problem is considered in particular due to its important applications in physics. Moreover, numerical experiments are presented to show the effectiveness of the proposed numerical method on specific physical problems such as heat conduction in an infinite plate and a fin.

## 1. Introduction

Fuzzy differential equations (FDEs) are utilized as a contribution to the improvement of the modelling problems in science and engineering. The difference of fuzzy approach is its suitable structure for uncertainty and subjective information on mathematical models of real world problems. As it is known to all, the mathematical models are used in various problems like population, quantum optics, gravity, control design, medicine etc. In most cases, physical problems are modelled by a boundary value problem. So, considering the boundary value problem with a fuzzy approach, gives us the fuzzy boundary value problems. A fuzzy boundary value problem may have vagueness in each part of the equation like initial condition, boundary condition etc. Due to its important applications, the fuzzy boundary value problems will be considered in the present talk. Then, applications of fuzzy boundary value problems on the flow of heat in an infinite plate and a fin will be presented.

According to the oldest sources on fuzzy derivatives, fuzzy differentiation started with the introduction of some basic properties of fuzzy set theory by Chang and Zadeh in their study [1]. Then, the paper of Dubois and Prade contributed to this definition in their paper [2].

In the literature, fuzzy differential equations have been studied by several different ways. The first and the most popular approach is using the Hukuhara differentiability for fuzzy-valued functions [3]. Unfortunately, this approach has a drawback: the solution becomes fuzzier as time goes by. Hence, the fuzzy solution behaves quite differently from the crisp solution. As an alternative to this definition, Seikkala [4] introduced the notion of fuzzy derivative as an extension of the Hukuhara derivative and fuzzy integral. To solve some drawbacks of generalization of fuzzy differential equations, Hllermeier [5] interpreted FDE as a family of differential inclusions.

The main shortcoming of using differential inclusions is that we do not have a derivative of a fuzzy-valued function. Then, the strongly generalized differentiability was introduced in [6] and studied in [7, 8]. But, the uniqueness of solutions for fuzzy differential equations is a problem under strongly generalized differentiability. Hence, we will recall and use H-differentiability concept which is preferred extensively [9-17] in the present paper.

Most of the problems in science and engineering require the solutions of a fuzzy differential equation (FDE) which are satisfied in fuzzy boundary values. A. Khastan and J. J. Nieto interpret a two-point boundary value problem for a second order fuzzy differential equation by using the generalized differentiability. So, the method of paper [18] is the selected approach of authors for handling the fuzzy differential equations in the present study.

On the other hand, numerical solutions of fuzzy differential equations draw great interest due to its important applications in physical problems. Numerical solution of a FDE is obtained now in a natural way, by extending the existing classical methods to the fuzzy case. Some numerical methods for FDE under the Hukuhara differentiability concept such as the fuzzy Adams-Bashforth, Euler, Adams-Moulton, Predictor-Corrector, Nystrom methods on FDEs appear in the literature [19-22]. Numerical methods for boundary value problems for linear second order FDEs are also considered in several studies [23, 24].

In the present study, the main aim is to solve a second-order fuzzy linear boundary value problem with a novel numerical method which is based on chasing method which came from the studies which are printed in [25] and [26]. Chasing method is based on the idea to establish a set of auxiliary problems that can be solved to find initial conditions at one of the boundaries. After the determination of initial conditions, the usual methods for solving initial boundary value problems can be applied.

The paper is organized as follows: In Section 2, the necessary preliminary information on fuzzy set theory and H-differentiability is given. In Section 3, fuzzy boundary value problems under generalized differentiability are presented and investigated. A new method for solving considered problem is introduced in Section 4. In Section 5, some boundary value problems for fuzzy heat equations are solved to illustrate effectiveness of the proposed method on several cases for physical applications.

## 2. Preliminaries

In the present section, some definitions and necessary notations on fuzzy set theory and fuzzy differentiability will be introduced.

### *2.1. Fuzzy set theory*

**Definition 2.1.** *Let $X$ be a nonempty set. A fuzzy set $u$ in $X$ is characterized by its membership function $u : X \to [0,1]$. Thus, $u(x)$ is interpreted as the degree of membership of an element $x$ in the fuzzy set $u$ for each $x \in X$.*

Let us denote by $\mathcal{F}$ the class of fuzzy subsets of the real axis (i.e., $u : R \to [0,1]$) satisfying the following properties:
$(i)$ $u$ is normal, that is, there exists $x_0 \in R$ such that $u(x_0) = 1$.
$(ii)$ $u$ is a convex fuzzy set (i.e. $u(\lambda x + (1-\lambda)y) \geq min\{u(x), u(y)\}$, for all $\lambda \in [0,1], x, y \in R$).
$(iii)$ $u$ is upper semicontinuous on $R$.
$(iv)$ $cl\{x \in R | u(x) > 0\}$ is compact where $cl$ denotes the closure of a subset.
Then $\mathcal{F}$ is called the space of fuzzy numbers. For each $\alpha \in (0,1]$ the $\alpha$-level set $[u]^\alpha$ of a fuzzy set $u$ is the subset of points $x \in R$ with

$$[u]^\alpha = [\underline{u}^\alpha, \overline{u}^\alpha] \tag{1}$$

where $\underline{u}$ and $\overline{u}$ are called lower and upper branch of $u$ respectively.
For $u \in E$ we define the length of $u$ as:

$$len(u) = \sup_{\alpha}(\overline{u}^{\alpha} - \underline{u}^{\alpha}). \tag{2}$$

A fuzzy number in parametric form is presented by an ordered pair of functions $(\underline{u}^{\alpha}, \overline{u}^{\alpha}), 0 \leq \alpha \leq 1$ , satisfying the following properties:
$(i)$ $\underline{u}^{\alpha}$ is a bounded nondecreasing left-continuous function of $\alpha$ over $(0,1]$ and right continuous for $\alpha = 0$.
$(ii)$ $\overline{u}^{\alpha}$ is a bounded nonincreasing left-continuous function of $\alpha$ over $(0,1]$ and right continuous for $\alpha = 0$.
$(iii)$ $\underline{u}^{\alpha} \leq \overline{u}^{\alpha}, 0 \leq \alpha \leq 1$.
For $u, v \in \mathcal{F}$ and $\lambda \in R$, the sum $u + v$ and the product $\lambda u$ are defined by $[u+v]^{\alpha} = [u]^{\alpha} + [v]^{\alpha}$ and $[\lambda u]^{\alpha} = \lambda[u]^{\alpha}$, for all $\alpha \in [0,1]$, where $[u]^{\alpha} + [v]^{\alpha}$ means the usual addition of two intervals (subsets) of $R$ and $\lambda[u]^{\alpha}$ means the usual product between a scalar and a subset of $R$. The metric on $E$ is defined by the equation

$$D(u,v) = \sup_{\alpha \in [0,1]} max\{|\underline{u}^{\alpha} - \underline{u}^{\alpha}|, |\overline{u}^{\alpha} - \overline{u}^{\alpha}|\}. \tag{3}$$

Here, $D$ is a Haussdorf distance of two intervals $[u]^{\alpha}$ and $[v]^{\alpha}$.

### *2.2. Hukuhara differentiability*

**Definition 2.2.** *Let $x, y \in \mathcal{F}$. If there exists $z \in \mathcal{F}$ such that $x = y + z$, then $z$ is called the H-difference of $x, y$ and it is denoted by $x \ominus y$.*

**Definition 2.3.** *Let $I = (0, l)$ and $f : I \to \mathcal{F}$ is a fuzzy function. We say that $f$ is differentiable at $t_0 \in I$ if there exists an element $f'(t_0) \in \mathcal{F}$ such that the limits*

$$\lim_{h \to 0^+} \frac{f(t_0+h) \ominus f(t_0)}{h}, \lim_{h \to 0^+} \frac{f(t_0) \ominus f(t_0-h)}{h}$$

*exist and are equal to $f'(t_0)$. Here, the limits are taken in the metric space $(\mathcal{F}, D)$.*

**Definition 2.4.** *Let $f : I \to \mathcal{F}$ and fix to $t_0 \in I$. We say that*
*$(i)$ $f$ is (1) differentiable at $t_0$, if there exists an element $f'(t_0) \in \mathcal{F}$ such that for all $h > 0$ sufficiently near to 0, there exist $f(t_0+h) \ominus f(t_0), f(t_0) \ominus f(t_0-h)$, and the limits (in the metric $D$) satisfy:*

$$\lim_{h \to 0^+} \frac{f(t_0+h) \ominus f(t_0)}{h} = \lim_{h \to 0^+} \frac{f(t_0) \ominus f(t_0-h)}{h} = F'(t_0) \tag{4}$$

*$(ii)$ $f$ is (2) differentiable at $t_0$, if there exists an element $f'(t_0) \in \mathcal{F}$ such that for all $h > 0$ sufficiently near to 0, there exist $f(t_0+h) \ominus f(t_0), f(t_0) \ominus f(t_0-h)$, and the limits (in the metric $D$) satisfy:*

$$\lim_{h \to 0^-} \frac{f(t_0+h) \ominus f(t_0)}{h} = \lim_{h \to 0^-} \frac{f(t_0) \ominus f(t_0-h)}{h} = F'(t_0) \tag{5}$$

If $f$ is $(n)$-differentiable at $t_0$, we denote its first derivatives by $D_n^{(1)} f(t_0)$, for $n = 1, 2$.
It is obvious that Hukuhara differentiable function has an increasing length of support. If the function doesn't have these properties then this function is not $H$-differentiable. To avoid this difficulty the authors in [11] introduced the following more general definition of derivative for fuzzy valued functions:

**Theorem 2.1.** *Let $f : I \to \mathcal{F}$ be fuzzy function, where $[f(t)]^\alpha = [\underline{f}_\alpha(t), \overline{f}_\alpha(t)]$ for each $\alpha \in [0,1]$.*
*(i) If $f$ is (1) differentiable in the first form, then $\underline{f}_\alpha$ and $\overline{f}_\alpha$ are differentiable functions and $[D_1^1 f(t)]^\alpha = [\underline{f}'_\alpha(t), \overline{f}'_\alpha(t)]$.*
*(ii) If $f$ is (2) differentiable in the first form, then $\underline{f}_\alpha$ and $\overline{f}_\alpha$ are differentiable functions and $[D_2^1 f(t)]^\alpha = [\overline{f}'_\alpha(t), \underline{f}'_\alpha(t)]$.*
Proof of Theorem 2.1 is given in [17].

The following theorem gives details on the generalized differentiability:
**Theorem 2.2.** *Let $D_1^1 f : I \to \mathcal{F}$ or $D_2^1 f : I \to \mathcal{F}$ be fuzzy functions, where $[f(t)]^\alpha = [\underline{f}^\alpha(t), \overline{f}^\alpha(t)]$ for $\forall\alpha \in [0,1]$. Then*
*(i) If $D_1^1 f$ is (1) differentiable, then $\underline{f}'^\alpha$ and $\overline{f}'^\alpha$ are differentiable functions and $[D_{1,1}^2 f(t)]^\alpha = [\underline{f}''_\alpha(t), \overline{f}''_\alpha(t)]$.*
*(ii) If $D_1^1 f$ is (2) differentiable, then $\underline{f}'^\alpha$ and $\overline{f}'^\alpha$ are differentiable functions and $[D_{1,2}^2 f(t)]^\alpha = [\overline{f}''_\alpha(t), \underline{f}''_\alpha(t)]$.*
*(iii) If $D_2^1 f$ is (1) differentiable, then $\underline{f}'^\alpha$ and $\overline{f}'^\alpha$ are differentiable functions and $[D_{2,1}^2 f(t)]^\alpha = [\overline{f}''_\alpha(t), \underline{f}''_\alpha(t)]$.*
*(iv) If $D_2^1 f$ is (2) differentiable, then $\underline{f}'^\alpha$ and $\overline{f}'^\alpha$ are differentiable functions and $[D_{2,2}^2 f(t)]^\alpha = [\underline{f}''_\alpha(t), \overline{f}''_\alpha(t)]$.*
Proof of Theorem 2.2 is presented in detail in [17].

## 3. Fuzzy Boundary Value Problem

We consider the two-point fuzzy boundary value problem which is defined as follows:

$$y''(t) = F(t, y(t), y'(t)), \; t \in J = [0, T] \tag{6}$$

with boundary conditions y(0)=$\gamma_\alpha$, $y'(T) = \lambda_\alpha$ where $y(t)$ is an unknown fuzzy function of crisp variable $t$, $f : J \times R^2 \to R$ is continuous and $\phi_1, \phi_2 \in R$.

Let $y(t)$ be a fuzzy-valued function on $J$ represented by $[\underline{y}(t;r), \overline{y}(t;r)]$. Let $y(t)$, $y'(t)$ are $H-$differentiable on $J$, then problem (6) can be transformed to the following BVPs systems by the method of Khastan and Nieto [17]: If $y(t)$ and $y'(t)$ are $1-$differentiable on $J$, then $y$ is called $(1,1)$-solution for problem (6), and obtained from

$$\begin{aligned} \underline{y}''(x,\alpha) &= F(x, \underline{y}(x,\alpha), \underline{y}'(x,\alpha)) \\ \overline{y}''(x,\alpha) &= F(x, \overline{y}(x,\alpha), \overline{y}'(x,\alpha)) \\ \underline{y}(a,\alpha) &= \underline{\gamma}_\alpha, \; \overline{y}(a,\alpha) = \overline{\gamma}_\alpha \\ \underline{y}'(b,\alpha) &= \underline{\lambda}_\alpha, \; \overline{y}'(b,\alpha) = \overline{\lambda}_\alpha. \end{aligned}$$

If $y(t)$ is $1-$differentiable and $y'(t)$ is $2-$differentiable on $J$, then $y$ is called $(1,2)$-solution for problem (6), and obtained from

$$\begin{aligned} \underline{y}''(x,\alpha) &= F(x, \overline{y}(x,\alpha), \overline{y}'(x,\alpha)) \\ \overline{y}''(x,\alpha) &= F(x, \underline{y}(x,\alpha), \underline{y}'(x,\alpha)) \\ \underline{y}(a,\alpha) &= \underline{\gamma}_\alpha, \; \overline{y}(a,\alpha) = \overline{\gamma}_\alpha \\ \underline{y}'(b,\alpha) &= \overline{\lambda}_\alpha, \; \overline{y}'(b,\alpha) = \underline{\lambda}_\alpha. \end{aligned}$$

If $y(t)$ is $2-$differentiable and $y'(t)$ is $1-$differentiable on $J$, then $y$ is called $(2,1)$-solution for problem (6), and obtained from

$$\begin{aligned}\underline{y}''(x,\alpha) &= F(x,\overline{y}(x,\alpha),\underline{y}'(x,\alpha))\\ \overline{y}''(x,\alpha) &= F(x,\underline{y}(x,\alpha),\overline{y}'(x,\alpha))\\ \underline{y}(a,\alpha) &= \underline{\gamma}_\alpha,\ \overline{y}(a,\alpha)=\overline{\gamma}_\alpha\\ \underline{y}'(b,\alpha) &= \underline{\lambda}_\alpha,\ \overline{y}'(b,\alpha)=\overline{\lambda}_\alpha.\end{aligned}$$

If $y(t)$ and $y'(t)$ are 2−differentiable on $J$, then $y$ is called $(2,2)$-solution for problem (6), and obtained from

$$\begin{aligned}\underline{y}''(x,\alpha) &= F(x,\underline{y}(x,\alpha),\overline{y}'(x,\alpha))\\ \overline{y}''(x,\alpha) &= F(x,\overline{y}(x,\alpha),\underline{y}'(x,\alpha))\\ \underline{y}(a,\alpha) &= \underline{\gamma}_\alpha,\ \overline{y}(a,\alpha)=\overline{\gamma}_\alpha\\ \underline{y}'(b,\alpha) &= \overline{\lambda}_\alpha,\ \overline{y}'(b,\alpha)=\underline{\lambda}_\alpha.\end{aligned}$$

Our strategy for solving problem (6) is based on the choice of derivative type in the fuzzy boundary value problem. According to the chosen derivative type, the fuzzy boundary value problem is converted into the corresponding system of boundary value problems. Finally, we consider the domain where the solution and its derivatives have valid level sets according to the differentiability type.

## 4. Chasing method

Chasing method for solving boundary value problems was developed by Gel'fand and Lokutsiyevskii and it was intoduced in detail by I.S.Berzin and N.F.Zhidkov [22, 23]. The method is based on the idea to establish a set of auxiliary problems that can be solved to find initial conditions at one of the boundaries. After the determination of initial conditions, the usual methods for solving initial boundary value problems can be applied.

In the present section, we will present the application of chasing method on a linear fuzzy boundary value problem. Firstly, the fuzzy second order differential equation

$$y''(x,\alpha) = p(x)y(x,\alpha)+q(x) \tag{7}$$

where $p(x), q(x)$ are known continuous functions is considered. The boundary conditions are given as

$$y'(a,\alpha) = \alpha_{00}y(a,\alpha)+\alpha_{10} \tag{8}$$

$$y'(b,\alpha) = \beta_{00}y(b,\alpha)+\beta_{10} \tag{9}$$

where $\alpha_{00}, \alpha_{10}, \beta_{00}, \beta_{10}$ are known constants.

We now consider a linear fuzzy first-order differential equation

$$y'(x,\alpha) = \alpha_0(x)y(x,\alpha)+\alpha_1(x) \tag{10}$$

and choose $\alpha_0(x)$ and $\alpha_1(x)$ so that $y$ still satisfies equation (7). Differentiating equation (10) with respect to $x$ and replacing $y'(x,\alpha)$ by the right-hand side of equation (10), we get

$$y''(x,\alpha) = (\alpha_0'(x)+[\alpha_0(x)]^2)y(x,\alpha)+(\alpha_1'(x)+\alpha_0(x)\alpha_1(x)). \tag{11}$$

From equation (7), it is seen that the following equations must be satisfied

$$\alpha'(x)+[\alpha_0(x)]^2 = p(x) \quad , \quad \alpha_0(a)=\alpha_{00} \tag{12}$$

$$\alpha_1'(x)+\alpha_0(x)\alpha_1(x) = q(x) \quad , \quad \alpha_1(a)=\alpha_{10} \tag{13}$$

Then, the two quantities $\alpha_0(b)$ and $\alpha_1(b)$ are obtained. From equation (10), we write

$$y'(b,\alpha) = \alpha_0(b)y(b,\alpha)+\alpha_1(b) \tag{14}$$

and equation (9) gives

$$y'(b,\alpha) = \beta_{00} y(b,\alpha) + \beta_{10}. \tag{15}$$

Since $\alpha_0(b)$ and $\alpha_1(b)$ are known quantities, equations (14) and (15) can be solved for

$$y(b,\alpha) = \frac{\beta_{10} - \alpha_1(b)}{\alpha_0(b) - \beta_{00}} \tag{16}$$

$$y'(b,\alpha) = \frac{\beta_{00}\alpha_1(b) - \beta_{10}\alpha_0(b)}{\beta_{00} - \alpha_0(b)}. \tag{17}$$

Thus, equation (7) is transformed into an initial value problem. Boundary conditions of problem (6) can be converted into the form of conditions (8) and (9). Also, problem (7) is a special case of problem (6). This allows us the combination of chasing method with fuzzy methodology on two point boundary value problems. The combination of chasing method with fuzzy methodology gives a new numerical method which is very effective on finding approximate solutions of different types of fuzzy boundary value problems. It is applicable to fuzzy boundary value problems with different boundary conditions. The effectiveness of the method on numerical solutions of fuzzy boundary value problems will be illustrated by the applications on physical problems in the next section.

## 5. Solutions for heat transfer problems

Heat transfer problems form an important part of the physical heat problems. Applications of them in engineering are including but not limited to resistance, cooling technologies and architecture. Due to its wide application area, we will present the application of the new constructed method on the following heat transfer problems.

### *5.1. Heat conduction in an infinite plate with heat generation*

In the present section, we firstly consider an infinite flat plate which separates a fluid at a temperature of $T_\infty$. Here, heat is generated within the plate at a constant rate, $q_s$.
To formulate the problem, the first law of thermodynamics can be written as

$$\{Aq_x\} + \{q_s A dx\} = \{A(q_x + dq_x)\}$$

where $A$ is the area.
Simplifying the above equation and introducing the Fourier law of conduction,

$$q_x = -k\frac{dT}{dx}$$

the first law of thermodynamics becomes

$$\frac{d^2T}{dx^2} + \frac{q_s}{k} = 0 \tag{18}$$

where total differentiation is used since the temperature is a function of $x$ only.
Since the temperature gradient must be zero at the plane midway between the two surfaces of the plate, it can be formulated as

$$\frac{dT(0)}{dx} = 0.$$

The second boundary condition is written in the usual way when the solid surface is in contact with a fluid at a different temperature with negligible radiation:

$$-k\frac{dT(l)}{dx} = \overline{h}[T(l) - T_\infty]$$

where $k$ is the heat conductivity, $\overline{h}$ is the convective heat transfer coefficient, $T_\infty$ is the temperature of the fluid away from the surface of the plate, and $l$ is half the thickness of the plate.
Introducing the dimensionless quantities

$$\xi = \frac{x}{l}, \quad \theta = \frac{T - T_\infty}{q_s l^2/k},$$

equation (17) becomes

$$\theta''(\xi,\alpha) = -1 \tag{19}$$

subject to the boundary conditions

$$\theta'(0,\alpha) = 0 \tag{20}$$

$$-\theta'(1,\alpha) = N_{bi}\theta(1,\alpha) \tag{21}$$

where $N_{bi}$ is the Biot number.
The exact solution of equation (18) is

$$\underline{\theta}(\xi,\alpha) = (\alpha - 1)[\frac{1}{2}(1-\xi^2) + \frac{1}{N_{bi}}], \tag{22}$$

$$\overline{\theta}(\xi,\alpha) = (1 - \alpha)[\frac{1}{2}(1-\xi^2) + \frac{1}{N_{bi}}]. \tag{23}$$

In Fig. 1 and 2, the exact solution and solutions of $(1,1)$ and $(1,2)$ systems are presented and for $(2,1)$ and $(2,2)$ solutions, we get the results that illustrated in Fig. 3 and 4 with $\alpha = 0$ and $N_{bi} = 0.5, 1.5, 3$.

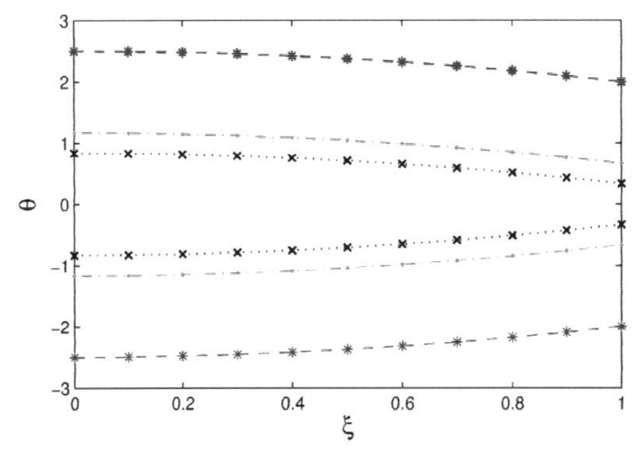


**Figure 1.** Solution $(1,1)$ of problem (18) for $N_{bi} = 0.5(- * -), 1.5(-\cdot-), 3(\cdot \times \cdot)$.

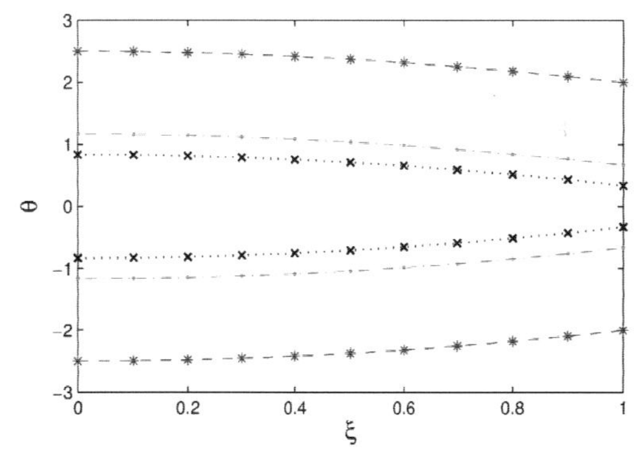


**Figure 2.** Solution $(1,2)$ of problem (18) for $N_{bi} = 0.5(- * -), 1.5(-\cdot-), 3(\cdot \times \cdot)$.

### *5.2. Heat transfer in a fin*

Here, it is a good approximation to assume that the temperature over every cross section perpendicular to the axis of the fin is uniform in the design of fins. The temperature at the base (x = 0) of the fin is known as Ts. If we consider an infinitesimal length $dx$ of the fin, taken as the control volume, conservation of energy can be written as

$$\{Aq_x\} = \{A(q_x + dq_x)\} + \{\overline{h}(Pdx)(T - T_\infty)\}$$

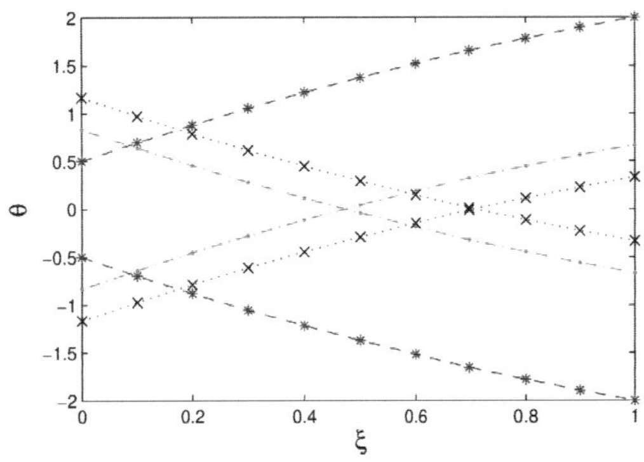


**Figure 3.** Solution $(2,1)$ of problem (18) for $N_{bi} = 0.5(- * -), 1.5(-.-), 3(\cdot \times \cdot)$.

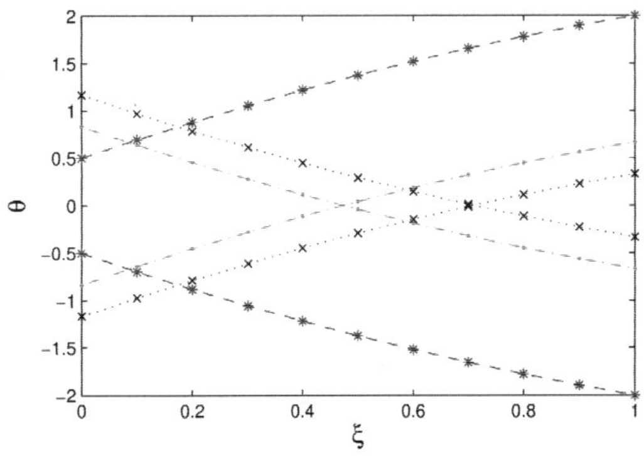


**Figure 4.** Solution $(2,2)$ of problem (18) for $N_{bi} = 0.5(- * -), 1.5(-.-), 3(\cdot \times \cdot)$.

where $A$ and $P$ are the cross-sectional area and the perimeter, respectively, and $\overline{h}$ is the convective heat transfer coefficient.
Simplifying the above equation and introducing the Fourier law of heating,

$$q_x = -k\frac{dT}{dx}$$

we get

$$\frac{d^2T}{dx^2} = m^2(T - T_\infty), \quad m = \frac{\overline{h}P}{kA}. \tag{24}$$

The temperature at the base ($x = 0$) of the fin is known as $T_s$, i.e. $T = T_s$. On the other surface, $x = L$, the plate is exposed to air and the only mode of heat transfer is by convection. Equating the conduction out of the solid to the convection into the fluid, we get

$$-k\frac{dT(L)}{dx} = \overline{h}[T(L) - T_\infty].$$

If the following dimensionless quantities are introduced

$$\xi = \frac{x}{l}, \quad \theta = \frac{T - T_\infty}{T_s - T_\infty},$$

equation (23) becomes

$$\theta''(\xi, \alpha) = \beta\theta(\xi, \alpha) \tag{25}$$

subject to the boundary conditions

$$\theta(0, \alpha) = 1 \tag{26}$$

$$-\theta'(1, \alpha) = N_{bi}\theta(1, \alpha) \tag{27}$$

where

$$\beta = m^2L^2, \quad N_{bi} = \frac{\overline{h}L}{k}.$$

The exact solution of equation (24) is

$$\underline{\theta}(\xi,\alpha)=(\alpha-1)[\frac{\sqrt{\beta}\cosh\sqrt{\beta}(1-\xi)+N_{bi}\sinh\sqrt{\beta}(1-\xi)}{\sqrt{\beta}\cosh\sqrt{\beta}+N_{bi}\sinh\sqrt{\beta}}], \tag{28}$$

$$\overline{\theta}(\xi,\alpha)=(1-\alpha)[\frac{\sqrt{\beta}\cosh\sqrt{\beta}(1-\xi)+N_{bi}\sinh\sqrt{\beta}(1-\xi)}{\sqrt{\beta}\cosh\sqrt{\beta}+N_{bi}\sinh\sqrt{\beta}}]. \tag{29}$$

In Fig. 5 and 6, the exact solution and solutions of $(1,1)$ and $(1,2)$ systems are presented for $\alpha=0$, $\beta=2$ and $N_{bi}=0.5,1.5,3$. For $(2,1)$ and $(2,2)$ solutions, we get the results that illustrated in Fig. 7 and 8 for $\alpha=0$, $\beta=1$ and $N_{bi}=0.5,1.5,3$.

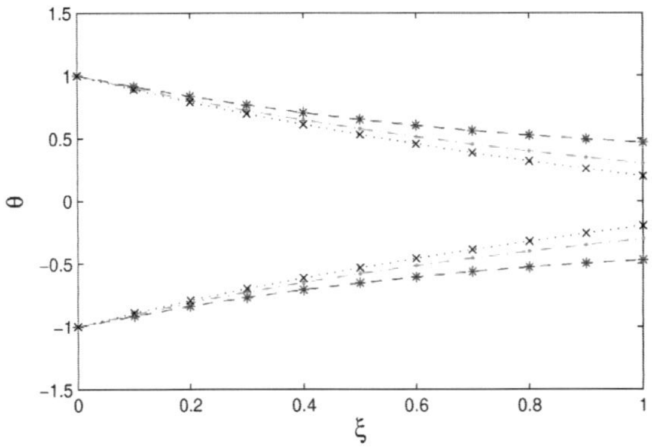


**Figure 5.** Solution $(1,1)$ of problem (24) for $N_{bi}=0.5(-*-),1.5(-.-),3(\cdot\times\cdot)$.

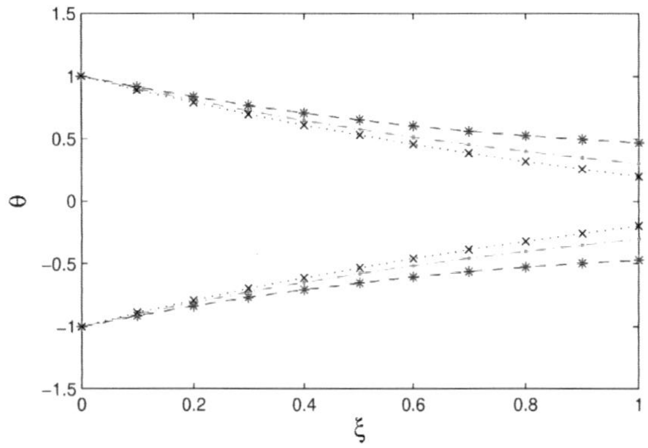


**Figure 6.** Solution $(1,2)$ of problem (24) for $N_{bi}=0.5(-*-),1.5(-.-),3(\cdot\times\cdot)$.

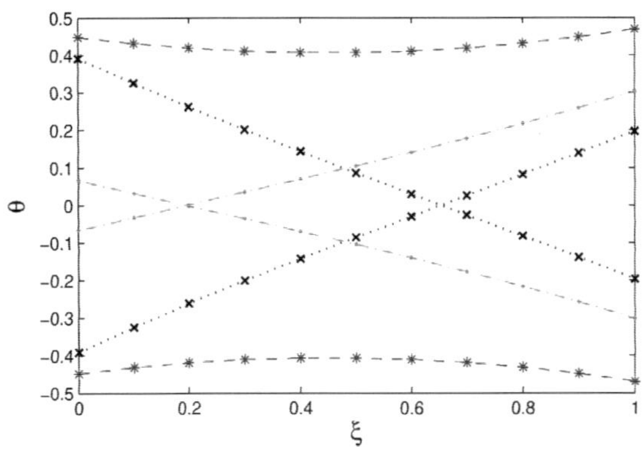


**Figure 7.** Solution $(2,1)$ of problem (24) for $N_{bi}=0.5(-*-),1.5(-.-),3(\cdot\times\cdot)$.

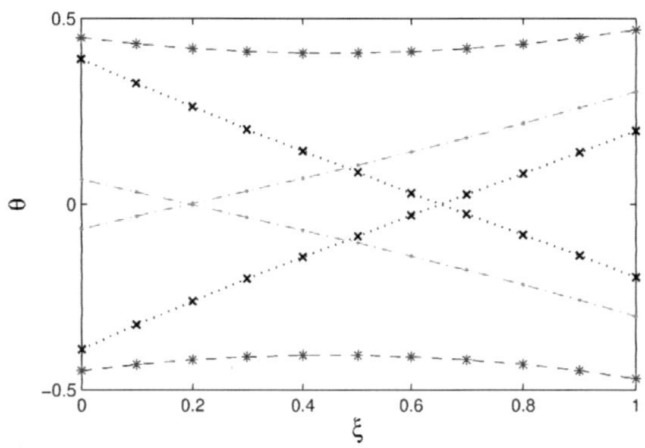


**Figure 8.** Solution $(2,2)$ of problem (24) for $N_{bi}=0.5(-*-),1.5(-.-),3(\cdot\times\cdot)$.

## 6. Conclusions

In the paper, a new numerical method has been introduced for fuzzy boundary value problems under generalized differentiability. This method offers a very interesting alternative for the conversion from a fuzzy boundary value problem to a fuzzy initial value problem. Another advantage of this method is the property that under certain conditions the solutions of the equations of forward and backward chasing methods increase slowly with independent variable. The proposed method can be applied to many physical problems with uncertainty and vagueness. In this paper, effectiveness of the proposed method is tested on heat transfer problems in an infinite plate and a fin.

## References

[1] Zadeh L A 1975 *Inform. Sci.* **8** (1975) 199-251
[2] Dubois D and Prade H 1982 *Fuzzy Sets and Syst.* **8** 225-233
[3] Kaleva O and Thomas F 1987 *Fuzzy Sets and Syst.* **24** 301-317
[4] Seikkala S 1987 *Fuzzy Sets and Syst.* **24** 319-330
[5] Hllermeier E 1997 *Int. J. Uncertain Fuzz.* **5** 117-137
[6] Stefanini L and Bede B 2009 *Nonlin. Anal.: Theory, Meth. Appl.* **71** 1311-1328
[7] Bede B, Rudas I J and Bencsik A L 2007 *Inf.Sci.* **177** 1648-1662
[8] Chalco-Cano Y and Roman-Flores H 2008 *Chaos and Sol. Fract.* **38** 112-119
[9] Chen M, Li D and Xue X 2011 *Fuzzy Sets and Syst.* **162** 67-78
[10] Diamond P 2000 *IEEE Trans. on Fuzzy Syst.* **8** 583-590
[11] Buckley J J and Thomas F 2000 *Fuzzy Sets and Syst.* **110** 43-54
[12] Diamond P and Watson P 2000 *Appl. Math. Lett.* **13** 31-35
[13] O'Reagan D, Lakshmikantam V and Nieto J 2003 *Nonlin. Anal.* **54** 405-415
[14] Ma M, Friedman M and Kandel A 1999 *Fuzzy Sets and Syst.* **105** 133-138
[15] Abbasbandy S and Allahviranloo T 2004 *Nonlin. Studies* **11** 117-129
[16] Puri M L, Ralescu D A 1983 *J. of Math. Anal. and Appl.* **91** 552-558
[17] Golayoglu A F and Koroglu C 2011 *CMES-Comput. Model. in Eng. Sci.* **75** 189-204
[18] Khastan A and Nieto J J 2010 *Nonlin. Anal.* **72** 3583-3593
[19] Palligkinis S C, Papageorgiou G and Famelis I T 2009 *Appl. Math. and Comput.* **209** 97-105
[20] Allahviranloo T, Ahmadi N and Ahmadi E, 2007 *Inf. Sci.* **177** 1633-1647
[21] Friedman M, Ma M and Kandel A 1999 *Fuzzy Sets and Syst.* **105** 35-48
[22] Khastan A and Ivaz K 2009 *Chaos and Sol. Fract.* **41** 859-868
[23] Allahviranloo T and Khalilpour K 2011 *World Appl. Sci. J.* **13** 2137-2147
[24] Khastan A, Bahrami F and Ivaz K 2009 *Boun. Val. Prob.* **395714** 1
[25] Berezin I S and Zhidkov N P 1965, *Computing methods* vol 2 (Oxford: Pergamon Press)
[26] Na T Y 1979 *Computational methods in Engineering Boundary Value Problems* (New York: Academic Press)

# A Comprehensive Study on Dielectric Properties of Volcanic Rock/PANI Composites

**M. Kılıç, Y. Karabul, M. Okutan and O. İçelli**
Department of Physics, Faculty of Science and Letters, Yıldız Technical University, Istanbul, Turkey

E-mail: oicelli@yildiz.edu.tr

**Abstract**. Basalt is a very well-known volcanic rock that is dark colored and relatively rich in iron and magnesium, almost located each country in the world. These rocks have been used in the refused rock industry, to produce building tiles, construction industrial, highway engineering. Powders and fibers of basalt rocks are widely used of radiation shielding, thermal stability, heat and sound insulation. This study examined three different basalt samples (coded CM-1, KYZ-13 and KYZ-24) collected from different regions of Van province in Turkey. Polyaniline (PANI) is one of the representative conductive polymers due to its fine environmental stability, huge electrical conductivity, as well as a comparatively low cost. Also, the electrical and thermal properties of polymer composites containing PANI have been widely studied.

The dielectric properties of Basalt/Polyaniline composites in different concentrations (10, 25, 50 wt.% PANI) have been investigated by dielectric spectroscopy method at the room temperature. The dielectric parameters (dielectric constants, loss and strength) were measured in the frequency range of $10^2$ Hz-$10^6$ Hz at room temperature. The electrical mechanism change with PANI dopant. A detailed dielectrically analysis of these composites will be presented.

## 1. Introduction

Basalt is the generic term for solidified volcanic lava that is dark colored and relatively rich in iron and magnesium, almost located every country in the world. Basalt rocks include a variety of crucial elements (Fe, Mg, Ca, K, P, and S) along with helpful elements (Si and Na) for plants. Also, they are intrinsically obtainable and environmentally sound as a soil amendment. There is growing interest in reinforcing polymer matrix composites with mineral basalt because of their moderate cost, high stiffness and strength, excellent corrosion and oxidation resistance, and heat resistance and thermal stability [1]. Conducting polymers breakthrough provides many new applications for polymeric components. The conducting polymers have been studied for the last 20 years, because of their essential physical attributes and potential applications in numerous electronic devices, e.g., organic field effect transistors, light emitting diodes, and chemical sensors [2].

Among the conducting polymers, Polyaniline is an excellent example of Π-conjugated polymer. The conductivity and I-V characteristics of Polyaniline depend on different experimental conditions like temperature, pH value, the concentration of oxidant and on the nature of dopant. Among different methods of synthesize of Polyaniline, the chemical oxidation and electrochemical synthesis are prominent [3]. It is one of the members of the intrinsically conducting polymer family. Due to possessing a high molecular weight. Also, PANI indicates high conductivity and good environmental stability in contrast to other conducting polymers [2].

The main goal of this study was to examine the effect of PANI on basalt and to evaluate the dielectric properties by varying the PANI concentration. The real and imaginary parts of the complex dielectric permittivity, dielectric strength and a.c. conductive mechanism of composites have been investigated by the dielectric spectroscopy method at the room temperature.

## 2. Experimental Procedure

### *2.1. Materials*

Basalt samples collected from Van city, Turkey. The basalts were taken from volcanic fields erupted from extensional fractures in different periods of time in the Pliocene to Quaternary. Petrology and volcanology of the volcanic regions are examined in detail by [4]. Chemical analyses of the basalt samples are taken by X-ray fluorescence (XRF) instrument. Operating conditions of the Philips PW-2400 XRF instrument were set at 60 kV and 50 mA. The results of chemical analysis of basalt samples are given in Table 1. According to the chemical analysis of basalt samples, $SiO_2$ has pure basic character [5].

**Table 1.** Chemical composition of basalt samples.

| Compound | CM-1 | KYZ-13 | KYZ-24 |
|---|---|---|---|
| $SiO_2$ | 41.668 | 47.790 | 47.086 |
| $TiO_2$ | 2.0800 | 1.3950 | 1.7910 |
| $Al_2O_3$ | 13.106 | 16.918 | 17.574 |
| $Fe_2O_3$ | 13.823 | 10.878 | 11.493 |
| MnO | 0.1920 | 0.1630 | 0.1720 |
| MgO | 9.7540 | 7.6190 | 8.6710 |
| CaO | 10.602 | 11.357 | 9.9660 |
| $Na_2O$ | 5.2610 | 3.1370 | 2.9340 |
| $K_2O$ | 1.7370 | 0.5190 | 0.1210 |
| $P_2O_5$ | 1.7770 | 0.2240 | 0.1920 |

Polyaniline (emeraldine base, average Mw ~5000) was purchased from Sigma-Aldrich. To investigate effects of PANI additive percentages (10.0, 25.0, 50.0 wt.%) on the electrical properties of basalt mechanically modified by PANI, Basalt/PANI composites were prepared by mixing of basalt samples with PANI. The prepared materials of composites were compression molded in a cold press at room temperature. Each one of these mixtures prepared as a pellet having 13 mm diameter, 0.500 g weight and 2 mm thickness.

### *2.2. Dielectric Measurements*

The dielectric measurements were carried out with a two-point probe arrangement. The measurements at room temperature with a high accuracy (0.17% typ.) of the samples were measured by an HP 4194A Impedance Analyzer. The dielectric parameters were measured in the frequency range of 100Hz–15 MHz, using an impedance analyzer interfaced with a computer.

## 3. Result and Discussion

Dielectric spectroscopy (DS) is a well-known and powerful technique which has been effectively used for the characterization of electrical properties of materials as a function of frequency and temperature [6,7]. The dielectric permittivity, energy loss and dielectric relaxation strength are important parameters for material science because these parameters used to decide on the suitability for a given application. The frequency dependence complex dielectric permittivity $\varepsilon^*(\omega)$ is given by

$$\varepsilon^*(\omega) = \varepsilon'(\omega) + i\varepsilon''(\omega) \quad (1)$$

Here, $\omega$, $\varepsilon'(\omega)$, and $\varepsilon''(\omega)$ are the angular frequency, the real and the imaginary part of the complex dielectric permittivity, respectively. The $\varepsilon'(\omega)$ and $\varepsilon''(\omega)$ for samples have been shown in Fig. 1. When the three types of basalts are doped by different weight percentage of PANI, the dielectric constant is decreased in low frequencies. Moreover, the $\varepsilon'(\omega)$ and $\varepsilon''(\omega)$ have large values in the low-frequency region and decreases rapidly in 10.0 wt.% PANI composites. Therefore the complex conjugates of dielectric can be written as [8],

$$\varepsilon'(\omega) - i\varepsilon''(\omega) = \varepsilon'_\infty + \frac{\varepsilon'_s - \varepsilon'_\infty}{1+(i\,\omega\,\tau)^{(1-\alpha)}} \tag{2}$$

In this expression, $\varepsilon'_s$ and $\varepsilon'_\infty$ are low and high frequency dielectric constants. In the dielectric spectroscopy technique, the dielectric relaxation strength $\Delta\varepsilon'$ is the difference between the dielectric values at low and high frequencies $\Delta\varepsilon'$ is given by;

$$\Delta\varepsilon' = \varepsilon'_s - \varepsilon'_\infty \tag{3}$$

The $\varepsilon'_s$, $\varepsilon'_\infty$ and $\Delta\varepsilon'$ values have been shown in Table 2. The highest value was obtained in CM-1/ 10.0 wt.% PANI sample in terms of dielectric strength parameters. It is also observed that the dielectric strength $\Delta\varepsilon'$ decreases with the increasing PANI doping concentrations.

**Table 2** The $\varepsilon'_s$, $\varepsilon'_\infty$ and $\Delta\varepsilon'$ values for the samples.

| **Sample** | $\varepsilon'_s$ | $\varepsilon'_\infty$ | $\Delta\varepsilon'$ |
|---|---|---|---|
| CM-1/ 10.0 wt.% PANI | 222.56 | 24.76 | 197.80 |
| CM-1/ 25.0 wt.% PANI | 83.13 | 30.59 | 52.54 |
| CM-1/ 50.0 wt.% PANI | 58.67 | 25.86 | 32.81 |
| KYZ-13/ 10.0 wt.% PANI | 148.94 | 28.40 | 120.54 |
| KYZ-13/ 25.0 wt.% PANI | 74.91 | 18.99 | 55.92 |
| KYZ-13/ 50.0 wt.% PANI | 42.45 | 9.49 | 32.96 |
| KYZ-24/ 10.0 wt.% PANI | 132.15 | 10.11 | 122.04 |
| KYZ-24/ 25.0 wt.% PANI | 98.55 | 30.74 | 67.81 |
| KYZ-24/ 50.0 wt.% PANI | 65.03 | 29.74 | 35.29 |

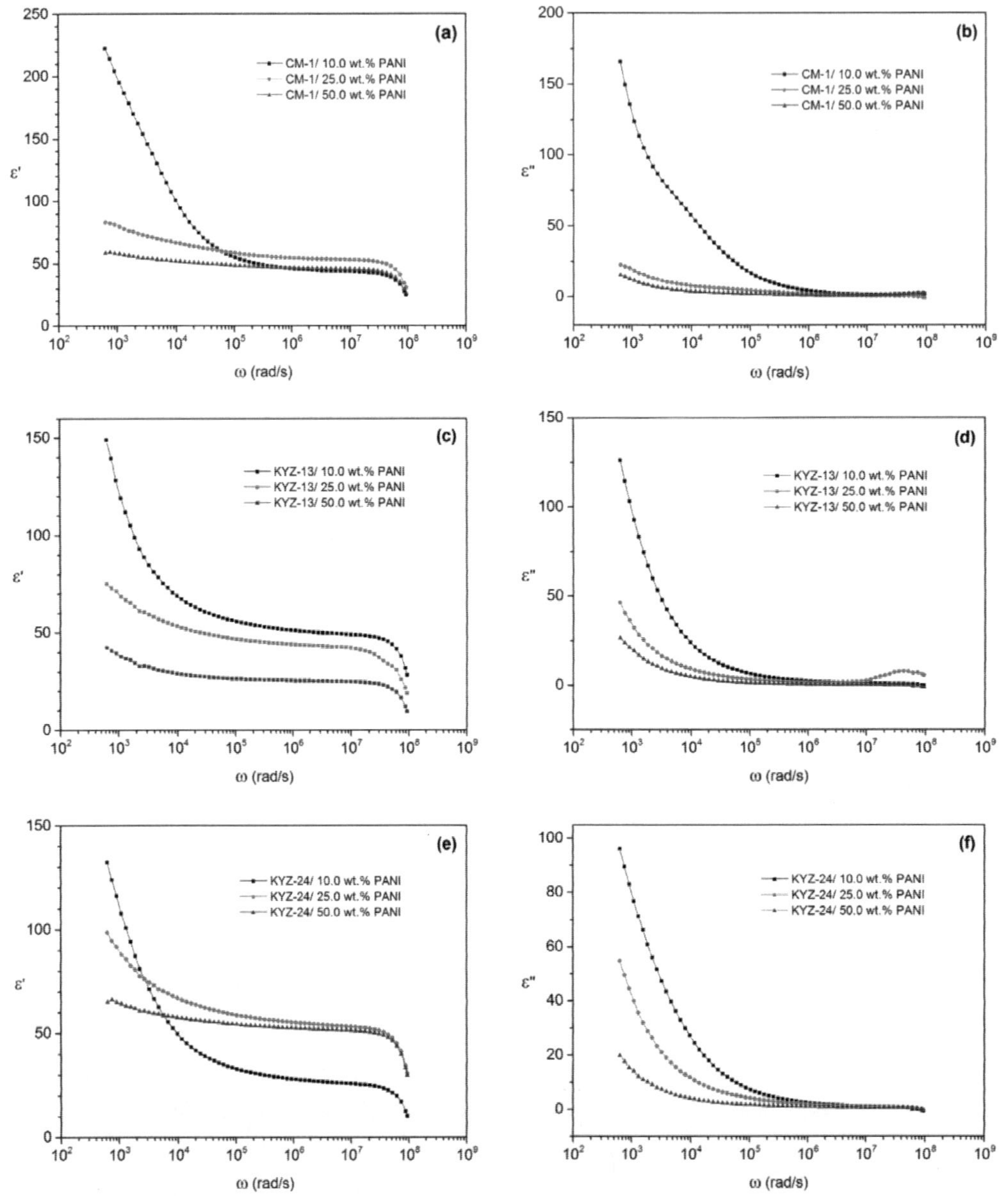


**Figure 1** The variation of (a, c, e) the real part and (b, d, f) the imaginary part of dielectric constant with angular frequency for the Basalt/PANI composites.

In dielectric spectroscopy, to characterize the electrical properties of materials, usually measurements of alternating current (a.c.) conductivity are preferred. The alternative current (a.c.) conductivity ($\sigma_{ac}$) was calculated using the empirical equation:

$$\sigma_{ac} = \omega \varepsilon_0 \varepsilon'' \qquad (4)$$

where, $\omega$, $\varepsilon_0$ and $\varepsilon''$ are the angular frequency, dielectric constant of free space, and imaginary part of the dielectric constant, respectively. The frequency dependent conductivity in solids is generally analyzed using Jonscher's power law [9]:

$$\sigma_{ac} = \sigma(0) + \sigma(\omega) = \sigma_{dc} + A\omega^{s} \tag{5}$$

Where, the first term is frequency independent conductivity ($\sigma(0)$) which is d.c. part of conductivity, and the second term is frequency dependent conductivity ($A\omega^s$) which is a.c. part of conductivity. Here, "A" is the coefficient, "s" represents the degree of interaction between mobile ions with the lattices around them and generally in the range of $0<s<1$ [10,11]. While $s \approx 0$ corresponds to D.C. conductivity [12], $s \approx 0.5$ corresponds to Correlated Barrier Hoping (CBH) Mechanism [13] and $s \leq 0.7 < 1$ conditions represent Quantum Mechanical Tunneling (QMT) model [14].

The frequency exponent, $s$ has been calculated by the slope of $\ln\sigma_{AC} = f(\ln\omega)$ function of the samples for two frequency regions (I. regions: ~ 100 Hz – 3.8 kHz, II. Regions: ~ 3.81 kHz – 15 MHz) given in Fig. 2. The results have been summarized in Table 3. It is clear that the curves tend to merge with a constant slope. The ionic conductivity mechanism all the samples correspond to the Quantum Mechanical Tunneling (QMT) model in the two frequency regions.

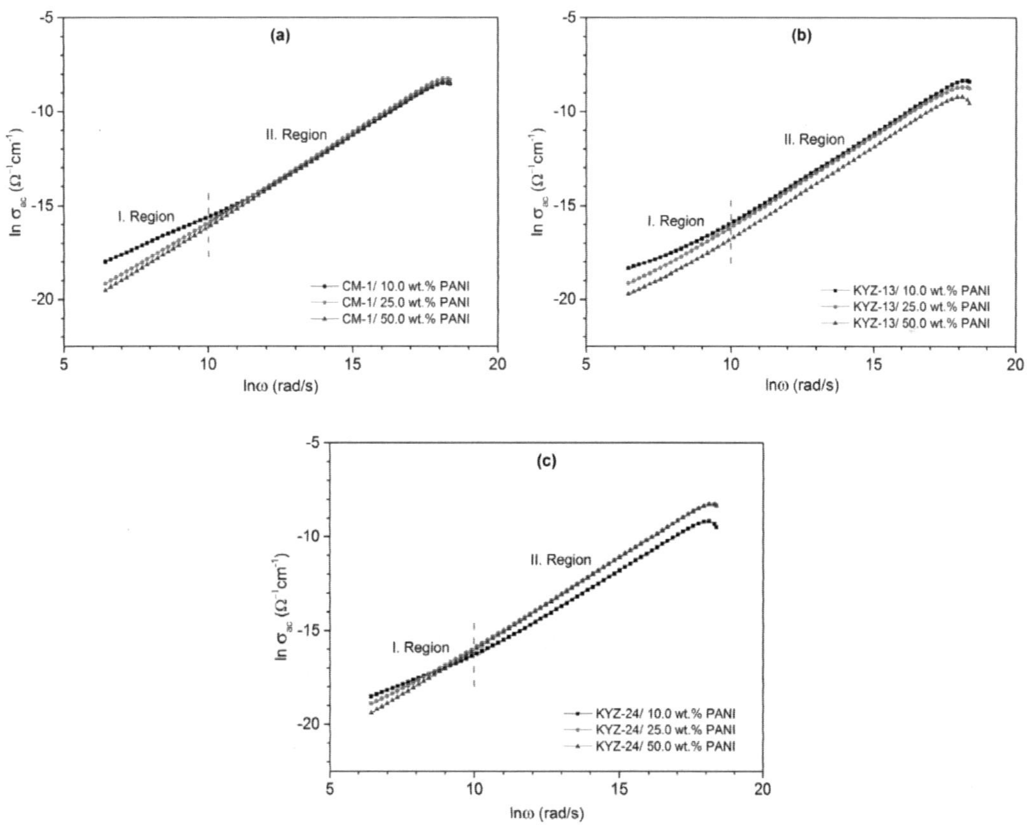


**Figure 2** The $\ln\sigma_{AC} = f(\ln\omega)$ curves for the Basalt/PANI composites.

**Table 3** The *s* parameters and conductivity mechanism for the samples.

| Sample | *s* parameter | |
|---|---|---|
| | **I. Region** | **II. Region** |
| CM-1/ 10.0 wt.% PANI | 0.676 | 0.925 |
| CM-1/ 25.0 wt.% PANI | 0.913 | 0.973 |
| CM-1/ 50.0 wt.% PANI | 0.945 | 0.982 |
| KYZ-13/ 10.0 wt.% PANI | 0.675 | 0.964 |
| KYZ-13/ 25.0 wt.% PANI | 0.845 | 0.964 |
| KYZ-13/ 50.0 wt.% PANI | 0.834 | 0.978 |
| KYZ-24/ 10.0 wt.% PANI | 0.617 | 0.928 |
| KYZ-24/ 25.0 wt.% PANI | 0.839 | 0.973 |
| KYZ-24/ 50.0 wt.% PANI | 0.943 | 0.982 |

## 4. Conclusion
This research represents the results of our investigation on the dielectric properties of three type Basalt/PANI composite using dielectric spectroscopy. All composites show that Pani additivities reduce the dielectric constant and strength at low frequencies. CM-1 with 10.0 wt% exhibits the highest strength. Also, these composite obey to mechanism of Quantum Mechanical Tunneling according to ionic conductivity.

## References
[1] T. Bhat, V. Chevali, X. Liu, S. Feih, A.P. Mouritz, Fire structural resistance of basalt fibre composite, Compos. Part A Appl. Sci. Manuf. 71 (2015) 107–115.
[2] R. Tas, M. Can, S. Sönmezoğlu, Preparation and characterization of polyaniline microrods synthesized by using dodecylbenzene sulfonic acid and periodic acid, Turkish J. Chem. 39 (2015) 589–599.
[3] C. Srinivas, D. Srinivasu, B. Kavitha, N. Narsimlu, K. Siva Kumar 5, Synthesis and Characterization of Nano Size Conducting Polyaniline, IOSR J. Appl. Phys. 1 (2012) 12–15.
[4] O. Vural, Vocanostratigraphy, petrology and magmatic evolution of the etrusk volcano and surrounding areas (North of Lake Van, Turkey), Yüzüncü Yıl University, 2011.
[5] Y. Karabul, L. Amon Susam, O. İçelli, Ö. Eyecioğlu, Computation of EABF and EBF for basalt rock samples, Nucl. Instruments Methods Phys. Res. Sect. A Accel. Spectrometers, Detect. Assoc. Equip. 797 (2015) 29–36.
[6] A.S. Nowick, A. V Vaysleyb, I. Kuskovsky, Universal dielectric response of variously doped $CeO_2$ ionically conducting ceramics, Phys. Rev. B. 58 (1998) 8398–8406.
[7] J. Suchanicz, The low-frequency dielectric relaxation $Na_{0.5}Bi_{0.5}TiO_3$ ceramics, Mater. Sci. Eng. B. 55 (1998) 114–118.
[8] K.S. Cole, R.H. Cole, Dispersion and Absorption in Dielectrics I. Alternating Current Characteristics, J. Chem. Phys. 9 (1941) 341.
[9] A.K. Jonscher, Dielectric Relaxation in Solids, Chelsea Dielectrics Press, London, 1983.
[10] A.K. Jonscher, C.D. Group., The universal dielectric response a review of data and their interpretation, (1979).
[11] K.S. Rao, P.M. Krishna, D.M. Prasad, T.S. Latha, C. Satyanarayana, Low frequency dielectric dispersion studies in ferroelectric ceramics of $Pb_{0.77}K_{0.26}Li_{0.2}Ti_{0.25}Nb_{1.8}O_6$, Indian J. Eng. Mater. Sci. 15 (2008) 215–223.
[12] P.R. Das, B. Pati, B.C. Sutar, R.N.P. Choudhury, Electrical properties of complex tungsten

bronze ferroelectrics; $Na_2Pb_2R_2W_2Ti_4V_4O_{30}$ (R = Gd, Eu), Adv. Mater. Lett. 3 (2012) 8–14.
[13] T. Tiwari, N. Srivastava, P.C. Srivastava, Ion Dynamics Study of Potato Starch + Sodium Salts Electrolyte System, Int. J. Electrochem. 2013 (2013) 1–8.
[14] S.R. Elliott, A.c. conduction in amorphous chalcogenide and pnictide semiconductors, Adv. Phys. 36 (1987) 135–217.